Charles Bendire

Life Histories of North American Birds with Special Reference to their

Breeding Habits and Eggs, with Twelve Lithographic Plates

Charles Bendire

Life Histories of North American Birds with Special Reference to their
Breeding Habits and Eggs, with Twelve Lithographic Plates

ISBN/EAN: 9783744750738

Printed in Europe, USA, Canada, Australia, Japan

Cover: Foto ©berggeist007 / pixelio.de

More available books at **www.hansebooks.com**

SMITHSONIAN

CONTRIBUTIONS TO KNOWLEDGE.

VOL. XXVIII.

CITY OF WASHINGTON.
PUBLISHED BY THE SMITHSONIAN INSTITUTION.
1892.

ADVERTISEMENT.

This volume forms the twenty-eighth of a series, composed of original memoirs on different branches of knowledge, published at the expense and under the direction of the Smithsonian Institution. The publication of this series forms part of a general plan adopted for carrying into effect the benevolent intentions of JAMES SMITHSON, Esq., of England. This gentleman left his property in trust to the United States of America, to found, at Washington, an institution which should bear his own name and have for its objects the "*increase and diffusion of knowledge among men.*" This trust was accepted by the Government of the United States, and an act of Congress was passed August 10, 1846, constituting the President and the other principal executive officers of the General Government, the Chief Justice of the Supreme Court, the Mayor of Washington* (and such other persons as they might elect honorary members), an establishment under the name of the "SMITHSONIAN INSTITUTION FOR THE INCREASE AND DIFFUSION OF KNOWLEDGE AMONG MEN." The members and honorary members of this establishment are to hold stated and special meetings for the supervision of the affairs of the Institution and for the advice and instruction of a Board of Regents, to whom the financial and other affairs are intrusted.

The Board of Regents consists of two members *ex officio* of the establishment, namely, the Vice-President of the United States, and the Chief Justice of the Supreme Court, together with twelve other members, three of whom are appointed by the Senate from its own body, three by the House of Representatives from its members, and six persons appointed by a joint resolution of both Houses. To this Board is given the power of electing a Secretary and other officers for conducting the active operations of the Institution.

To carry into effect the purposes of the testator the plan of organization should evidently embrace two objects: one, the increase of knowledge by the addition of new truths to the existing stock; the other, the diffusion of knowledge, thus increased, among men. No restriction is made in favor of any kind of knowledge; and hence each branch is entitled to, and should receive, a share of attention.

---

* This office has been abolished.

The act of congress establishing the Institution directs, as a part of the plan of organization, the formation of a library, a museum, and a gallery of art, together with provisions for physical research and popular lectures, while it leaves to the Regents the power of adopting such other parts of an organization as they may deem best suited to promote the objects of the bequest.

After much deliberation the Regents resolved to divide the annual income into two equal parts—one part to be devoted to the increase and diffusion of knowledge by means of original research and publications; the other half of the income to be applied in accordance with the requirements of the act of Congress to the gradual formation of a library, a museum, and a gallery of art.

The following are the details of the parts of the general plan of organization provisionally adopted at the meeting of the Regents, December 8, 1847.

DETAILS OF THE FIRST PART OF THE PLAN.

I. To increase Knowledge.—*It is proposed to stimulate research by offering rewards for original memoirs on all subjects of investigation.*

1. The memoirs thus obtained to be published in a series of volumes in a quarto form and entitled "Smithsonian Contributions to Knowledge."

2. No memoir on subjects of physical science to be accepted for publication which does not furnish a positive addition to human knowledge resting on original research; and all unverified speculations to be rejected.

3. Each memoir presented to the Institution to be submitted for examination to a commission of persons of reputation for learning in the branch to which the memoir pertains, and to be accepted for publication, only in case the report of this commission is favorable.

4. The commission to be chosen by the officers of the Institution, and the name of the author as far as practicable concealed, unless a favorable decision be made.

5. The volumes of the memoirs to be exchanged for the transactions of literary and scientific societies, and copies to be given to all the colleges and principal libraries in this country. One part of the remaining copies may be offered for sale; and the other carefully preserved, to form complete sets of the work, to supply the demand from new institutions.

6. An abstract, or popular account, of the contents of these memoirs to be given to the public through the annual report of the Regents to Congress.

II. To increase Knowledge.—*It is also proposed to appropriate a portion of the income annually to special objects of research, under the direction of suitable persons.*

1. The objects and the amount appropriated to be recommended by counselors of the Institution.

2. Appropriations in different years to different objects, so that in course of time each branch of knowledge may receive a share.

3. The results obtained from these appropriations to be published, with the memoirs before mentioned, in the volumes of the Smithsonian Contributions to Knowledge.

4. Examples of objects for which appropriations may be made—

(1) System of extended meteorological observations for solving the problem of American storms.

(2) Explorations in descriptive natural history, and geological, mathematical, and topographical surveys to collect material for the formation of a physical atlas of the United States.

(3) Solution of experimental problems, such as a new determination of the weight of the earth, of the velocity of electricity and of light, chemical analyses of soils and plants, collection and publication of articles of science accumulated in the offices of Government.

(4) Institution of statistical inquiries with reference to physical, **moral, and** political subjects.

(5) Historical researches, and accurate surveys of places celebrated in American history.

(6) Ethnological researches, particularly with reference to the different races of men in North America; also explorations and accurate surveys of the mounds and other remains of the ancient people of our country.

I. To diffuse Knowledge.—*It is proposed to publish a series of reports giving an account of the new discoveries in science and of the changes made from year to year in all branches of knowledge not strictly professional.*

1. Some of these reports may be published annually, others at longer intervals, as the income of the Institution or the changes in the branches of knowledge may indicate.

2. The reports are to be prepared by collaborators eminent in the different branches of knowledge.

3. Each collaborator to be furnished with the journals and publications, domestic and foreign, necessary to the compilation of his report; to be paid a certain sum for his labors, and to be named on the title-page of the report.

4. The reports to be published in separate parts, so that persons interested in a particular branch can procure the parts relating to it without purchasing the whole.

5. These reports may be presented to Congress for partial distribution, the remaining copies to be given to literary and scientific institutions and sold to individuals for a moderate price.

The following are some of the subjects which may be embraced in the reports:

I. PHYSICAL CLASS.

1. Physics, including astronomy, natural philosophy, chemistry, and meteorology.
2. Natural history, including botany, zoölogy, geology, etc.
3. Agriculture.
4. Application of science to arts.

II. MORAL AND POLITICAL CLASS.

5. Ethnology, including particular history, comparative philology, antiquities, etc.
6. Statistics and political economy.
7. Mental and moral philosophy.
8. A survey of the political events of the world, penal reform, etc.

III. LITERATURE AND THE FINE ARTS.

9. Modern literature.
10. The fine arts, and their application to the useful arts.
11. Bibliography.
12. Obituary notices of distinguished individuals.

II. To DIFFUSE KNOWLEDGE.—*It is proposed to publish occasionally separate treatises on subjects of general interest.*

1. These treatises may occasionally consist of valuable memoirs translated from foreign languages, or of articles prepared under the direction of the Insti-

tution, or procured by offering premiums for the best exposition of a given subject.

2. The treatises to be submitted to a commission of competent judges, previous to their publication.

DETAILS OF THE SECOND PART OF THE PLAN OF ORGANIZATION.

This part contemplates the formation of a library, a museum, and a gallery of art.

1. To carry out the plan before described, a library will be required, consisting, first, of a complete collection of the transactions and proceedings of all the learned societies in the world; second, of the more important current periodical publications and other works necessary in preparing the periodical reports.

2. The Institution should make special collections, particularly of objects to verify its own publications. Also a collection of instruments of research in all branches of experimental science.

3. With reference to the collection of books other than those mentioned above, catalogues of all the different libraries in the United States should be procured, in order that the valuable books first purchased may be such as are not to be found elsewhere in the United States.

4. Also catalogues of memoirs and of books in foreign libraries and other materials should be collected for rendering the Institution a center of bibliographical knowledge, whence the student may be directed to any work which he may require.

5. It is believed that the collections in natural history will increase by donation as rapidly as the income of the Institution can make provision for their reception, and therefore it will seldom be necessary to purchase any article of this kind.

6. Attempts should be made to procure for the gallery of art casts of the most celebrated articles of ancient and modern sculpture.

7. The arts may be encouraged by providing a room, free of expense, for the exhibition of the objects of the Art Union and other similar societies.

8. A small appropriation should annually be made for models of antiquity, such as those of the remains of ancient temples, etc.

9. The Secretary and his assistants during the session of Congress will be required to illustrate new discoveries in science and to exhibit new objects of art; distinguished individuals should also be invited to give lectures on subjects of general interest.

In accordance with the rules adopted in the programme of organization, each memoir has been favorably reported on by a commission appointed for its examination. It is, however, impossible in most cases to verify the statements of an author; and therefore neither the commission nor the Institution can be responsible for more than the general character of a memoir.

OFFICERS

OF THE

SMITHSONIAN INSTITUTION.

BENJAMIN HARRISON,

PRESIDENT OF THE UNITED STATES.

Ex-officio PRESIDING OFFICER OF THE INSTITUTION.

MELVILLE W. FULLER,

CHIEF JUSTICE OF THE U. S. SUPREME COURT.

CHANCELLOR OF THE INSTITUTION.

SAMUEL P. LANGLEY,

SECRETARY OF THE INSTITUTION.

G. BROWN GOODE,

ASSISTANT SECRETARY.

IX

MEMBERS EX-OFFICIO OF THE INSTITUTION.

x

REGENTS.

SMITHSONIAN CONTRIBUTIONS TO KNOWLEDGE.

—840—

LIFE HISTORIES

OF

NORTH AMERICAN BIRDS

WITH SPECIAL REFERENCE TO

THEIR BREEDING HABITS AND EGGS,

WITH

TWELVE LITHOGRAPHIC PLATES.

BY

CHARLES BENDIRE, CAPTAIN, U. S. ARMY (RETIRED),

Honorary Curator of the Department of Oölogy, U. S. National Museum,
Member of the American Ornithologists' Union.

WASHINGTON:
GOVERNMENT PRINTING OFFICE.
1892.

II

ADVERTISEMENT.

The following memoir by Capt. Charles Bendire having been published at the joint expense of the Smithsonian fund and of the printing appropriation of the U. S. National Museum, two separate editions are issued, one forming a portion of the series of "Smithsonian Contributions to Knowledge," and the other appearing as a Special Bulletin of the U. S. National Museum.

In accordance with the rule adopted by the Smithsonian Institution, the work has been submitted for examination to a commission consisting of Messrs. George N. Lawrence and William Brewster. The memoir, having been recommended for publication by these gentlemen, is herewith presented as a work of original research, illustrating more particularly the oölogy and breeding habits of North American birds.

S. P. LANGLEY,
Secretary.

SMITHSONIAN INSTITUTION,
Washington, May, 1892.

III

TABLE OF CONTENTS.

GALLINACEOUS BIRDS.

Family TETRAONIDÆ. Grouse, Partridges, etc.

Family PHASIANIDÆ. Pheasants, etc.

Family CRACIDÆ. Curassows and Guans.

TABLE OF CONTENTS.

PIGEONS OR DOVES.

Family COLUMBIDÆ.   Pigeons.

BIRDS OF PREY.

Family CATHARTIDÆ.   American Vultures.

Family FALCONIDÆ.   Vultures, Falcons, Hawks, Eagles, etc.

INTRODUCTION.

This work on the Life Histories of North American Birds is based largely upon the collections in the U. S. National Museum. It was the wish of the late Prof. Spencer F. Baird that a comprehensive work on this subject should be published, bringing together the great advances in our knowledge made during the past few years. This wish was also shared by Prof. Langley and Dr. Goode, and with their concurrence the present work has been written.

Since the publication by the Smithsonian Institution in 1857 of an initial volume on North American Oölogy, by the late Dr. T. M. Brewer—which work was not finished owing to lack of material—and of Baird, Brewer and Ridgway's "History of North American Birds" in 1874, no systematic and comprehensive work on the oölogy of this country has appeared. Large collections have been brought together during the last three decades, and great advances, only rendered possible by the more general interest that the subject has attracted, have been made.

It is not intended that this work shall consist merely of descriptions of nests and eggs. Special attention has been given to the life history, the migratory and breeding ranges, and food of each species. In this connection the latest information, including the field notes made by myself and others and hitherto unpublished, has been freely used.

Although involving considerably more labor and a certain amount of repetition, I treat each species and subspecies separately, and endeavor to define the "breeding range" of each as accurately as possible. This method is to some extent open to criticism, and especially so where a species is divided into several geographical races between the boundaries of whose ranges a neutral zone exists in which they intergrade. On account of the limited knowledge we possess of many of our birds, I am well aware that the information given under this head is more or less imperfect, but this is irremediable in many instances at present.

The present volume relates only to land birds. The classification given in the Code and Check List of the American Ornithologists' Union has been followed, and the synonymy and nomenclature used in this list have also been adopted, with the emendations that have been made up to date.

When it is not expressly stated to the contrary, the type specimens figured have been presented to the U. S. National Museum collection by the gentlemen whose names are given. Without mentioning each by name, my thanks are due to the many friends whose assistance has so greatly aided me in the preparation of this volume and added to its completeness, and whose coöperation will, I hope, be continued until the work is finished.

The original water-color drawings from which the plates have been reproduced were made by Mr. John L. Ridgway, of Washington, D. C., to whose skill and painstaking care the excellence of the illustrations is largely due. The chromolithographic reproductions of these plates were made by the Ketterlinus Printing Company, of Philadelphia, Pa., and it gives me pleasure to say that they are as faithful copies of the original drawings as it is possible to make.

THE AUTHOR.

LIFE HISTORIES OF NORTH AMERICAN BIRDS.

By CHARLES BENDIRE, *Captain U. S. Army (retired).*

GALLINACEOUS BIRDS.

Family TETRAONIDÆ. GROUSE, PARTRIDGES, ETC.

1. Colinus virginianus (LINNÆUS).

BOB WHITE.

Tetrao virginianus LINNÆUS, Systema Naturæ, ed. 10, 1, 1758, 161.
Colinus virginianus STEJNEGER, Auk, II, January 1885, 45.
(B 471, C 389, R 480, C 571, U 289.)

GEOGRAPHICAL RANGE: Eastern United States and southern Ontario, Canada; west to eastern Minnesota, Nebraska, Kansas, Indian Territory, and eastern Texas; south to Georgia, Alabama, and other Gulf States.

This species, one of the most widely distributed of our game birds, is better known throughout the Northern and Middle States as the Quail, and under the name of Partridge or Virginia Partridge in the South. It is found everywhere, more or less abundantly in suitable localities within the United States, east of the Missouri and Mississippi Rivers, excepting in Florida, where it is replaced by the Florida Bob White, and in the northern portions of the New England States. In these, north of Massachusetts, it is rare, but occurs occasionally in the southern portions of Vermont and New Hampshire, and less frequently in Maine. In northern New York it is very rare. West of the Mississippi it occurs in Louisiana, eastern Texas, the eastern part of the Indian Territory, Arkansas, Missouri, Kansas, and the greater portion of Nebraska, where it has advanced beyond the central part of the State.

Mr. W. M. Wolfe, of Kearney, Nebraska, informs me that the Bob Whites are becoming more and more abundant, and are now common as far west as Ogallala. Dr. T. E. Wilcox, surgeon U. S. Army, writes me to the same effect from Fort Niobrara, Nebraska, saying it is steadily advancing westward and is now to be found 30 miles west of this post. It is also gradually advancing northward.

Mr. W. W. Cooke states that in Minnesota it has followed up the settlements, and in the eastern part of the State has reached the line of the Northern

1

Pacific Railroad, about latitude 46°; also that in South Dakota it is abundant, and has advanced to latitude 44° 30'.[1]

North of the United States Mr. T. McIlwraith gives it as a permanent resident in southern Ontario, Canada.[2]

At the present time the Bob Whites are most abundant in the Central and some of the Southern States. They have also been successfully introduced in various localities in the West. According to information received from Mr. Denis Gale, of Gold Hill, Boulder County, Colorado, it is now well established along the South Platte River and its tributaries in the vicinity of Denver, Colorado, and is known to occur also in portions of northern New Mexico.

As early as 1872 Prof. J. A. Allen stated in the American Naturalist that these birds had recently been introduced in the Great Salt Lake Valley, Utah, and that in the summer of 1871 young had been raised and gave promise of multiplying rapidly and becoming thoroughly naturalized. At the present time they are common in various parts of Utah, and Professor Allen's predictions have been fully verified. According to Mr. H. K. Taylor the Bob Whites are quite abundant about Gilroy, California.[3]

In the vicinity of Boisé City, Idaho, a few pairs were turned out some time in 1875. In the fall of 1878 I found them abundant between that point and Snake River, all along the Boisé River, and in 1882 they had spread to the west side of Snake River, fully 50 miles from where they were first liberated. Dr. T. E. Wilcox, U. S. Army, who first noticed them there, says, "I never saw coveys so large and numerous as I found them about Boisé. Cover and food, as well as climate, are all favorable."[4]

They are also quite abundant now in portions of the Willamette Valley, Oregon, as well as on several islands in Puget Sound, Washington. In fact, they are well adapted for introduction into any country where the climate is not too severe in winter, and where suitable food and shelter are to be found, they seem to thrive and adapt themselves to the surroundings.

Excepting, perhaps, in its extreme northern range, the Bob Whites are residents, and breed wherever found. They are partial to more or less open country. Fields and pastures, interspersed with small bodies of woodland, country roads, bordered by brush and briar patches, as well as the edges of meadow and lowlands, are its favorite abiding places. In southern Louisiana they are very partial to the borders of hammock land and open pine woods.

They are never found in large packs; each covey generally keeps to itself, and rarely moves far from the place where it was raised. The mating season commences in April, when the coveys or such portions of them remaining begin to break up, each pair selecting a suitable nesting site. Nidification begins usually about May 1; in the Southern States somewhat earlier, and in

[1] Report on Bird Migration of the Mississippi Valley, Bull. 11, U. S. Dept. of Agric., Div. Economic Ornith., p. 102.
[2] Birds of Ontario, pp. 140, 141.
[3] Ornithologist and Oölogist, Vol. IV, 1889, p. 93.
[4] Auk, Vol. II, 1885, p. 315.

the more northern portions of their breeding range it is often delayed until June. The nest is always placed on the ground and is generally a very simple affair. A saucer-shaped cavity is excavated (occasionally quite a deep one) alongside a patch of overhanging weeds or a tall bunch of grass. Again, it may be placed under a small bush or in a briar patch, by the side of a fence, in cultivated fields or pastures, and even in gardens close to houses; and in the South, "Cotton rows" are favorite nesting sites. This cavity is lined with dry grasses or bits of grain stubble. The nest is generally well hidden, arched over naturally by overhanging vines, bushes or weeds, and usually open on one side. Occasionally a nest is arched over artificially, but in most cases, where there is no natural cover existing, no dome is attempted.

Judge John N. Clark, of Saybrook, Connecticut, writes me of having seen a male Bob White at work constructing a domed nest. He says: "In May, 1887, while on a hill back of my house one morning, I heard a Quail whistle, but the note, which was continually repeated, had a smothered sound. Tracking the notes to their source, I found a male Bob White building a nest in a little patch of dewberry vines. He was busy carrying in the grasses and weaving a roof, as well as whistling at his work. The dome was very expertly fashioned, and fitted into its place without changing the surroundings, so that I believe I would never have observed it, had he kept quiet." Another nest, found by Mr. G. E. Beyer, of New Orleans, Louisiana, was entirely constructed of pine needles, arched over, and the entrance probably a foot or more from the nest proper.

In North Carolina, according to Mr. R. B. McLaughlin, the Bob Whites preferred to nest in sedge-fields, so very common in that region, and nearly all the nests observed by him were placed near paths and roads. The favorite materials used for lining the nest were the long dry blades from the sedge tussocks.

Capt. B. F. Goss, of Pewaukee, Wisconsin, has found these birds nesting in the open prairies, miles from timber and brush of any kind; but such instances are rather unusual.

Among uncommon nesting sites the following deserves mention: Mr. Lynds Jones, of Grinnell, Iowa, found a nest of a pair of Bob Whites under the edge of a bridge, which contained nine eggs. It had been placed under a plank in the road, and during a heavy rainstorm was flooded and deserted.

Prof. Robert Ridgway, of the Smithsonian Institution, found a Bob White's nest containing fresh eggs, on October 16, and Mr. H. C. Munger, of Jefferson City, Missouri, publishes in Forest and Stream, of March 6, 1879, a still more remarkable find. He writes as follows:

"JEFFERSON CITY, MISSOURI, *February* 6, 1879.

"Editor FOREST AND STREAM:

"I noticed an article in a local paper here yesterday, stating that a gentleman while out hunting in Calaway County, a county adjoining this, in the month of January, found a Quail's nest with fifteen or sixteen eggs, and the mother bird sitting on the nest. After she flew off the nest, he examined it carefully, the bunch of grass covering it being filled with ice and frozen solid, leaving just space enough under it for the bird and her nest and a place of exit. A few days after finding it he and other parties went to examine it again. This time they found the bird still sitting on the nest but frozen to death. A portion of the eggs had been hatched, but the young were also frozen. Was this not a very singular occurrence? I should have been somewhat skeptical in regard to it if I had not met with very nearly a similar case while out quail shooting four years ago this winter, in company with a venerable sportsman, Mr. Pratt, of this place. Our dogs made a point. We flushed a single bird after considerable kicking around in the grass and snow, and found she had been sitting on her nest containing three apparently fresh eggs; but alas, she never returned to finish her maternal duties. It was too late when we found the cause of her reluctant flight."

The Bob White is unquestionably the most prolific of all our game birds, the number of eggs laid varying from twelve to eighteen to a clutch. Fifteen may be considered a fair average. As many as thirty-seven eggs have been found in one nest, unquestionably the product of two, or even three, hens. In such large sets the eggs are always placed in layers or tiers, the small or pointed ends usually toward the center. An egg is laid daily till the set is completed.

The late Dr. T. M. Brewer states that he "never found less than twenty-four eggs in a nest, and from that to thirty-two."[1] If the eggs are all laid by a single bird, which I think is doubtful, such large sets as Dr. Brewer mentions may possibly be accounted for in the following manner: In Massachusetts and in other portions of its northern range the Bob Whites probably rear but one brood, and lay a larger number of eggs to a set than they do in the Middle and Southern States, where the fact seems to be pretty well established that two and even three broods are sometimes raised during a favorable season; parents with young of three different sizes having been met with now and then, which would tend to substantiate this assertion. Incubation lasts about twenty-four days, in which duty the male is said to assist, at times at least.

Mr. Lynds Jones, who has had excellent opportunities to study the habits of the Bob White, writes me: "The female is seldom seen during the nesting season, while the male attracts our attention with his loud and fearless call, usually uttered from some fencepost or other elevated position. If driven from this, he darts into the grass or shrubbery and there repeats his call. I never succeeded in flushing the female at such times; she is shy and coy, while the male is bold and fearless. While I have never flushed the male from the nest, I have frequently found him near it. If the nest is disturbed while the set of eggs is still incomplete, the birds usually abandon it; but should incubation be somewhat advanced they will return and hatch their brood. The male

---

[1] History North American Birds, Vol. III, p. 472.

is very attentive to the setting hen, often making excursions into the grass after food, apparently for her benefit."

That the male Bob White takes the whole duty of incubation upon himself, should some accident befall the female, which unfortunately happens only too often, is conclusively proven by the following statement, from Dr. William C. Avery, of Greensboro, Alabama, who writes me as follows: "In June, 1886, while on a visit to Dr. J. M. Pickett, of Cedarville, Alabama, this gentleman informed me of having seen a male Bob White incubating; he had visited the nest at various times during the day, and on different days, and always found the male on the nest. Wishing to be an eye witness of so interesting a phenomenon, I rode several miles with the doctor to see this male Bob White on his nest. There we found him, faithfully warming his treasures, but not into life. The eggs were never hatched. Dr. Pickett went frequently to the place, until long after the period of incubation had elapsed, and finding that the eggs would not hatch, he destroyed them, to prevent the useless occupation of the nest by the male. The female had probably been dead some hours, and the eggs were cold before the male took the nest, hence they did not hatch. How different is he in his nature from some other gallinaceous birds, which only seek the female when impelled by sexual desire. I know no other bird that will take the nest and faithfully brood upon the eggs when the female has been killed."[1]

These birds are very sociable in disposition, and, when not constantly disturbed or shot at, become quite tame and may frequently be seen about dwelling houses, barns, and in gardens, especially during the late fall, winter, and early spring. As soon as the young are hatched they become more shy and retiring. The young leave the nest as soon as hatched, and have been seen running about with pieces of the shell sticking to them. They are faithfully cared for by both parents, who make use of all sorts of artifices, such as feigning lameness and fluttering along just out of reach of the intruder, to lure him away from the young brood: the young scattering, in the mean time, and hiding in the grass and under leaves at the danger signal of the parents, and remaining quiet until called together again by either of them, as soon as all danger is passed. When they are about two or three weeks old, the male takes charge of the first brood, while the female begins to lay her second clutch of eggs. This is usually a smaller one than the first, averaging only about twelve eggs. The young are at first exclusively fed on insect food, and later on small seeds, grains, and berries.

Aside from insects of various kinds, the favorite food of the Bob White consists of buckwheat, wheat, rye, oats, the seed of the locust, wild pease, tick-trefoil (*Desmodium*), smartweeds (*Polygonum*), sunflower, and bitterweed, the partridge berry (*Michella*), wintergreen (*Gaultheria*), nannyberries (*Viburnum*),

---

[1] The fact that the male Bob White takes occasionally the entire duties of incubation on himself, should the female be killed, appears to be not an unusual occurrence with this species, at least two similar instances having come under the observation of other parties.

wild grapes, and other small berries. In the late fall they often feed on the seeds of the skunk cabbage, acorns of different kinds, as well as on beech-nuts.

Mr. W. M. Wolfe, of Kearney, Nebraska, writes me: "Here, the male takes the young to the wheat fields and stubble early in July; at first, they return to the brush for the night, but as soon as harvesting fairly commences they spend all their time in the fields, huddling together at night in the open. Here they form a circle with their heads out and crowd close together. The male remains outside the ring and close at hand. The female, after raising her second brood, takes the chicks to the stubble as soon as they are able to fly. The broods unite in September, and all care on the part of the parents ceases soon after, though they all remain together until the following spring."

Aside from the many enemies that the Bob White has to contend with during the breeding season, the mowing machine is probably one of the greatest factors of destruction, as many brooding birds and eggs are annually destroyed through its agency.

The males commence singing about May 1; their song is the well-known "Bob White," or "Ah, Bob White." One of their love notes may be translated as "Pease most ripe," another call as "No more wet," or "More wet." A shrill "wee-tech" is used as a note of warning, and one to assemble when the covey has dispersed resembles "Quoi-hee, quoi-hee." A subdued clucking when undisturbed, and a rapidly repeated twitter when suddenly surprised, are frequently used as well.

In the fall, in certain portions of the country, these birds, while not actually migratory, leave the localities where they raised their broods for others, possibly on account of the desire for some particular kind of food. Mr. G. E. Beyer writes me that in the vicinity of Madisonville, Louisiana, the Bob Whites leave the hammock lands in the fall and retreat considerable distances into the open pine woods, along small water courses, returning only when nesting time approaches.

The eggs of the Bob White vary from a round ovate to subpyriform in shape, are dull white in color, slightly glossy, and often partially stained a buffy yellow by contact with the grass or soil on which they lie. The shell is smooth and remarkably strong and thick for the size of the egg. Their average size is about 30 by 24 millimetres, the largest egg in the U. S. National Museum collection measuring 32.5 by 25, the smallest 26 by 22.5 millimetres.

The type specimen, No. 12786, Pl. 1, Fig. 1, selected from a set of ten eggs, was collected by Dr. William Wood at East Windsor Hill, Connecticut, June 14, 1866.

2. Colinus virginianus floridanus (Coues).

FLORIDA BOB WHITE.

Ortyx virginianus var. *floridanus* Coues, Key to North American birds, 1872, 237.
Colinus virginianus floridanus Stejneger, Auk, II, January, 1885, 45.
(B —, C 389a, R 480a, C 572, U 289a.)

GEOGRAPHICAL RANGE: Florida, except the extreme southern portion.

This somewhat smaller and darker race is found only in Florida. Dr. W. L. Ralph, who has enjoyed excellent opportunities of studying the habits of the Florida Bob White, and is well known as a reliable and careful observer, writes me as follows: "It is still common throughout the northern and central parts of the State, and probably in the southern portions as well, but they are not nearly so abundant as formerly, owing to the persecution they receive from northern visitors and negroes, and to the want of efficient game laws. They are very tame and confiding, and when not molested prefer to live near man, probably on account of greater security from the attacks of beasts and birds of prey. They become much attached to the localities where they breed, and seldom wander far from these, even when much persecuted. I have known cases where they were hunted day after day until their number was reduced to two or three birds to each covey, yet those which were left could always be found at their old places of resort. The localities they like best are open woods grown up with saw palmettos or low bushes, or fields with woods near them, and they are particularly fond of slovenly cultivated grounds that have bushes and weeds growing thickly along their borders."

The pairing season commences early. Mr. J. F. Menge writes me: "In Lee County, Florida, it nests sometimes as early as February 15. A dry secluded spot is selected for a nesting site, usually under a saw palmetto or low bush in open woods or in a field thickly grown up with grass and weeds."

Mr. W. E. D. Scott states, "The Florida Bob White is abundant in the vicinity of Tarpon Springs, and breeds in numbers in early April. At least two broods are raised, as I have found birds but a few days old in the first week in July."

Their general habits are similar to those of its northern relative. Their food consists of insects of various kinds, small seeds, and cabbage-palm berries, and their various notes resemble those of the Bob White. Mr. Scott heard males singing as early as January 19, 1889, and February 7, 1890. The average number of eggs laid varies from eleven to thirteen. Mr. Menge has found as many as twenty-three in one nest, however. Two and probably three broods are raised in a season. The eggs resemble those of *Colinus virginianus* in every respect. The average size of twenty-three specimens is 30 by 24 millimetres, the largest measuring 31.5 by 24.5, the smallest 28.5 by 23 millimetres. None are figured.

3. Colinus virginianus texanus (LAWRENCE).

TEXAN BOB WHITE.

Ortyx texanus LAWRENCE, Annals Lyceum Natural History, **N. Y.**, VI, April, 1853, 1.
Colinus virginianus texanus STEJNEGER, Auk, II, January, 1885, 45.
(B 472, C 389b, R 480b, C 573, U 289b.)

GEOGRAPHICAL RANGE: Southern and western Texas, and northeastern Mexico north to western Kansas.

The Texan Bob White is a resident of the greater part of Texas, excepting the so-called Staked Plains in the northwestern part of the State. In eastern Texas it intergrades with *Colinus virginianus*. It is most abundant in the central parts of the State. Its range northward extends well into the Indian Territory and it has also been taken in western Kansas, where it is rare, however. In its general habits it does not differ materially from *Colinus virginianus*.

Capt. P. M. Thorne, Twenty-second Infantry, U. S. Army, writes me: "During a month I spent on the road between Forts Duncan and McIntosh, Texas, I found this species common, and so unsophisticated that a covey would not even squat when my dog pointed them; they would move on slowly, chattering to each other, evidently talking the matter over. At Fort Duncan it seemed odd that I could always find them mornings and evenings close to the bank of the Rio Grande on the American side, and as soon as flushed they flew straight over into Mexico. The river here is over half a mile wide. At Fort Clark, Texas, I have taken them nearly full grown on July 29, and found them barely able to fly as late as September 20."

Mr. George B. Sennett records them as common at Lomita, in the lower Rio Grande Valley, coming into the inclosures of the ranch at all times and feeding about the corncribs with Blackbirds and Pigeons.

Mr. William Lloyd, of Marfa, Texas, informs me, "The Texan Bob White is a bird of the lowlands, and is not found above an altitude of 2,000 feet. Their food consists of small berries, acorns, grain, buds and leaves of aromatic herbs and small shrubs, varied with occasional beetles, grasshoppers, and ants, especially the winged females, of which they seem to be very fond. They are very insuspicious, and their low notes, uttered while feeding, attract a good many enemies. I have seen foxes on the watch, and the Marsh Harrier perched in a clump of grass on the lookout, waiting for them to pass. But the many large rattlesnakes found here are their worst enemies. One killed in May had swallowed five of these birds at one meal; another, a female evidently caught on her nest and a half dozen of her eggs; a third, four Bob Whites and a Scaled Partridge. The young are also greatly affected and many killed by heavy rains in June and July; numbers perish then from cold and protracted wet weather. When alarmed by a Hawk sailing overhead they run under the mother for protection, as domestic chickens do."

Mr. J. A. Singley, of Giddings, Texas, writes me: "During the very hot weather of the summer this species is always to be found under the large detached live oaks standing in the prairies. It is cool there in the hottest weather, and the birds know it. They are easily approached then, and are often killed by the dozen at such times by so-called sportsmen, though they are in poor condition and molting plumage. The favorite nesting site of the Texan Bob White is a bunch of sedge grass. A slight cavity is made in the center, this is lined with a few straws and arched over with similar material. Sometimes a covered way or tunnel leads to the entrance of the nest. Occasionally a nest is placed under a bush and not covered or arched."

Two broods are usually raised in a season, and even three at times. The average number of eggs to a set is about fifteen. Full sets of eggs have been found as early as May 3, and again late in September. Mr. Singley met with a covey of young just hatched September 22. He also found five eggs of this species in the nest of a domestic hen, not over 30 yards from the house. A nest with nineteen eggs when first found, contained ten additional ones five days later, proving conclusively that it was occupied by two birds. Only one incubated, however. Mr. H. P. Attwater found as many as thirty-three eggs in a nest of the Texan Bob White on May 14, 1889, near San Antonio, Texas. This set is now in the collection of Mr. Samuel B. Ladd, of Westchester, Pennsylvania. While examining this collection I was shown a very peculiar set of nine eggs, taken on May 16, 1889, by the above-mentioned collector, near the same place. These eggs, while of the usual shape and color, are all more or less spotted and streaked with pale reddish brown and lilac shell markings, principally about the larger end. The nest from which they were taken was found in a cornfield.

The eggs of the Texan Bob White are in no way different from those of *Colinus virginianus*. The average measurement of fifty-nine specimens in the U. S. National Museum collection is 30 by 24 millimetres; the largest egg measuring 31.5 by 24.5, the smallest 27.5 by 22 millimetres. On account of the similarity of these eggs to those of the preceding subspecies none are figured.

4. Colinus virginianus cubanensis GOULD.

CUBAN BOB WHITE.

Ortyx cubanensis GOULD, Monograph of the Odontophorinæ, 1850, Pl. 2.
Colinus virginianus cubanensis RIDGWAY, Manual North American Birds, 1887, p. 188.
(B —, C —, R —, C—, U 289c.)

GEOGRAPHICAL RANGE: Cuba and southwestern Florida.

This slightly smaller and darker colored bird than *Colinus virginianus floridanus*, is found in limited numbers in southwestern Florida, south of Lake Okeechobee and Tampa Bay. Specimens from Miami, Dade County, on the east coast, are intermediate between this and the Florida Bob White. Dr. Jean Gundlach reports it as common on the Island of Cuba, and says: "It is not found at any time in the heavy forests, but along their outskirts, in the bushes and undergrowth. During the mating season the male perches on one of the lower branches of an isolated tree, or some other elevated position, and calls to the female. This note resembles its generic name '*Ortyx*' or '*Ortys*.' Another call is somewhat similar to the low grunting of the Guinea pig. Nidification begins in March and lasts till May. The female deposits about sixteen white eggs in a slight excavation of the ground, sparingly lined with grasses and always covered with overhanging plants."[1]

According to Dr. Juan Vilaró, professor of natural history, University of Havana, Cuba, "The Cuban Bob White lays from ten to eighteen eggs; these are usually deposited, between the months of April and July, in a slight cavity of the ground sheltered by vegetation. They feed on small fruits, seeds, and leafy shoots. The young, if alarmed, raise the feathers of the head and upper part of the neck, spread out the tail and wings a little, and run in various directions to hide, reassembling again in obedience to the call of the parent birds. The male assists in the duties of incubation. This bird is locally known as '*Codorniz*.'"

The average measurement of nine eggs in the U. S. National Museum collection from Cuba is 30.5 by 24 millimetres. The largest egg measures 31.5 by 25.5, the smallest 29.5 by 23.5 millimetres. They are indistinguishable from the eggs of the preceding subspecies, and none are figured on this account.

5. Colinus ridgwayi BREWSTER.

MASKED BOB WHITE.

Colinus ridgwayi BREWSTER, Auk, II, April, 1885, 199.
(B —, C —, R —, C --, U 291.)

GEOGRAPHICAL RANGE: Sonora, Mexico, and southern Arizona.

This species, only recently added to our avifauna, was first described by Mr. William Brewster in the Auk (Vol. II, 1885, p. 199), from a specimen taken by Mr. F. Stephens, August 11, 1884, about 18 miles southwest of the little town of Sasabe, in Sonora, Mexico. Fully a year previously, however, Mr. Herbert Brown, of Tucson, Arizona, obtained some of these birds within our border, which he sent East, where several ornithologists examined and pronounced them to be Grayson's Bob White (*Colinus graysoni*), a Mexican species, which proved to be a mistake, as the specimens sent were identical with the bird subsequently described by Mr. Brewster. The credit of discovering the Masked Bob White clearly belongs to Mr. Brown, and what little information we possess about its range within the limits of the United States and the habits of this species is principally due to his patient inquiry and personal investigation. He gives its range as follows: "The Masked Bob White is found in the country lying between the Barboquivari Range in Arizona and the Gulf coast in Sonora, more especially between the Barboquivari and the Plomoso, where this species is quite abundant. They are also found on the

[1] Journal für Ornithologie, Cabanis, 1856, p. 337.

Sonoita Creek, about 60 miles north of the Sonora line. From the Sonoita Valley they range in a westerly direction fully 100 miles, and through a strip of country not less than 30 miles in width within Arizona Territory. Very probably they may go beyond this, both to the east and west. The habits of the Masked Bob White, so far as we know them, appear to resemble very closely those of the common Quail, only slightly modified by the conditions of their environment. They utter the characteristic call of ‘Bob White’ with bold, full notes, and perch on rocks or bushes while calling. They do not appear to be a mountain bird, but live on the mesas (table lands) in the valleys, and possibly in the foothills.

“The Masked Bob White was, three years since, abundant in the neighborhood of Bolle's Well, a stage station on the Quijotoa road, near the northern end of the Barboquivari Range, 29 miles southwest of Tucson and about 40 miles north of the Mexican boundary line. As the station was then comparatively new the grass thereabouts was high and these Quail could be had for the taking; but now that the stock has eaten away the grass the birds have not for a year or more been seen about the place. On the road from Bolle's Well west to the Coyote Range (about 25 miles) these Quail were frequently to be met with, but the teamsters and travelers have killed or frightened them off. One of the former assured me that he had killed as many as five at one shot. Ten miles south of Bolle's, in the Altar Valley, we came across a small covey, perhaps a dozen in all. The bright deep chestnut breast plumage of the males looked red in the sun and gave the birds a most magnificent appearance. We secured but one, a male, the rest secreting themselves in the tall sacaton grass, which at this point was between 4 and 5 feet high, and as we had no dog we did not follow them. Our next place to find them was on the mesa southeast of the peak, where we camped to hunt for them, but they were scarce here, and we managed to secure but few.

“In addition to their ‘Bob White’ they have a second call of ‘Hoo-we,’ articulated and as clean cut as their ‘Bob White.’ This call of ‘Hoo-we’ they use when scattered, and more especially when separated toward nightfall. At this hour I noted, that, although they occasionally call ‘Bob White,’ they never repeated the first syllable, as in the daytime they now and then attempted to do. In body they are plumpness itself; in this respect, considering size, they overmatch the Arizona Quail (*Callipepla gambeli*) with which I compared them. In actual size of body, however, the latter is the larger. Of three stomachs of this species examined, one contained a species of mustard seed, a few chaparral berries, and some six or eight beetles and other insects, ranging in length from a half inch down to the size of a pin head. The second was similarly provided, but contained, in lieu of mustard seed, a grasshopper fully an inch in length. These two were taken on the mesa. The third, from a bird taken in the valley, contained about twenty medium-sized red ants, several crescent-shaped seeds, and a large number of small, fleshy, green leaves.”

¹ Extracted from “Arizona Quail Notes,” Forest and Stream, December 31, 1885. The name of “C. ridgwayi” is substituted by me for C. graysoni, where used in above article.

Lieut. H. C. Benson, Fourth Cavalry, U. S. Army, found the Masked Bob White fairly abundant near Campos and Bacuachi, Sonora, securing a number of specimens in the summer of 1886, and he writes me that they only frequented cultivated fields there, where wheat and barley had been raised. He also found another species of Partridge associated with these birds near Campos, which probably crosses our border also. This Partridge was recently described by Mr. R. Ridgway, in Forest and Stream of March 3, 1887, and named "*Callipepla elegans bensoni*," in honor of its discoverer.

Nothing positive has been known about the nest and eggs of the Masked Bob White till the present season, though one of their nests containing six eggs was found some years ago. They were allowed to remain in hope of seeing the number increased, and when visited at another time they had hatched. The nest as described to Mr. Brown was a shallow excavation alongside a tuft of grass. The eggs were white and unspotted. In the spring of 1890 Mr. Brown succeeded in obtaining one of these eggs cut from the oviduct of the female, as well as a set of eleven eggs said to belong to this species. These eggs were found early in May in a similar situation in the vicinity of Arivaca. The egg of the Masked Bob White is white, unspotted, subpyriform in shape, and their average measurement is 31 by 25 millimetres.

Since the foregoing was written, Mr. Otho C. Poling writes me from Fort Huachuca, Arizona, as follows: "I first met with the Masked Bob White on May 24, 1890, in a series of low foothills extending off to the northwest of the Huachuca Mountains, and ending in a somewhat higher range of hills called the Canella Range, being the northwestern termination of the Huachucas. Most of these hills are thickly grown up with juniper, cedar, manzanita, stunted oaks, pines, and with a heavy growth of grass. While riding along one of these grassy ridges I flushed what I supposed was a "Fool Quail," *Cyrtonyx montezumæ*. I at once staked my horse, and on shooting the bird I found it to be a male of the Masked Quail. I continued the search and had spent a half hour or more, when, as I passed within 2 feet of a mescal plant, *Agave americana*, I suddenly flushed the female from underneath it. I first shot her and then expected to find the nest; but after searching for several hours and failing, I gave it up. The female contained an egg which was fully developed and would probably have been laid within half an hour. I searched for the nest of this species on the two succeeding days as well, but made no further discoveries. The single egg of *Colinus ridgwayi* obtained by me is pure white and measures 31 by 24 millimetres.

"The Canella Range is about 25 miles north of the Mexican border. Two more specimens of the Masked Quail, both females, were shot about June 10, in the Huachuca Mountains, near the middle of the range, at an elevation of about 6,000 feet. The birds were found in a cañon about 15 miles north of the border, on the northeastern slope of the range."

From the foregoing it would appear that the Masked Bob White is confined to a narrow strip of country along our southwestern border, and is nowhere as common as the Gambel's and Scaled Partridges, which are found in the same regions. The eggs appear to be indistinguishable from those of the eastern Bob White, and no specimen is figured on that account.

6. Oreortyx pictus (Douglas).

MOUNTAIN PARTRIDGE.

Ortyx picta DOUGLAS, Transactions of the Linnæan Society, XVI, 1829, 143.
Oreortyx pictus BAIRD, Birds of North America, 1858, 642.
(B 473, C 390, R 481, C 574, U 292.)

GEOGRAPHICAL RANGE: Pacific coast districts, from Santa Barbara, California, north to Washington.

This handsome bird is the largest of the American Partridges. It is better known on the Pacific coast by the name of "Mountain Quail." Its range is not an extensive one, as it is only found along the western slopes of the Coast Range in California from about latitude 34° northward, and throughout western Oregon, where it has a slightly more easterly range than in California, typical specimens having been taken near Mount Hood. In regard to its occurrence in the new State of Washington, Prof. O. B. Johnson, of the University of Washington, Seattle, Washington, writes me as follows: "Twenty years ago this species was found but little north of the Willamette Valley, Oregon, but they gradually worked down the south side of the Columbia River toward Astoria, and in 1872 I was informed that some of these birds, shot at Kalama, Washington, were the first seen north of the Columbia. A crate of trapped birds sent to the Seattle market, were some time afterward purchased by the Young Naturalists' Society and set free. These have since multiplied nicely, and others have been sent to Whitby Island, 40 miles north of Seattle, where, I understand, they are also doing well. A covey wintered in a barn lot with the hens, just at the outskirts of Seattle this winter."

It is only within the last twenty years that this Partridge has obtained a permanent footing in Washington, and while a few birds may have crossed the Columbia River near Kalama, the majority were introduced, quite a number having been liberated near Vancouver Barracks and other localities as well. This species is affected by climatological conditions, and is only to be found in the moist mountainous regions along the coast, where the rainfall is heavy. In the dryer regions of the interior it is replaced by a paler race, *Oreortyx pictus plumiferus*.

The Mountain Partridge is a constant resident wherever found, and is quite common in portions of its range, especially about Fort Gaston, California. Its habits are similar to those of the Plumed Partridge, which will be more fully described. Professor Johnson writes me, "the males have a sharp challenge note, which they utter with the beak pointing straight up and with wings dropped, while standing on a fencepost, broken limb, or upturned root. It is a sharp whistle of but one note, and may be imitated by whistling the word "querk" in middle E, and is audible for a long distance."

Nidification commences about the middle of May, and ordinarily but one brood is raised. The nest is placed on the ground, alongside or under an old log, or on side hills under thick bushes and clumps of ferns, occasionally along the edges of clearings, grain fields, or meadows. A nest found May 27, 1877, near Coquille, Oregon, containing six fresh eggs, was well concealed under a bunch of tall ferns, in a tract of timber killed by a forest fire. Another, taken in Ukiah Valley, Mendocino County, California, June 2, 1883, by Mr. C. Purdy, contained twelve fresh eggs. This nest was found under a bush of poison oak among a lot of dry leaves on a steep hillside. The average number of eggs laid by this Partridge is about ten, most of the sets containing from eight to twelve. An occasional nest contains as many as sixteen, but such large sets are rare.

The eggs vary from a pale cream color to a rich creamy buff, and are unspotted. In shape they are short ovate, and very much resemble unmarked eggs of the Ruffed Grouse although of smaller size. They are indistinguishable in shape, size, and color from the eggs of the next subspecies, and for that reason none are figured. The average size of nineteen specimens in the U. S. National Museum collection is 34.5 by 26 millimetres, the largest egg measuring 36 by 26.5, the smallest, 34 by 25 millimetres.

7. Oreortyx pictus plumiferus (GOULD).

PLUMED PARTRIDGE.

Ortyx plumifera GOULD, Proceedings Zoölogical Society, 1837, 42.
Oreortyx pictus var. *plumiferus* RIDGWAY, in History North American Birds, III, 1874, 476.
(B —, C —, R 481a, C —, U 292a.)

GEOGRAPHICAL RANGE: From the west side of the Cascade Range in northern Oregon (except near the coast) south, along both sides of the Sierra Nevada and the southern coast ranges of California (south of latitude 34° only) to northern Lower California.

The Plumed Partridge, a bird as handsome as the preceding, inhabits the interior mountain regions from the southern border of California northward through middle western Oregon, as well as parts of western Nevada, approaching the seacoast in the extreme southern portion of its range only. It is everywhere known as the "Mountain Quail," and deserves this name far more than the preceding subspecies, reaching much higher altitudes than the former. On the eastern slopes of the Sierra Nevada it has been observed from Walker's Pass, near Fort Tejon, northward. Personally, I have often met with it on Mount Kearsarge, in Inyo County, California, where it reached an altitude of 10,000 feet in summer. Mr. Robert Ridgway noticed this subspecies near Carson City, and in the Comstock Mountains near Pyramid Lake, Nevada, and Dr. A. K. Fisher obtained it in the Argus Range of mountains in south-eastern California, the most easterly known point of its range. It is essentially a bird of the mountains, where it is more partial to the open pine forests and rocky ridges, covered with chaparral and undergrowth, than to the densely timbered portions of the ranges. I have always found it a shy and retiring bird, usually to be found only in small coveys, and on being alarmed trusting more to the legs than wings for escape. While equally abundant on both sides of the Sierra Nevada, and common enough on the western slopes of the Cascade Mountains in Oregon, it does not seem to occur on the east side of the latter range; at any rate no specimens have been brought to the notice of naturalists from such localities, and as far as known to me, unless recently introduced, it does not occur in Washington. It is a resident and breeds wherever found, excepting in the higher portions of the mountains frequented as summer haunts, from which the birds retire on the approach of winter.

Mr. Charles A. Allen, of Nicasio, California, a careful and trustworthy observer, writes me as follows: "I find this Partridge all through the Sierras. In the spring many go up to the snow line, returning in the fall below the point of snowfall. These vertical migrations are performed *entirely on foot*, unless streams must be crossed, when they take to their wings, but alight at once on gaining the opposite side, and continue their travels on foot."

The mating season begins in the latter part of March and the beginning of April, according to latitude and altitude. The call note of the male is a clear whistle, like "whu-ié-whu-ié," usually uttered from an old stump, the top of a rock, or a bush. When alarmed, a note like "quit-quit" is used. In the higher mountains but a single brood is raised; but in the lower foothills they rear two broods occasionally, the male caring for the first one while the female is busy hatching the second.

I met with a brood of young birds, perhaps a week or ten days old, near Jacksonville, Oregon, on June 17, 1883. The male, which had them in charge, performed the usual tactics of feigning lameness, and tried his very best to draw my attention away from the young, uttering in the mean time a shrill sound resembling "Quaih-quaih," and showed a great deal of distress, seeing I paid no attention to him. The young, already handsome and active little creatures, scattered promptly in all directions, and the majority were most effectually hidden in an instant. As nearly as I was able to judge they numbered eleven. I caught one, but after examining it turned it loose again. The feathers of the crest already showed very plainly.

Their food consists of insects, the buds and tender tops of leguminous plants, small seeds, and berries of various kinds. The nest, simply a slight depression in the ground scratched out by the bird, and lined perhaps with a few dry leaves, pine needles, grasses, and usually a few feathers lost by the hen while incubating, is sometimes placed alongside an old log, at other times under low bushes or tufts of weeds, ferns, and, when nesting in the vicinity of of a logging camp, a favorite site is under the fallen tops of pine trees that

have been left by wood-choppers, the boughs of which afford excellent cover
for the nest.

Mr. L. Belding found a deserted nest of this species in a cavity of the
trunk of a standing tree near Big Trees, California, but in this locality they
nest oftener in thickets of the rock rose or the tar-weed, and according to his
observations they do not desert their nests for slight cause, like the Bob White
or the California Quail. The eggs vary from eight to fourteen in number,
averaging usually about eleven, and are indistinguishable from those of the
preceding subspecies. They are generally more or less stained by contact with
the lining of the nest and the soil.

Mr. Loren W. Green, of the U. S. Fish Commission, reports the Plumed
Partridge as quite common near Baird, Shasta County, California, and he found
a nest containing nineteen eggs, seeing also a brood of twenty young birds. In
that vicinity they raise two broods a season. The earliest date on which eggs
were found by him was April 15, and the latest August 15. Occasionally a
nest is placed on top of old decayed tree stumps. Rattlesnakes are very common in this locality, and these birds have adopted such sites, probably from the
fact that they afford them better protection from such enemies. A nest taken
by Mr. Green on May 24, 1886, near Redding, California, contained twelve
fresh eggs; these were placed in a slight excavation on a hillside under a
small bush, and well concealed from view. Another, found May 5, 1885, in
the Volcan Mountains, San Diego County, California, by Mr. F. W. Paine, at
an altitude of about 5,000 feet, contained eight slightly incubated eggs. It was
placed among a lot of dry leaves in a pine and fir forest.

Mr. A. M. Ingersoll, of San Diego, California, found a nest of the Plumed
Partridge containing ten eggs, under a mass of excelsior and other shavings
that had lodged against a brace of a snowshed, less than 15 feet from the railroad track on the Union Pacific Railroad in the Sierra Nevada Mountains.

Mr. F. Stevens found a full set of eggs of this subspecies in southern
California as early as April 7.

The ground color of the eggs of the Plumed Partridge varies from pale
cream to a reddish buff, and in shape they are short ovate. Incubation lasts
about twenty-one days, and an egg is laid daily until the set is complete. The
shell is smooth and slightly glossy, and usually more or less stained by contact
with the lining of the nest or the soil.

The average measurement of sixty-six specimens in the U. S. National
Museum collection is 34.5 by 26.5 millimetres. The largest egg of the series
measures 38 by 28, the smallest 32 by 25.5 millimetres. The type, No. 10049,
(Pl. 1, Fig. 3), selected from a set of ten eggs, is one of the darkest colored specimens; it was taken June 7, 1866, near Downieville, California, by Mr. William
Veille. No. 18187 (Pl. 1, Fig. 2), one of the palest colored specimens of the
series, showing the other extreme, was collected by Mr. L. Belding, June 10,
1880, in Bear Valley, Alpine County, California. This nest had been abandoned and contained only six eggs when found.

8. Oreortyx pictus confinis ANTHONY.

SAN PEDRO PARTRIDGE.

Oreortyx pictus confinis ANTHONY, Proceedings California Academy Sciences, 2d ser.,
ii, October 11, 1889, 74.
(B —, C —, R —, C —,U 292b.)

GEOGRAPHICAL RANGE: San Pedro Mountains, Lower California (and southward?).

This recently described subspecies differs from *Oreortyx pictus plumiferus* in its grayer upper parts and thicker bill.

Mr. A. W. Anthony, who discovered this new race, writes me that the breeding range of the San Pedro Partridge extends from the foothills along the base of the San Pedro Mountains, Lower California, to the tops of the highest peaks, estimated at about 12,500 feet. It is not common below an altitude of 2,500 feet. He says: "The first call notes are heard about the last of February, a soft-tuned, far-reaching 'chay-chay,' as mellow and sweet as the notes of a flute. The alarm note when surprised is a soft, loud, 'ch-ch-ch-c-c-ca, ch-c-c-ca,' which, as the danger becomes more evident, becomes harder, sounding like 'kee-kee-ke-ea,' which is taken up by each member of the covey as they gradually draw away, and long after the last of the band has disappeared the soft chorus floats back from the manzanitas and lilacs which clothe the hills where this subspecies is found. Frequently a loud 'pit-pit' is heard as they take wing, but they are more silent when on the wing than *Callipepla californica*. A few pairs bred about my camp at Valladares Creek, 6 miles from the base of the range and about 2,500 feet above the sea level, but nearly all of the flocks that wintered along the creek at this point disappeared by March, leaving only an occasional pair, which sought the shelter of the manzanitas high up on the hillsides, from whence their clear, mellow notes were heard morning and evening, so suggestive of cool brooks and rustling pines, but so out of place in the hot barren hills of this region.

"The only eggs I have taken were found at my camp on the Valladares Creek, March 29, 1889. The nest, placed in the midst of thick manzanita chaparral high up on a hillside having a northern exposure, consisted of a mere hollow under a manzanita bush, lined, or rather filled, with dry leaves of the lilac and manzanita, and contained but a single egg. The female, which I shot as she left the nest, was about to deposit a second, which I secured upon skinning her." Mr. Anthony further states that the eggs of the San Pedro Partridge resemble those of the Plumed Partridge in shape and color, being creamy white and unspotted. The two specimens measure 36 by 28 and 37 by 27 millimetres, respectively.

26957—Bull. 1——2

9. Callipepla squamata (Vigors).

SCALED PARTRIDGE.

Ortyx squamatus Vigors, Zoölogical Journal, v, 1830, 275.
Callipepla squamata Gray, Genera of Birds, iii, 1846, 514.
(B 476, C 393, R 484, C 577, U 293.)

Geographical Range: Northern Mexico and contiguous border of United States, from western Texas, through New Mexico, to southern Arizona.

The Scaled Partridge, usually called the Blue Quail and also the White Top-knot Quail, is a constant resident of southwestern Texas from about latitude 28° northward along the valley of the Rio Grande, as well as of a considerable portion of New Mexico and southern Arizona, extending south into Mexico. Specimens have been taken as far north as Chico Springs, Colfax County, New Mexico, where Mr. Thurlow Washburne reports them as fairly common in rocky places wherever water is found. Bonham, Fannin County, Texas, is the most eastern point at which it has been noted, and in Arizona the mountains in the vicinity of the Colorado River probably mark its western boundary. It has been traced at least 90 miles west by southwest from Tucson to Wood's Station, and probably extends still further west in this direction. It does not appear to occur north and east of the White and Mogollon Mountains. They are very common in some portions of southern Arizona. Mr. Herbert Brown writes me from Tucson as follows: "I have seen this bird, both far away from, and in the immediate vicinity of, water, on the open valleys and plains, and also in the rough foothills of the mountains. In the Altar Valley, west and southwest of Tucson, I have seen large numbers of them, but as the foothills of the mountains are approached they give way to Gambel's Quail entirely. They are also at all times very numerous in the Sierritas about 50 miles south of Tucson, as well as west of the Catalina Mountains."

According to my own observations the Scaled Partridge is found most abundantly on the high plateaus bordering on the principal streams of the regions under consideration, reaching an altitude of from 1,500 to nearly 7,000 feet. It shuns timbered country, and in southern Arizona, where I have frequently met with these birds, they seemed to me to prefer the most barren and driest portions of that scantily watered Territory. I invariably found them back in the foothills and mesas, from 2 to 5 miles distant from the river beds, which are generally dry for the greater part of the year.

These barren and rocky foothills and table lands are covered in places with a dry, harsh vegetation, consisting of different species of cacti, stunted yuccas, catclaw-mimosa, creosote, and dwarf sage bushes, where the soil is so parched that scarcely anything else will flourish, and where nearly every shrub is covered with sharp spines or thorns; such places I found to be the favorite home of the Scaled Partridge. Many times have I seen coveys miles from water, and it appeared to me that, judging from the kind of country it inhabits by preference, this universally needed fluid was scarcely required by these birds. From the repelling nature of the country it generally frequents it is naturally hunted very little; still, I found them exceedingly shy and wary and very difficult to approach, far more so than Gambel's Partridge. It prefers to trust to its legs for escape almost entirely, and is generally successful, being an expert and swift runner, dodging in and out among the bushes with great ease and dexterity, consequently they are soon lost to sight. The covey generally follow a leader, Indian-file fashion. Their habits seem, however, to differ somewhat in other sections.

Mr. E. W. Nelson furnishes me with the following observations about this species: "In New Mexico I have found the Scaled Partridge abundant on the Jornada del Muerto, and thence north along the valley of the Rio Grande to the vicinity of Santa Fé and Cañoncito. Near the latter place the birds occur sparingly in the lower portions of the cedar and piñon belt, but as a rule the Scaled Partridge is preëminently a species of the open arid plains.

"In many instances I have found them far from water, but they make regular visits to the watering places. On the Jornada del Muerto and on Santa Fé Creek I found them frequenting the open plains, away from the water in the middle of the day, and in the vicinity of the water late in the afternoon. At this time they are often seen in company with Gambel's Quail amongst the bushes and coarse grass or weeds bordering the water courses. They are very difficult to flush, owing to the rapidity with which they run through the bushes or other vegetation. When flushed, they scatter and only fly a short distance, when they alight and run on as before. As soon as the alarm is over the old birds reunite the flock by a low call note.

"The latter part of summer and early fall they gather into coveys, often containing several broods, as I observed in 1882, in the valley of the Gila River, near Clifton, Arizona. At this season they frequented the low bare hillsides, or the now dry water courses and the fields adjoining these, associating with Gambel's Partridge. They are easily trapped in the fall and winter, and many are caught by the natives and taken to the markets of the larger towns of New Mexico and Arizona."

Mr. William Lloyd writes me from Marfa, Texas, as follows: "The Blue Quail loves a sandy table land, where they spend considerable time in taking sand baths. I have often watched them doing so, pecking and chasing each other like a brood of young chickens. Good clear water is a necessity to them. They are local, but travel at least 3 miles for water. In the evenings they retire to the smaller ridges or hillocks and their calls are heard on all sides as the scattered covey collects. Several times I have seen packs numbering sixty to eighty, but coveys from twenty-five to thirty are much oftener noticed. The mating season begins sometimes as early as March, and after the female commences laying, generally about six weeks later, the male at about sundown every fine evening, mounts a convenient bush or rock and calls his mate, which approaches noiselessly and they disappear together.

The young are wary, and crouch in or under the smallest tufts of grass, while the parents remain in full sight. During the middle of the day they frequently alight in trees, usually large oaks, but they roost on the ground at night."

Mr. A. W Anthony says: "I think this species, like *Callipepla californica vallicola*, is largely influenced by the seasons in regard to its nesting, although perhaps not to the same extent. In southwestern New Mexico the season of 1886 was a very dry one, no rain falling during the spring and but very little during the preceding winter. As a consequence the vegetation of the deserts was unusually scarce. The Scaled Partridges remained in flocks till very late, and in many cases did not break up at all. Young birds were very scarce during the fall, and none were noted, I think, till after August 10. The only nest that came under my notice that season was taken on July 31. It was located on a high, rocky hillside, in a slight hollow between two rocks, and slightly lined with dry grass. A large dead *Agave* (American aloe) had fallen over the nest, hiding it so effectually that its discovery was accidental, the female being flushed by my brushing against the dry leaves of this plant. It contained eight fresh eggs. I have taken chicks not over two weeks old in Grant County, New Mexico, as late as October 10, while young of the year taken the same day could not be distinguished from the adults by the closest scrutiny."

I believe two and even three broods are occasionally raised in a season, the male assisting in the care of the young, but not in incubation. This lasts about twenty-one days. Full sets of the eggs of the Scaled Quail have been taken early in April and others late in September. The nest is always placed on the ground, as far as known, usually under the shelter of a yucca or small bush, overgrown with grama grass, and in the Rio Grande Valley, New Mexico, occasionally in a wheatfield. In southern Arizona Mr. O. C. Poling has found the Scaled Partridge nesting in corn and grain fields, in alfalfa meadows and potato patches, as well as on almost barren flats, where only a few scattering bushes but a few inches high grew. Here the nesting season began about May 1, and eggs were found as late as July 15.

Their food consists of small seeds, grain when procurable, berries of various kinds, the tender tops of plants, small beetles, ants, and grasshoppers. A young female of this species, probably about ten days old, and taken by Mr. Herbert Brown in Altar Valley, Arizona, September 27, 1885, may be described as follows: Upper parts chestnut-brown, each feather with medial T-shaped white markings. Tail-feathers ashy, barred with alternate pale black and white bars. Upper parts of the breast buffy-grayish, each feather about the neck with a V-shaped lighter area. Belly rusty buff, transversely barred with brown. Sides of head reddish buff, crest brownish, each feather with a narrow medial stripe of white.

According to Mr. Lloyd their call note sounds something like a lengthened "chip-churr, chip-churr;" the same, only more rapidly repeated, is also given when alarmed, and a guttural "oom-oom-oom" is uttered when worried or chased by a Hawk. The young utter a plaintive "peep-peep," very much like young chickens. Like the rest of the Partridge tribe they are able to run about as soon as hatched.

Mr. W. H. Cobb, of Albuquerque, New Mexico, informs me "that in this part of the Territory this species is a sort of semi-migrant. The greater portion of the birds move to the higher mesas and foothills of the mountains to breed, and during the cold weather return to the river bottoms where, in favorable localities, a few remain throughout the year."

Mr. W. E. D. Scott found this species in great abundance in a little valley west of the Santa Catalina Mountains, which the road from old Camp Grant to Tucson, Arizona, crosses. Here, he says, they associated with Gambel's Quail, apparently on the most friendly terms.

The number of eggs to a set ranges usually from nine to sixteen, generally about eleven or twelve, and an egg is deposited daily. Occasionally a larger set is found. The shells of these eggs are very thick and without lustre. The ground color varies from a very pale creamy white to a pale buff. The markings are sharp and well defined in most cases, varying from mere pin points, scarcely perceptible to the naked eye, to the size of No. 12 shot. These spots are usually round and of equal size, and pretty evenly distributed over the entire egg. Occasionally a set is marked with somewhat more irregular, as well as larger, spots or blotches, resembling certain types of eggs of *Callipepla gambeli*, but these markings are always paler colored and not so pronounced. They vary in color from a pale reddish brown or ochraceous to a vinaceous buff and fawn color in different sets.

The average measurement of twenty-eight specimens in the U. S. National Museum collection is 32.5 by 25 millimetres; the largest egg of the series measuring 34 by 27, the smallest 30.5 by 25 millimetres. In shape they vary from short ovate to subpyriform.

As there is practically no difference in the eggs of this species and of the Chestnut-bellied Scaled Partridge, the eggs figured as typical of the latter would also answer for the present species under consideration. If there is any difference it seems to be in size only, and even this is doubtful.

The type specimen, No. 23165 (Pl. 1, Fig. 4), selected from an incomplete set of six eggs, was collected by First Lieut. H. C. Benson, Fourth Cavalry, U. S. Army, near Fort Huachuca, Arizona, August 6, 1886, and represents one of the palest colored eggs of the series.

No. 23776 (Pl. 1, Fig. 5), from a set of twelve eggs collected by Lieut. M. H. Barnum, Third Cavalry, U. S. Army, near Marathon, Texas, June 22, 1889, represents one of the heavier marked eggs of this species, and the remaining two types figured under the next subspecies are still better marked and represent the extremes.

10. Callipepla squamata castanogastris BREWSTER.

CHESTNUT-BELLIED SCALED PARTRIDGE.

Callipepla squamata castanogastris BREWSTER, Bulletin Nuttall Ornithological Club,
VIII, January, 1883, 34.
(B —, C —, R —, C —, U 294c.)

GEOGRAPHICAL RANGE: Lower Rio Grande Valley in Texas, south to San Luis
Potosí, eastern Mexico.

This well-marked subspecies is easily distinguished from the Scaled Partridge, being a much richer and darker colored bird than the latter. Its range seems to be a very restricted one, and is confined to the Lower Rio Grande Valley in Texas and eastern Mexico. It has been taken near Fort Brown, Texas, by Asst. Surg. James C. Merrill, U. S. Army, where it is rare, and it extends northwestward from this locality at least to Eagle Pass, Texas. According to Mr. George B. Sennett, the foothills of the Rio Grande, about 100 miles back from the coast, mark the eastern limit of this bird.

"Mr. C. W. Beckham reports this subspecies from Mineral City, about 50 miles northwest from Corpus Christi, Texas."[1]

According to Mr. J. A. Singley, it seems to be very common about Ringgold Barracks and Rio Grande City, Texas, from which points it does not extend more than 50 miles into the interior. He reports them as abundant in the hilly country near Rio Grande City, and at the Las Cuevas Rancho, 15 miles south, and as rare at and below Hidalgo, Texas.

The general habits of the Chestnut-bellied Scaled Partridge as well as its food are very similar to those of the preceding subspecies. The mating and nesting season, however, commences somewhat earlier. Full sets of eggs have been taken near Rio Grande City, and at Camargo on the Mexican side of the river opposite, as early as March 11, and from that time up to July 10. Two broods are unquestionably raised in a season. Mr. Thomas H. Jackson, of West Chester, Pennsylvania, gives the average number of eggs laid by this species as fifteen, based on data taken from twenty-seven sets. The largest number found in one nest was twenty-three. I am indebted to him, as well as to Mr. George B. Sennett, for the loan of a number of specimens for examination. Their nests are always placed on the ground; a slight hollow in the sand is scratched out by the bird, usually under a clump of weeds or grass, or a prickly-pear bush. They are very slightly lined with dry grasses. The shape and color of their eggs are very similar to those of *Callipepla squamata*.

From the material before me it would appear that more of the eggs of this subspecies are plainly and distinctly spotted than is the case with the former, and they also average a trifle smaller. The average measurement of seventy-seven specimens examined is 31 by 24 millimetres. The largest egg of the series in

[1] Proceedings U. S. National Museum, Vol. X, 1887, p. 656.

the U. S. National Museum collection measures 34 by 25, the smallest 25 by 21 millimetres.

The type specimen, No. 24021 (Pl. 1, Fig. 7), selected from a set of thirteen eggs, was taken June 21, 1890, near Camargo, Mexico, and purchased from Mr. Thomas H. Jackson, of West Chester, Pennsylvania.

An egg (figured on Pl. 1, Fig. 6), from a set of fifteen collected near Rio Grande City, Texas, was borrowed from, and is now in the collection of, the above-mentioned gentleman. These two eggs represent the heavier marked types, others resemble the two eggs figured under the preceding subspecies so much that they are practically indistinguishable. Specimens marked like the four eggs figured can be found among the eggs of either subspecies.

11. Callipepla californica (Shaw).

CALIFORNIA PARTRIDGE.

Tetrao californicus Shaw, Naturalists' Miscellany, 1797 (?), Pl. cccxlv.
Callipepla californica Gould, Monograph Odontophorinæ, 1850, Pl. xvi.
(B 474, C 391, R 482, C 575, U 294.)

GEOGRAPHICAL RANGE: Coast region of California, Oregon, Washington, and Vancouver Island, British Columbia.

This handsome and well-known western game bird, commonly called Valley or Top-knot Quail, is an inhabitant of the coast region of California from about latitude 34° northward along the coast of Oregon, the new State of Washington, and some of the islands adjacent thereto, including Vancouver Island, British Columbia. In Washington and the islands of Puget Sound it was originally introduced, however, and according to Dr. Suckley, one of the pioneer naturalists of the Northwest coast, this was first done by Governor Charles H. Mason and Mr. Goldsborough as early as 1857, when two lots were turned out on the prairies near Puget Sound, and by the following winter they had increased largely.[1]

Prof. O. B. Johnson, of Seattle, Washington, states: "This species is very common now on Whitby Island, in Puget Sound, which seems to be especially suitable to these birds, owing to the extensive prairies and open oak parks found thereon."

Their favorite haunts are the undergrowth and thickets along water courses, brush-covered side hills, and cañons, frequenting the roads, cultivated fields, vineyards, and edges of clearings to feed. It is a constant resident, and breeds wherever found.

The mating time commences early in March, sometimes later, depending on the season. Then the large packs into which this species gathers in the fall of the year break up gradually, each pair of birds selecting a suitable nesting site. In the more densely settled portions of California this Partridge is by no means as common now as it was a decade ago, when it was not unusual to

---

[1] History North American Birds, 1874, Vol. III, p. 481.

see packs numbering five hundred and more together, while now, at least near the larger cities, coveys even of fifty birds are rarely seen. In localities where not constantly harassed and hunted the California Partridge becomes surprisingly tame and confiding, in fact almost domesticated; and under such circumstances many nest close to houses and outbuildings and in the shrubbery of gardens adjoining human habitations.

Mr. Charles A. Allen, of Nicasio, Marin County, California, writes me as follows: "I found a nest of this species under a bunch of Snowberry bushes (*Symphoricarpus*), not 30 feet distant from my house, containing twenty-one eggs. I watched this nest daily, and two weeks after finding it the eggs hatched. The female was still on the nest, and the little heads of the young were peeping out all around her. They became very tame, ran all around the yard, and took but little notice of the members of my family when going among them. I think the average number of eggs laid by this species is about fourteen. While the female is incubating, the male usually mounts some old stump, a dead limb, or fencepost in the vicinity of the nest, and every few seconds utters a long-drawn note not unlike 'whää-whää.'"

Mr. W. Otto Emerson, of Haywards, California, states: "I have never known the male to assist in the duties of incubation, but he will make his appearance twice a day near the nesting site. First at break of day, when he gives his call note, "kuck-ku, kuck-ku;" the female then comes off to feed an hour, and the same is repeated at dusk. During the past summer, 1889, I found a nest of this species in a pile of brush in the chicken yard, and in May, 1880, a nest was found in a similar situation, within 15 feet of our front door. Both carriages and persons were passing nearly every hour within 4 feet of this nest, and the bird was frequently disturbed, but did not seem to mind it much. Another nest was placed in the short grass alongside of a highway, with nothing at all to conceal it. The food of the downy young consists of insects, small seeds of various plants, and chickweed. When alarmed, the old bird gives one or two notes of warning and flies away. The young, when still too small to fly, hide quickly under anything in the shape of a leaf or in the grass, and lie close to the ground till the danger is passed, when they are called together again by the parent."

The nest of the California Partridge is but a very flimsy affair at best. Any place alongside of a rock, log, or an old stump, under a pile of brush, small bush, or a bunch of weeds or grass will answer. Occasionally the eggs are laid in a perfectly open situation without any attempt at concealment whatever, and now and then a hen's nest in the chicken house is used for this purpose. The site once selected, and it does not seem to be a difficult matter to please them in this respect, a slight hollow is scratched out by the bird, and this is sparingly lined with any convenient material near at hand, usually bits of grass. As incubation advances a few feathers drop from the setting hen and work in among the eggs. These, I believe, are constantly turned and rearranged from day to day. Incubation is variously stated to last from twenty-one to twenty-eight days. I believe the first-mentioned period is nearer correct.

Mr. Walter E. Bryant, well known as an excellent ornithologist, writes as follows regarding some unusual nesting sites of the California Partidge: "Essentially a ground-building species, but several cases have come under my notice of its nesting in trees, upon the upright end of a broken or decayed limb, or at the intersection of two large branches. A few years ago a brood was hatched in and safely conducted away from a vine-covered trellis at the front door of a popular seminary. How the parents managed to get the tender young down to the ground is not known."[1]

The young run about as soon as hatched. Usually but one brood is raised, occasionally two. In the latter case the male takes charge of the young when they are about three weeks old, the female then laying the eggs for the second. Downy young have been observed as early as May 20 in the southern portions of their range, and some broods are undoubtedly hatched still earlier. In the fall, when the young are full grown and able to shift for themselves, they collect in large packs, a number of coveys associating together until spring. They are much shyer then and more difficult to approach. The usual call note, when one of these packs become scattered, is a rather unmusical "ca-āpe, ca-āpe," the last syllable drawn out; another note, like "kā-kāāh," is also used on such occasions.

From twelve to sixteen eggs seems to be the average number laid. The largest number found in a nest of which I have any reliable record is twenty-one, but undoubtedly more are occasionally found when two hens lay in the same nest. Their ground color is usually creamy white, but now and then a decidedly buff-colored set is found. The markings vary from fine dots, usually well rounded and of various sizes, to irregular outlined spots and blotches of different shades of dark chestnut brown, olivaceous drab, and golden russet, generally pretty evenly scattered over the entire egg. In shape and thickness of shell they resemble the eggs of *Colinus virginianus*.

The average measurement of forty-eight specimens in the U. S. National Museum collection is 32 by 25 millimetres. The largest egg of this series measures 35 by 26, the smallest 30 by 24 millimetres.

Of the three type specimens figured, which are selected to show the variations and different styles of markings, Nos. 17922 and 21109 (from the Bendire collection) were obtained near Santa Cruz, California, on June 13, 1874, and July 21, 1877 (Pl. 1, Figs. 8 and 10), and No. 23912 (Pl. 1, Fig. 9) was taken near Haywards, California, on April 21, 1883, by Mr. W. Otto Emerson.

---

[1] Bulletin California Academy of Sciences, II, 1887. p. 451.

12. Callipepla **californica** vallicola RIDGWAY.

VALLEY PARTRIDGE.

Callipepla californica vallicola RIDGWAY, Proceedings **U. S. National Museum, VIII,** 1885, 355.

(B —, C —, R —, C —, U 294*a*.)

GEOGRAPHICAL RANGE: From western and southern Oregon, except near the coast, south through western Nevada and the interior of California to Cape St. Lucas, Lower California.

This race, a paler and grayer-colored bird than the preceding, is an inhabitant of the drier interior valleys and foothills of the mountains ranging from Cape St. Lucas, Lower California, throughout central California east of the Coast Range, through western Oregon. It occurs on both sides of the Sierra Nevada, at least as far north as the head of Owen's River, and eastward to the western border of Death Valley, California, where I met with this subspecies in 1867. In southeastern California, according to Dr. Elliott Coues, "It reaches nearly to the Colorado River, following along the course of the Mojave to the spot where it sinks in the desert, there meeting the western extension of the range of *Callipepla gambeli.*"[1]

In southwestern Oregon it does not appear to occur anywhere on the eastern slopes of the Cascade Mountains, north of Fort Klamath, unless recently introduced. It is common in the upper and middle parts of the Willamette Valley, while in the lower part of this valley it intergrades with *Callipepla californica*. It has been transplanted to Utah, in the vicinity of Ogden, as well as in various parts of Nevada, where it is now found in suitable localities along the entire western border of this State, from Carson and Reno, along the west shore of Pyramid Lake, to the northern end of Warner Valley, Oregon. It is a resident and breeds wherever found. A few years ago the Valley Partridge was exceedingly abundant about Fort Bidwell, in the extreme northeastern part of the State of California.

Mr. A. C. Lowell writes me from there as follows: "These birds are unable to stand the severe cold of this region, especially when accompanied by a heavy fall of snow. In the winter of 1887-'88 about 2 feet of snow fell, followed by three very severe nights in which the thermometer reached 28° below zero. This killed most of these birds. In the following fall I heard of but three or four coveys of Quail within a radius of 60 miles where thousands had been the year before. They ranged from the north end of Warner Valley south to Reno, Nevada, and were especially numerous in Buffalo Cañon and along the west shore of Pyramid Lake. They were very common up to the summits of the Warner Mountains, which attain here an altitude of about 6,000 feet, though the cañons and water courses found along those slopes were their favorite resorts. I have never seen or heard of

---

[1] Birds of the Northwest, 1874, p. 449.

a covey of these Quail down in the cultivated fields of the valleys. Here, at least, they prefer to live exclusively on the brush-covered hillsides."

That the Valley Partridge differs very much in its general habits in certain localities has long been known to me, still I did not for an instant suppose that I would meet with this species in a place like Fort Klamath, Oregon. Anyone at all acquainted with these birds would certainly not look for them in this locality, and I was greatly surprised to find a covey here in November, 1882. The post, situated in the upper Klamath Valley, is nearly surrounded by large pine forests. In winter the snowfall is generally quite heavy and the summer climate is variable, usually cool and damp, frost occurring sometimes every month in the year. To make sure of their identity I shot two of these birds, and found them plump and in excellent condition. The remainder, about a dozen in number, seemed to stand the winter well, the thermometer falling more than once considerably below zero, and in the summer of 1883 I noticed two coveys of half-grown birds. They were excessively shy at all times, living, as a rule, in the more open pine timber, and when disturbed flying at once into the densest growth of young pines and hiding in the trees. I never saw them on the open meadow or valley lands. I am certain that they were not introduced here, and as they are often known to travel long distances on foot, I believe they followed up the wagon road to the post, and this ending at that point, they settled down permanently, the mountains by which the fort is hemmed in barring a further advance. On the lower Klamath River they are common enough, but the character of the country is quite different there, and eminently suited to these birds, which can not be said of the locality where the post is situated.

The Valley Partridges found along the coast in southern California are intermediate between the two races, while the birds found in Lower California are typical *Callipepla californica vallicola*.

Mr. A. W. Anthony writes me about the Lower California birds as follows: "I found the Valley Partridge very common in the mountains of Lower California, up to an altitude of about 9,000 feet. Both in southern and Lower California I was told by the Indians and native Mexicans that during very dry seasons the Valley Quail did not nest, but remained in large flocks during the entire summer. This statement I was able to verify by personal observations during the summer of 1887. These birds were seen by me in large flocks throughout the spring and summer months, and only two or three broods of young were noticed. Birds taken during April, May, and June showed but little development of the ovaries. Should the winter rains, however, be sufficient to insure an abundance of seeds and grasses, the coveys begin to break up early in March, and from every hill in the land the loud challenge of the male is heard. The call notes of this subspecies are quite varied, frequently the same bird changing his call six or seven times within half an hour.

"A call heard frequently during the nesting season, and which seems to be a challenge from the male, is a clear, loud 'thee-hooo,' or 'queh-ōōō' (stress

being laid on the last syllable). This is usually uttered when the bird is
perched on the topmost lobe of a cactus or other commanding point, and is
repeated at intervals of two or three minutes for hours. The call from the male
to his mate is a soft-tuned 'ah-wah' or 'ah-hooh,' tender and clear, and is
heard during the entire nesting season. The alarm note of both sexes is a low
'quit-quit-quit,' usually uttered as the birds cluster under the shelter of a
bush before they begin to scatter and run; and it seems to be rather more of a
discussion as to the extent of the danger and the best method of avoiding it
than the note which is heard a moment later when they have decided that the
case is serious and requires a prompt retreat. Then a sharp 'chip-chip,' or 'pip,
pip-pip' is uttered by each bird as it dashes forward a few feet before taking
wing or till hidden in the nearest manzanita thicket. By far the most common
call at all seasons is one resembling 'ca-ra-ho,' repeated four or five times,
and the accent shifted from one syllable to another as suits the fancy of the performer. This note is often heard when the covey has been suddenly surprised,
and sounds very much like an angry remonstrance against the intrusion.
Sometimes when the covey is scattered, or the old bird is calling her brood
together, a call something like 'ca-raw' is used."

Mr. William Proud, who is quite familiar with the habits of the Valley
Partridge, writes me from Butte County, California, regarding them as follows:
"Hundreds of these birds roost every night in the shrubbery around my house.
Some of them are very tame, feeding among the chickens and coming on the
veranda. They appear to know that they are protected. They mostly roost in
thick brush, and on the ground when the brush is not at hand. In early seasons they begin to pair in the last week of February, but the time varies somewhat according to the season. During this period there is considerable fighting
among the males for the favor of the coveted female. This is kept up until they
are suitably mated and the nesting season arrives. This usually begins here
about the last week in March, when the pairs scatter among the shrubbery
along the banks of creeks and in adjacent ravines, along hedge rows and brush
fences and on the borders of cultivated fields. The earliest nest I ever found
was on March 15, and on April 15 I met young birds probably a couple of days
old. I consider fourteen eggs to be about the average number laid by these
birds, and have found as many as twenty-four in a nest. The large sets I attribute to other hens laying in the nest, probably young birds which have failed to
make preparation for their own eggs. On May 21 my dog pointed a Valley
Partridge on her nest which contained twenty-two eggs, and every one hatched.

"During incubation the male is very attentive and watchful, usually
taking an elevated position near the nest, where with crest erect and tail
spread he bids defiance to all intruders, uttering an oft-repeated 'whew-whew-
whew.' When the brooding hen leaves the nest to feed, should he be absent
from the post of duty, her cry of 'tobacco, tobacco,' very plainly given,
brings him up at once. In fact, their call notes are very varied. I frequently
heard an old cock call out at night 'ah-hooh, ah-hooh,' the first note in a
low key.

"As soon as the young are hatched, they immediately leave the nest, keeping under cover as much as possible. Should the brood be disturbed, the old birds will run and flutter along the ground to draw the attention of the dog, or whatever may have frightened them, to themselves and away from the young. In about ten days these can fly a short distance. The Valley Partridge feeds on insects and the young and tender leaves of clover and green pease, later, on grain and various small seeds; in the fall they eat wild grapes and are also very partial to the seeds of the amaranth, also those of *Mentzelia lævicaulis.* Here only one brood is raised in a season, and incubation, as nearly as I can ascertain, lasts about twenty-eight days."

The nests and eggs of the Valley Partridge are similar in every respect to those of the California Partridge, and the number of eggs usually laid is about the same. In southern California they often nest under small juniper bushes and in prickly-pear or cactus patches. Usually but one brood is raised, but under favorable circumstances two are not uncommon. Mr. Anthony's statement that the Valley Partridge does not nest in exceptionally dry seasons in portions of its range has been fully verified by me through other observers, and appears to be a well-established fact.

None of the eggs of this Partridge are figured, as they are indistinguishable from those of the preceding. A number of the eggs from Cape St. Lucas of this subspecies average a trifle smaller than California and Oregon specimens, reducing the average measurement somewhat.

Seventy-six specimens in the U. S. National Museum collection average 31 by 24 millimetres. The largest egg of the series measuring 34 by 25, the smallest 28 by 23 millimetres.

13. Callipepla gambeli (NUTTALL).

GAMBEL'S PARTRIDGE.

Lophortyx gambeli "NUTTALL." GAMBEL, Proceedings Academy Natural Sciences, Phila., 1843, 260.
Callipepla gambeli GOULD. Monograph Odontophorinæ, 1850, Pl. XVII.

(B 475, C 392, R 483, C 576, U 295.)

GEOGRAPHICAL RANGE. Northwestern Mexico and contiguous portions of United States from western Texas to southern California, north to southern Nevada and southern Utah.

The home of this graceful and interesting species includes that portion of southern California commonly known as the "Great American Desert." Here Gambel's Partridge reaches the most western point of its range, near San Gorgonio Pass, in San Bernardino County, California, where it overlaps that of the Valley Partridge and hybrids are found;[1] thence it ranges eastward through Arizona and the greater part of New Mexico into western Texas. To the north

---

[1] Auk, Vol. II, 1885, p. 247.

it is found in southwestern Utah and the Death Valley region of southern Nevada, as well as in parts of northern New Mexico, where Dr. C. J. Newberry, jr., met with it a few miles south of Santa Fé. South it extends into western Mexico. It is a resident, and breeds wherever found.[1]

In southern Arizona, along the valley of the Gila River, it used to be exceedingly abundant before the day of railroads, and is yet, I presume. In those days Gambel's Partridge was one of the most pleasing sights to the weary traveler over Arizona's hot and dusty plains, where springs and even stagnant water holes were few and far between, and stretches of 50 miles without water were not unusual. The presence of these handsome little game birds always indicated that this much-needed fluid, poor as it often might be, was not far off, and this cheered you, for which reason alone, if for no other, their appearance was doubly welcome. Numerous wells along the principal highways and railroads have changed all this now, and a journey through Arizona to-day has lost about all its terrors, and can be made in comparative comfort and even luxury.

For one of the most exquisite pieces of word painting of Arizona, as it appeared thirty years ago, and at the same time giving an exceedingly interesting and accurate account of the life-history of Gambel's Partridge, I refer the reader to an article in the Ibis of January, 1866, entitled "Field Notes on *Lophortyx gambeli*, by Elliott Coues, M. D.," which will not fail to prove attractive to the most critical observer. Concerning the relative abundance of Gambel's Partridge in Arizona at present, Mr. Herbert Brown, of Tucson, writes me as follows: "There is no diminution in their numbers; if anything, they have multiplied in proportion to the extent of increased cultivation. I have been told that some farmers on the Salt and Gila Rivers, about Florence and Phœnix, poisoned them as a nuisance, and in a 'game bill,' introduced in the Arizona legislature in 1885, Partridges had to be stricken out from protection before the bill could pass."

Wherever water is found Gambel's Partridge is common throughout southern Arizona up to an altitude of 5,000 feet; and in New Mexico, Mr W. H. Cobb, of Albuquerque, informs me of meeting with young fledglings in the pine forests at an altitude of 8,000 to 9,000 feet. In 1872 I found this species very abundant near my camp on Rillitto Creek, the present site of Fort Lowell, 7 miles northeast of Tucson. During the winter and early spring coveys of these birds might be seen almost daily, feeding and dusting themselves in the immediate vicinity of my camp, and especially on the wagon roads leading to it. They frequented these mostly in the mornings and occasionally in the evenings, the birds scratching about in the sand and dusting themselves like domestic fowls. They appeared **very** sociable, and were constantly calling to each other as the scattered covey moved from

[1] Lieut. Robert C. Van Vliet, U. S. Army, tells me that he tried to introduce this species in the vicinity of Fort Union, New Mexico, liberating fifty of these birds in February, 1874. They all disappeared within a year. The birds met with by Dr. Newberry, **near** Santa Fé, may have been stragglers or descendants of this lot.

place to place. This note resembled the grunting of a sucking pig more than anything else, and it is rather difficult to reproduce the exact sound in print. Any of the following syllables resembles it, "quoit," "oit," "woét," uttered rapidly but in a low tone. During the mating and breeding season, the former commencing usually in the latter part of February, the latter about the first week in April and occasionally later, according to the season, the male frequently utters a call like "yuk-käe-ja, yuk-käe-ja," each syllable distinctly articulated and the last two somewhat drawn out. A trim, handsome, and proud-looking cock, whose more somber-colored mate had a nest close by, used an old mesquite stump, about 4 feet high, and not more than 20 feet from my tent, as his favorite perch, and I had many excellent opportunities to watch him closely. Standing perfectly erect, with his beak straight up in the air, his tail slightly spread and wings somewhat drooping, he uttered this call in a clear strong voice every few minutes for half an hour or so, or until disturbed by something, and this he repeated several times a day. I consider it a call of challenge or of exultation, and it was taken up usually by any other male in the vicinity at the time. During the mating season the males fight each other persistently, and the victor defends his chosen home against intrusion with much valor. It is a pleasing and interesting sight to watch the male courting his mate, uttering at the time some low cooing notes, and strutting around the coy female in the most stately manner possible, bowing his head and making his obeisance to her. While a handsome bird at all times, he certainly looks his best during this love-making period. The alarm note is a sharp discordant "crïèr, crïèr," several times rapidly repeated, and is usually uttered by the entire covey almost simultaneously. Although they nested abundantly in close proximity to my camp, I saw but a single brood of birds that were probably not more than a day or two old. Small as these were, they nevertheless managed to run and hide so quickly in the undergrowth in which I found them that I failed to catch one for closer examination. The hen tried to draw me away by the usual devices, and showed considerable anxiety. Half-grown birds were much more frequently met with by me, and not until they are well able to fly do they make excursions in the more open country, away from the tangled undergrowth and vine-covered chaparral of the creek bottoms. Their food, like that of the other species of this genus, consists of insects of various kinds, especially grasshoppers and ants, small seeds, grain when obtainable, the tender leaves and buds of leguminous plants, and berries. In the early fall and winter they pack, and from two to five hundred may, at times, be met with on favorite feeding grounds.

During the intense heat of the Arizona summers Gambel's Quail, like most other birds, prefers to remain in the shady and cool spots in the creek bottoms, frequently perching in the trees, and I believe the majority of these birds spend the nights in them as well. They take to trees very readily at all times. The nesting season of 1872, compared with subsequent ones, was an unusually late one, and though I searched carefully for the nests of this species during both April and May, I failed to find a single one before May 29. This contained ten fresh eggs. During June I found a number, however, also two in July, and one as late as August 17. I believe two broods are regularly raised in a season. Incubation, as near as I was able to learn, lasts from twenty-one to twenty-four days, and does not begin until all the eggs are laid, and these are deposited daily.

The nest of Gambel's Partridge is simply a slight oval-shaped hollow, scratched out in the sandy soil of the bottom lands, usually alongside of a bunch of "sacaton," a species of tall rye grass, the dry stems and blades of last year's growth hanging down on all sides of the new growth and hiding the nest well from view. Others are placed under, or in a pile of, brush or drift brought down from the mountains by freshets and lodged against some old stump, the roots of trees, or other obstructions on some of the numerous islands in the now dry creek beds, refreshing green spots amid a dreary waste of sand. (It is perhaps as well to mention that many of the so-called creeks in Arizona are dry for about ten months of the year, the water sinking below the sand for a foot or or two, but running below this through the coarser gravel, digging being necessary in order to reach it.) These so-called islands are always covered with a luxurious vegetation, and it is in this that most of the Partridges nest. According to my observations only a comparatively small number resort to the cactus and yucca covered foothills and mesas some distance back, where the nests are usually placed under the spreading leaves of one of the latter-named plants. If grain fields are near by they nest sometimes amidst the growing grain in these, and should the latter be surrounded by brush fences, these also furnish favorite nesting sites.

Among the nests observed by me two were placed in situations above ground. One of these was found June 2 on top of a good-sized rotten willow stump, about 2½ feet from the ground, in a slight decayed depression in its center, which had, perhaps, been enlarged by the bird. The eggs were laid on a few dry cottonwood leaves, and were partly covered by these. Another pair appropriated an old Road-runner's nest, *Geococcyx californianus*, in a mesquite tree, about 5 feet from the ground, to which apparently a little additional lining had been added by the bird. The nest contained ten fresh eggs when found on June 27, 1872.

Mr. Herbert Brown found a pair of these birds occupying a newly-made nest of a Palmer's Thrasher, *Harporhynchus curvirostris palmeri*, in which seven eggs had been deposited. This nest was placed in and near the top of a cholla cactus about 4 feet from the ground. He says: "My first impression was that an Indian had probably placed them there, but I was soon convinced to the contrary, as I found it impossible to get my head near the nest without first breaking down a part of the cholla with the barrels of my gun. The eggs were fresh and finely marked."[2]

---

[1] Forest and Stream, June 4, 1885.

Birds resorting to nesting sites in trees or cacti have undoubtedly lost their eggs or small young on former occasions, and learned from experience that such a situation is in many respects a safer one.

During the nesting season of 1872, I found, upon a second visit, that several incomplete sets of eggs belonging to this species had been destroyed or removed. The numerous large snakes of various kinds, especially the rattlesnake, must be counted among the worst of their enemies.

On one occasion I found a Gambel's Partridge's nest in the side of a sandbank. A portion of this had been washed away by a former freshet and a sod of grass having been undermined thereby fell over it, being still firmly held in place by its roots. The bird had scratched out a hole in the sandy bank behind this sod and deposited her eggs therein, and it appeared to me to be an extremely well-selected nesting site. It proved otherwise, however, for a few days later, when passing by the spot again, I put my hand in the cavity, the contents of which were not visible without raising the sod, I came in contact with something cold which I at first supposed to be a snake; and being curious to see what it really was and not able to dislodge it, I raised the sod with a stick and found a land terrapin taking its ease in the nest. Not the sign of an egg remained, neither were any broken shells visible. Whether the reptile had eaten the eggs or not I was unable to decide, as I found no remains of them in the stomach. That reptiles of various kinds are not adverse to an egg diet is shown by the following instance kindly furnished by Mr. Herbert Brown. A Gila monster, *Heloderma suspectum*, had been caught alive near Tucson, Arizona, on April 14, 1850, and was placed in a packing box for safekeeping over night. Next morning five eggs were found in the box with the occupant. Two of these were forwarded to me for identification by Mr. Brown, who wished to know if they were the eggs of this reptile or of Gambel's Partridge, he surmising the latter. There was no difficulty in solving this problem, for the shells of the eggs, although considerably injured, plainly showed the peculiar markings of the egg of Gambel's Partridge, and even the shape, leaving no possible doubt that they were the product of one of these birds and not of the Gila monster, which had probably swallowed them whole on the day it was caught and thrown them up during the night.

The nests of Gambel's Partridge are lined usually, but very slightly, with bits of dry grasses or leaves, and often contain no lining whatever, the eggs lying on the dry, sandy soil. These usually number from ten to twelve in a set, but occasionally double these numbers are found, which are unquestionably the product of more than one hen. I have several times found ordinary-sized sets placed in two layers, one egg on top of the other, the cavity being in such cases deep, narrow and not rounded.

A set found by me June 20, 1872, contained nineteen fresh eggs, evidently laid by two different birds, as the eggs showed two radically different

26957—Bull. 1——3

and distinct types of markings. These were likewise placed in two layers. In the hot Gila River Valley in southern Arizona, nidification commences in some seasons by the middle of March.

Mr. John Swinburne informs me of finding a full set of eggs on March 19, near Phœnix, in Maricopa County. In the vicinity of Tueson they lay somewhat later. The earliest date at which eggs of this species have been found there, according to the observations of Mr. Brown, is April 4, usually about the latter part of this month and the beginning of May, the nesting season continuing into August and sometimes even to September.

The eggs of Gambel's Partridge are short ovate in form, and the ground color varies from a dull white to a creamy white and pale buff color. The eggs are spotted, clouded, and blotched, sometimes very heavily, with irregular markings or blotches, and again with well-defined and rounded spots of dark seal-brown and écru drab. Diffused over these blotches is found a peculiar purplish or pinkish bloom, difficult to describe, resembling somewhat the rich bloom found on blue grapes and various kinds of plums when first picked. These markings, when touched by water or moisture of any kind, change radically, becoming seal brown, or chestnut brown of different shades, according to the variable amount of pigment on the shell of the egg. Carefully blown specimens will retain this peculiar bloom for years, and some eggs collected by me and now deposited in the U. S. National Museum, one of which is figured, show this as plainly to-day as when they were first taken, fully eighteen years ago. Eggs of *Callipepla gambeli* are, as a rule, more heavily spotted than those of the two California Partridges, and the color of the markings in the majority of specimens is decidedly different. The peculiar golden russet shade so often present in the eggs of the latter is almost entirely wanting here, and is replaced by darker and more bluish brown tints.

The average measurement of ninety-seven specimens in the U. S. National Museum collection is 31.5 by 24 millimetres, the largest egg of the series measuring 34 by 26, the smallest 28.5 by 24 millimetres. The type specimens, No. 16480 (Pl. 1, Fig. 11), selected from a set of ten eggs, taken June 14, 1872, and No. 21116, two eggs selected from a set of nineteen (Pl. 1, Figs. 13 and 14), one showing the peculiar bloom before mentioned, and the other a decided difference in the style of markings, taken June 20, 1872, near Rillitto Creek, Arizona (Bendire collection), were found by the writer. No. 23938 (Pl. 1, Fig. 12), from a set of ten eggs, was taken by Mr. Herbert Brown at the Laguna, near Tucson, Arizona, May 19, 1889.

14. Cyrtonyx montezumæ (Vigors).

MASSENA PARTRIDGE.

Ortyx montezumæ Vigors, Zoölogical Journal, v, 1830, 275.
Cyrtonyx montezumæ Stejneger, Auk, ii, January, 1885, 46.
(B 477, C 394, R 485, C 578, U 296.)

Geographical range: Western and central Mexico, from Mazatlan and valley of Mexico, north to western Texas, New Mexico, and Arizona.

This handsome and peculiarly marked Partridge, better known in western Texas as the "Black" or "Black-bellied" Quail, and in Arizona as the "Fool" Quail, inhabits the rough mountainous regions of the last-mentioned Territory north to at least the vicinity of Fort Whipple, which, as far as known at present, marks the western limit of its range, and where it was first obtained by the well-known ornithologist, Dr. Elliott Coues. Thence it extends eastward through New Mexico, north to about latitude 36°, where Capt. William L. Carpenter, Ninth Infantry, U. S. Army, observed it in the upper Rio Grande Valley, near Taos. It is also found in suitable localities in the intervening country, in a southeasterly direction, throughout portions of western and southwestern Texas. Mr. Dresser's specimen, obtained in the Bandera Hills, about 40 miles northwest of San Antonio, marks about the most easterly known point of its range. According to Mr. William Lloyd, it ranges south from the Llano Estacado and mountainous regions of western Texas to the Sierra Madre Mountains in Sonora, Chihuahua, and Sinaloa, and the mountains in Jalisco in northwestern Mexico, inhabiting regions from an altitude of 4,000 to 9,000 feet. In the less elevated parts of its range it is a constant resident and breeds, but at the higher altitudes it is only a summer visitor, retiring to the lower foothills on the approach of winter.

Although sixty years have passed since the Massena Partridge was first described by Vigors, nothing absolutely reliable was known about the nest and eggs of this bird up to the season of 1890. Not a single positively identified egg was to be found in any of the larger and well-known oölogical collections of the country, and up to the time of this writing no description of them has been published. This is rather remarkable when the extensive range which this species occupies within our borders is considered, and also the fact that in many localities it is by no means rare. Nevertheless the Massena Partridge, next to the Lesser Prairie Hen, *Tympanuchus pallidicinctus*, is still one of the least-known game birds of the United States.

Mr. William Lloyd writes me from Marfa, Texas, that "the favorite resorts of the Massena Partridge are the rocky ravines or arroyas that head well up in the mountains. They quickly, however, adapt themselves to changed conditions of life and are now to be seen around the ranches picking up grain and scratching in the fields. In the vicinity of Fort Davis,

Texas, they have been exceptionally numerous and may frequently be seen sitting on the stone walls surrounding grainfields in Limpia Cañon. In Mexico I have seen them several times living contentedly in cages. In Mesquite Cañon they are the only Partridge found, and in June and July, 1887, I spent some time there trying principally to locate the nest and eggs of this species. I found a single egg in a depression at the roots of a tasaca cactus, presumably belonging to this species. It was white, without any markings whatever. While there I was informed by two different parties living in the vicinity that each of them had found a nest the previous year, 1886, containing eight and ten eggs respectively, which they had eaten. They described the eggs as being white in color. Both said that the nest was simply a slight hollow, one under a small shin-oak bush, the other alongside a sotol plant. The call note of this bird is a low murmuring whine, more like that of the rock-squirrel, *S. grammurus*, than a bird, and it can be heard quite a distance. I can not imitate it in syllables. They are very fond of acorns, mountain laurel, arbutus, cedar, and other berries, and range in coveys from eight to twelve."

Capt. Platt M. Thorne, Twenty-second Infantry, U. S. Army, writes me: "I found the Massena Partridge common at both Forts McKavett and Clark, Texas, where they apparently liked the same kind of ground as the Texan Bob White, yet the lines of their habitat seem mysteriously restricted for some reason. Can it be that their food is peculiar? All the stomachs I have examined (fall birds) contained little else than large quantities of white shiny bulbous roots, rounded at both ends, and about the size of French pease. I regret now that I never forwarded any of these roots, that it might be determined what they were. You are aware how well these birds are adapted to scratching and I have an idea that this root food might account for their restricted distribution. I also found them abundant on a divide near Nueces River, but I never saw any within 20 miles of the Rio Grande."

Lieut. Robert C. Van Vliet, Tenth Infantry, U. S. Army, also met with the Massena Partridge in western Texas and northern New Mexico (Fort Union), usually along the sides of rocky ravines. He tells me that they were fairly common, and that their food (at least during the fall and early winter), consisted almost entirely of a small angular brownish-looking bulb, with a white kernel, the root of a short grass, their crops containing scarcely anything else excepting small particles of gravel. He often saw where they had scratched out holes to the depth of 2 inches in search of these roots, and such evidences were always abundant in localities frequented by these birds. Their call note is a clear "dsiup-chiur." He rarely saw coveys consisting of more than eight birds. Polecats seem to be one of their principal enemies.

Capt. William L. Carpenter, U. S. Army, states: "I have observed this species in the Rio Grande Valley, near Taos, New Mexico, and more frequently on the headwaters of the Black and White Rivers, where it undoubtedly breeds, and I have often looked for their nests unsuccessfully. In the spring and summer they are usually found in pairs; the balance of the year they range all through the White Mountain region of Arizona above an altitude of 4,000 feet, in coveys, but these are never numerous, and usually small in size. They are probably more subject to the attacks of predatory animals than any other species, owing to their confiding disposition, which has given them the name of 'Fool' Quail. I once stopped my horse, when about to step on one, and watched it for some time without creating alarm. After admiring it for several moments, squatting close to the ground within a yard of the horse, and watching me intently, but apparently without fear, I dismounted, and almost caught it with my hat, from under which it fluttered away. The flight, which is remarkably rapid, is accompanied by a peculiar clucking."

According to Mr. John Swinburne, of St. John's, Apache County, Arizona, the favorite localities frequented by this species during the breeding season are thick live-oak scrub and patches of rank grass, at an altitude of from 7,000 to 9,000 feet. He says: "Here they are summer residents only, descending to much lower altitudes in winter. They lie very close at all times, allowing one to almost step on them before they move. I have seen this species on the White Mountains during the breeding season, and saw young birds of the year shot there. Even the adults seem very stupid when suddenly flushed, and after flying a short distance, alight and attempt to hide in most conspicuous places. I have seen men follow and kill them by throwing stones."

Mr. E. W. Nelson writes me as follows: "In September, 1882, I found this bird rather common near Chloride, and Fairview, New Mexico. Old birds with half-grown young were found late in the afternoon each day in the roads leading down the bottoms of open brush-bordered cañons that extend down the flanks and foothills of the Black Range in this vicinity. A small stream was usually found in these, which disappeared in the sand a mile or two below on reaching the open barren country.

"The Massena Partridges were commonly found dusting themselves in the roads, and usually stood and watched our approach until we were within a few yards, and then flew into the bordering thicket and laid very close. When a covey was surprised among the grass they arose at our feet and scattered in every direction, but never went very far, and while flying off they would utter low notes of alarm, sounding like 'chuk-chuk-chuk.' I also found them not uncommon in the Santa Rita Mountains of southern Arizona in July, 1884. Here they occupied the live-oak belt below the lower limit of the pines. On the northeastern slopes of the White Mountains, near Springerville, Arizona, a pair has raised a brood during several successive seasons at the lower edge of the pine forest, at an altitude of about 7,500 feet. After the young are hatched they are often led up among the pines to an altitude of between 8,000 and 9,000 feet, where I have seen them.

"The birds breeding along the northern limit of their habitat migrate southward in October. In southern Arizona the same result of a warmer winter climate is obtained by descending the flanks of the mountains. The summer range of this species is just above and bordering that of Gambel's Quail in parts of Arizona and New Mexico. The fact that Gambel's Quail changes its range but little in winter results in these birds being found very frequently occupying the same ground at this season. I have never seen the Massena Partridge in coveys larger than would be attributed to a pair of adults with a small brood of young. Frequently a pair raise but three or four, and I do not remember having ever seen more than six or seven of these birds in a covey."

Personally I met with this species several times in the foothills and cañons of the Santa Rita, Patagonia, and Huachuca Mountains in southern Arizona in the early part of August, 1872, while scouting after hostile Indians, but had no time then to study their habits nor to look for their nests. A small covey of young, less than half grown, were seen by me on August 14 in a cañon of the Patagonia Mountains, about 12 miles from Camp Crittenden, and an addled egg was picked up from an abandoned nest under a small yucca in the same vicinity by one of my packers, whose attention was drawn to the place by seeing several broken egg shells lying about the yucca, and dismounting to investigate he found the egg under the bush and concealed by it, which he handed to me some two hours afterwards. The nest, he said, was within 5 feet of the trail I had previously passed over. While not absolutely certain of the identity of this egg I always felt confident that it belonged to this species, and since I have had an opportunity of examining the eggs taken during the season of 1890 I have no further doubt of it. The egg in question is ovate in shape, differing in this respect from all the eggs of the genus *Colinus* I have ever seen, which are usually rounded ovate, or subpyriform. The egg is pure white in color, the shell is smooth and close grained, and it measures 32 by 23 millimetres.

Mr. Otho C. Poling writes me that he found the Massena Partridge in parts of the Whetstone, the Santa Rita, Patagonia, and Huachuca Mountains of southern Arizona, where they were fairly common. He says: "During most of the year the Massenas remain in coveys of from four to a dozen birds in number, and even at the height of the nesting season I have several times found coveys of half a dozen together, while I have shot pairs in the month of February.

"On June 12, 1890, I shot a female and found a fully developed egg in her oviduct which would have been laid soon. It measures 30.5 by 25 millimetres, and is pure white in color. In another female, shot the same day, the ovary contained small ova about the size of No. 6 shot, which would not have been laid for some weeks. On July 15 I found my first productive nest. I was climbing up a steep mountain side on the northeast of the Huachuca Mountains, some 10 miles north of the border, when at an elevation of about 8,000 feet I flushed the female almost directly under my feet and shot it. The hillside was covered in places with patches of pines and aspens, as well as with low bushes and grasses. The nest was directly under a dead limb which was grown over with dead grass, and so completely hidden that until I had removed the limb and some of the grass it was not discernible at all. The nest was sunken in the ground, and composed of small grass stems, arched over, and the bird could only enter it by a long tunnel leading to it from under the limb and the grass growing around it. The eggs were eight in number and naturally white, but they were badly stained by the damp ground, their color being now a brownish white. They were almost hatched. The female must have remained on them all the time to have caused such uniform incubation and preserved the eggs from spoiling by the excessive dampness.

"On July 27 I met with a female and brood of about a dozen young. The entire family was in view when I at first saw them crossing an old trail. They at once entered some dense bushes, and I failed to capture or even see any of them again. The young were probably about a week old. On August 31 I discovered another brood, about a dozen in number, which were but a few days out of the nest. I secured one of the young which must have been hatched late in the month."

Mr. G. W. Todd writes me as follows: "I first met with the Massena Partridge in Bandera County, Texas, in 1883, where they very scarce, and I learned but little of its habits for a long time. They are very simple and unsuspicious, and apparently live so much in such barren and waste places that they do not see enough of man to make them afraid. On seeing a person they generally squat at once, or run a little way and hide. They will hardly fly until one is almost on them, but when they finally do fly they go much further than either the Texan Bob White or the Scaled Partridge, and on alighting they run rapidly for a little distance and then squat again, generally flushing easier the second time. It is rare to see more than six together; two or three are more often met with. In the fall of 1886 I found a covey of five on a wet and misty day, and killed three of them with a Winchester rifle before the remaining two flew. I never found their nest nor met with small young until this year. I saw but a single young bird this season, and this seemed to be entirely alone. They are not very abundant here, and are always found in the most barren places, among rocks and wastes, where even the prickly pear is stunted, and no bush grows over 3 feet high. When scared they utter a kind of whistling sound, a curious combination between a chuckle and a whistle, and while flying they make a noise a good deal like a Prairie Hen, though softer and less loud, like 'chuc-chuc-chuc' rapidly repeated.

"The only nest of this species I have ever seen was situated under the edge of a big bunch of a coarse specie of grass, known as 'hickory grass.' This grass grows out from the center and hangs over on all sides until the blades touch the ground. It is a round, hard-stemmed grass, and only grows on the most sterile soil. According to my observations the Massena Partridge is seldom seen in other localities than where this grass grows. I was riding at a walk up the slope of a barren hill when my horse almost stepped on a nest, touching just the rim of it. The bird gave a startled flutter, alighting again within 3 feet of the nest and not over 6 feet from me; thence she walked away with her crest slightly erected, uttering a low chuckling whistle until lost to view behind a Spanish bayonet plant (yucca), about 30 feet off. I was riding a rather unruly horse, and had to return about 30 yards to tie him to a yucca, before I could examine the nest. This was placed in a slight depression, possibly dug out by some animal, the top of the nest being on a level with the earth around it. It was well lined with fine stalks of wire-grass almost exclusively, the cavity being about 5 inches in diameter and 2 inches deep. At the back, next to the grass, it was slightly arched over, and the overhanging blades of grass hid it entirely from sight. The nest was more carefully made than the average Bob White's nest, and very nicely concealed."

The eggs, ten in number, were fresh when found, pure white in color, rather glossy, and the majority of them are more elongated than those of the Bob White. A few of these eggs resemble those of the latter somewhat in shape, but the greater number are distinctly ovate and much more glossy. Some are slightly granulated, and corrugations converge from near the middle to the small end.

This set of eggs of the Massena Partridge is now in Mr. Thomas H. Jackson's collection, at West Chester, Pennsylvania, who has kindly allowed me to examine them and figure one. They were taken by Mr. G. W. Todd, near the head of Turkey Creek, in Kinney County, Texas, June 22, 1890, and are, as far as I am aware, the first fully identified eggs of this species that have been found.

Mr. Todd has kindly sent me a couple of skins and stomachs of these birds. The latter, according to the report of Dr. C. Hart Merriam, in charge of the Division of Ornithology, U. S. Department of Agriculture, contained principally cactus seeds (*Opuntia*), a few bits of cactus prickles, a lot of finely ground vegetable matter with a trace of insects, and a large amount of coarse sand, mainly iron ore.

The average measurement of the eight eggs found by Mr. Poling is 32 by 24 millimetres. The largest egg of Mr. Jackson's set measures 33 by 24.5 millimetres. This is figured on Pl. 1, Fig. 15, the smallest measuring 30 by 23.5 millimetres; they average 31.5 by 24 millimetres.

15. Dendragapus obscurus (Say).

DUSKY GROUSE.

Tetrao obscurus Say, Long's Expedition, II, 1823, 14.
Dendragapus obscurus Elliot, Proceedings Academy Natural Sciences, Philadelphia, 1864, 23.

(B 459, C 381, R 471, C 557, U 297.)

Geographical range: Southern Rocky Mountains, from central Arizona and New Mexico, north to southeastern Idaho and central Wyoming, east to southwestern South Dakota, west to northeastern Nevada.

With our present limited knowledge it is rather a difficult matter to define accurately the range of the Dusky Grouse from that of the two subspecies, the "Sooty" and "Richardson's" Grouse; this can only be done approximately as yet. The three forms are well known and rank as the finest of game birds, and next to the Sage Fowl are the largest Grouse found within the United States.

Beginning with the northern range of the Dusky Grouse as well as I can define it, this includes a small portion of southeastern Idaho, where it intergrades with *D. obscurus fuliginosus,* thence eastward through Wyoming and western South Dakota (Black Hills), south and west through northeastern Nevada (East Humboldt Mountains), Utah, central and western Colorado, as well as northern and central Arizona and nearly the whole of New Mexico, excepting the extreme southern portion south of the Rio Mimbres, which marks the most southern limit of its range.

It is more or less a common resident in suitable localities, *i. e.,* the outer borders of the timbered mountain regions of the States and Territories mentioned, and breeds wherever found. It is best known as the Blue Grouse, and is also called Pine Grouse and Pine Hen.

Mr. Denis Gale, of Boulder County, Colorado, a careful and reliable observer, writes me as follows: " Here in Colorado the Dusky Grouse ranges from an altitude of about 7,000 feet to timber line. Having once selected a place to raise a brood they do not stray far from the neighborhood. Water at no great distance is always kept in view. The lower gulches and side hills are mostly chosen for their summer homes. During the mating season if you are anywhere near the haunts of a pair you will surely hear the male and most likely see him. He may interview you on foot, strutting along before you, in short hurried tacks alternating from right to left, with widespread tail tipped forward, head drawn in and back and wings dragging along the ground, much in the style of a turkey gobbler. At other times you may hear his mimic thunder overhead again and again in his flight from tree to tree. As you walk along he leads, and this reconnoitering on his part, if you are not familiar with it, may cause you to suppose that the trees are alive with these Grouse. He then takes his stand upon a rock, stump, or log, and in the manner already described distends the lower part of his neck, opens his frill of white, edged with the darker feather tips, showing in its center a pink narrow line describing somewhat the segment of a circle, then with very little apparent motion he performs his growling or groaning, I don't know which to call it, having the strange peculiarity of seeming quite distant when quite near, and near when distant; in fact, appearing to come from every direction but the true one. The first time I heard the sound I concluded it was the distant laboring of one of our small mountain sawmills wrestling in agony with some cross-grained saw-log. It appeared to me like it.

"As near as I can judge by meeting with the young broods, these birds nest at the lowest points about May 15, at the highest about the beginning of June. The number of chicks seen by me in a brood ranged from three to eight. The young in the downy stage are beautiful, delicate little objects. Upon one occasion I met with a covey which had just been hatched; they were quite nimble, and with the exception of one which I caught they hid themselves with great address. Until I released the little prisoner the female showed great distress, clucking in the most beseeching manner, accompanied with suitable gestures, similar to but more tender and graceful than those of our domestic hen. She stood within 6 or 7 feet of me pleading her cause and easily won it. In her beautiful summer dress of brown, handsomely plumed as she was, she looked very interesting.

"In a single instance only, with a brood about ten days old, have I noted the presence of both parents. Perched upon a fallen tree the male seemed to be on the lookout, while the female and young were feeding close by. This seeming indifference of the male while the brood is very young, allowing his mate to protect them, if he really is always near at hand, looks very strange, and yet it may be the case, since he is generally with the covey when the young are well grown. Directly the young are able to travel, the hen Grouse leads them to some desirable opening skirting the timber, or gulch where bearberries, wild raspberries, gooseberries, and currants, as well as grasshoppers, worms, and grubs are abundant, managing them just as the domestic hen does her brood. The young grow rapidly, and when about two weeks old can do a little with their wings; then instead of hiding on the ground they flush and endeavor to conceal themselves in the standing timber. Until almost fully grown they are very foolish; flushed, they will tree at once, in the silly belief that they are out of danger, and will quietly suffer themselves to be pelted with clubs and stones till they are struck down one after another. With a shotgun, of course the whole covey is bagged without much trouble, and as they are, in my opinion, the most delicious of all Grouse for the table, they are gathered up unsparingly."

Mr. John Swinburne informs me that in southeastern Arizona the Dusky Grouse frequents thick spruce and fir timber, and is generally found at an altitude of about 9,000 feet. He says: "If found on the ground they almost invariably fly into the nearest tree and sit there, moving their heads from side to side, gazing at the intruder first with one eye then with the other. I have shot at them repeatedly with the rifle and pistol before they flew from the branch on which they had settled. Broods of young, on being disturbed, scatter and hide, the old bird flying into a neighboring tree. These broods usually number from eight to ten."

A nest found near Fort Garland, Colorado, is described by Mr. H. W. Henshaw as follows: "A nest found June 16 contained seven eggs on the point of hatching. The nesting site was a peculiar one, being in an open glade, where the grass had been recently burned off. The nest proper was a slight collection of dried grass placed in a depression between two tussocks, there apparently having been no attempt made at concealment."[1]

The Dusky Grouse raises but a single brood a season, and, as a rule, the nest is well concealed. A slight depression is scratched out by the bird, alongside an old log, under a small thick bush or a tall bunch of grass; this is slightly lined with pine needles, bits of dry grass, or whatever suitable material is most convenient to the site selected. The number of eggs to a set varies from seven to ten, rarely more, although they have generally been credited with larger numbers, up to fifteen. Such large sets are very exceptional, and eight or nine are the numbers most often found. An egg is deposited daily, and incubation does not commence till the set is completed. Nidification begins usually about the middle of May, and varies somewhat, both according to season and altitude. Incubation lasts, as nearly as I can determine, from eighteen to twenty-four days. The eggs resemble in shape, size, and markings those of the Sooty Grouse; and as the U. S. National Museum collection contains a much better series of this race, showing considerable variation both in the ground color and the markings, and as the same differences would unquestionably be found in an equal number of the eggs of the Dusky Grouse, I have had only a series of the former figured. The average size of the few specimens in the U. S. National Museum collection is 50.5 by 35 millimetres.

16. Dendragapus obscurus fuliginosus Ridgway.

SOOTY GROUSE.

Canace obscura var. *fuliginosa* Ridgway, Bulletin Essex Institute, v, December, 1873, 199.
Dendragapus obscurus fuliginosus Ridgway, Proceedings U. S. National Museum, VIII, 1885, 355.

(B —, C 381b, R 471a, C 559, U 297a.)

GEOGRAPHICAL RANGE: Northwest Coast Mountains, from California north to Alaska (Sitka), east to western Nevada, western Idaho, and middle British Columbia.

The Sooty Grouse, as fine a game bird as the preceding, is an inhabitant of the mountains of the Northwest. It has been taken as far north as Portage Bay, Alaska, near latitude 60°, and probably reaches farther in this direction

---

[1] Explorations and Surveys west of 100th meridian, Wheeler, 1873, p. 92.

wherever good-sized timber is found. South it ranges through British Columbia, Washington, Oregon, and the greater portion of California, to about latitude 35° (vicinity of Fort Tejon in the southern Sierra Nevada). East it is found to the western slopes of the Bitter Root Mountains in Idaho, intergrading in the northern and central portions of this State with *Dendragapus obscurus richardsoni*. It also occurs in western Nevada, and is fairly abundant in suitable localities throughout its range, at altitudes varying from 2,500 to 9,000 feet. It is a constant resident and breeds wherever found. All the Sooty Grouse from Alaska, and I presume from the Northwest coast generally, are much darker, almost a sooty black, than specimens from eastern Washington and Oregon, which resemble the Dusky Grouse much more in their general coloration than the northern bird.

Personally I have met with the Sooty Grouse in various sections of the Pacific coast, such as Mount Kearsarge in Inyo County, near the headwaters of the King and Kern Rivers, California, and in numerous localities in Oregon, Washington, and Idaho.

The following account of this species is taken from an article of mine published in the Auk (Vol. VI), January, 1889:

"I first met with the Sooty Grouse on Craig's Mountain, near Fort Lapwai, Idaho, on the Nez Percé Indian Reservation, and was told by both trappers and Indians that these birds did not remain there during the winter, in which belief I consequently shared at that time. I was also told that when a covey had been located in a tree, by being careful always to shoot the bird sitting lowest, the whole lot might be successfully secured. This may be so, but somehow it always failed with me; usually after the second shot, often even after the first, and certainly at the third, the remaining birds took wing, and generally flew quite a distance before alighting again, nearly always placing a deep cañon between themselves and me.

"At Fort Lapwai, in the early fall of 1870 and of 1871, on two or three occasions I found a few of these birds feeding with large packs of the Sharp-tailed Grouse. This must, however, be considered an unusual behavior, as I never noticed it anywhere else subsequently, although both species were equally abundant in other localities where I met them frequently in after years. The favorite locations to look for the Sooty Grouse during the spring and summer are the sunny, upper parts of the foothills, bordering on the heavier timbered portions of the mountains, among the scattered pines and the various berry-bearing bushes found in such situations, and along the sides of cañons. According to my observations these birds are scarcely ever found any distance within the really heavy timber. In the middle of the day they can usually be looked for with success amongst the deciduous trees and shrubbery found along the mountain streams in cañons, especially if there is an occasional pine or fir tree mixed amongst the former. The cocks separate from the hens after incubation has commenced, I believe, and keep in little companies, of from four to six, by themselves, joining the

young broods again in the early fall. At any rate, I have more than once
come upon several cocks in June and July without seeing a single hen
amongst them. High rocky points near the edges of the main timber, amongst
juniper and mountain mahogany thickets, are their favorite abiding places at
that time of the year. The young chicks are kept by the hen for the first
week or two in close proximity to the place where they were hatched, and not
until they have attained two weeks' growth will they be found along the
willows and thickets bordering the mountain streams. Their food consists at
first principally of grasshoppers, insects, and tender plant tops, and later in the
season of various species of berries found then in abundance everywhere, as
well as the seeds of a species of wild sunflower, of which they seem to be very
fond. It is astonishing how soon the young chicks learn to fly, and well, too,
and how quickly they can hide and scatter at the first alarm note of the mother
bird, which invariably tries by various devices to draw the attention of the
intruder to herself and away from her young. A comparatively small leaf, a
bunch of grass, anything in fact will answer their purpose; you will scarcely
be able to notice them before they are all securely hidden, and unless you
should have a well-trained dog to assist you, the chances are that you will
fail to find a single one, even when the immediate surroundings are comparatively open. After the young broods are about half grown, they spend the
greater portion of the day, and I believe the night as well, among the shrubbery in the creek bottoms, feeding along the side hills in the early hours of the
morning and evening. During the heat of the day they keep close to the
water, in shady trees and the heavy undergrowth. They walk to their feeding
grounds, but in going to water they usually fly down from the side hills.

"The love note of the cock has a very peculiar sound, hard to describe.
It can be heard at almost any hour of the day in the spring, often in the
beginning of March, when there is still plenty of snow to be found, and it
is kept up till well into the month of May. It is known as hooting or booming. The cocks when engaged in this amusement may be found perched on
horizontal limbs of large pine or fir trees, with their air sacks inflated to the
utmost, wings drooping, and the tail expanded. They then present a very
ludicrous appearance, especially about the head. When at rest these air
sacks, of a pale orange yellow color in the spring, are only noticeable by
separating the feathers on the neck and upper parts of the breast, but when
inflated they are the size of a medium orange and somewhat resemble one
cut in halves. This call is repeated several times in rapid succession,
decreasing gradually in volume, but can at any time be heard at quite a
distance. It appears to be produced by the sudden forcing of a portion of
the air in the sack through the throat, and is quite misleading as to the exact
locality whence uttered, the birds being expert ventriloquists.

"I have frequently tried in vain to locate one while so engaged, where
there were but few trees in the vicinity; and although I searched each one
through carefully, and with a powerful field glass to assist me, I had to give

it up, completely baffled. It is beyond my power to describe this love call accurately. Some naturalists state that it resembles the sound made by blowing into the bunghole of an empty barrel; others find a resemblance to the cooing of a pigeon, and some to the noise made by whirring a rattan cane rapidly through the air. The latter sound comes nearer to it in my opinion than anything else. The closest approach to it I can give in letters is a deep guttural 'mubum,' the first letter scarcely sounded.

"The accounts of the nesting habits of the Sooty Grouse are somewhat vague, the number of eggs to a set being variously given as from eight to fifteen. I have personally examined quite a number of the nests of this Grouse between May 6, 1871, and June 25, 1883. The largest number of eggs found by me in a set was ten in two instances; three sets contained nine each, seven sets contained eight each, and five sets seven eggs or less. The last were probably incomplete, although some of these sets were advanced in incubation. I think that eight eggs is the ordinary number laid by these birds.

"Eggs may be looked for from April 15 to the latter part of May, according to altitude. The earliest date on which I observed eggs of this Grouse was April 18, 1877, when a set was found by Lieut. G. R. Bacon, First Cavalry, containing seven fresh specimens. The nest was placed on the ground among the roots of a willow bush growing under a solitary pine tree in a small ravine 5 miles northwest of Camp Harney, Oregon. The nest was composed entirely of dry pine needles picked up in the vicinity.

"A nest found by me April 22, 1877, about 4 miles west of Camp Harney, was placed under the roots of a fallen juniper tree, in a grove of the same species, growing on an elevated plateau close to the pine belt. This nest was well hidden, a mere depression in the ground, and composed of dry grasses, a few feathers from the bird's breast, and dry pine needles. The nine eggs were about half way imbedded in this mass, and nearly fresh.

"As a rule, most of the nests found by me were placed in similar situations, under old logs or the roots of fallen trees, and generally fairly well hidden from view, and amongst the more open pine timber along the outskirts of the forest proper. Occasionally, however, a nest may be found some little distance from timber, and in the lower parts of mountain valleys. I found such a nest on April 26, 1878, among some bunches of tall rye-grass, in a comparatively open place, and within a yard of Cow Creek, a small mountain stream about 4 miles east of Camp Harney. There was no timber of any size, only small willow bushes, within 2 miles of this nest, which was placed under one of these rye-grass bunches, and the bird sat so close that I actually stepped partly on her and broke two of the eggs in doing so. This nest contained eight slightly incubated eggs. It was composed of dead grass and a few feathers.

"The most exposed nest, without any attempt at concealment whatever, that came under my observation, I found on June 8, 1876, on the northern slope and near the summit of the Cañon City Mountain, in Grant County, Oregon, at an altitude of about 6,800 feet. I was returning from escort duty to Cañon City, and sent the party with me around by the stage road which wound in zigzag turns up the steep mountain, while with one of my men I took a much shorter, but far steeper, Indian trail which intersected the wagon road again on the summit.

"Near this intersecting point the trail passed through a beautiful oval-shaped mountain meadow of about an acre in extent, near the summit of which stood a solitary young fir tree. No other trees were growing nearer than 30 yards from this one. The meadow itself was covered with a luxuriant growth of short, crisp mountain grass and Alpine flowers, altogether as lovely a spot to take a rest as could well be found. Arriving at this point, and knowing that the party would not be along for more than half an hour at least, I dismounted and unsaddled my horse to let him have a roll and a good chance at the sweet mountain grass, of which opportunities he was not slow to take advantage. Throwing the saddle in the shade made by the little fir I laid down to take a rest myself. I had a fine setter dog with me who had been ranging along both sides of the trail and who came up wagging his tail just as I had settled myself comfortably. Rock, my setter, had approached perhaps within 2 feet of me at a pretty brisk lope, when all of a sudden he came to an abrupt halt, fairly freezing and stiffening in his tracks, and made a dead point alongside of me. I could not understand at first what this meant; even my horse thought it worth the while to stop eating, and with his ears pointed forward was looking in the same direction. Rock, was fairly trembling with excitement, but kept to his point. Jumping up quickly I looked to the right and the rear, thinking that perhaps a rattlesnake might be coiled up in the grass, and saw at once the cause of my dog's strange behavior. It was only a poor Sooty Grouse sitting within 3 feet of me on her nest, containing two chicks and seven eggs on the point of hatching. It was as touching a sight as I had ever seen; the poor bird, although scared nearly to death, with every feather pressed close to her body, and fairly within reach of the dog, still persisted in trying to hide her treasures; and her tender brown eyes looked entreatingly on us rude intruders, and if eyes can speak hers certainly pleaded most eloquently for mercy. She let me almost touch her before she fluttered off her nest, feigning lameness, and disappeared in the undergrowth. Counting the eggs and examining one of the chicks, which apparently had only left the shell a few minutes before, I at once vacated the vicinity and took up a position some 50 yards in an opposite direction from that the bird had taken, to watch further proceedings. The grass was so short that it did not hide the bird, which, after waiting, perhaps ten minutes, came slowly creeping and crouching toward the nest and covered the eggs again. I did not disturb her further, and hope that although her selection of a nesting site so thoroughly exposed was not judicious, she may have succeeded in rearing her brood in safety. None of the eggs in the nest touched each other; they were all about half covered or imbedded in the material out of which the nest was made—dry grass, pine and fir needles, and a few of the bird's feathers, presumably plucked out by herself."

A very good description of the booming or hooting of the Sooty Grouse is published in Forest and Stream, May 23, 1889, by a correspondent signing himself "Stanstead." The article was sent from Vancouver Island, British Columbia, May 4, an extract from which reads as follows: "While driving near the city with the veteran shot, R. Maynard, we saw a pair of Blue Grouse quite near the trail, and the cock bird gave us a most entertaining exhibition of the charms that he displays in wooing his mate. Like a turkey cock he strutted about with his wings trailing on the ground, his tail feathers erect and spread-out fanlike to their fullest extent, his neck distended, and on each side of his neck the feathers were turned out so as to resemble a pair of round white rosettes, nearly 3 inches in diameter, with an oblong red spot in their center where the skin of the neck was exposed. His head seemed to be crowned with a fiery red comb. Excepting the rosettes, he was in appearance a miniature turkey-gobbler. Every few seconds he would strut up to his demure but sleek-looking mate, puff out his neck, and with a jerky movement of his head, utter his boom or hoot, 'boom, boom, boom.' As he grew more and more demonstrative in his actions, his modest mate flew up to an overhanging limb to escape his familiarities, and we drove away, leaving him still strutting on the ground underneath the tree where his mate sat perched. The comb, I should judge, was produced by the spots over the eyes becoming enlarged and inflamed with passion."

According to the observations made by Capt. T. E. Wilcox, assistant surgeon U. S. Army, in the vicinity of Lake Chelan, Washington, in the latter part of August, 1883, the Sooty Grouse will pack at times and gather in large coveys, though not to the same extent as the Pinnated and Sharp-tailed Grouse. He also writes me: "I once caught an old Grouse with a fishhook. I had my rod on my shoulder and suddenly came upon a covey, about the size of Quails, and caught one with my hands. This made the old bird frantic; she attacked me, and alighting on my rod, the hook pierced her foot. I was pulling her in, when my leader broke and she flew off. Of course I released her chicken. I killed a male in the Boisé Mountains, December 2, 1879, which weighed 3 pounds 10 ounces, but some killed by me in the Cascade Mountains seemed to be much larger. While on Lake Chelan in 1883, hunting white goats, I flushed a covey of Grouse, and here heard for the first time the call note of the female for her young. It was low, but distinct, something like that by the Bob White, just before it flushes. At this time, last of August, the birds were well grown. I have always found these birds near water. In 1881, while going to Indian Valley, Idaho, I rode past some, one being near enough to touch with a switch I had in my hand, yet they all walked out of the trail as quietly as domestic fowls would have done, and then resumed their dusting."

But one brood is raised in a season. Incubation lasts, according to different observers, from eighteen to twenty-four days. Females seem to predominate in numbers, but I do not think that these birds are polygamous. Their ordinary note resembles the cackling of the domestic hen very much. The Indian name of the Sooty Grouse on the Northwest coast is "*Tyhee-Callaw-Callaw*," "Chief Bird."

According to my own observations, made in various portions of Oregon, Washington, and Idaho, the usual number of eggs laid by the Sooty Grouse is about eight, and occasionally as many as ten are found in a set. Prof. O. B. Johnson, of the University of Washington, Seattle, Washington, informs me, however, that he found as many as sixteen eggs in a nest, and gives the average number from eight to twelve. The former, I think, will as a rule, come nearer to the correct average.

The eggs are ovate in shape, and the ground color varies from pale cream to a cream-buff, the latter being more common. In a single set before me it is a pale cinnamon. The eggs are more or less spotted over their entire surface with fine dots of chocolate or chestnut brown; these spots vary considerably in size in different sets, ranging from the size of No. 3 shot to that of mustard seed. These markings are generally well rounded, regular in shape, and pretty evenly distributed over the entire egg. They never run into irregular and heavy blotches, such as are frequently found in the eggs of the Canada Grouse, *Dendragapus canadensis*, which approach the pattern usually found among those of the Willow Ptarmigan, *Lagopus lagopus*, much nearer than the former. All of these markings can be readily washed off, as well as the overlying ground color, while they are still quite fresh, leaving the shell a delicate pale creamy white. In fact this coloring matter rubs off very readily, and occasionally fresh eggs will not stand even a good wiping. An egg is usually deposited daily and incubation does not begin until the set is completed, the male taking apparently no part in this duty nor in the care of the young after they are hatched.

The average size of ninety-six specimens in the U. S. National Museum collection is 48.5 by 34.5 millimetres. The largest egg of the series measures 52 by 37, the smallest 45 by 32.5 millimetres.

The type specimens show the different variations found in the eggs of the Sooty Grouse, and are all from the Bendire collection, having been collected by the author. No. 21073 (Pl. 1, Fig. 16), is from a set of ten, taken near Camp Harney, Oregon, May 10, 1876; No. 21074 (Pl. 1, Fig. 17), from a set of eight collected on the Canyon City Mountain, Grant County, Oregon, June 8, 1876; No. 21079 (Pl. 1, Fig. 18), from a set of nine, taken May 10, 1877, near Camp Harney, Oregon, and No. 21080 (Pl. 1, Fig. 19), from a set of seven eggs found near Fort Klamath, Oregon, and taken May 22, 1883. The majority of the eggs of this sub-species resemble the specimen figured on Pl. 1, Fig. 18, more than the other types.

17. Dendragapus obscurus richardsonii (Sabine).

RICHARDSON'S GROUSE.

Tetrao richardsonii, "Sabine MS.," Douglas. Linnæan Transactions, XVI. iii, 1829, 141.

Dendragapus obscurus richardsonii Ridgway. Proceedings U. S. National Museum, VIII, 1885, 355.

(B —, C 381*a*, R 471*b*, C 558, U 297*b*.)

GEOGRAPHICAL RANGE: Northern Rocky Mountains, mainly on eastern slopes, from southern Montana, northeastern Idaho, and eastern British Columbia, north into British America (Liard River).

This distinctly marked race of Dusky Grouse inhabits the timbered regions along the eastern slopes of the Rocky Mountains, from southern Montana and contiguous parts of Idaho northward, through the interior of British North America, to about latitude 61°. It was in the latter vicinity (Liard River) that Mr. J. Lockhart, of the Hudson Bay Company, obtained the most northern specimens of this bird that are in the U. S. National Museum collection.

Like the preceding, it is a resident, and breeds wherever found, and its habits are similar to those of the Dusky Grouse. In northern Wyoming and the eastern parts of central Idaho this Grouse intergrades with its more southern relative, and in northeastern Idaho and western Montana it does the same with the Sooty Grouse. It is a common enough bird in suitable localities throughout the mountainous portions of Montana, especially in the Big Horn Mountains and along the headwaters of the Musselshell River, where I personally met with them. Still, in some sections of this State, apparently quite suited to these birds, where an abundance of good-sized pine timber is found, they are entirely wanting. I have been unable to account for this fact, or to ascertain a good reason therefor, as plenty of good water and an abundance of food is to be found thereabouts.

Mr. Robert S. Williams, of Great Falls, Montana, writes me: "On June 21, 1885, while crossing over the almost bare summit of a small knoll in the foothills of the Belt Mountains, I suddenly almost ran into a brood of young Richardson's Grouse, which had evidently been hatched out but a very short time. The young, about ten in number, were closely huddled together, the old bird standing by their side, with head up, and eyes fairly blazing at the unexpected intruder. I was almost within reach of them, but neither old or young made a single motion or uttered a sound, while I stood watching them for several moments; and I left them in the same position.

"I have often met with coveys a little older, but have never seen the parent bird attempt to draw off the attention of any one by the feints so cunningly carried out by the Ruffed Grouse. These birds feed largely on grasshoppers when such are abundant."

Like the Sooty Grouse, after the young are fairly grown, these birds spend the greater portion of the late summer and autumn along the creek bottoms, fringed with dense thickets of cottonwoods, and many berry-bearing bushes, and at such times they become exceedingly fat[1]. I have seen them fully 10 miles away from any pine timber at this time of the year, and occasionally quite a distance from timber of any kind. Their nesting habits, as far as known, as well as the eggs, are similar in every respect to those of the Dusky and Sooty Grouse. The latter seem to average a trifle smaller, the mean being 47 by 34 millimetres. The largest specimen measures 51 by 34.5, the smallest 43 by 33.5 millimetres. This apparent difference in size can scarcely be taken into account, and is due, no doubt, to the small number (eleven specimens) in the U. S. National Museum collection; the majority of these, all from one set, laid probably by a young bird, are very small, and they reduce the general average considerably. As these eggs are indistinguishable from those of the preceding race none are figured.

18. Dendragapus canadensis (Linnæus).

CANADA GROUSE.

Tetrao canadensis Linnæus, Systema Naturae, ed. 10, I, 1758, 159.
Dendragapus canadensis Ridgway, Proceedings U. S. National Museum, VIII, 1885, 355.
(B 460, C 380, R 472, C 555, U 298).

GEOGRAPHICAL RANGE: Northern North America east of the Rocky Mountains, from the northern portions of the New England States, New York, Michigan, and Minnesota northwestward to Alaska (reaching coast at Kadiak, St. Michael, etc.).

The breeding range of the Canada Grouse, or the Spruce Partridge, extends from northwestern Alaska (Kowak or Putnam River) southeastward throughout British North America from ocean to ocean, south to central Minnesota, northern Wisconsin, northern Michigan, northern New York, and northern New England. It must, however, be considered as rather a rare summer resident within the United States, excepting northern Minnesota, where it is said to be common in the immense forests of the northeastern parts of the State, and extending westward to the edge of the prairie at White Earth.[2]

The Canada Grouse is usually resident, and breeds wherever found. At times, however, it is partially migratory during the winter; probably due more to lack of suitable food than to cold, as it has been found in considerable numbers, during the severest kind of weather, as far north as latitude 67°. Its favorite abiding places are the dense thickets of tamarack, *Larix americana*, also called hackmatac, and in groves and swamps of evergreen woods.

Mr. L. M. Turner, in his manuscript on the Birds of Labrador and Ungava, makes the following statement: "The mating season occurs in this locality (Fort Chimo) in the latter part of April or early May. It is said that the

[1] According to Dr. C. Hart Merriam, this Grouse feeds largely on the berries of *Arctostaphylos uva-ursi* and *Ribes cereum*, besides green leaves of the willow and other bushes.
[2] Bulletin 11, Dept. of Agriculture, Bird Migration Mississippi Valley, 1888, p. 103.

weather at this season may influence the pairing of this species for two or three weeks later. The males exercise much intrigue to secure the object of their choice for the season, although I have reason to suspect that some of these birds retain their mate for more than one season, as I have frequently found a pair together in the depth of winter, these two being the only ones of the kind to be found in the vicinity.

"Laying begins about the 5th of June, and incubation about the 12th. The young are hatched in about seventeen days. Young birds about five days old were obtained June 28, and others, able to fly, were secured July 10. Through the exertions of Miss Lizzie Ford I was enabled to secure two sets of eggs of this species. The nest consisted merely of a few stalks and blades of grass, loosely arranged among the moss of a higher spot, under the drooping limbs of a spruce, situated in a swamp. A few feathers from the parent bird were also in the nest. The number of eggs in this nest was seven, all quite fresh. A second set, also of seven eggs, was found in a similar situation, and near the location of the nest previously described.

"The food of the Spruce Partridge consists of the tender terminal buds of the spruce, and in winter this seems to be their only food. In a great number of birds examined during that season this was the only substance found in their gizzards, mixed at times with an astonishing quantity of gravel. I was surprised to find these stones of such uniformity of size and material. Crystallized quartz fragments, in certain instances, formed alone the triturating substance, and rarely were there fragments of granite or other stones. In fact many of the birds had not a discolored stone in their gizzard. In the spring and summer months these birds consume quantities of berries of *Empetrum* and *Vaccinium*."

Mr. J. W. Banks, of St. John's, New Brunswick, writes me: "Mr. James Lingley, an old backwoodsman and close observer, found two nests of the Canada Grouse, one on May 4, which was partially hidden under the trunk of a fallen tree. He killed the female with a stick of wood, not knowing she had a nest close by. On picking her up he found an egg she had just laid, and looking around found the nest with seven eggs. May 20 he found a second nest. This was placed between two small fir bushes that grew quite close together, and contained thirteen eggs. In both cases the nests were composed of dried leaves. He also describes the drumming of the male during the mating season, as follows: 'After strutting back and forth for a few minutes, the male flew straight up, as high as the surrounding trees, about 14 feet; here he remained stationary an instant, and while on suspended wing did the drumming with the wings, resembling distant thunder, meanwhile dropping down slowly to the spot from where he started, to repeat the same thing over and over again. The only food he noticed them take was the needles of the fir.'"

On the other hand, Mr. J. H. Yarnall, who has examined the crops of a great number of these birds "never found anything in them but the needles of the hackmatack."

Mr. Manly Hardy, of Brewer, Maine, a reliable and careful observer, writes me as follows: "I have been over every part of this State where this bird is likely to be abundant, east from Penobscot, from the sea to the North Corner Monument, but I have always found the Canada Grouse very scarce everywhere. Five once and six at another time are the largest number I ever saw together. I have many times traveled a month, and sometimes two months constantly in the woods, where they ought to be, without seeing over one or two.

"A Micmac Indian, whom I consider reliable, tells me of having seen a pack of many thousands somewhere east of Halifax, Nova Scotia, on which their whole village lived for weeks, moving after them when they moved. The males greatly preponderate over the females, at least two to one. They feed almost entirely on the needles of spruce and fir, also hackmatack and berries in summer. They show a preference for some fir trees over others, as I have seen them return to the same tree until it was nearly stripped. When disturbed, they always take to the trees, walking about in them, from one branch to another. My father, who had opportunities to see them drum, told me they drummed in the air while descending from a tree. They would fly up on a tree, then start off and drum on the way to the ground, like a Quaker grasshopper. When on the ground they scratch a great deal more than other Grouse do."

Another description of the drumming is as follows: "The Canada Grouse performs its 'drumming' upon the trunk of a standing tree of rather small size, preferably one that is inclined from the perpendicular, and in the following manner: Commencing near the base of the tree selected, the bird flutters upward with somewhat slow progress, but rapidly beating wings, which produce the drumming sound. Having thus ascended 15 or 20 feet it glides quietly on wing to the ground and repeats the manœuvre. Favorite places are resorted to habitually, and these 'drumming trees' are well known to observant woodsmen. I have seen one that was so well worn upon the bark as to lead to the belief that it had been used for this purpose for many years. This tree was a spruce of 6 inches diameter, with an inclination of about 15 degrees from the perpendicular, and was known to have been used as a 'drumming tree' for several seasons. The upper surface and sides of the trunk were so worn by the feet and wings of the bird or birds using it for drumming, that for a distance of 12 or 15 feet the bark had become quite smooth and red as if rubbed."[1]

Mr. Watson L. Bishop, of Kentville, Nova Scotia, has succeeded in domesticating the Canada Grouse, and he has published several very interesting accounts of their habits as observed by him, in the "Forest and Stream," giving its many readers a great deal of new and valuable information about the life-history of these birds, a portion of which I extract. He says: "As the nesting season approaches I prepare suitable places for them by placing spruce

---

[1] Birds of Maine, Everett Smith, Forest and Stream, February 8, 1883, p. 26.

boughs in such a way as to form cozy little shelters, where the birds will be pretty well concealed from view. I then gather up some old dry leaves and grass and scatter it about on the ground near where I have prepared a place for the nest. The bird pays no attention to this until she wants to lay. She will then select one of these places, and, after scratching a deep cup-shaped place in the ground, deposit in it her eggs. When the hen is on the nest she is continually making a kind of cooing sound, which I have never heard them make on any other occasion. If there should be sufficient material within easy reach of the nest the bird will sometimes cover the eggs up, but not in all cases.

"No nesting material is taken to the nest until after three or four eggs are laid. After this number has been deposited, the hen after laying an egg, and while leaving the nest, will pick up straws, grass, and leaves, or whatever suitable material is at hand, and throw it backward over her back as she leaves the nest, and by the time the set is complete, quite a quantity of this litter is collected about the nest. She will then sit in her nest and reach out and gather in the nesting material and place it about her, and when completed the nest is very deep and nicely bordered with grass and leaves.

"So strong is the habit, or instinct, of throwing the nesting materials over the back, that they will frequently throw it away from the nest, instead of toward it, as the hen will sometimes follow a trail of material that will turn her 'right about' so that her head is toward the nest, but all the time she will continue to throw what she picks up over her back. This, of course, is throwing the material away from the nest. Discovering her mistake, she will then 'right about face' and pick up the same material that an instant before was being thrown away, and throw it over her back again toward the nest.

"The way they will steal eggs from one another would do credit to a London pickpocket. Two hens had their nests near together, perhaps 2 feet apart, and as each hen laid every other day, one nest would be vacant while the other would be occupied. The hen that laid last would not go away until she had stolen the nest egg from the other nest and placed it in her own. I once saw a hen attempt to steal an egg from another nest that was 20 feet away. She worked faithfully at it for half an hour or more, but did not succeed in moving the coveted egg more than about 8 feet, it being up hill. The egg so frequently got away from her and rolled back a foot or more each time, that she at last got disgusted, and gave up the task. I had no fear of getting the sets mixed, as each was so different in color and shape from the other.

"On going to the pen one evening I found one of the hens on the nest, and I knew she was beginning to sit, as all the others had gone to roost. Slipping my hand under her I found three eggs, the nest egg, the one just laid, and the one stolen from the other nest. I picked two of them up and held them before her, when she all at once placed her bill over the one held between my thumb and forefinger, and tried to pull it out of my hand; I did not let her have it, however, and she immediately stepped upon the side of the nest and placing her bill over the remaining egg, drew it up out of the nest and pushed it back out of sight, as much as to say 'you have two, and that is all you can have.' I must confess that it was with great reluctance I took these eggs from her, she pleaded so hard for them.

"Anyone who has seen eggs of the Canada Grouse only after they have lain in the nest until the whole set is complete, can have no idea of the beauty of a fresh-laid egg. I have now in my collection about eighty of these eggs, all perfect specimens.

"The male bird begins to strut in March. I remember very well the first time I saw one strutting. I had obtained the bird in the fall, and he used to sit about, bunched up almost in a round ball, as the female did, until one morning, when I went to feed them I found him strutting. His attitude was so different that one would scarcely have known it was the same bird. I went in the house and told my wife to come and see him, remarking that whether the female laid any eggs or not, I was well paid by this sight for all my trouble. I was so interested in seeing him strut that I had the photographer bring his camera in and take some stereoscopic views of him while strutting.

"I will describe as nearly as I can his conduct and attitude while strutting: The tail stands almost erect, the wings are slightly raised from the body and a little drooped, the head is still well up, and the feathers of the breast and throat are raised and standing out in regular rows, which press the feathers of the nape and hind neck well back, forming a smooth kind of cape on the back of the neck. This smooth cape contrasts beautifully with the ruffled black and white feathers of the throat and fore breast. The red comb over each eye is enlarged until the two nearly meet over the top of the head. This comb the bird is able to enlarge or reduce at will, and while he is strutting the expanded tail is moved from side to side. The two center feathers do not move, but each side expands and contracts alternately with each step as the bird walks. This movement of the tail produces a peculiar rustling, like that of silk. This attitude gives him a very dignified and even conceited air. He tries to attract attention in every possible way, by flying from the ground up on a perch, and back to the ground, making all the noise he can in doing so. Then he will thump some hard substance with his bill. I have had him fly up on my shoulder and thump my collar. At this season he is very bold, and will scarcely keep enough out of the way to avoid being stepped on. He will sometimes sit with his breast almost touching the earth, his feathers erect as in strutting, and making peculiar nodding and circular motions of the head from side to side; he will remain in this position two or three minutes at a time. He is a most beautiful bird, and shows by his actions that he is perfectly aware of the fact.

"As the spring and summer advance the food given these Canada Grouse must be changed with the season, and it is only with a perfect knowledge of their wants and with constant care, that they can be safely carried through the heat of the summer and the moulting season. In the nesting season the females are very quarrelsome, and at this time more than two or three cannot be kept in the same pen, but in July they may be all turned together again, and they will agree very well until the following March."[1]

The Canada Grouse breeds early. Eggs now in the U. S. National Museum collection have been taken by Mr. B. R. Ross, of the Hudson Bay Company, near Fort Simpson, British North America, north of latitude 62°, as early as May 23. But a single brood is raised in a season. The number of eggs to a set varies from nine to thirteen, rarely more, usually about eleven, and in exceptional cases as many as sixteen. An egg is deposited every other day, and incubation does not begin till the clutch is completed. In form the eggs vary from ovate to elongate ovate. Their ground color, which is only superficial, is also very variable, ranging from a pale creamy buff to a decided reddish buff or pale cinnamon, and again to brownish buff with intermediate shades. The eggs are irregularly spotted and blotched with reddish brown or burnt umber. The spots vary considerably in size and shape, but are never close enough together to hide the ground color. An occasional specimen is but very slightly marked, and now and then one may be entirely unspotted.

The average measurement of fifty specimens in the U. S. National Museum collection is 43.5 by 31.5 millimetres, the largest egg measuring 48 by 33, the smallest 41 by 31 millimetres.

Of the type specimens selected to show the variations in color and markings, No. 22367 (Pl. 1, Fig. 20), was taken near Whale River, Ungava Bay, June 3, 1883; Nos. 22398 and 22399 (Pl. 1, Figs. 21 and 22), near Fort Chimo, Northeast Territory, Dominion of Canada, both on July 1, 1884.

These eggs were all collected by Mr. L. M. Turner while on duty as United States signal observer at Fort Chimo, and No. 24024 (Pl. 1, Fig. 23), is from a set of thirteen, laid in confinement in the spring of 1890, and purchased from Mr. W. L. Bishop, Kentville, Nova Scotia. The set from which this specimen is selected is much richer colored than any of the eggs taken from these birds in a wild state, and may be partly caused by the food they received in captivity.

19. Dendragapus franklinii (DOUGLAS).

FRANKLIN'S GROUSE.

Tetrao franklinii DOUGLAS, Transactions Linnæan Society, XVI, iii, 1829, 139.
Dendragapus franklinii RIDGWAY, Proceedings U. S. National Museum, VIII, 1885, 355.
(B 461, C 380*a*, R 472*a*, C 556, U 299.)

GEOGRAPHICAL RANGE: Northern Rocky Mountains (chiefly north of the United States) and west to the Coast ranges.

The breeding range of Franklin's Grouse, which still remains one of the rarest birds in the ornithological collections of the United States, extends from about latitude 60°, in southern Alaska, but along the coast only, south through British Columbia and Washington, to northern Oregon, where it reaches its southern limit at about latitude 45°. Eastward it ranges through the higher mountains of northern and central Idaho, and northwestern Montana to the Belt range. In Alaska, north of latitude 60°, it is replaced by *Dendragapus canadensis*, which likewise reaches the coast here, and the present species occupies but a comparatively small portion of this extensive territory. Within the United States, Franklin's Grouse is perhaps most common in suitable localities throughout northern and central Idaho, in the almost impenetrable and densely timbered mountain ranges bordering the headwaters of the north and south forks of the Clearwater, and the tributaries of the Salmon River. Throughout this region this species is known as the "Fool Hen;" an eminently proper and well-deserved name, it being entirely unsuspicious, allowing itself frequently to be knocked off the trees with sticks or stones, and it can often be caught by hand.

My friend, Dr. T. E. Wilcox, U. S. Army, says: "The cocks of this species are fearless and pugnacious, refusing to flee from man, and even attacking an intruder. I have been able to get within 3 or 4 feet before they would hop to another branch or twig. I always found them near running water or along the borders of high marshes. Its flight is not noisy like that of other Grouse or Partridges."

Mr. George Bird Grinnell says: "When alarmed or uneasy, Franklin's Grouse, as well as the Dusky Grouse, has the habit of erecting the feathers of the neck just below the head. This is done very commonly, and gives the bird a very odd appearance. It is analogous to the habit of the Ruffed Grouse, which, under the influence of certain emotions, erects the black ruff, and as does the Pinnated Grouse its little falciform feathers on the neck."

While stationed at Fort Lapwai, Idaho, from 1868 to 1871, I saw these birds on several occasions, and learned a good deal about them from packers and trappers. In those days the town of Lewiston, situated at the junction of the Snake and Clearwater Rivers, 11 miles west of Fort Lapwai, was the main supply depot for the various mining camps in northern Idaho, and every pound of freight for the mines had to be carried there on pack mules. The main trails to Oro Fino, Florence, and Warrens, the three principal mining centers at that time, passed right through the garrison, and it was no unusual sight for a half dozen pack trains, numbering a couple of hundred mules, to pass by there in a day. The route followed by these trains passed, for a portion of the way at least, over as rough and rugged a country as can be found anywhere, up one mountain and down another; some places being so rocky and steep that it seemed impossible for the heavily laden mules to keep their footing, and the underbrush so dense and thick on either side that it was almost impenetrable. The few narrow mountain valleys met with were no better. The melting of the deep snows rendered them nearly bottomless during the greater portion of the summer, making them a shaking, trembling quagmire in which the poor mules floundered up to their bellies in mud and mire. If you desired to become acquainted with the habits of Franklin's Grouse your inclination might be gratified in such localities as here described, viz, along the edges of wet or swampy mountain valleys, the so-called "Canas prairies," or the borders of the numerous little streams found in such regions among groves or thickets of spruce and tamarack. Few naturalists have as yet been sufficiently interested to invade their favorite haunts. They are also quite abundant on the Lolo trail over the Bitter Root Mountains, from the Nez Percé Indian Reservation to Missoula, Montana. I have met with them here as well as in the Salmon River Mountains, south of Mount Idaho, at an altitude of from 6,000 to 9,000 feet, during the Nez Percé campaign in the summer of 1877, but had no time then to observe their habits closely.

In the summer of 1881 I found a single covey, numbering about ten birds, in the low, flat and densely timbered region between the southern end of Pend d'Oreille Lake (the old steamboat landing) and Lake Cœur d'Alène, Idaho, at an altitude not exceeding 3,500 feet, I should think. I bagged three of these birds, and was quite surprised to find them in such a locality. As far as I have been able to learn, they usually occurred only at altitudes from 5,000 to 9,000 feet, and scarcely ever left the higher mountains. They were scratching in the dust on the trail I was following, and simply ran into the thick underbrush on each side, where they were quickly hidden.

Franklin's Grouse is a constant resident wherever found, and abundant enough in certain localities. Large numbers are yearly killed by both Indians and packers; in fact, this Grouse seems to furnish the latter their principal fresh-meat supply during the summer months, and they are by no means unpalatable at this time, as they feed more or less on various berries and grasshoppers, and not so much on the buds and leaves of the spruce and tamarack, as at other seasons of the year.

According to the best information obtainable, but a single brood is raised in a season, and their actions and drumming during the mating season are similar to those of the Canada Grouse. Nidification begins during the latter part of May or the beginning of June, depending somewhat on altitude and the season. The nesting habits and number of eggs laid to a set appear to be similar to those of the former species. There are no full sets of eggs in the U. S. National Museum collection.

Among an extremely interesting collection of birds' nests and eggs, made by Mr. R. MacFarlane, chief factor of the Hudson Bay Company, near Stewart Lake, New Caledonia District, British Columbia, during the season of 1889, and throwing much light on the distribution of a number of species found in this little known and practically unexplored territory, are two incomplete sets of eggs of this bird.

Three eggs of Franklin's Grouse and one of the Canadian Ruffed Grouse were found in one nest by an Indian near Babine, in the latter part of May, 1889, and a second nest, also containing three eggs, was brought to Mr. MacFarlane with the parent, by another Indian, who found it near Fort St.

James, on June 9 of the same year. The nest was merely a slight depression in the ground, and was lined with a few decayed leaves. Two of the eggs contained well-formed embryos and the third was addled.

Through the kindness of Mr. W. E. Traill, in charge of one of the Hudson Bay Company posts in British Columbia, parts of three sets of these rare eggs, fifteen in number, were collected during the season of 1890; taken on May 20, 27, and 30, respectively. The nests were shallow depressions in the moss-covered ground, lined with bits of dry grass, and were placed at the borders of spruce thickets. The eggs were fresh when found. They resemble those of the Canada Grouse in shape, color, and markings, but average a trifle smaller.

The average size of twenty-three specimens in the U. S. National Museum collection is 42 by 31 millimetres; the largest egg measuring 45 by 32.5, the smallest 38.5 by 30 millimetres. As they are similar to those of the preceding species, none are figured.

20. Bonasa umbellus (LINNÆUS).

RUFFED GROUSE.

Tetrao umbellus LINNÆUS, Systema Naturæ, ed. 12, I, 1766, 275.
Bonasa umbellus STEPHENS, General Zoölogy, XI, 1819, 300.
(B 465, C 385, R 473, C 565, U 300.)

GEOGRAPHICAL RANGE: Eastern United States, west to edge of Great Plains (?); north to Massachusetts (lowlands), Minnesota, southern Ontario, Canada; south to northern South Carolina and northwestern Georgia (uplands), Tennessee, Arkansas, etc.

The typical Ruffed Grouse or Partridge of the Northern States and the Pheasant of the South, inhabits and breeds throughout the wooded sections of the eastern United States, from Massachusetts westward, through New York, Pennsylvania, Ohio, Michigan, Wisconsin, Minnesota, and the southeastern portions of North and South Dakota, thence south through southeastern Nebraska and Missouri, the mountainous regions of Arkansas, eastern Tennessee, western North Carolina, northeastern Alabama, northwestern Georgia, and northern South Carolina, as well as in the remaining States included within the boundaries mentioned. Throughout its southern range the Ruffed Grouse is mostly confined to the mountain regions, and is seldom if ever found in the lowlands during the breeding season. In the New England States north of Massachusetts it intergrades with *B. umbellus togata*, the majority of the specimens found throughout southern Maine, New Hampshire, Vermont, and northern New York being scarcely referable to either form, birds found in the high lands approaching the Canadian Ruffed Grouse, while those in the valleys, are nearer typical *Bonasa umbellus*. The Ruffed Grouse found in southern Ontario, Canada, are referable to this race.

It is generally a resident and breeds wherever found, ranking with the Bob White in importance as a game bird. The Ruffed Grouse is naturally tame and unsuspicious, and let it once realize that it is protected it becomes almost as much at home in the immediate vicinity of man as a domestic fowl, and quickly learns to know its friends. At the fine country residence of the Hon. Clinton L. Merriam, near Locust Grove, New York, especially during the winter, it is not an unusual sight to see several of these handsome birds unconcernedly walking about the shrubbery surrounding his home, and even coming on the veranda of the house to feed. They, like many other animals about the place, have learned that here at least they are among friends, and plainly show their full confidence in them. Even during the mating season a cock Grouse may frequently be seen in the act of drumming within 50 yards of some of the outbuildings.

How different are the habits of these birds from those of the Ruffed Grouse as we usually see them. From the almost constant persecution they are subjected to throughout the year, in the more thickly settled portions of the United States at least, they have become a most cunning and extremely wary bird, and it takes a quick eye as well as steady nerves to arrest its swift and powerful flight when once on the wing and bring it to bag.

Notwithstanding the army of sportsmen, who leave this bird but little rest during the open season, and the great number annually snared, the numerous four-footed enemies it has to contend with during the breeding season, including cats, mink, weasels, foxes, and squirrels, as well as crows and birds of prey—like a few of the hawks and owls, which destroy either the eggs or young—and natural causes, such as wet and cold seasons, which are also exceedingly destructive to the newly-hatched young, this noble game bird seems, nevertheless, to hold its own fairly well over the greater portion of its range, and while they may be scarce one season, in the next they may be comparatively common.

The Ruffed Grouse is partial to an undulating and hilly country, one well wooded and covered with considerable undergrowth, interspersed here and there with cultivated fields and meadow lands. In the southern portions of its range, this bird is confined to the more mountainous and Alpine regions, being seldom found far away from such places, excepting in the late fall. As winter approaches, the coveys leave their feeding grounds in the mountains and repair to more congenial haunts along the edges of the neighboring valleys.

The mating season occasionally commences early in February, but usually about the beginning of March, when the familiar drumming of the male may be frequently heard, though the bird is not often seen. This drumming of the Ruffed Grouse has been often described, and many different theories have been advanced as to how the sound is produced. It is generally conceded now by most naturalists, including such well-known ornithologists as Brewster, Merriam, and Henshaw, that the sound is produced by the outspread wings of the bird being brought suddenly downward against the air, without striking anything.

Mr. Manly Hardy, of Brewer, Maine, well known as a reliable student of nature and a careful observer, describes the drumming as follows:

"The cock Grouse usually selects a mossy log, near some open hedge, clearing, or woods road, and partly screened by bushes, where he can see and not be seen. When about to drum he erects his neck feathers, spreads his tail, and, with drooping wings, steps with a jerking motion along the log for some distance each way from his drumming place, walking back and forth several times and looking sharply in every direction; then, standing crosswise, he stretches himself to his fullest height and delivers the blows with his wings fully upon his sides, his wings being several inches clear from the log. After drumming he settles quietly down into a sitting posture, and remains silently listening for five or ten minutes, when, if no cause for alarm is discovered, he repeats the process."

The drumming place is resorted to by the male from year to year. It may be a log, a rock, an old stump, or when such are not available, a small hillock is made to answer the purpose equally as well. While this drumming can not be considered a love note, as it may be heard almost every month in the year, and sometimes in the night as well as in the daytime, yet it must undoubtedly have some attraction for the female, and I think is performed as a sign of bodily vigor and to notify her of his whereabouts. Occasionally it causes a jealous rival to put in an appearance also, when a rough-and-tumble fight ensues. The female is seldom seen near the drumming place.

No game bird is more courageous than the Ruffed Grouse in the defense of its young; and the various tactics made use of, such as feigning injury, and fluttering along the ground just out of reach, are well known and often successful.

By many persons the Ruffed Grouse is considered polygamous, and while I can not actually disprove this assertion I doubt it very much.

The nest, like that of all the Grouse family, consists of a slight hollow scratched out at the base of a standing tree, a rock, under or alongside an old log, the fallen top of a tree, a brush pile, an old fence corner, or in the tangled undergrowth and thickets near a stream. Usually it is well and securely hidden, and placed in a secluded locality. Now and then, however, a nest will be found in quite an exposed and unlikely place, without any pretense at concealment. I have a photograph of such a one before me now, showing the bird on the nest. It was placed amongst a lot of fallen leaves, alongside the trunk of a tree, apparently a spruce, and close to a fence, in quite an open place.

Mr. Lynds Jones, of Grinnell, Iowa, found a nest of the Ruffed Grouse in a hollow stump, and Mr. C. M. Jones, of Eastford, Connecticut, found one in a swamp, on a little cradle knoll, surrounded by water. Mr. William N. Colton, of Biddeford, Maine, records a nest found between the stems of three young birches, fully 8 inches from the ground.

The nest itself is a very slight affair, and does not take long to construct. It is lined with a little dry grass, dead leaves, pine needles, or whatever is most conveniently found in the immediate vicinity of the nesting site.

Occasionally the Ruffed Grouse breeds very early, even in the more northern portions of its range. I have reliable records of full sets of eggs found in central New York as early as April 1 and April 2. Usually, however, the beginning of May is the breeding season of this species. If the bird is disturbed on the nest and the eggs are handled before the complete number has been laid and incubation fairly begun, it will frequently abandon its nest. The male leaves his mate as soon as she commences to sit, and apparently does not join the family again until the young are nearly fully grown. Incubation lasts from twenty-four to twenty-eight days, and but a single brood is raised in a season. If there are exceptions to this rule they are rare.

When incubation is somewhat advanced the Ruffed Grouse is loath to leave her eggs, and will allow herself to be very closely approached, relying on her color and motionless attitude for protection.

Mr. Lynds Jones writes me that he once stepped directly over a sitting bird without knowing it until the bird flew off behind him. Mr. A. S. Johnson, of Hydeville, Vermont, relates a similar experience, as follows: "I stood within 2 feet of a Ruffed Grouse sitting on her nest, which did not as much as wink till I stooped over closer to see how near she would let me approach. Then she slipped off the nest and skulked off 4 or 5 rods, stopping then to watch what I was going to do. The nest contained ten eggs. I passed by the spot several times after this and saw the bird on the nest each time, but did not disturb her."

The young are able to run about as soon as out of the shell and are cared for by the mother as a hen manages her brood. Their food at first consists almost entirely of insects (such as ants, beetles, small larvæ, and grasshoppers) and worms. When a little older they are taken to old wood roads for the double purpose of feeding on berries and such grain as is found among the droppings of horses, and more especially to take dust baths in order to free themselves from vermin. The cluck of the mother resembles that of the common barnyard fowl, only it is more subdued. When suddenly alarmed, a shrill squeal is given by the female; this, according to Dr. William L. Ralph, resembles very much the whining of a young puppy; and while the parent faces the intruder with every feather raised, the young hide quickly under anything in the vicinity that may afford protection, and they remain there perfectly quiet until called together again by their parent.

Till about half grown the Ruffed Grouse roosts with her young on the ground, afterward in trees. They do not pack at any time of the year, but remain in coveys, or what is left of these, seldom more than six or eight birds being found together.

During the summer and fall the food of the Ruffed Grouse is quite varied. Dr. A. K. Fisher, of the Department of Agriculture, Washington, District of

Columbia, writes me on this subject as follows: "The Ruffed Grouse is very fond of grasshoppers and crickets as an article of diet, and when these insects are abundant it is rare to find a stomach or crop that does not contain their remains. One specimen, shot late in October, had the crop and stomach distended with the larvæ of *Edema albifrons*, a caterpillar which feeds extensively on the leaves of the maple. Beechnuts, chestnuts, and acorns of the chestnut and white oaks are also common articles of food. Among berries early in the season, the blackberries, blueberries, raspberries, and elderberries are eaten with relish, while later in the year the wintergreen (*Gaultheria*), partridge berry (*Mitchella*), with their foliage, sumach berries (including those of the poisonous species), cranberries, black alder (*Ilex*), dogwood (*Cornus*), nannyberries (*Viburnum*), and wild grapes form their chief diet. In the fall the foliage of plants often forms a large part of their food, that of clover, strawberry, buttercup, wintergreen, and partridge berry predominating.

"A fine male, shot at Lake George, New York, November 1, 1889, had the crop and stomach distended with the leaves of the peppermint. In the winter these birds feed on the buds of trees, preferring those of the apple, ironwood, black and white birch, and poplar."

The number of eggs to a set varies from eight to fourteen; about eleven may be called a fair average. If the first set is destroyed, a second and usually a smaller one is laid. Sets of sixteen eggs or over are of rare occurrence, but I have a reliable record of one numbering twenty-three eggs. Mr John T. Paintin, of Coralville, Johnson County, Iowa, found this set May 26, 1886, near the Iowa River, 10 miles north of Iowa City. He was walking along in the timber, and in stepping over a rotten log almost stepped upon the Grouse. The eggs were carefully counted and the number found to be twenty-three; they were almost hatched, and were not disturbed.

In form they are ovate, or short ovate, their ground color varying from milky white to pinkish buff. About one-half of the eggs in the U. S. National Museum collection are more or less spotted with rounded dots, varying in size from a No. 4 shot to mustard seed or dust shot. These markings vary from pale reddish brown to drab color, and none of the eggs are heavily marked.

The average measurement of forty-four specimens in the U. S. National Museum collection is 38.5 by 30 millimetres, the largest egg of the series measuring 40 by 32, the smallest 33 by 25 millimetres.

As there is practically no difference in the eggs of the geographic races of the Ruffed Grouse, the type specimens figured have been selected with the object of showing as nearly as possible the variations both in ground color and markings, irrespective of race, similar specimens being sure to be found in a sufficiently large series of each form.

The type specimen of *Bonasa umbellus* (No. 23308, Pl. 2, Fig. 1), selected from a set of eight eggs collected by Mr. C. W. Richmond, near Harper's Ferry, West Virginia, May 30, 1885, represents one of the lightest colored specimens in the entire series, and is perfectly plain colored and unspotted.

21. Bonasa umbellus togata (Linnæus).

CANADIAN RUFFED GROUSE.

Tetrao togatus Linnæus, Systema Naturæ, ed. 12, 1766, 275.
Bonasa umbellus togata Ridgway, Proceedings U. S. National Museum, VIII, 1885, 355.

(B —, C —, R —, C, — U 300*a*.)

GEOGRAPHICAL RANGE: British Columbia, Washington and Oregon, excepting the coast districts, and from Idaho north and eastward to James Bay (Moose Factory), northern and central Maine and Nova Scotia; south occasionally in the mountains of New England and northern New York.

This race inhabits and breeds in the wooded districts from the mouth of the St. Lawrence River, westward through central and northern Maine and thence throughout the British possessions to the eastern slopes of the Cascade Range in Washington and Oregon, as far south at least as Fort Klamath, close to the boundary line of California. On the western slopes of the Bitter Root Mountains it reënters the United States, and is the typical form found throughout northern and middle Idaho, Oregon, and Washington east of the Cascades. Thence it ranges northward along the eastern spurs of the Fraser River and Cariboo Mountains to Fort St. James, Stewart Lake, New Caledonia district, in British Columbia, where it is common to about latitude 56°, and probably still further north in this direction.

In the central Rocky Mountain region the range of the Canadian Ruffed Grouse is locally intercepted by the southern extension of that of *B. umbellus umbelloides*, the latter being more of an Alpine form, and seemingly restricted to the mountainous sections.

The habits of this race are very similar to those of the common Ruffed Grouse. Throughout Canada and the British possessions it is better known by the name of Partridge and Birch Partridge.

Mr. Ernest E. Thompson, of Toronto, Canada, has kindly placed his field notes on this race at my disposal, and I make the following extracts from them: "Every field man must be acquainted with the simulation of lameness, by which many birds decoy or try to decoy intruders from their nests. This is an invariable device of the Partridge, and I have no doubt that it is quite successful with the natural foes of the bird, indeed it is often so with man. A dog, as I have often seen, is certain to be misled and duped, and there is little doubt that a mink, skunk, raccoon, fox, coyote, or wolf, would fare no better. Imagine the effect of the bird's tactics on a prowling fox; he has scented her as she sets, he is almost upon her, but she has been watching him, and suddenly with a loud 'whirr' she springs up and tumbles a few yards before him. The suddenness and noise with which the bird appears causes the fox to be totally carried away; he forgets all his former experience, he never thinks of the eggs, his mind is filled with the thought of the wounded bird almost within his reach; a few more bounds and his meal will be secured. So he springs and springs, and very nearly catches her, and in his excitement he is led on, and away, till finally the bird flies off, leaving him a quarter of a mile or more from the nest.

"If instead of eggs the Partridge has chicks, she does not await the coming of the enemy, but runs to meet and mislead him ere yet he is in the neighborhood of the brood; she then leads him far away, and returning by a circuitous route, gathers her young together again by her clucking. When surprised she utters a well-known danger signal, a peculiar whine, whereupon the young ones hide under logs and among grass. Many persons say they will each seize a leaf in their beaks and then turn over on their backs. I have never found any support for this idea, although I have often seen one of the little creatures crawl under a dead leaf. On July 3, 1884, while exploring in the Carberry spruce bush, Manitoba, with a friend, we passed a tree at whose roots was a Partridge's nest, but we would not have discovered it had not the mother pursued us some 20 feet and begun a vigorous attack on our legs, whereupon we turned and found the nest. It was just at that critical moment when the young were coming out. Those that were hatched, some six or eight, hid so effectually within a space of 6 feet that no sign of them could be seen. After their first rush, and once hidden, they ceased their plaintive 'peeping' and maintained a dead silence. Meanwhile the mother was sorely distressed, running about our feet with drooping wings, whining grievously, in such entire forgetfulness of herself and in such agony of anxiety for her young, that the hardest hearted must have pitied her and have felt constrained to leave her in peace, as we did."

Mr. Manly Hardy states: "The young run as soon as they chip the egg. If disturbed when only a few days old, the hen immediately flies at the intruder, making a loud noise, often striking him in the face or breast. The young usually drop where they are, remaining perfectly motionless. The parent throws herself on her breast and kicks herself along with her feet, aided by her spread wings, making a loud squealing noise. She goes just fast enough so that the pursuer can not quite get his hand on her, recovering, in a rod or two, to seem only broken-winged, and a short distance further on suddenly darting off. If one keeps quiet, in a short time she returns to the vicinity and calls her chicks, who come out of their hiding places and rejoin her. I have once seen the old cock with the brood, and on this occasion he gallantly defended the rear, until the rest made good their escape. He stood with wings raised and tail spread, ready to fight the intruder. I have seen the young fly into a tree when still in the yellow down; and when not larger than a Pine Grosbeak they will fly long distances, giving the alarm note of 'quit, quit,' just like an old bird. The young, a few days old, are shyer than the wariest adults. The noise made by the Ruffed Grouse in flying 'is made on purpose' to alarm others in the vicinity; they can fly as quietly as any bird if they choose.

26957—Bull. 1——5

"The males never congregate during the breeding season or after, and I never but once saw two adult males within one-fourth of a mile of each other between April and September. I consider that the drumming is not a call to the female, as they drum nearly or quite as much in the fall as in the spring, and I have heard them drumming every month in the year. I have never seen the least evidence that the Ruffed Grouse is polygamous."

Besides the various foods mentioned in the previous article, the Canadian Ruffed Grouse, according to Mr. Hardy, feeds not alone on the poplar buds, but also on the hard old leaves. He writes me: "I have killed one with its crop filled with such leaves on the 20th of August, and they eat them continuously, until the last have fallen in late October. They do this when other food is abundant. Buds of willow, yellow and white birch, hophornbeam, thorn plums, rosehips, leaves of tame sorrel, of the rock polypod, fungus from birch trees, the seeds of touch-me-nots (*Impatiens fulva*), wild raisins, and highland cranberries (both species of *Viburnum*) form also a part of their bill of fare. They seem to be especially fond of beechnuts. I have a record of finding seventy-six in one bird's crop and over sixty in another."

Personally, I have met with this bird quite frequently in various portions of Oregon and Washington, as well as in the north of Idaho, where it was especially abundant and exceedingly tame and unsuspicious. On the trail from Fort Lapwai, Idaho, to Fort Colville, Washington, in 1869 to 1871, I have seen, more than once, over fifty of these birds in a day's travel, without looking for them. Coveys of from eight to twelve were frequently met lying in any dusty place on the trail, taking sun baths and scratching around like chickens. When closely approached they would hop up or fly into the nearest tree or bush and remain there perfectly unconcerned, and I have seen them knocked down with sticks and stones.

On one of these trips, in the beginning of June, 1870, I saw a Ruffed Grouse, with a brood of young, attack an Indian dog that had attached itself to our party, and drive him off. We were riding through a little aspen thicket, some 10 miles north of the Spokane River, when the dog suddenly ran on the bird with her brood. She certainly looked the very incarnation of fury, every feather on her body was standing on end, as she fairly flew at the dog, perfectly reckless of consequences; but was so nimble and quick in her movements that she escaped all harm, and actually compelled the dog, by various peckings on the legs and head, to turn tail and run. At the same time she uttered a sharp, hissing sound of defiance rather than fear, which reminded me more of the hissing and spitting of an angry cat than anything emanating from a bird.

The nesting habits of the Canadian Ruffed Grouse, as well as the eggs, are in every respect similar to those of typical *Bonasa umbellus*. Mr. J. W. Banks, of St. Johns, New Brunswick, writes me: "Here with us a very common nesting place is what is called a fallow. This is a piece of woods chopped down in the fall, to be burned when sufficiently dry, usually in the latter part of May or early in June. Being composed chiefly of spruce and fir, it burns very rapidly. I found two nests (or rather the remains, for the eggs were badly scorched) in one of these burnt fallows, and a few feet from each nest the bones of the mother Grouse. A farmer acquaintance told me of finding a nest of this bird, which contained ten eggs, in a fallow he was about to burn, and knowing of another nest with an equal number of eggs, the thought occurred to him to put the eggs in the nest of the other bird that would not be endangered by the fire, and watch developments. He had the satisfaction of knowing that the eggs were hatched."

A nest of this Grouse was found by Mr. R. MacFarlane, of the Hudson Bay Company, near Fort St. James, British Columbia, May 16, 1889. It contained eight nearly fresh eggs, and was placed close to the foot of a pine tree in a slight depression scratched out by the bird. It was sparingly lined with grass, dry leaves, and a few feathers, and situated near a small lake; the female was snared on the nest. Judging from the number of skins of this Grouse, sent on at the same time, it must be quite common there.

But one brood is raised in a season. Incubation lasts from twenty-four to twenty-eight days, and does not begin until the clutch is completed, an egg, I believe, being deposited daily. The number of eggs to a set varies from eight to fourteen, rarely more. In form and color these are indistinguishable from those of the former subspecies. In size they average a trifle larger. The mean measurement of thirty-nine specimens in the U. S. National Museum collection is 40 by 31 millimetres, the largest egg of the series measuring 44 by 33, the smallest 37 by 29 millimetres. The type specimen, No. 4772 (Pl. 2, Fig. 2), selected from a set of eight, one of the darkest colored and most distinctly marked eggs of the entire series, was obtained by J. R. Willis, near Halifax, Nova Scotia, June, 1861.

22. Bonasa umbellus umbelloides (Douglas).

GRAY RUFFED GROUSE.

Tetrao umbelloides DOUGLAS, Transactions Linnæan Society, XVI, 1829, 148.
Bonasa umbellus var. umbelloides BAIRD, Birds of North America, 1858, 925.
(B 465*, C 385a, R 473a, C 566, U 300b.)

GEOGRAPHICAL RANGE: Rocky Mountain region of the United States and British America, north to Alaska, east to Manitoba.

The Gray Ruffed Grouse, the lightest colored of the forms of *Bonasa*, in which the gray tints strongly predominate over all others, inhabits the central Rocky Mountain system, from latitude 65° (Kaltag Mountains, near the head of Norton Sound) and the valley of the Yukon River in Alaska, south and southeast along the Yukon and Mackenzie Rivers, through British North America, eastern Idaho, Montana, western North Dakota, Wyoming, Utah, and Colorado. Like the preceding, it is generally a resident and breeds wherever found.

This well-marked and easily recognized subspecies, within the United States inhabits the dense undergrowth usually found along the sides of cañons and the clear mountain streams running through these, from an altitude of 7,000 to 10,000 feet, and, excepting in the fall and winter, it is rarely seen in the lower foothills or plains. Considering the isolated localities it inhabits, where it is seldom molested by man, it is an extremely shy bird, much more so than the Canadian Ruffed Grouse, and is not nearly so abundant as the latter. It habits are similar; and, besides the usual food used by the members of this family, in the late fall it feeds, to a great extent, on the leaves and fruit of a species of wild plum, growing in abundance along the foothills of the Big Horn Mountains in Montana, where, at that time of the year, it is often found associated with the Sharp-tailed Grouse, and not uncommon. The "ruffs," instead of being of the usual dark color, are, in an occasional specimen, of a beautiful bronze or coppery hue.

The nesting habits also, as well as the eggs of the Gray Ruffed Grouse, are in no way different from those of the preceding subspecies.

Mr. Robert S. Williams, of Great Falls, Montana, writes me: "I found a nest of this subspecies July 3, 1889; it was placed under the trunk of a fallen cottonwood tree, which rested about a foot from the ground. Otherwise the nest was not concealed in any way. The eggs, eleven in number, were evidently about to hatch, and I did not disturb them. Visiting the nest the succeeding day, the old bird let me climb over the fallen trunk above her without leaving the eggs."

Mr. W. H. Dall, U. S. Coast Survey, found the Gray Ruffed Grouse nesting near Nulato, Alaska, in May, and a set of eggs were found in an old willow stump. The average measurement of twenty-nine eggs in the U. S. National Museum collection is 40.5 by 30 millimetres. The largest egg of the series measures 43 by 31.5, the smallest 38 by 30 millimetres. The type specimen, (No. 22830, Pl. 2, Fig. 3), was taken May 18, 1886, by Mr. Ernest E. Thompson, near Carberry, Manitoba. It is of a pure rich cream color and unspotted.

23. Bonasa umbellus sabini (Douglas).

OREGON RUFFED GROUSE.

Tetrao sabini Douglas, Transactions Linnæan Society, XVI, iii, 1829, 137.
Bonasa umbellus var. sabinei Coues, Key to North American Birds, 1872, 235.
(B 466, C 385b, R 473b, C 567, U 300c.)

GEOGRAPHICAL RANGE: Coast Mountains of northern California, Oregon, Washington, and British Columbia.

The range of the Oregon Ruffed Grouse, the darkest and handsomest race of the genus Bonasa, is restricted to the wooded portions of country between the western slopes of the Coast Range and the Pacific Ocean, as well as the islands adjacent thereto. It is found from about latitude 57°, in the vicinity of Sitka, Alaska, south through western British Columbia, western Washington, western Oregon, and northwestern California, it having been taken near Humboldt Bay. Like the preceding, it is a constant resident and breeds wherever found, its general habits differing in no particular from those of its allies. In central Washington and Oregon it intergrades with the Canadian Ruffed Grouse, the majority of specimens approaching closer to the last-mentioned race.

According to Dr. Suckley, owing to the mildness of the season in the vicinity of Fort Steilacoom, the males commence drumming as early as January, and in February they are heard to drum throughout the night. In the autumn they collect in great numbers in the crab-apple thickets near the salt marshes at the mouths of the rivers emptying into Puget Sound. There they feed for about six weeks on the ripe fruit of the northwestern crab-apple, the *Pyrus rivularis* of Nuttall.[1]

Nidification begins about the middle of April and lasts sometimes till late in June. April 14 is the earliest date I have on which eggs have been found— a record given me by Prof. O. B. Johnson, of the Washington University, Seattle, Washington.

The number of eggs to a set varies from seven to thirteen, rarely more. A small set of six, partly incubated, were collected for me near North Saanich, Vancouver Island, British Columbia, June 28, 1876; probably a second laying, the first brood having been destroyed. The nest, a slight hollow in the ground scratched out by the bird, was placed under the fallen branches of a spruce tree. The cavity was lined with dead leaves and spruce needles, as well as a few feathers. This nest was found close to a small creek and was well concealed. Mr. A. W. Anthony found a nest in a similar situation near Beaverton, Oregon, on May 16, 1885. It contained seven eggs and incubation had commenced. A single brood is usually reared in a season.

The average measurement of twenty specimens in the U. S. National Museum collection is 41 by 30.5 millimetres, the largest egg of the series measuring 44 by 31.5, the smallest 38 by 29 millimetres. The type specimen (No. 6886, Pl. 2, Fig. 4) was taken by Mr. James Hepburne, near Victoria, British Columbia, in the spring of 1862.

24. Lagopus lagopus (Linnæus).

WILLOW PTARMIGAN.

Tetrao lagopus Linnæus, Systema Naturæ, ed. 10, i, 1758, 159.
Lagopus lagopus Stejneger, Proceedings U. S. National Museum, viii, 1885, 20.
(B 467, 470, C 386, R 474, C 568, U 301.)

Geographical range: Northern portions of northern hemisphere, south in winter, in America to Sitka, Alaska, the British provinces, and occasionally within the northern border of the United States.

The breeding range of the Willow Ptarmigan, or Willow Grouse, is confined to the Arctic regions of America, the so-called fur countries, seldom

---
[1] History of North American Birds, i-74, B. B. and R., Vol. iii, p. 454.

extending further south than latitude 55°, and then only in the eastern portions of its range, in Labrador and the shores of Hudson Bay. In winter these birds are partly migratory, and are sometimes found in considerable numbers as far south as latitude 50°, and stragglers on rare occasions have been taken within the northern borders of the United States. According to Richardson, considerable numbers remain in the wooded tracts, as far north as latitude 67°, even in the coldest winters.

Mr. E. W. Nelson states: "In the northern portions of their respective range these Grouse are summer residents, frequenting the extensive open country and being most abundant along the barren seacoast region of Bering Sea and the Arctic coast; but in autumn, the last of August and during September, they unite in great flocks and migrate south to the sheltered banks of the Kuskokwim and Yukon Rivers, and their numerous tributaries. In early spring as the warmth of the returning sun begins to be felt, they troop back to their breeding grounds once more.

"During a large portion of the year these birds form one of the most characteristic accompaniments of the scenery in the northern portion of Alaska. During the winter season these birds extend their range south to Sitka and Kadiak, from whence specimens in white plumage are in the U. S. National Museum collection.

"Toward the end of March, as the small bare spots commence to show on the tundra, the Eskimo say, this will bring the Ptarmigan from the shelter of the interior valleys, and their observation proves true.

"At St. Michael these birds commence their love-making according to the character of the season. In some years by the 1st of April their loud notes of challenge are heard; but the recurrence of cold weather usually puts a temporary stop to their proceedings. About the 5th or 15th of this month the first dark feathers commence to appear about the heads and necks of the males. During some seasons the males make scarcely any progress in changing their plumage up to the middle of May, when I have frequently seen them with only a trace of dark about the head and neck. In the spring of 1878 the first males were heard calling on the 26th of April, and on April 27, in 1879, the males were just commencing to moult, showing a few dark feathers, but the seasons were unusually late. In autumn the change frequently commences the last of September, and by the first of October it is well under way, the winter moult being completed towards the end of this month.

"At the Yukon mouth in the evening of May 24, these Ptarmigan were heard uttering their hoarse notes all about. As we were sitting by the tent my interpreter took my rifle, and going off a short distance worked a lump of snow to about the size of one of these birds. Fixing a bunch of dark-brown moss on one end of the snow to represent the bird's head, he set his decoy upon a bare mossy knoll; then retiring a short distance behind the knoll he began imitating the call of the male until a bird came whirring along, and taking up the gauntlet lit close by its supposed rival and fell a victim to the ruse.

"The note used by the native in this instance was a peculiar nasal 'yak-yak-yak-yak.' This was made by placing his hands over his mouth and closing the nose with thumb and finger. At this time the males were continually pursuing each other or holding possession of prominent knolls, frequently rising thence 5 to 10 yards in the air, with quick wing strokes, and descending with stiffened wings with the tips curved downward. While ascending they uttered a series of notes which may be represented by the syllables 'kû-kû-kû-kû,' which is changed as the bird descends to a hard rolling 'kr-r-r-r-,' in a very deep guttural tone, ending as the bird reaches the ground. Frequently a pair would fly at each other full tilt, and a few feathers would be knocked out, the weaker bird quickly taking flight again, while the victor rises, as just described, and utters his loud note of defiance and victory. On other occasions, when the birds are more evenly matched, they fight fiercely until the ground is strewn with feathers.

" By May 24 almost all these birds are paired, but some did not complete their nuptials until the first few days in June. This Grouse takes but a single mate in northern Alaska, and I am informed by the natives of Unalaska that the same is the case with the Rock Grouse found on the Aleutian Islands, nor have I ever known of the Ptarmigan assembling in numbers about any special meeting place to carry on their love affairs; they scatter about as previously mentioned, being seen singly here and there on prominent knolls over the flat country. Early in June, rarely so early as the last of May, the first eggs are laid; by June 20 and 25 the downy young are usually out, and when approached the female crouches close to the ground amongst her brood. When she sees it is impossible to escape notice, she rolls and tumbles away as though mortally injured, and thus tries to lead one from her chicks. The young at the same time try to escape by running away in different directions through the grass. At this season the female and male both moult and assume a plumage which differs considerably. The young are fledged and on the wing at varying dates through July, and are nearly full grown by the 1st to the 10th of August. They are handsome little creatures in brown and yellow down, with a chestnut cap and black lines down the back. A few days after birth the young begin to show traces of the first full plumage upon their breasts. * * *

"In nesting, these birds usually gather a few grasses and dry leaves, and with them they loosely line a shallow depression which is situated on the side of some slight knoll or dry place on the open grass and moss covered tundra."[1]

Mr. L. M. Turner, in his manuscript on the birds of Labrador and Ungava, makes the following statement regarding this species: "In the spring these birds repair, as the snow melts, to the lower grounds and prepare for the nuptial season. About the 10th of April they may be heard croaking or barking on all sides. A male selects a favorable tract of territory for the location of the nest, and endeavors to induce a female to resort to that place. He usually selects the highest portion of the tract, whence he launches into the

[1] Extracts from Report upon Natural History Collections made in Alaska, 1877-1881, Nelson, pp. 132-135.

air uttering a barking sound of nearly a dozen separate notes, thence sails or flutters in a circle to alight at the place whence he started, or to alight on another high place, from which he repeats the act while flying to his former place. Immediately on alighting, he utters a sound similar to the Indian word *chū-xwan* (what is it?) and repeats it several times, and in the course of a few minutes again launches in the air. Early in the morning hundreds of these birds may be heard, continuing until near 11 o'clock, when the bird then becomes silent until after 3 o'clock, when he again goes through the same performance, though with less vigor than in the morning. In the course of a few days a female may be found in the vicinity. The actions of the male are now redoubled, and woe be to any bird of his kind which attempts to even cross his chosen locality. Battles ensue which for fierceness are seldom equaled by birds of larger size.

"In the vicinity of Fort Chimo the nesting of this species begins during the latter part of May. The nest is usually placed in a dry spot among the swamps or on the hillsides where straggling bushes grow. The nest is merely a depression in the mosses, and contains a few blades and stalks of grass, together with a few feathers from the parent bird, which is now in the height of the moult from the winter to the summer plumage.

"The first eggs obtained were two, on June 1, 1884, this being the earliest record at Fort Chimo. The number laid for a set varies greatly in different localities. At Fort Chimo, seven to nine is the usual number, although in exceptional instances as many as eleven and rarely thirteen may be found. While I was at St. Michael (Norton Sound, Alaska) I frequently found nests containing as many as fifteen and several times found seventeen. I was there informed that over twenty eggs had been taken from a single nest. On neither side of the continent did I hear that more than one female deposited eggs in the same nest. I can affirm that a clutch of seven eggs may be taken, and, if the nest be not disturbed, the female will deposit nearly the same number again. These may again be taken, and not over three eggs will be deposited, and if disturbed a third time she will lay no more unless she selects a new location, which, of course, would be difficult to ascertain.

"I can not speak accurately on the subject, but think that seventeen days are required to incubate the eggs. On the 20th of June I obtained a young bird of this species, which was less than forty-eight hours out of the shell. This was the earliest record. Thousands of these young must perish annually, either from the cold rains, or from their parents being killed for food. The Indians consider the downy young of the Ptarmigan a special delicacy, even it taken from the shell; the bird serves in lieu of an oyster. I once had occasion to require the services of several Indian women to blow some eggs, which, during a pressure of other work, I had no time to do. I set them to work and frequently went to see if the work was progressing satisfactorily. I observed a pile of birds without, and some with feathers on, lying on a board. I inquired why they were being reserved. An old woman picked up one of the birds by the leg, and throwing back her head opened her mouth and indicated the purpose plainer than words could tell. After the middle of August the birds have acquired a good size, and are then feeding on berries of various kinds. They then are quite tender, of nearly white flesh, and when properly prepared form a pleasant food for the table. The young birds of the year attain their full growth by the 1st of November."

Mr. R. MacFarlane, chief factor of the Hudson Bay Company, who is exceedingly well qualified to speak about the Willow Ptarmigan, says: "This species is exceedingly abundant in the neighborhood of Fort Anderson, on the Lower Anderson River, and in the wooded country to the eastward. It is not, however, common in the Barren Grounds, especially from Horton River to Franklin Bay, where it is replaced by *L. rupestris* The nest is invariably on the ground, and consists of a few withered leaves placed in a shallow cavity or depression. The female sometimes leaves it only when almost trodden under foot, in fact several were swooped upon and caught thereon by hand. They usually begin to lay about the end of May or the beginning of June. The process of moulting, or the gradual assumption of their summer plumage, commences a week or two earlier. The female lays from seven to ten, twelve, and occasionally as many as thirteen, eggs, which I find was the greatest number recorded, and we had reason to know that some, at least, of the nests were used by Ptarmigan several seasons in succession. When very closely approached as stated, the female would frequently flutter off, sometimes spreading her wings and ruffling her feathers, as if to attack or frighten away intruders, and at other times calling out in distressed tones, and acting as if she had been severely wounded.

"In one instance where an Indian collector had found a nest which contained seven eggs, he placed a snare thereon; but on returning to the spot a few hours afterwards he was surprised to find that six of the eggs had disappeared in the interim, and as no eggshells were left behind (the male escaped) they were in all probability removed by the parents to a safer position. The male bird is generally not far away from the nest, and his peculiarly hoarse and prolonged note is frequently heard, the more especially between the hours of 10 p. m. and 2 a. m. Both, however, displayed great courage and devotion in protecting from capture their young, which we often encountered on our return coast trips.

"About the end of September, during October, and early in November *L. lagopus* assembles in great flocks, but during the winter it was seldom that more than two or three dozen were ever noticed in single companies. They are, however, most winters very numerous in the neighborhood of Fort Good Hope and other Hudson Bay Company posts in the Mackenzie River district; but as the spring sets in they begin to migrate northward. It is very doubtful if many breed to the south of latitude 68°, at least in the valley of the Anderson."[1]

[1] From R. MacFarlane's Manuscript on Land and Water Birds Nesting in British North America.

The food of the Willow Ptarmigan, during the early spring and summer, consists principally of the buds and tender leaves of the various species of birch and willows found in that region, and several kinds of berries, such as arbutus, cranberry, and whortleberry, as well as insects of different species, of which they find an abundant supply during the short summer season.

All observations made on the habits of the Ptarmigan during the breeding season tend to show that the male is equally devoted, and shows a strong attachment for the young, assisting in taking care of them, and displaying as great a solicitude for their safety as the female, differing in this respect from most of the Grouse family, by whom the care and protection of the young is apparently almost entirely left to their mates.

The nests of the Willow Ptarmigan are, as a rule, not particularly well hidden, and judging from the large number of eggs of this species in the U. S. National Museum collection, procured principally by Mr. R. MacFarlane, of the Hudson Bay Company, near Anderson River Fort, in about latitude 68°, they must be exceedingly abundant at this point.

The average number of eggs to a set is from seven to eleven, and but one brood is raised in a season. The eggs vary in shape from ovate to elongate ovate. The ground color ranges from cream color to a pronounced reddish buff, with several intermediate shades. In some specimens it is very clearly seen, in others it is almost completely obscured by the heavy confluent blotches and markings. The latter vary from well-defined and nearly even-sized spots of different sizes to confluent and clouded blotches, and smears of various shades of dark reddish and clove brown, completely obscuring the ground color in some instances. All this coloring matter can be readily removed in a freshly-laid egg, leaving the shell a pale creamy white, and they show an almost endless variation in shape, color, and size. All the specimens in the U. S. National Museum collection were taken in the month of June, the majority about the middle of this month.

The average measurement of two hundred and fifty specimens in the U. S. National Museum collection is 43 by 31 millimetres. The largest egg of this series measures 47 by 33.5, the smallest 39.5 by 28 millimetres, and a runt specimen 20 by 17.5 millimetres.

The types selected to show the variations in the styles of markings and coloration were obtained as follows: No. 6023 (Pl. 2, Fig. 5), from an incomplete set of three, taken by G. Bannister near Whale River, Ungava Bay, Labrador, June, 1862; No. 9251 (Pl. 2, Fig. 6), from a set of seven, June 29, 1863; No. 10689 (Pl. 2, Fig. 7), from a set of seven, June 20, 1865; all the latter being from the region east of Anderson River Fort, British North America, and collected by R. MacFarlane. No. 16461 (Pl. 2, Fig. 8), from a set of eight, taken June 20, 1872, by W. H. Dall, U. S. Coast Survey, on Popof Island (one of the Shumagin Group), Alaska Peninsula; No. 17042 (Pl. 2, Fig. 9), from a set of six, taken June 3, 1874, near St. Michael, Alaska, by L. M. Turner, U. S. Signal Service; and No. 21364 (Pl. 2, Fig. 10), from a set of ten, taken June 11, 1880, at St. Michael, Alaska, by E. W. Nelson, U. S. Signal Service.

25. Lagopus lagopus alleni Stejneger.

ALLEN'S PTARMIGAN.

Lagopus alba alleni Stejneger, Auk, 1, 1884, 369.
Lagopus lagopus alleni Stejneger, Proceedings U. S. National Museum, VIII, 1885, 20.
(B —, C —, R —, C —, U 301a.)

Geographical range: Newfoundland.

According to Dr. L. Stejneger this newly described subspecies is similar to *Lagopus lagopus*, but distinguishable by having the shafts of both primaries and secondaries black, the wing feathers and even some of the coverts marked and mottled with the same color.

Dr. C. Hart Merriam refers to this bird in the Ornithologist and Oölogist (Vol. VIII, No. 6, 1883, p. 43) as the Common or Willow Ptarmigan, and says: "It is still an abundant resident in Newfoundland, even in the vicinity of St. Johns, and thousands of them are killed annually on the peninsula of Avalon alone. It frequents rocky barrens, feeding on seeds and berries of the stunted plants that thrive in these exposed situations."

In his notes on the "Zoölogy of Newfoundland," Henry Reeks, esq., F. L. S., makes the following statement, which unquestionably applies to this race : "The Willow Grouse is called ' Partridge' by the settlers, and it frequents beds of alder and dwarf birch in swampy places, especially on the borders of lakes and rivers. It breeds on the ground among stunted black spruce, in rather drier situations."[1]

The breeding range of this well-marked race of the Willow Ptarmigan seems, as far as at present known, to be confined to the island of Newfoundland, where it is a resident. I have been unable to find any description of the eggs, which undoubtedly are indistinguishable from those of *Lagopus lagopus*.

26. Lagopus rupestris (Gmelin).

ROCK PTARMIGAN.

Tetrao rupestris Gmelin, Systema Naturæ, I, ii, 1788, 751.
Lagopus rupestris Leach, Zoölogical Miscellany, II, 1817, 290.
(B 468, C 387, R 475, C 569, U 302.)

Geographical range: Arctic America in general, southeastward to the Gulf of St. Lawrence (Anticosti), except the northern extremity of the peninsula of Labrador, and region thence northward, Greenland and the Aleutian Islands.

The breeding range of the Rock Ptarmigan extends through Arctic North America, from the Alaska Peninsula and Bering Strait, along the Arctic coast,

---

[1] Zoölogist, second series, 1869, Vol. IV, p. 1747.

southeast through the Barren Grounds, to the west coast of Hudson Bay, the Northeast Territory, and southern Labrador, and possibly Anticosti Island, in the Gulf of St. Lawrence.

Mr. E. W. Nelson states that "this beautiful Ptarmigan is a common resident of the Alaskan mainland, and unlike the common White Ptarmigan it frequents the summits of the low hills and mountains during the summer season, where it remains until the severe weather of early winter forces it down to the lower elevations and under the shelter of the bush-bordered ravines and furrows marking the slopes."[1]

There is evidently but little difference in the general habits of this species and those of the common Willow Ptarmigan, except that it frequents higher altitudes during the breeding season.

We are indebted to Mr. R. MacFarlane for nearly all we know about the breeding habits, nests, and eggs of this interesting species. He says: "This Ptarmigan is not near so plentiful as *L. lagopus*, and we only met with it in any considerable numbers from Horton River, Barren Grounds, to the shores of Franklin Bay. Very few nests were found to the eastward of that river, or on the coast or 'barrens' of the Lower Anderson. Its nest is similar but it lays fewer eggs than *L. lagopus*, as nine proved to be the rarely attained maximum among an aggregate record of sixty-five nests. The usual number was six or seven, and there were some which held only four and five eggs. It was no easy matter, however, to find the nests of this species, as the plumage of the birds and the color of the eggs both strongly resembled the neighboring vegetation. At the same time the female sat so very closely that more than one was caught on the nest, and I recollect an instance where the female bird, on the very near approach of our party, must have crouched as much as possible in the hope that she might not be noticed, which would have happened had not one of the smartest of our Indian collectors caught a glance of her eye. Although lots of male 'Rockers' were observed on our summer trips, feeding and otherwise disporting themselves in the 'barrens,' yet comparatively few nests were obtained, and, except in 1862, not one well-identified example was discovered west of Horton River, but during the winter scores of *L. rupestris* were met with in the forest country east of Fort Anderson."[2]

The "Barren Grounds," so often referred to in connection with the breeding grounds of numerous birds, are thus described by Mr. R. MacFarlane in a paper entitled, "On an Expedition down the Begh-Ula or Anderson River:" "The belt of timber which at Fort Anderson[3] extends for over 30 miles to the eastward, rapidly narrows and becomes a mere fringe along the Anderson River, and disappears to the northward of the sixty-ninth parallel of latitude. The country is thickly interspersed with sheets of water varying in size from mere small ponds to small and fair sized lakes. In traveling northeast toward

[1] Report on the Natural History Collections made in Alaska 1877-1881, Nelson, p. 136.
[2] From R. MacFarlane's Manuscript on Land and Water Birds Nesting in British North America.
[3] Established on Anderson River in 1861, and abandoned in 1866. Approx. lat. 68° 3′.

Franklin Bay, on the Arctic coast, several dry, swampy, mossy, and peaty plains were passed before reaching the 'Barren Grounds' proper. The country thence to the height of land between the Anderson and the deep gorge-like valley through which the Wilmot Horton River (MacFarlane River of Petitot's map) flows, as well as from the 'crossing' of the latter to the high plateau which forms the western sea-bank of Franklin Bay, consists of vast plains or steppes of a flat or undulating character, diversified by some small lakes and gently sloping eminences, not dissimilar in appearance to portions of the Northwest prairies. In the region here spoken of, however, the ridges occasionally assume a mound-like hilly character, while one or two intersecting affluents of the Wilmot Horton flow through valleys in which a few stunted spruce, birch, and willows appear at intervals. On the banks of one of these, near its mouth, we observed a sheltered grove of spruce and willows of larger growth, wherein moose and musk oxen had frequently browsed. We met with no more spruce, nor any traces of the moose to the eastward, and I doubt if many stragglers range beyond latitude 69° north.

"The greater part of the Barren Grounds is every season covered with short grasses, mosses, and small flowering plants, while patches of sedgy or peaty soil occur at longer or shorter distances. On these, as well as along the smaller rivulets, river and lake banks, Labrador tea, cranberries, and a few other kinds of berries, dwarf birch, willows, etc., grow. Large, flat spaces had the honeycombed appearance usually presented in early spring by land which has been turned over in autumn. There were few signs of vegetation on these, while some sandy and many other spots were virtually sterile. * * * These Barren Grounds are chiefly composed of a peaty, sandy, clayey, or gravelly soil, but stones are rare and rock *in situ* (limestone?) was encountered only two or three times on the line of march from the woods to the coast."[1]

This description will give the reader a good idea of the summer home of the Rock Ptarmigan; and while its food differs probably but slightly from that of the Willow Ptarmigan, it must necessarily be restricted to a much smaller variety. Their nests, usually placed among the dwarf brush or sedge-covered patches of the tundras on these barrens, are much harder to find than those of the latter, and the U. S. National Museum is almost entirely indebted to the indefatigable Mr. R. MacFarlane for the handsome series of eggs of this species in the collection, all of which, with the exception of a single set, were obtained by him.

Nidification begins about the middle of May in Alaska, and correspondingly later in the Barren Grounds, usually from June 15 to July 10. But a single brood is raised in a season. The number of eggs to a set varies from six to ten, rarely more, and usually but seven or eight are laid. These are ovate or short ovate in form, resembling the eggs of *Lagopus lagopus* considerably both in color and markings, but they average smaller. The majority are readily distinguished from those of the latter, the markings as a rule being

[1] Canadian Record of Science, January, 1890, pp. 52, 53.

smaller and better defined, and seldom running into indistinct and irregular blotches as is frequently the case in the eggs of that species.

The ground color ranges from a pale cream to a decided yellowish-buff, and in many specimens this is entirely hidden by a vinaceous rufous suffusion. The spots and blotches range from a dark clove brown to a dark claret red, with paler colored edgings; they are of various sizes, from the size of a buck-shot to that of No. 10 shot, and are irregularly distributed over the egg.

The average measurement of ninety nine specimens in the U. S. National Museum collection is 42 by 30 millimetres. The largest egg in this series measures 44 by 32.5, the smallest 39 by 29 millimetres.

The types selected to show the different styles of markings are as follows: No. 7642 (Pl. 2, Fig. 11), from an incomplete set of three eggs, taken near Franklin Bay, Arctic coast, June 26, 1883; No. 9268 (Pl. 2, Fig. 12), from a set of eight, taken near Anderson River, Arctic America, June 10, 1863; No. 9273 (Pl. 2, Fig. 13) from an incomplete set of four, same locality, taken July 7, 1863; and No. 9284 (Pl. 2, Fig. 14), from a set of six, same locality, June 3, 1863; all having been collected by Mr. MacFarlane. No. 14997 (Pl. 2, Fig. 15) is from a set of ten eggs, taken in the Gens-du-large or Romanzof Mountains, Alaska, by Mr. James McDougall, of the Hudson Bay Company, in the latter part of May, 1869.

27. Lagopus rupestris reinhardti (Brehm).

REINHARDT'S PTARMIGAN.

Lagopus reinhardi (err. typ.) Brehm, Lehrbuch europäischer Vögel, 1823, 440.
Lagopus rupestris reinhardti Blasius, List European Birds, 1862, 16.
(B —, C —, R —, C —, U 3020.)

GEOGRAPHICAL RANGE : Greenland, islands on western side of Cumberland Gulf, and northern extremity of Labrador (Ungava).

The breeding range of Reinhardt's Ptarmigan, as known at present, includes both shores of Baffin Bay, Davis Strait, and Hudson Strait, ranging well up into the Arctic circle. It is a common bird in Greenland, and a number of its eggs collected in the vicinity of Sukkertoppen are in the U. S. National Museum collection.

It is not at all rare in the northern portions of Labrador, and Mr. L. M. Turner, of the U. S. Signal Service, makes the following statement regarding this subspecies, in his "Notes on the Birds of Labrador and Ungava:"

"This Ptarmigan is known to the white people as the Rock Grouse, or simply as 'Rocker.' In the southern portion of Labrador these Ptarmigan are not very numerous, but become so as the more northern and elevated portions of the country are reached. They prefer more open ground *'and rarely straggle even into the skirts of the wooded tracts.'* The hilltops and 'barrens' (hence often called Barren Ground Bird) are their favorite resorts. As these tracts are more extensive in the northern portions of Labrador and Ungava, these birds are there very abundant. During the summer months they are quite scarce in the vicinity of Fort Chimo, retiring to the interior and the hills of George River for that season. In the month of May the nuptial season arrives and is continued until about June, when nesting and laying begin. The birds are by this time scattered, each pair now taking possession of a large tract of stunted vegetation, among which they make their nest and rear their young. I was never able to procure the eggs of this species. Only young birds a few days old were brought to me, and some of larger size.

"As before stated the mating season begins in May, and during this period the male acts in the strangest manner to secure the affection of his chosen mate. He does not launch high in air and croak like the Willow Ptarmigan, but runs around his prospective bride with tail spread, wings either dragging like those of the common Turkey, or else his head and neck stretched out, and breast in contact with the ground, pushing himself in this manner by the feet, which are extended behind. The male at this time ruffles every feather of his body, twists his neck in various positions, and the supraorbital processes are swollen and erect. He utters a most peculiar sound, something like a growling 'kurr-kurr,' and as the passion of the display increases the bird performs the most astonishing antics, such as leaping in the air without effort of wings, rolling over and over, acting withal as if beside himself with ardor.

"The males engage in most desperate battles; the engagement lasts for hours, or until one is utterly exhausted, the feathers of head, neck, and breast strewing the ground. A manoeuvre is for the pursued bird to lead the other off a great distance and suddenly fly back to the female, who sits or feeds as unconcerned as it is possible for a bird to do. She acts thoroughly the most heartless coquette, while he is a most passionately devoted lover. He will rather die than forsake her side, and often places himself between the hunter and her, uttering notes of warning for her to escape, while attention is drawn to him, who is the more conspicuous.

"When the young are with the parents they rely upon their color to hide themselves among the nearly similar vegetation from which they procure their food. I am certain I have walked directly over young birds which were well able to fly. If the parent birds are first shot, the entire number of young may be secured, as they will not fly until nearly trodden upon, and then only for a few yards, where they may easily be seen. I have found on two occasions an adult female with a brood of thirteen young. All of the flocks were secured without trouble. At other times only three or four young would be found with both parents. The young are very tender when first hatched; no amount of most careful attention will induce them to eat, and after only a few hours' captivity they die. I could never keep them alive above twelve hours. The changeable weather, sudden squalls of snow or rain, must be the death of scores of these delicate creatures. Their note is a soft piping 'pe-pe-pe,' uttered several times, and has the same sound as that of the young of the Bob White, *Colinus virginianus.*

"In the young birds just hatched, and up to the age of three weeks, it is difficult, if possible at all, to distinguish them to a certainty from the young of *Lagopus lagopus* of the same age. They are slightly darker, and the lower parts have a greenish tinge to the down instead of yellow, as in the young of *L. lagopus.* Although I have preserved a great number of these young birds I would still hesitate to assert to which of the two species they belong. After the age of three weeks they may be easily distinguished by the bill. By the 10th of August the wing quills have begun to show the winter plumage. The first primary is then white and nearly half its normal length, with the second and third showing considerable development. The bird is at this time about the size of a Bob White (*Colinus virginianus*)."

The food of Reinhardt's Ptarmigan during the summer consists of insects as well as various leaves and berries, such as those of the crowberry (*Empetrum nigrum*), whortleberries, the tender leaves of the dwarf birch and white birch (*Betula alpestris*) as well as the buds, willow buds, and sorrel. Mr. Ludwig Kumlien shot a specimen near Cumberland Sound, whose crop was crammed full of sphagnum moss. They are usually met with in small coveys from six to ten birds, rarely more.

But a single brood is reared in a season. The eggs are usually deposited during the month of June, and the sets vary from six to fourteen, very rarely more. They are absolutely indistinguishable from those of the Rock Ptarmigan *Lagopus rupestris;* in fact, the average measurement of thirty-three specimens in the U. S. National Museum collection, from Greenland, corresponds exactly with that of the preceding species, giving an average of 42 by 30 millimetres. The largest egg of the series measures 44 by 32, the smallest 40 by 29 millimetres. The majority of these eggs were collected near Godthaab and Sukkertoppen, Greenland, and are a gift of Governor E. Fencker.

None are figured, as they are exactly like the eggs of *Lagopus rupestris.*

28. Lagopus rupestris nelsoni Stejneger.

NELSON'S PTARMIGAN.

Lagopus rupestris nelsoni STEJNEGER, Auk, I, 1884, 226.
(B —, C —, R —, C —, U 302*b*.)

GEOGRAPHICAL RANGE: Island of Unalaska and adjacent islands in the Aleutian Chain.

The types of this comparatively new race were taken by Mr. Nelson, at Unalaska, one of the Aleutian Islands; he reports it as common there, frequenting the mountain tops and slopes, and breeding in June.

Mr. Turner, in his "Contributions to the Natural History of Alaska," refers to this subspecies in his article on *Lagopus rupestris*, from which it had not been separated when his account was written. He says:

"On some of the islands it is extremely abundant, among those may be mentioned Unalaska, Unimak, Akutan, and Akun. It is a resident where found, and, among the islands, rarely leaves its native island. At Unalaska they seem to prefer the high rocky ledges, but everywhere come down to the low narrow valleys to roost and rear their young. They rarely assemble in large flocks; a dozen to twenty individuals usually compose a flock. The season begins in the early part of May, and is continued for about mating three weeks, by which time the site for the nest is chosen, usually amidst the tall grasses at the mouth of a wide valley, or else on the more open tundra among the moss and scanty grass.

"The nest of this bird is composed of a few stalks of grass and the feathers that may fall from the mother's breast. The nest is a very careless affair, and often, near the completion of incubation, the eggs will lie on the bare ground surrounded by a slight circle of grass stalks that have apparently been kicked aside by the mother impatient of her task. The number of eggs varies from nine to seventeen, eleven being the usual number. The exact date of incubation was not determined by me. The young are able to follow the mother as soon as they are hatched.

"As this bird never collects into large flocks, I always supposed the flocks seen in winter were the parents with the brood reared the previous summer."

There are no eggs of Nelson's Ptarmigan in the U. S. National Museum collection, neither have any of the ornithologists, who met with this subspecies, described them. There is every reason to presume, however, that they are indistinguishable from the eggs of the Rock Ptarmigan.

29. Lagopus rupestris atkhensis (Turner).

TURNER'S PTARMIGAN.

Lagopus mutus atkhensis Turner, Proceedings U. S. National Museum, v, July 29, 1882, 227, 230.
Lagopus rupestris atkhensis Nelson, Cruise of the Corwin, 1883, 56c+82.
(B—, C—, R—, C—, U 302c.)

Geographical range: Atka Island, Aleutian Chain.

Mr. E. W. Nelson states: "Among the specimens secured by Mr. L. M. Turner, during his residence in the Aleutian Islands, are four Ptarmigans, which, upon examination, prove to represent a well-marked geographical race of *L. rupestris*. His specimens were secured June 7 and May 29, upon Atka Island, at the extreme western end of the Aleutian Chain. They are found upon this island, and undoubtedly upon those adjoining. * * * It is undoubtedly to this race that Mr. Dall refers in his 'Contribution to the

Ornithology of the western end of the Aleutian Chain,' when he speaks of finding nine much incubated eggs on June 21, at Attu Island, and chicks which were hatched at Kiska July 8."[1]

Mr. Turner himself writes: "When I first obtained these birds I was struck with the greater size, and also, with the shape of the bill, and greater length of the claws when compared with the mainland bird. This bird frequents the lowlands and hills of the western islands of the Aleutian Chain. They are quite plentiful on Atka, Amchitka, and Attu Islands. The nest is built amongst the rank grasses at the bases of hills and the lowlands near the beach. It is carelessly arranged with a few dried grass stalks and other trash that may be near. The eggs vary from eleven to seventeen, and are darker in color than those of the *L. rupestris* and but slightly inferior in size to those of *L. lagopus*. A number of eggs of this species were procured, but broken in transportation; hence, I can give no measurements of them.

"The general habits of this species are those of the other species. At Attu they frequent the higher elevations, probably on account of the great number of foxes (*Vulpes lagopus* BAIRD) which occur on that island and have but little to subsist on. The natives of Attu assert that this species of Ptarmigan occurs on Agattu Island, and that it is quite numerous there, probably on account of the absence of foxes."[2]

Nothing further is known of the nesting habits of Turner's Ptarmigan, and no specimens of their eggs have as yet found their way into the U. S. National Museum collection.

30. Lagopus welchi BREWSTER.

WELCH'S PTARMIGAN.

Lagopus welchi BREWSTER, Auk, II, April, 1885, 194.
(B —, C —, R —, C —, U 303.)

GEOGRAPHICAL RANGE: Newfoundland.

This newly described species is based on specimens obtained during May and June, 1883, by Mr. George O. Welch, in whose honor it has been named.

Mr. Brewster states: "The colors in the male of this Ptarmigan are confused and blended to such a degree that a detailed description, however carefully drawn, fails to do them justice. The general effect is that of a dark, grayish-plumbeous bird (colored not unlike the Oregon form of the Dusky Grouse), plentifully besprinkled with fine dots of pepper-and-salt color. * * *

"According to Mr. Welch these Ptarmigan are numerous in Newfoundland, where they are strictly confined to the bleak sides and summits of rocky hills and mountains in the interior. Unlike the Willow Grouse of that island, which in winter wander long distances and frequently cross the Gulf to Lab-

[1] Report on the Natural History Collections made in Alaska 1877-1881, Nelson, p. 139.
[2] Contributions to the Natural History of Alaska, 1886, Turner, p. 156.

rador, the Rock Ptarmigan are very local, and for the most part spend their lives on or near the hills where they are reared."[1]

In his notes on the "Zoölogy of Newfoundland," in the Zoölogist (second series, 1869, Vol. IV, p. 1747), Henry Reeks, esq., refers to the present species, under the name of Rock Ptarmigan, as follows: "A truly Alpine species in Newfoundland; rarely found below the line of stunted black spruce, except in the depths of winter, when they descend to the lowlands and feed on the buds of dwarf trees, sometimes in company with the Willow Grouse; but I never saw this species perch on trees. It is called by the settlers the 'Mountain Partridge.'"

The nest and eggs of this species have not as yet, so far as I am aware, been described, but are presumably similar to those of *Lagopus rupestris*.

31. Lagopus leucurus SWAINSON.

WHITE-TAILED PTARMIGAN.

Lagopus leucurus SWAINSON, Fauna Boreali Americana, II, 1831, Pl. 63.
(B 469, C 388, R 476, C 570, U 304.)

GEOGRAPHICAL RANGE: Alpine summits of Rocky Mountains; south to New Mexico; north into British America (as far as Fort Halkett, Liard River); west to higher ranges of Oregon, Washington, and British Columbia.

The breeding range of the White-tailed Ptarmigan, within the United States at least, is only found on or near the summits of the higher mountain ranges, and apparently always above timber line. It extends from Alaska southward through western North America, reaching its most southerly point in northern New Mexico (vicinity of Taos), where Dr. B. J. D. Irwin, U. S. Army, obtained specimens near Cantonment Burgwyn. It has also been met with in eastern Idaho, Montana, Wyoming and Colorado, where, in suitable localities, it is by no means rare.

On the Pacific coast it is reported common in the mountains of British Columbia and the Olympic and Cascade Ranges in Washington, especially on Mounts Baker, Ranier, and St. Helen's. In Oregon it is reported from Mounts Hood and Jefferson, and, according to Indian testimony, it occurs as far south as Diamond Peak, 60 miles north of Fort Klamath, Oregon. I know of no record, however, that this species has actually been taken in Oregon. The Washington records are given on the authority of Prof. O. B. Johnson, of Washington University, Seattle, Washington, as well as of other correspondents.

Mr. W. M. Wolfe, of Kearney, Nebraska, writes me of having found the White-tailed Ptarmigan in the Wind River Mountains, Wyoming, and in the Bitter Root Mountains of Idaho and Montana.

It is a resident and breeds wherever found, rarely leaving the mountain summits, even during the severest winter weather, and then only descending

[1] Brewster, Auk, II, April, 1885, pp. 194, 195.

2,000 or 3,000 feet at most, seldom being found at a lower altitude than 8,000 or 9,000 feet at any time. In the Rocky Mountain region it is generally known by the very appropriate name of the "White" or "Snow" Quail.

Mr. George Bird Grinnell has kindly furnished me the following information about this species: "I have found the White-tailed Ptarmigan in considerable numbers in Colorado, Montana, and British Columbia. Where I have seen them they have always been above timber line. Although on a few occasions I have met with these birds in the late summer when the young were little more than half grown and the broods were still together, my experience with them has been chiefly in the autumn when hunting mountain sheep and white goats high up among the summits of the ranges. At this season of the year they are usually found in small numbers, from two to a half dozen being the ordinary size of the flocks. Last year, however (October, 1889), I came across a pack of these birds in the Cascade Mountains of British Columbia, where there were twenty-five or thirty together. In the autumn the birds are generally rather wild, and if nearly approached become quite uneasy and run about, holding the tail elevated and looking very much like a white Fan-tail Pigeon. At this season the only cry that I have heard is a sharp cackle like that of a frightened hen. This the bird begins to utter a short time before it takes wing, and continues it for quite a little while after having begun to fly.

"On the high plateaus where this bird is found the wind often blows with a tremendous sweep and is almost strong enough to throw down a man. When such a wind is blowing the Ptarmigan dig out for themselves little nests or hollows in the snow banks, in which they lie with their heads toward the wind and quite protected from it.

"Often on the rocky slopes where there is no snow they may be seen lying crouched on the ground behind rocks or small stones, with their heads directed to the quarter from which the wind blows. If startled from such a place they all take wing at once, looking like a flock of white Pigeons, and fly for a short distance, but as soon as they touch the ground again they throw themselves flat on it behind the most convenient shelter. Among the high mountains of the St. Mary's Lake region in Montana I have seen birds of this species which were pure white by the 20th of September. On the other hand, in the Cascade region of British Columbia, they have still a good many brown spots in early October. I presume the change of plumage varies with the locality and often with the individual bird."

Mr. A. W. Anthony writes me as follows: "In southern Colorado, where I have met with this species, nesting must begin some time from the first to the middle of June, as I have found young birds but an hour or so from the egg, from July 1 to the 18th. The nests I have seen were located in the loose rocky débris of steep hillsides, a simple depression in the short fine grass which grows in small patches between the rocks above the timber line

Although utterly devoid of protection from bush or shrub, so nearly does the sitting bird resemble the gray bowlders which surround her on every side that the discovery of the nest is due largely to accident. When incubating it is nearly impossible to flush the bird, according to my experience. Twice have I escaped stepping upon a sitting Ptarmigan by only an inch or so, and once I reined in my horse at a time when another step would have crushed out the life of a brood of nine chicks but an hour or so from the egg. In this case the parent crouched at the horse's feet, and, though in momentary danger of being stepped on, made no attempt to escape until I had dismounted and put out my hand to catch her. She then fluttered to the top of a rock a few feet distant, and watched me as I handled the young, constantly uttering low anxious protests. The chicks were still too young to escape, mere little awkward bunches of down that stumbled and fell over one another when they attempted to run.

"Miners in whom I place confidence have told me that they have lifted sitting Ptarmigan from the nest and handled the eggs, while the bird stood but a few feet distant watching her treasures and uttering an occasional squeak like a sitting hen. One, which had her nest near the trail between the cabin and the mine, was annoyed in this way so often that she would attempt to regain the nest while the eggs were being handled, and had to be frequently pushed aside; she never failed to peck at the hand and utter her protesting 'k-r-rrr' whenever any one attempted to touch her, and made no attempt to fly away.

"I have never heard of a nest at a lower elevation than one I found in Saguache County, Colorado, which was not over 200 feet above timber line. I think that they usually nest above 12,000 feet. Judging from the broods of young I have flushed in August, I consider nine about the average number."

Mr. Drew, in his field notes on the Birds of San Juan County, Colorado, makes the following statement about this species: "Very common; breeds. They are found above timber line in summer, where they feed on the leaves and flowers of the marsh marigold, *Caltha leptosepala.* * * * They are usually quiet during the day, but active and noisy in the evening, making a cackling like Prairie Chickens. * * * They have from eight to ten young at a brood."[1]

Mr. Dennis Gale writes me as follows: "Irrespective of season, as a general rule, a single bird will not flush unless urged to it. During the summer months this is especially noticeable; they will only move out of your way when directly in your path, and close upon them, by short tacks right and left, sidling off from you, at each tack changing sides, moving quickest on the short run just before slowing up for the turn. Two or more together are much more likely to flush, and if alarmed while flying will utter a quick repeated 'köck, köck,' very like the note uttered by *Pediocætes phasianellus campestris* under similar circumstances.

---

[1] Bulletin Nuttall Ornithological Club, Vol. vi, 1881, p. 141.

"As near as I can decide, they nest about the middle of June and hatch out their young about the middle of July. It is very rarely that a female is seen from the beginning of June until noticed in company with her brood. At this season I have frequently met with the males singly, and sometimes as many as five together; and I do not think that they take any share in the duties of incubation.

"I met with two broods, one with, I think, seven chicks just hatched out, and the other of five, nearly two weeks old. The latter showed no white; they had, in both cases, a general gray appearance; the newly hatched brood was in the downy phase. There was a disposition, clearly proven with the chicks of both broods, to hide when the hen signaled danger; but some of the older ones flushed and flew at least 50 yards. The females were very tame and would not flush; in fact they could not be induced by mild treatment to leave the place where the young had hidden. They walked around me so close that I could have touched them with my hand, and showed a marked concern for the safety of their broods, clucking in a manner very similar to our domestic hen. I am of the opinion that this species is very much less numerous now than they were ten years ago, and I believe the conditions generally favor a yearly contraction in their numbers; at least this will be the case wherever their summer range is available for stock to graze over."

The crop of a bird of this species, kindly sent me by Mr. Gale for examination as to the nature of its food, was filled with the buds and catkins of a species of birch, *Betula glandulosa*.

The number of eggs to a set are variously stated at from four to fifteen. From eight to ten may be considered a fair average. Notwithstanding the fact that the nests of the White-tailed Ptarmigan are said to have been found repeatedly, but very few of the eggs of this species have as yet found their way into collections. The U. S. National Museum has an incomplete set of four, taken by Mr. A. D. Wilson, one of the chiefs of the Hayden Geological Survey, in the San Juan Mountains, southern Colorado, at an altitude of 12,300 feet. He told me that he accidentally stumbled on the nest, while crossing a rocky mesa, on the morning of July 15, 1875. The nest was placed between a couple of lichen-covered rocks, and contained, if he remembered rightly, five or six nearly fresh eggs. The female skulked off as he was in the act of stepping over her, and hid amongst the rocks some 20 feet away.

Dr. Elliott Coues, in an article "On the breeding habits, nest, and eggs of the White-tailed Ptarmigan," describes this nest as follows: "The nest in its present state measures scarcely 5 inches in diameter by about an inch in depth. It thus seems rather small for the size of the bird, but is probably somewhat compressed in transportation. The shape is saucer-like, but with very litle concavity of surface. The bottom is decidedly and regularly convex in all directions, apparently fitting a considerable depression in the ground. The outline is to all intents circular. The nest is rather closely matted, the material interlacing it in all directions, and retains considerable consistency.

The material is chiefly fine dried-grass stems; with these are mixed, however, a few small leaves and weed tops and quite a number of feathers. The latter, evidently those of the parent birds, are imbedded throughout the substance of the nest, though more numerous upon its surface, where a dozen or so are deposited; there may have been some loose ones lost in handling."[1]

A set of these rare eggs has recently been obtained by Mr. Thomas H. Jackson, of West Chester, Pennsylvania, who kindly allowed me to examine it, and placed all the information regarding it at my disposal. This set, containing but four eggs, in which incubation had already begun, was taken by Mr. Evan Lewis, in the vicinity of the Chicago Lakes, in Clear Creek County, Colorado, on June 19, 1890, at an altitude of about 12,200 feet. The nest itself was but a slight hollow in the ground, lined with a few small twigs, blades of grass, and a few feathers. It was about such a nest as a Bantam hen would make.

Mr. Lewis says: "The bird did not leave the nest until I stepped within a foot of it; then she strutted around, dragging her wings, very much like a Turkey does. When I returned to get the eggs, she allowed me to stroke her with my hand, and was about as tame as an average hen is when sitting. Foxes are very numerous around here, so that I did not dare to leave the eggs to see if others would be laid. I saw several of these birds, both males and females, the latter always between 3 and 5 o'clock p. m. The location of this nest, just above timber line, on the level top of a ridge, near isolated patches of dwarf willows, made me think they always nested in such places; but one sitting bird I saw feeding started up the mountain, running a short distance, then flew about a thousand feet, and after resting a few seconds repeated its flight and disappeared over the top of the mountain.

"I met a covey of young Ptarmigan about July 17, 1886, near the top of the mountain, at an altitude of about 13,000 feet. They were not very shy, and my companion and I counted them. I am not quite positive as to the number, but am under the impression there were nine or fifteen. I judged them to be nearly two weeks old. I ran after one, which tried to creep under a large rock, and I readily caught it. The old bird flew around my head and came close enough to knock my hat off, and as soon as we were about 100 feet away she began to call her flock together. I never saw more than two adult birds together, and should two males meet they immediately commence fighting, till one finally drives the other away."

The shape of the White-tailed Ptarmigan's eggs is an elliptical ovate. Their ground color varies from a creamy buff to a pale reddish or salmon buff. The markings are few, generally small in size and well defined. Some eggs, however, are much more heavily spotted than others, and in these the markings are more irregular and in the shape of blotches. These markings vary from reddish brown to chocolate brown.

---

[1] Bulletin U. S. Geological Surveys of the Territories, 2d series, v, 1875, p. 3.

Compared with other eggs of the Grouse family, they resemble far more the eggs of *Dendragapus* than *Lagopus*. None of the markings run into each other, as is often the case in eggs of the different species of *Lagopus*, and they are not nearly so heavily spotted. As in all Grouse eggs, these markings are entirely superficial. The three perfect specimens in the U. S. National Museum collection measure 43 by 30, 42.5 by 29, and 44 by 30 millimetres. Mr. Jackson's specimens are a trifle larger, and measure 46 by 30.5, 44.5 by 30.5, 44 by 31, and 44 by 30.5 millimetres.

The type specimen, No. 17200 (Pl. 2, Fig. 16), from an incomplete set, was collected by Mr. A. D. Wilson, of the Hayden Geological Survey, on July 15, 1875, in the San Juan Mountains, southern Colorado; the second (Pl. 2, Fig. 17), from a set of four, now in Mr. Thomas H. Jackson's collection, and kindly loaned for figuring, was taken June 19, 1890, in Clear Creek County, Colorado, by Mr. Evan Lewis.

32. Tympanuchus americanus (REICHENBACH).

PRAIRIE HEN.

Cupidonia americanus REICHENBACH. Systema Avium, 1852, p. xxix; based on Vollst. Naturg. Hühnen., Pl. 217, Figs. 1896-1898.
Tympanuchus cupido americanus RIDGWAY, Manuscript.
(B 464, C 384, R 477, C 563, U 305.)

GEOGRAPHICAL RANGE: Prairies of Mississippi Valley; south to Louisiana and Texas; west to northern Indian Territory, middle Kansas, Nebraska, and eastern North and South Dakota; east to Kentucky, Indiana, northwestern Ohio, southeastern Michigan, and southwestern Ontario, Canada; north to southern Manitoba.

The breeding range of the Pinnated Grouse or Prairie Hen extends over the prairie country of the Mississippi Valley, from southeastern Texas and Louisiana; north to Manitoba to about latitude 50°, vicinity of Winnipeg; east to western Ontario, Canada, southeastern Michigan (Monroe County), and northwestern Ohio, where they are rare now—a few still exist in Kentucky; west, they range to eastern North and South Dakota, throughout Nebraska, eastern and central Kansas, and the northern portion of the Indian Territory; south, at least to Fort Reno. They are not uncommon in suitable localities in northwestern Indiana, central Illinois, Missouri, and Wisconsin, and very abundant in Iowa, Minnesota, eastern Kansas, Nebraska, and the eastern portion of the two Dakotas. The range of this species is rapidly contracting along its eastern border, and equally rapidly extending both north and westward, where it is following the settlements. It is partly migratory in the northern portions, and resident from the central portions of its range south.

Mr. W. W Cooke writes on this subject as follows: "The Prairie Chicken is commonly said to be a resident bird, and so it is in the larger part of its range, but in Iowa a regular though local migration takes place. This has been men-

tioned by former writers, and in the spring of 1884 a special study was made
of the matter. Many observers unite in testifying to the facts in the case, and
what is still more important, there is not a dissenting voice. One of the
observers does not exaggerate when he says: 'Prairie Chickens migrate as
regularly as a Canada Goose.' Summing up all the information received, the
facts of the case are as follows: In November and December large flocks of
Prairie Chickens come from northern Iowa and southern Minnesota to settle for
the winter in northern Missouri and southern Iowa. This migration varies in
bulk with the severity of the winter. During an early cold snap immense
flocks come from the northern prairies to southern Iowa, while in mild, open
winters the migration is much less pronounced. During a cold wet spring the
northward movement in March and April is largely arrested on the arrival of
the flocks in northern Iowa, but an early spring, with fair weather, finds them
abundant in the southern tier of counties in Minnesota, and many flocks pass
still further north. The most remarkable feature of this movement is found in
the *sex* of the migrants. It is the females that migrate, leaving the males to
brave the winter's cold. Mr. Miller, of Heron Lake, Minnesota, fairly states the
case when he says: 'The females in this latitude migrate south in the fall and
come back in the spring, about one or two days after the first Ducks; and they
keep coming in flocks of from ten to thirty for about three days, all flying
north. The Grouse that stay all winter are males."[1]

The mating season begins early, about the beginning of March, and the
packs sometimes commence to break up while the ground is still covered with
snow.

Judge John Dean Caton describes the love-making of the male Pinnated
Grouse as follows: "The spring of the year is the season of courtship with
them, and it does not last all the year round as it does with humans, and
they do it in rather a loud way, too; and instead of taking the evening, as
many people are inclined to do, they choose the early morning. Early in
the morning you may see them assemble in parties, from a dozen to fifty
together, on some high dry knolls, where the grass is short, and their goings
on would make you laugh. The cock birds have a loose patch of naked
yellow skin on each side of the neck just below the head, and above these
on either side, just where the head joins the neck, are a few long black
feathers, which ordinarily lay backward on the neck, but which, when excited,
they can pitch straight forward. Those yellow naked patches on either side
of the neck cover sacs which they can blow up like a bladder whenever
they choose. These are their ornaments, which they display to the best
advantage before the gentler sex at these love feasts. This they do by blowing up these air sacs till they look like two ripe oranges, on each side of
the neck, projecting their long black ears right forward, ruffling up all the
feathers of the body till they stand out straight, and dropping their wings
to the ground like a Turkey cock. Now they look just lovely, as the coy

[1] Bulletin II, Dept. of Agriculture, Report on Bird Migration in the Mississippi Valley, 1888, p. 105.

timid maidens seem to say, as they cast side glances at them, full of admiration and of love.

"Then it is that the proud cock, in order to complete his triumph, will rush forward at his best speed for two or three rods through the midst of the love-sick damsels, pouring out as he goes a booming noise, almost a hoarse roar, only more subdued, which may be heard for at least 2 miles in the still morning air. This heavy booming sound is by no means harsh or unpleasant; on the contrary, it is soft and even harmonious. When standing in the open prairie at early dawn listening to hundreds of different voices, pitched on different keys, coming from every direction and from various distances, the listener is rather soothed than excited. If this sound is heavier than the deep key notes of a large organ, it is much softer, though vastly more powerful, and may be heard at a much greater distance. One who has heard such a concert can never after mistake or forget it.

"Every few minutes this display is repeated. I have seen not only one, but more than twenty cocks going through this funny operation at once, but then they seem careful not to run against each other, for they have not yet got to the fighting point. After a little while the lady birds begin to show an interest in the proceedings by moving about quickly a few yards at a time, and then standing still a short time. When these actions are continued by a large number of birds at a time, it presents a funny sight, and you can easily think they are moving to the measure of music.

"The party breaks up when the sun is half an hour high, to be repeated the next morning and every morning for a week or two before all make satisfactory matches. It is toward the latter part of the love season that the fighting takes place among the cocks, probably by two who have fallen in love with the same sweetheart, whose modesty prevents her from selecting between them."[1]

Nesting follows quickly after the birds are once paired, but as a rule they seem to show very poor judgment in the selection of the sites. Immense numbers of nests are annually destroyed, either by fire in dry seasons or water during wet ones, not taking the many other enemies into consideration at all, and it is safe to compute the loss of eggs alone, from the first two mentioned causes, at 50 per cent. Many nests with eggs are also yearly plowed up.

On the prairies they generally select unburnt places to nest in, where the old grass is thick; others prefer the borders of large marshes, where, during a wet season, they are almost certain to be destroyed by water. The nest is simply a slight excavation, alongside of some slough, in a fence corner among tall grass or a clump of weeds, or in cultivated fields or meadows, and again on open prairies, where the grass is very short. If there is plenty of material at hand the nest is often quite thickly lined, but on burned prairie very little lining is used, as no effort is made by the hen to bring material from a distance. Apparently no particular attempt is made to conceal the nest; the bird sits so

---

[1] Forest and Stream, March 29, 1883, p. 165.

close and harmonizes so well with the surroundings that she often escapes observation when in plain view.

Asst. Surg. J. C. Merrill, U. S. Army, found a nest of the Prairie Hen about half a mile from Fort Reno, Indian Territory, on May 20, 1890, which he describes as follows: "The nest was placed in a tussock of tall prairie grass growing on sloping ground on the open prairie. It was composed of dried grass blades, well matted together, and a few feathers from the parent. The nest was open on the northeast side of the tussock, and directly opposite was a narrow opening, or rather a tunnel, through the grass. On the two occasions upon which the bird was flushed from the nest she sat with her head toward this tunnel, through which she left the nest and skulked off. The eggs, fourteen in number, were found in one layer, and arranged without system, the smaller ends pointing in all directions. Incubation was far advanced."

This set and the female thereof are now in the U. S. National Museum collection.

Laying begins in the southern portions of its range sometimes as early as the latter part of March, and further north fully a month to six weeks later. Next to the Bob White I consider the Pinnated Grouse one of the most prolific of our game birds, laying from eleven to fourteen eggs on an average to a set; sets of twenty and more eggs have been repeatedly found and are not especially rare. Mr. Horace A. Kline, of Vesta, Johnson County, Nebraska, reports in the Ornithologist and Oölogist (August, 1882, p. 150), "that during this year he had seen two nests containing twenty-one eggs each," and gives the average of a large number examined as fourteen. He also states that "one of the most destructive agents to the nests of these valuable birds is the prairie fire. Many of the stockmen do not burn their hay ground until the middle of May, and so thousands of eggs are destroyed every year. In passing over one of these burned fields I counted five nests containing seventy-eight eggs on about one acre of ground."

Now and then a nest of this species is found above ground. Probably the bird had lost her first brood and learned wisdom from her former experience. Mr. P. H. Smith, jr., writes me that he found a nest of this species near Greenville, Bond County, Illinois, containing five eggs, on the top of an old hay-stack, 6 feet from the ground.

Mr. J. W. Preston, of Baxter, Iowa, informs me that a number of years ago he frightened a Prairie Hen from her nest of eggs in a marsh that was subject to overflow; the nest was entirely submerged and the bird was incubating the cold eggs. Not eight feet distant, on a tussock, a Marsh Harrier was caring for her clutch of eggs. Strange neighbors!

As a rule but one brood is raised in a season, but occasionally nests with fresh eggs are found in July and even in August, which seems to indicate that now and then a second brood may be reared; if this is actually the case, it is exceptional, unless the first eggs have been destroyed.

Incubation lasts from three to four weeks. The male does not assist in this duty, but keeps to himself. The young leave the nest as soon as hatched, and are cared for by the female alone. Their food at first consists almost, if not entirely, of insects, and when grasshoppers are plenty, as they frequently are in the northern parts of their breeding range, they subsist almost exclusively on them. Later they frequent the grain fields and feed on the different cereals as well as other small seeds and berries. The female is much devoted to her young, and will act similarly to the Ruffed Grouse in trying to attract the attention of the intruder to herself and away from the chicks, which hide quickly in the grass at the first intimation of danger which the parent may give.

By the latter part of August most of the broods are well grown and able to care for themselves. Two or three coveys pack together then, and later in the fall packs numbering fully five hundred and more may be seen where these birds are common.

The eggs of the Prairie Hen, as previously stated, number on an average from eleven to fourteen to a set, and if the first clutch is destroyed, which is unfortunately too often the case, a second and smaller set is laid. The color of the eggs varies from pale cream to vinaceous and olive buff, as well as light brown and clay color. Scarcely two sets are alike in this respect. The majority of the eggs of this species in the U. S. National Museum collection, are faintly but regularly spotted with fine pin points of reddish brown, in some instances scarcely perceptible to the naked eye. In a few sets in the series the markings show plainly, but none are larger than a No. 6 shot, and but few that size. The spots are pretty regular in size, and well defined, even when very small.

The eggs are ovate; a few short-ovate in shape. The average size is 43 by 32.5 millimetres. The largest egg in one hundred and two specimens measures 46 by 34, the smallest 40 by 30 millimetres. A runt egg of this species in the collection measures but 18 by 15 millimetres.

The type-specimen (No. 3103), from a set of eleven eggs (Pl. 2, Fig. 18), was collected by J. W. Tolman, near Winnebago, Illinois, May, 1860. No. 14517, from a set of sixteen eggs (Pl. 2, Fig. 19), was collected by G. and C. Blackburn, in Buchanan County, Iowa, May 22, 1868. This is the heaviest marked set in the series. No. 21102, from a set of eleven (Pl. 2, Fig. 20), collected for myself by Capt. B. F. Goss, Pewaukee, Wisconsin, May 20, 1876, is one of the darkest sets in the series.

33. Tympanuchus cupido (LINNÆUS).

HEATH HEN.

Tetrao cupido LINNÆUS, Systema Naturæ, ed. 10, I, 1758, 160.
Tympanuchus cupido RIDGWAY, Proceedings U. S. National Museum, VIII, 1885, **355**.
(B 464, part; C 384, part; R 477, part; C 563, part; U 306.)

GEOGRAPHICAL RANGE: Island of Martha's Vineyard, Massachusetts.

The breeding range of the Heath Hen is, at present, limited to the island of Martha's Vineyard, Massachusetts, where these birds are strictly protected.

Mr. William Brewster says: "They were formerly found at various points in eastern Massachusetts, southern Connecticut, Long Island, New Jersey, and Pennsylvania; perhaps also southern New England, and the Middle States generally. A *woodland* species, inhabiting scrubby tracts of oak and pine. * * * The general differences between this bird and its Western representative, *T. americanus*, are difficult of adequate definition, for the reason that they consist largely in shades of color rather than in markings. Its small size, short tarsus, acutely lance-pointed feathers of the neck-tufts, white-tipped scapulars, general reddish coloration above and restricted light markings beneath, are, however, readily appreciable and apparently constant characters. * * *

"The Heath Hen (I use the vernacular name by which it was known to our forefathers) is still common on Martha's Vineyard, where it is mainly, if not exclusively, confined to the woods, haunting oak scrub by preference and feeding largely on acorns. Being strictly protected by law, but few are probably killed. I am told by one of the Boston marketmen, however, that he has had as many as twenty from the Vineyard in a single season. He also says that they average nearly a pound less in weight than Western specimens, and on this account do not sell as readily.

"The bird is not found on the neighboring island of Naushon, despite statements by recent writers to that effect, nor is there any good evidence that it ever occurred there. There is also no reason to believe that the stock on Martha's Vineyard has been vitiated by the introduction of Western birds. It is simply the last remnant of a once more or less widely distributed race, preserved in this limited area partly by accident, partly by care. According to the best testimony available the colony is in no present danger of extinction."[1]

From a more recent article on this species published by Mr. Brewster, based on information gathered by him during a visit to Martha's Vineyard in July, 1890, I extract the following: "Throughout Martha's Vineyard the Heath Hen (locally pronounced hêth'n, as this Grouse is universally called) is well known to almost every one. Even in such seaport towns as Cottage City and Edgartown most of the people have at least heard of it, and in

the thinly settled interior it is frequently seen in the roads or along the
edges of the cover by the farmers, or started in the depths of the woods by
the hounds of the rabbit and fox hunters.

"Its range extends, practically, over the entire wooded portion of the
island, but the bird is not found regularly or at all numerously outside an
area of about 40 square miles. This area comprises most of the elevated
central portions of the island, although it also touches the sea at not a few
points on the north and south shores. In places it rolls into great rounded
hills and long irregular ridges, over which are scattered stretches of second-
growth woods, often miles in extent, and composed chiefly of scarlet, black,
white, and post oaks from 15 to 40 feet in height. Here and there, where
the valleys spread out broad and level, are fields which were cleared by the
early settlers more than a hundred years ago, and which still retain sufficient
fertility to yield very good crops of English hay, corn, potatoes, and other
vegetables. Again, this undulating surface gives way to wide, level, sandy
plains, covered with a growth of bear, chinquapin, and post-oak scrub, from
knee to waist high, so stiff and matted as to be almost impenetrable; or to
rocky pastures, dotted with thickets of sweet fern, bayberry, huckleberry,
dwarf sumac, and other low-growing shrubs.

"Clear, rapid trout brooks wind their way to the sea through open
meadows, or long narrow swamps wooded with red maples, black alders,
high huckleberry bushes, andromeda, and poison dogwood, and overrun with
tangled skeins of green briars.

"At all seasons the Heath Hens live almost exclusively in the oak woods,
where the acorns furnish them abundant food, although, like our Ruffed
Grouse, they occasionally at early morning and just after sunset venture out
a little way in the open to pick up scattered grains of corn or to pluck a few
clover leaves, of which they are extremely fond. They also wander to some
extent over the scrub-oak plains, especially when blueberries are ripe and
abundant. In winter, during long-continued snows, they sometimes approach
buildings, to feed upon the grain which the farmers throw out to them. A
man living near West Tisbury told me that last winter a flock visited his
barn at about the same hour each day. One cold snowy morning he counted
sixteen perched in a row on the top rail of a fence near the barnyard. It is
unusual to see so many together now, the number in a covey rarely exceeding
six or eight, but in former times packs containing from one to two hundred
birds each were occasionally met with late in the autumn.

"Only one person of the many whom I questioned on the subject had
ever seen a Heath Hen's nest. It was in oak woods, among sprouts at the base
of a large stump, and contained either twelve or thirteen eggs. The date, he
thought, was about June 10. This seemed late, but I have a set of six eggs
taken on the Vineyard July 24, 1885, and on July 19, 1890, I met a blueberry
picker who only the day before had started a brood of six young, less than
half grown. These facts prove that this bird is habitually a late breeder.

"The farmers about Tisbury say that in spring the male Heath Hen makes a booming or tooting noise. This, according to their descriptions must resemble the love notes of the Western Pinnated Grouse. About sunrise, on warm still mornings in May, several birds may be sometimes heard at once, apparently answering one another.

"During my stay at Martha's Vineyard, I obtained as many estimates as possible of the number of Heath Hens which are believed to exist there at the present time. My most trustworthy informants were, creditably, averse to what was apparently mere idle guessing; but when I questioned them, first as to the extent of the region over which the birds ranged, and next as to how many on the average could be found in a square mile within this region, they answered readily enough, and even with some positiveness. As already stated the total present range of the Heath Hen covers about 40 square miles. The estimates of the average number of birds per mile varied from three to five, giving from one hundred and twenty to two hundred birds for the total number. These estimates, it should be stated, relate to the number of birds believed to have been left over from last winter. If these breed freely and at all successfully, there should be a total of fully five hundred, young and old together, at the beginning of the present autumn. When one considers the limited area to which these birds are confined, it is evident that within this area they must be reasonably abundant. I was assured that with the aid of a good dog it was not at all difficult to start twenty-five or thirty in a day, and on one occasion eight were killed by two guns. This, however, can be done only by those familiar with the country and the habits of the birds."[1]

The only eggs of the Heath Hen in any collection, as far as known to me, are the set of six referred to above in Mr. Brewster's article, and now in his cabinet. These were taken July 24, 1885, at Martha's Vineyard, Massachusetts, and brought unblown to C. J. Maynard, from whom they were procured. They contained large embryos, and were saved with considerable difficulty.

The specimen figured (No. 23945, U. S. National Museum collection, Pl. 3, Fig. 2) is one of these eggs. It is creamy buff in color, with a slight greenish tint, ovate in form, and unspotted. It measures 44 by 33 millimetres.

---

[1] Forest and Stream, September 25, 1890.

34. Tympanuchus pallidicinctus (Ridgway).

LESSER PRAIRIE HEN.

Cupidonia cupido var. *pallidicincta* RIDGWAY, Bulletin Essex Institute, v, December, 1873, 199.
Tympanuchus pallidicinctus RIDGWAY, Proceedings U. S. National Museum, VIII, 1885, 355.

(B —, C 384*a*, R 477*a*, C 564, U 307.)

GEOGRAPHICAL RANGE: Southwestern parts of Kansas and western Indian Territory, western (and southern?) Texas.

The breeding range of the Lesser Prairie Hen, a smaller, paler-colored species than *T. americanus*, is not as well known as could be desired, and as far as our present knowledge goes includes southwestern Kansas, the western parts of Indian Territory as well as portions of northwestern and perhaps southern Texas. The latter locality is based on the statement of Asst. Surg. James C. Merrill, U. S. Army, who says, in his Notes on the Ornithology of Southern Texas (pp. 159, 160): "I am informed by a person perfectly familiar with the bird that the Prairie Chicken is occasionally seen on the prairies of Miradores ranch, which is about 30 miles north of the fort (referring to Fort Brown, Texas), and a few miles from the coast. This is probably about the southernmost point in the range of this bird." This statement is further confirmed by Lieut. Col. Lawrence S. Babbitt, U. S. Ordnance Corps, who writes me under date of March 18, 1890, as follows: "The Prairie Hen is not found in the immediate vicinity of San Antonio, Texas, but exists in great numbers south and southeast and in limited numbers north and west from here, all at about an average distance of 100 miles from the above mentioned locality."

It is possible, however, that this species may only be a winter resident in southern Texas. Mr. William Lloyd, in his Notes on the Birds of Western Texas, states: "*Tympanuchus pallidicinctus*. Lesser Prairie Hen. Winter visitor; seen in October and November in Concho County and also in winter on Middle Concho in Tom Green County. Abundant near Colorado City on the Texas and Pacific Railroad. I believe this record extends the range to the southwest. Westward it was abundant to the foothills of the Davis Mountains. Said to have been driven from the Pan Handle counties by the numerous prairie fires."[1]

The nesting habits of this species are undoubtedly similar in every respect to those of *Tympanuchus americanus*. Mr. C. S. McCarthy found it breeding abundantly 40 miles west of Fort Cobb, in the Indian Territory, he taking not less than three nests with eggs on June 1, 1860. Presumably but one brood is raised in a season. The number of eggs to a set is probably about the same as that of the previously mentioned species. The three sets in the

[1] Auk, Vol. IV, 1887, p. 187.

U. S. National Museum collection, contain seven eggs each, and are probably incomplete. They are somewhat lighter colored than the eggs of the common Prairie Hen and almost unmarked, but this is perhaps not constant. Their shape is ovate.

The ground color varies from pale creamy white to buff. The markings, which are all very fine, not larger than pin-points, are lavender colored. More than two-thirds of the eggs are unspotted, and all look so till closely examined. In size they average a trifle smaller than the eggs of the Prairie Hen. The mean measurement of twenty-two specimens in the U. S. National Museum collection is 42 by 32.5 millimetres. The largest egg measures 43.5 by 33.5, the smallest 41 by 31 millimetres.

The type specimen, No. 4011 (Pl. 3, Fig. 1), was obtained June 1, 1860, 40 miles west of Fort Cobb, Indian Territory, by C. S. McCarthy. This set contained seven eggs.

35. Pediocætes phasianellus (LINNÆUS).

SHARP-TAILED GROUSE.

Tetrao phasianellus LINNÆUS, Systema Naturæ, ed. 10, I, 1758, 160.
Pediocætes phasianellus ELLIOT. Proceedings Academy Natural Sciences, Phila., 1862, 403 (*nec* BAIRD, 1858, qui subsp. *columbianus*).
(B —, C 383, R 478, C 561, U 308.)

GEOGRAPHICAL RANGE: Interior of British America, east of Rocky Mountains, about James Bay (Moose Factory), and the western shore of Hudson Bay, northern Manitoba ; north at least to Fort Simpson, Mackenzie River, Northwest Territory.

The breeding range of the Sharp-tailed Grouse extends from about latitude 52° north, and westward through British America to the eastern slopes of the Rocky Mountains, as far north as latitude 69°, and probably still farther within the Arctic circle. Mr. C. P. Gaudet, of the Hudson Bay Company, found it breeding in latitude 68°, near Fort Good Hope, in the Mackenzie River Basin. Its northeastern range is not well defined, but it probably reaches the northern shores of Hudson Bay, and breeds possibly as far south as Moose Factory, James Bay, about latitude 51°, 40'. It seems to be especially abundant in the vicinity of Great Slave Lake, between latitude 61° and 63°; most of the eggs in the U. S. National Museum collection coming either from Forts Rae, Providence, or Resolution, all three posts situated on different parts of this lake.

Comparatively little is as yet known about the breeding habits of this subspecies. There is no reason to suppose, however, that they differ materially from those of its more southern relatives, which will be fully described. The Sharp-tailed Grouse is said to inhabit the wooded districts of the fur countries, as well as the borders of the extensive prairies or tundras near the numerous lakes found throughout that region, and it is probably more or less migratory in the winter.

But a single brood is raised in a season. Nidification begins, sometimes at least, extremely early with this species, eggs having been found May 1, 1863, by Mr. L. Clarke, of the Hudson Bay Company, at Fort Rae, in latitude 63°. These must have been laid long before the ice and snow disappeared from the surrounding country. Mr. R. MacFarlane also took a nest containing nine eggs, on May 15, 1884, near Fort Providence. According to this gentleman, the Sharp-tailed Grouse breeds also in the pine forests on both sides of the Lockhart and Upper Anderson Rivers, where one or two nests were taken, but the eggs were afterward lost.

The number of eggs to a set varies from seven to fourteen, and their ground color from a fawn color with a vinaceous rufous bloom, to chocolate, tawny, and olive brown in different specimens. The majority of the eggs are finely marked with small, well-defined spots of reddish brown and lavender, resembling the markings found on the eggs of *Tympanuchus americanus*, only they are much more distinct. Compared with the eggs of the two southern subspecies *P. phasianellus columbianus* and *P. phasianellus campestris*, they usually are very much darker colored, even the palest specimens being darker than the heaviest marked eggs of either of the two subspecies. These markings are entirely superficial, and when removed leave the shell a creamy white in some cases and a very pale green in others. In shape they are usually ovate. The average measurement of thirty-four eggs in the U. S. National Museum collection is 44.5 by 32 millimetres. The largest egg of this series measures 48 by 33, the smallest 42 by 30 millimetres.

Of the type specimens selected to show the variations, No. 7619 (Pl. 3, Fig. 3), from an incomplete set of seven, was obtained May 10, 1863, near Fort Rae, Great Slave Lake, by Mr. L. Clarke, jr., of the Hudson Bay Company; No. 7620 (Pl. 3, Fig. 4), from an incomplete set of six, taken June 1, 1863, by the same gentleman, in the same locality; and No. 22503 (Pl. 3, Fig. 5), a single egg, taken May 16, 1885, near Fort Providence, Great Slave Lake, was obtained from Mr. R. MacFarlane, also of the Hudson Bay Company.

36. Pediocætes phasianellus columbianus (ORD).

COLUMBIAN SHARP-TAILED GROUSE.

Phasianus columbianus ORD, Guthrie's Geography, 2d Am. ed., II, 1815, 317.
Pediœcetes phasianellus var. *columbianus* COUES, Key to North American Birds, 1872. 234.

(B 463, C 383a, R 478a, C 562, U 308a.)

GEOGRAPHICAL RANGE: Northwestern United States; south to northeastern California, northern Nevada, and Utah; east to Montana and Wyoming; west to Oregon and Washington; north, chiefly west of Rocky Mountains, through British Columbia, to central Alaska (Fort Yukon).

The Columbian Sharp-tailed Grouse inhabits the grass-covered plains of the Northwest. Its breeding range extends from eastern Montana and Wyoming, westward through northern and central Utah, the whole of Idaho, eastern and central Oregon and Washington, south to northern Nevada and northeastern California, along the eastern slopes of the Siskiyou Mountains. The latter, as well as the eastern spurs of the Cascades, forms a barrier to its westward extension in Oregon and Washington, and it is here, at Fort Klamath, Oregon, that this bird reaches the most westerly point of its range. North it is found throughout eastern British Columbia, on both sides of the Rocky Mountains, and it has been taken as far north as Fort Yukon, Alaska.

The habits of the Columbian Sharp-tailed Grouse, also known as the Spike-tail and the Prairie Chicken, are very similar to those of its eastern relative, *P. phasianellus campestris*. It is one of the most abundant and best known game birds of the Northwest, inhabiting the prairie country to be found along the foothills of the numerous mountain chains intersecting its range; seldom venturing into the wooded portions for any distance, and then only during the winter months, when it is partially migratory in certain sections.

According to my own experience the Columbian Sharp-tail breeds more frequently on the sheltered and sunny slopes of the grass-covered foothills of the mountains than in the lower valleys and creek bottoms. At Fort Lapwai, Idaho, this Grouse was exceedingly common about twenty years ago, but it is much less so now. It then gathered into large packs during the late fall and winter, frequently numbering two hundred and more. These kept together until about the beginning of March, when they commenced to break up. The "dancing" indulged in during the mating season, and which will be fully described in the succeeding article, began at Fort Lapwai (the only place where I had the opportunity of witnessing it) usually between the 1st and 10th of March, and by the end of that month most of the birds were paired and had selected their nesting sites. Nidification began usually from about April 15, to May 1, according to the season. I found a set of fifteen eggs, which had been sat upon about a week or ten days, on April 22, 1871. Some birds must have laid earlier still, as it was no uncommon sight to find fully grown birds by July 10. All the nests of this species which I examined were invariably well concealed and rather difficult to find. You might search daily for a couple of weeks and be unsuccessful in finding a nest, and again you might stumble on two or three on the same day. A bunch-grass covered hillside, with a southerly exposure, seemed to be a favorite nesting site with this Grouse at Fort Lapwai, while at Camp Harney, Oregon, they confined themselves during the breeding season to the sage brush covered plains of the Harney Valley, interspersed here and there with a low grassy swale, nesting along the borders of these, where the grass attained a heavier growth.

The nest, like that of all the Grouse, is always placed on the ground, usually close alongside some tall bunch of coarse grass, which hides it completely from view. Even if it did not, the female harmonizes in color so thoroughly with her surroundings that she is not apt to be noticed, unless she should leave her nest, which she does not do very readily, as she is a very close sitter. A slight hollow, usually scratched out on the upper side of a bunch of grass, if the nest is placed on a hillside, is fairly lined with dry grass, of which there is ordinarily an abundance to be found in the vicinity, and this constitutes the nest. A few feathers from the lower parts of the bird are usually mixed in among the eggs, each one of which is often imbedded about two-thirds in its own mould and does not touch the others. Once only did I find the eggs placed on top of each other, eight in the lower and five in the upper layer.

Incubation lasts about twenty-one days, the female attending to this duty exclusively, the males keeping by themselves, usually in small parties of from three to five, frequenting the higher hills and edges of the table lands in the vicinity of the nests. I do not believe that this Grouse is polygamous.

At Camp Harney I have found eggs of this subspecies as late as June 18, and as a rule they nested fully from four to six weeks later there than at Fort Lapwai. The female is exceedingly devoted to her young brood, and I have seen one boldly attack my dog, who accidentally happened to run into a young covey about a week old, while I was riding along one of the tributaries of Lapwai Creek, in the latter part of May, 1871.

But one brood is raised in a season. The young are active, handsome little creatures, and able to use their legs at once on leaving the shell. They are at first fed mostly on insects, young grasshoppers and crickets forming the principal portion of their bill of fare. The former are always abundant and easily obtained; later, when the young are able to fly, the mother leads them to the creek bottoms, where they find an abundance of berries and browse. They are especially fond of the seeds of the wild sunflower, which grows very abundantly in some places, and when these are ripe, many of these birds can be found in the vicinity where these plants grow.

The habits of the Columbian Sharp-tailed Grouse vary very materially in different portions of the country where I have met with them. At Fort Klamath, Oregon, where they are rather rare, I have found them inhabiting decidedly marshy and swampy country, and keeping close to, if not in the edges of, the pine timber throughout the year. At Fort Custer, Montana, this Grouse, during the winter, was much more arboreal than terrestrial in its habits, moving around on the limbs of the large cottonwood trees as unconcernedly as on the ground; spending in this way almost all their time, except when feeding. At Harney, Oregon, and Lapwai, Idaho, they might be frequently seen in small trees and bushes which grow along the creeks, but scarcely ever in large trees, of which there was an abundance. Here, they uttered very few notes at any time, while at Fort Custer I have frequently heard them cackling in the tall cottonwoods which grew along the Big Horn River bottom, before I had approached within several hundred yards of them, evidently giving notice to other birds in the vicinity of my coming.

This fine game bird is decreasing very rapidly throughout its range. It does not seem to prosper in the vicinity of man, and as the country is becoming more and more settled, it recedes before civilization. As it is not a particularly shy bird, it falls an easy victim to the gunner.

In Oregon, Washington, and Idaho, where it used to be exceedingly abundant a decade ago, it is every year becoming rarer, and at the present rate of decrease it will not be long before it will be numbered among the game birds of the past, at least in all fertile portions of the country, retaining only a precarious foothold in the more sterile sections of these States where the lands are too poor and rocky to be successfully cultivated.

From eleven to fourteen eggs are laid to each set, rarely more. These are usually short ovate in shape, and very small for the size of the bird. The ground color varies from creamy buff to pale olive brown. An occasional specimen has a pale vinaceous bloom overlying the ground color. The majority of the eggs are slightly spotted with reddish brown; the markings, for the most part, are very fine, the spots varying from mere pin points to the size of No. 6 shot. All these markings are superficial and easily rubbed off on a freshly laid specimen.

The average measurement of seventy-two specimens in the U. S. National Museum collection is 43 by 32 millimetres. The largest egg in this series measures 46.5 by 34.5, the smallest 39 by 31 millimetres.

Of the types, No. 9139 (Pl. 3, Fig. 6), from an incomplete set of four, was collected May 29, 1862, near Fort Yukon, Alaska, by Mr. J. Lockhart, of the Hudson Bay Company; No. 21103 (Pl. 3, Fig. 7), selected from a set of fifteen, was taken April 22, 1871, near Fort Lapwai, Idaho; and No. 21106 (Pl. 3, Fig. 8), from a set of eleven eggs, taken June 18, 1876, near Camp Harney, Oregon. The last two are from the Bendire collection.

37. Pediocætes phasianellus campestris RIDGWAY.

PRAIRIE SHARP-TAILED GROUSE.

Pediocætes phasianellus campestris RIDGWAY, Proceedings Biological Society, Washington, II, April 10, 1884, 93.
(B —, C —, R —, C —, U 308b.)

GEOGRAPHICAL RANGE: Plains and prairies of the United States; north to Manitoba; east to Wisconsin and northern Illinois; west to eastern Colorado; south to eastern New Mexico.

This recently described subspecies differs from *Pediocætes phasianellus columbianus* in its rather lighter and much more ochraceous coloration above, in having the black bars narrower and less regular, and in the V-shaped markings of the lower parts being much less distinct. It is thus described by Mr. R. Ridgway, and from types coming from Illinois and the Rosebud River, Montana.

The breeding range of the Prairie Sharp-tailed Grouse extends from northern Illinois, west through southern Wisconsin, northwestern Iowa, middle and western Kansas, through eastern Colorado to northeastern New

Mexico, north through western Nebraska, eastern Wyoming, eastern Montana, Minnesota, and the two Dakotas, and extending north of our border into southeastern Assiniboia and southern Manitoba. Specimens taken along the southern portions of these provinces are typical *P. phasianellus campestris;* and those from middle Manitoba northward to about latitude 52° are intermediate between this and *P. phasianellus.*

The eastern range of this Grouse is becoming rapidly restricted. In Illinois they are very rarely found now. Col. N. S. Goss reports them as becoming rare in Kansas; and the case is the same in Wisconsin and Iowa. Mr. Denis Gale writes: "The Prairie Sharp-tailed Grouse was quite plentiful fifteen years ago on the plains about Denver, Colorado. They are seldom met with now; the last I saw was in the winter of 1886. In 1885, I met one of these birds far up in the foothills at an elevation of over 8,000 feet. Unlike the Prairie Hen, *Tympanuchus americanus,* grain and corn fields have but few attractions for these birds; this, and the stamping out by cattle of the whole country's surface, supplemented by the pot hunter's shotgun to secure a toothsome morsel in and out of season, no doubt accounts for their present scarcity."

Mr. W. M. Wolfe, of Kearney, Nebraska, says: "The Prairie Sharp-tailed Grouse, *P. phasianellus campestris,* was formerly abundant in central Nebraska. Now it has retired before civilization, and the Pinnated Grouse has taken its place. Cold winters, notably that of 1885, drive it back into thinly settled regions. In northwestern Nebraska, where both species are still found, they not infrequently mingle in winter, but are bitter enemies in warm weather. Then they have no occasion to be together, for the Sharp-tailed Grouse always prefers its natural food, tender twigs and insects, while the Pinnated Grouse must have grain. A Sharp-tail never loves a wheat field so well as when there is an abundance of grasshoppers to be found there."

Mr. George Bird Grinnell has kindly furnished me some notes on the habits of the Sharp-tailed Grouse, which are mostly referable to this race. He writes me as follows: "The Sharp-tailed Grouse, which, in certain sections, is called "Speckled Belly" and "Willow" Grouse, I have found in various years almost everywhere west of the Mississippi River, east of the Sierra Nevadas, and north of the Platte River. In the old days it used to be very common all along the Platte and the Loup Rivers in Nebraska, and in the country which lies between these two streams. I have also found it nearly as abundant in the mountains, sometimes even late in the autumn, coming upon single birds or a considerable brood, far up toward the edge of timber in the most narrow wooded ravines. This species is partly migratory, and there is the very greatest difference in the habits of the bird in summer and winter. As soon as the first hard frosts come in the autumn the birds seem to take to the timber, and begin to feed on the buds of the willow and the quaking aspen. At this time they spend a large portion of their time in the trees, and are very wild. In the Shirley Basin, in western Wyoming, a locality where I have never seen any of these birds in summer, they are abundant in winter. At this season they live in quaking aspen thickets along the mountains, and there I have seen hundreds of them roosting on top of a big barn which stands just at edge of a grove of quaking aspen timber. It was always easy in the morning, just after sunrise, to step out of the house, and, with a .22-caliber rifle, shoot off the heads of as many of these birds as were needed for eating for the next two or three days.

"I have only one note on these birds which seems particularly worth mentioning, and of this I spoke in my report to Col. William Ludlow, on the birds noticed during a reconnoissance to the Black Hills of Dakota, in 1874, which was published by the Engineer Bureau of the War Department. The Sharp-tailed Grouse has a cry which is unlike that of any other Grouse with which I am familiar, although something very similar has been observed in the case, I think, of one of the Ptarmigans. On the plains of Dakota, in 1874, having scattered a brood of Sharp-tailed Grouse, consisting of a mother and a dozen well-grown young, I sat down to wait for them to get together. The mother had flown to the top of a hill not far off, where she sat on the ground in plain sight, and after a few moments began to call to the young, which immediately answered her from the different points where they had taken refuge. The call of the mother and the young was a guttural, raucous croak, which quite closely resembled the croaking of a raven at a little distance. I plainly saw the old bird utter its note, and subsequently followed up the calls uttered by more than one of the young ones, until I started them and killed one or two as they flew. I do not know that this cry of the Sharp-tailed Grouse has been noted by any other observer."

Mr. Ernest E. Thompson has also kindly placed some of his notes on this race, made in southern Manitoba, at my disposal, and I make the following extracts from them: "The Sharp-tailed Grouse, while eminently a prairie bird in the summer time, usually retires to the woods and sandhills on the approach of winter, but in the spring, before the snow is gone, they again perform a partial migration and scatter over the prairies, where alone they are to be found during the summer. They are very shy at all times, but during the winter the comparatively heedless individuals have been so thoroughly weeded out by their numerous enemies, that it requires no slight amount of stalking to get within the range of a flock in the springtime.

"The advent of the Grouse on the still snow-covered plains might prove premature, but that they find a good friend in the wild prairie rose (*Rosa blanda*), which is abundant everywhere; and the ruddy hips, unlike most fruits, do not fall when ripe, but continue to hang on the stiff stems until they are dislodged by the coming of the next season's crop. On the 'Big Plain' stones of any kind are unknown, and in nearly all parts of Manitoba gravel is unattainable during the winter, so that the Sharp-tails and other birds, that require these aids to digestion, would be at a loss, were it not that the friendly rose also supplies this need; for the hips, besides being sweet and nutritious, contain a number of small angular hard seeds which answer perfectly the purpose of the gravel.

"To illustrate the importance of this shrub in this regard, I append a table of observations on the contents of crops and gizzards of Grouse killed during the various months as indicated:

January—Rose-hips, browse, and *Esquisetum* tops.
February—Rose-hips and browse.
March—Rose-hips and browse.
April—Rose-hips and browse of birch and willow.
May—Rose-hips and sand-flowers (*Anemone patens*).
June—Rose-hips, grass, grasshoppers, and *Procaria costalis.*
July—Rose-hips, seeds of star-grass, and *P. costalis.*
August—Rose-hips, grass, strawberries, and *P. costalis.*
September—Rose-hips, grass, berries, and *P. costalis.*
October—Rose-hips, grass, and various berries.
November—Rose-hips, birch and willow browse, and berries of arbutus.
December—Rose-hips, juniper berries, and browse.

"This is of course a mere list of staples, as in reality nothing of the nature of grain, fruit, leaves, or insects comes amiss to this nearly omnivorous bird, but it illustrates the importance of the rose-hips, which are always obtainable, as they grow everywhere, and do not fall when ripe. In the course of my experience I have examined some hundreds of gizzards of the Prairie Chicken, and do not recollect ever finding one devoid of the stony seeds of the wild rose.

"After the disappearance of the snow, and the coming of warmer weather, the chickens meet every morning at gray dawn in companies of from six to twenty, on some selected hillock or knoll, and indulge in what is called 'the dance.' This performance I have often watched, and it presents the most amusing spectacle I have yet witnessed in bird life. At first the birds may be seen standing about in ordinary attitudes, when suddenly one of them lowers its head, spreads out its wings nearly horizontally and its tail perpendicularly, distends its air sacs and erects its feathers, then rushes across the 'floor,' taking the shortest of steps, but stamping its feet so hard and rapidly, that the sound is like that of a kettledrum; at the same time it utters a sort of bubbling crow, which seems to come from the air sacs, beats the air with its wings and vibrates its tail, so that it produces a loud, rustling noise, and thus contrives at once to make as extraordinary a spectacle of itself as possible. As soon as one commences, all join in, rattling, stamping, drumming, crowing, and dancing together furiously; louder and louder the noise, faster and faster the dance becomes, until at last as they madly whirl about, the birds leap over each other in their excitement. After a brief spell the energy of the dancers begins to abate, and shortly afterward they cease, and stand or move about very quietly, until they are again started by one of their number leading off.

"The whole performance reminds one so strongly of a 'Cree dance' as to suggest the possibility of its being the prototype of the Indian exercise. The space occupied by the dancers is from 50 to 100 feet across, and as it is returned to year after year, the grass is usually worn off, and the ground trampled down hard and smooth. The 'dancing' is indulged in at any time of the morning or evening in May, but it is usually at its height before sunrise. Its erotic character can hardly be questioned, but I can not fix its place or value in the nuptial ceremonies. The fact that I have several times noticed the birds join for a brief 'set-to' in the late fall, merely emphasizes its parallelism to the drumming and strutting of the Ruffed Grouse, as well as the singing of small birds.

"The whole affair bears a close resemblance to the manœuvring of the European Ruff, and from this and other reasons I am inclined to suspect the Sharp-tail of polygamy. When the birds are disturbed on the hill they immediately take wing and scatter, uttering as they rise their ordinary alarm note, a peculiar vibratory 'cack, cack, cack.' This is almost always uttered simultaneously with the beating of the wings, and so rarely, except under these circumstances, that at first I supposed it was caused by the wings alone, but since then I have heard the sound both when the birds were sailing and when they were on the ground, besides seeing them fly off silently. They have also a call, a soft, clear whistle of three slurred notes, E, A, D, and a sort of grunt of alarm, which is joined in by the pack as they fly off. Their mode of flight is to flap and sail by turns every 40 or 50 yards, and so rapid and strong are they on the wing that I have seen a chicken save itself by its swiftness from the first swoop of a Peregrine Falcon, while another was seen to escape by flight from a Snowy Owl.

"The nest of this subspecies is placed in the long rank grass, under some tuft that will aid in its concealment, and is usually not far from a tract of brush land or other cover. It is little more than a slight hollow in the ground, arched over by the grass. The eggs, usually fourteen, but sometimes fifteen or sixteen, in number, are very small for the size of the bird. Immediately before expulsion they are of a delicate bluish green; on being laid they show a purplish, grape-like bloom; after a few days they become deep chocolate brown, with a few dark spots. After a fortnight has transpired they are usually of a dirty white. This change is partly due to bleaching and also to the scratching they receive from the mother's bill and feet.

"Incubation lasts about twenty-one days. The young when first hatched are covered with golden yellow down, and are spotted with black above. This covering assists them materially in hiding when they squat in the grass. At the age of six weeks they are fully feathered, and at two months fully grown. Although still under guidance of the mother at this time, there are usually not more than six or seven young ones left out of the original average brood of fourteen, which shows the number of chicks which fall a prey to their natural enemies, while many sets of eggs also are destroyed by the fires which annually devastate the prairies. As the fall advances, they gather more and more into flocks and become regular visitors to the stubble fields, and in consequence regular articles of diet with the farmer."

The nesting habits are similar to those of the Columbian Sharp-tailed Grouse, and the average number of eggs about the same, from eleven to fourteen. Their favorite nesting sites, according to Mr. W. M. Wolfe, are usually along upland thickets or the edges of timber, near streams, and he also states that occasionally two broods are raised in a season. He found one nest on the bare ground under the shelter of a rock.

Dr. T. E. Wilcox, assistant surgeon, U. S. Army, writes me from Fort Niobrara, Nebraska: "In the spring of 1889 a nest of this species was found, and the bird was incubating. Neatly coiled among the eggs was a fox blacksnake, *Coluber vulpinus*, the death of which, perhaps, insured a successful result of the effort to rear offspring. Here they nest usually in the sandhills, remote from water. Wild grapes, rose-hips, plums, sand-cherries, besides grasses, the tops of plants, grasshoppers, and other insects afford them food."

The nesting season varies according to latitude. In the more southern portions of its range it begins in April, usually about the latter half of this month, and it is protracted to the middle of June occasionally, in the more northern localities.

Incubation lasts about three weeks, and usually but a single brood is raised. The eggs, in shape and color, are exact counterparts of those of the Columbian Sharp-tailed Grouse. They can not be distinguished from each other with certainty. The average measurement of thirty-eight specimens in the U. S. National Museum collection is 42.5 by 31.5 millimetres. The largest egg measures 46 by 32.5, the smallest 40 by 30.5 millimetres.

The type specimen, No. 16645 (Pl. 3, Fig. 9), selected from an incomplete set of three eggs, was taken by Dr. Elliott Coues, near Pembina, North Dakota, June 6, 1873; No. 22829 (Pl. 3, Fig. 10), from a partial set of nine eggs, was collected by Mr. Ernest E. Thompson, in May, 1886, near Carberry, Manitoba.

38.　Centrocercus urophasianus (BONAPARTE).

SAGE GROUSE.

Tetrao urophasianus BONAPARTE, Zoölogical Journal, III, 1827, 213.
Centrocercus urophasianus SWAINSON, Fauna Boreali Americana, II, 1831, 497, Pl. 58.
(B 462, C 382, R 479, C 560, U 309.)

GEOGRAPHICAL RANGE: Sagebrush-covered plains of the interior, principally within the United States from North Dakota, and southern Assiniboia to Washington, and casually(?) to southern British Columbia; south to northern New Mexico, Utah, and Nevada, west to Oregon and California; east to Colorado, Nebraska, South and North Dakota.

The home of the Sage Grouse is found on the dry sagebrush-covered plains and table lands of the western parts of the United States. Its breeding range extends northward to our boundary, latitude 49°, from western

North Dakota, through Montana, Idaho, and Washington, to the eastern slopes of the Cascade Range, thence south through that portion of Oregon east of the Cascades, northeastern California east of the Sierra Nevada, through Nevada, Utah, Colorado, and northwestern New Mexico; thence eastward again through western Nebraska, Wyoming, and western South Dakota. According to Prof. J. Macoun, it is not rare on some of the tributaries of the Upper Missouri in southern Assiniboia, about 30 miles north of the boundary.

The Sage Grouse, next to the Wild Turkey, is the largest of the game birds found within the limits of the United States, full-grown males attaining not infrequently a weight of 8 pounds. The female is much smaller, rarely weighing more than 5 pounds. This species usually is a resident throughout the year wherever found, but in portions of its range is partly migratory. This has been questioned by some writers, but I have positive proof that in the upper Sylvies Valley, in the Blue Mountains of Oregon, where they breed very abundantly at an altitude of from 6,000 to 6,500 feet, every bird leaves the region at the approach of winter, migrating to the lower Harney Valley, and remaining there until the return of spring. The reason for this migration is easily explained. During the winter these birds feed almost entirely on the leaves of the sage (*Artemisia*), which usually grows to the height of 2 or 3 feet in some of the richer valley lands. In the upper Sylvies Valley the snow generally covers these sage bushes entirely, even during a comparatively mild winter, hence their principal food supply is cut off, and the birds are necessarily compelled to migrate in order to find something to eat. During the greater portion of the year their food is much more varied than is generally supposed, but the leaves of the sage are always more or less used at all seasons. From personal observation I know that the seed tops of various grasses and leguminous plants, as well as berries of different kinds, grasshoppers, and crickets (*Anabus simplex*), are consumed to a considerable extent during the summer months.

Capt. William L. Carpenter, Ninth Infantry, U. S. Army, writes me on this subject as follows: "The Sage Grouse does not feed on the *Artemisia* as exclusively as reported, only resorting to this diet, which renders it so objectionable for the table, in the season when nothing else is obtainable. In summer, its principal food (in Wyoming and Colorado) is the leaves, blossoms, and pods of the different species of plants belonging to the genus *Astragalus*, and *Vicia*, commonly called wild pease, which are always eagerly sought for and consumed in great quantities. At this season dissection has shown me also grasshoppers, crickets, and a few of the smaller beetles."

Mr. Robert S. Williams, Great Falls, Montana, states: "I think these birds are seldom seen far from *Artemisia* tracts; but once I scared up a flock of six or eight in the middle of a mountain meadow, among tall grass. I obtained one of the birds, and its crop was filled with the blossoms of a species of golden-rod (*Solidago rigida*)."

Mr. George H. Wyman makes the following statement in Forest and Stream, August 29, 1889: "The Sage Cock will eat the leaves from sagebrush when it cannot get berries or grain, but it will go farther for a morning feed from a wheat field than any bird I know, except the Wild Goose. I have killed Sage Fowl with stomachs filled with ripe wheat picked up the same morning, in places where none was to be had nearer than 8 miles, and in fact with no cultivation of any kind nearer in any direction. They fly long distances in search of food, but return to roost in the same place at night, generally on some steep hillside, free from shrubs or high grass."

From the foregoing statements it will be seen that the food of the Sage Grouse, during the summer months at least, is quite varied. The fact that the stomachs of these birds are soft, and unlike, in this respect, that of all gallinaceous birds, is well known. No doubt on this account it feeds mostly on leaves and the tender tops of various plants, as well as insects; still there is no reason to doubt that where grain is obtainable, which is not often the case where these birds are found, it will also resort to it as an article of food.

It is a hardy bird, taking kindly to the higher altitudes in the mountain parks, as well as to the hottest and most barren portions of the alkali-covered valleys, as long as they support a scanty growth of *Artemisia*, which seems to be a positive necessity to the existence of this Grouse.

Mr. A. C Lowell, of Fort Bidwell, California, writes me that "Sage Fowl are much more numerous with us some years than in others, and I believe this bird is a victim of a grub which I have several times found in the walls of the abdominal cavity. These grubs are about 1¼ inches long, flat, with a head much broader than its body; they look like the large white worms which are often found cutting shallow paths under the bark of decaying logs."

The mating season begins early in March, and sometimes even in the latter part of February, in fact long before the snow has disappeared. While not at any time what might be called a graceful bird when on the ground, the Sage Cock, during this season, when actively engaged in his courtship, is unquestionably the most comical-looking bird I have ever seen, and it would be hard to say what he most resembles.

Directly west, and about half a mile from Camp Harney, Oregon, on a rocky table land, sparsely covered with stunted patches of sage, a number of these birds wintered regularly, and early one morning in the first week of March, 1877, I had the long-wished-for opportunity to observe the actions of a single cock while paying court to several females near him; and I presume he did his very best. His large, pale yellow air sacs were fully inflated, and not only expanded forward, but apparently upward as well, rising at least an inch above his head, which, consequently, was scarcely noticeable, giving the bird an exceedingly peculiar appearance. He looked decidedly top-heavy and ready to topple over at the slightest provocation. The few long, spiny feathers along the edges of the air sacs stood straight out, and the grayish white of the upper parts showed in strong contrast with the black of the breast.

His tail was spread out fan-like, at right angles from the body, and was moved from side to side with a slow quivering movement. The wings were trailing on the ground. While in this position he moved around with short, stately, and hesitating steps, slowly and gingerly, evidently highly satisfied with his performance, uttering, at the same time, low, grunting, guttural sounds, somewhat similar to the purring of a cat when pleased, only louder. This was kept up for some ten minutes. After having regained his usual attitude it was hard to believe that this was the same bird I had seen but a few minutes before.

During the winter the birds pack, and I have more than once seen from fifty to a hundred together. By the 10th of March they are already pretty well scattered, and many are paired by this time. In the vicinity of Camp Harney nidification usually begins about the middle of April. I have found a full set of nine slightly incubated eggs of this species on April 7, 1877, and the first egg must have been laid before March 25. Fresh specimens were also taken by me as late as June 2, probably a second laying, the first having been destroyed. But one brood is raised in a season.

The nest is always placed on the ground, in a slight depression, usually under the shelter of a small sage bush. I have found several, however, some little distance from sage brush flats, alongside and sheltered by a bunch of tall rye grass (*Elymus condensatus?*), near the borders of small creeks. The nest is usually very poorly lined, and in fact the eggs frequently lay on the bare ground without any lining whatever, and are often found in quite exposed situations. I found such a one on May 11, 1875. My notes read as follows: "I stumbled accidentally on this nest. It was placed within a yard of a much-used Indian trail, in a very exposed position, so much so that I saw the eggs while still 5 yards off. There really was no nest, simply a mere depression scratched out by the bird on the south side of a very small sage bush, which afforded no concealment or protection from rain whatever. The bush itself was not over a foot and a half high, growing on a rocky plateau about 3 miles east of Camp Harney. A few feathers were scattered among the eggs which laid on the bare ground, and were separated from each other by bits of grass and dry leaves of the sage. One of the eggs was nearly covered with dirt and almost buried out of sight. The set contained eight eggs, and these were nearly hatched. They were cold when found, and the nest had evidently been abandoned for some days."

As a rule the Sage Hen is a very close sitter, and is loath to leave her nest at any time. I have almost stepped on them before they would quit their eggs.

Capt. William L. Carpenter, U. S. Army, writes me: "I found a nest at Fort Bridger, Wyoming, where this species is numerous, June 1, with nine fresh eggs. I was standing alongside a sage bush watching butterflies; several times looking down carelessly without seeing anything unusual, when happening again to glance at the foot of the bush, in the very place before observed, I saw the winking of an eye. Looking more intently a grayish mass was discerned blending perfectly with the color of the bush, which outlined itself into the form of a Sage Hen not 2 feet from my foot. She certainly would have been overlooked had not the movement of her eyelids attracted my attention. I stood there fully five minutes admiring the beautiful bird, which could have been caught in my butterfly net, then walked back and forth, and finally passed around the bush to observe it from behind. Not until then did it become frightened and fly away with a loud cackling. The nest was a depression at the foot of a sage bush, lined with dead grass and sage leaves. The spot was marked and visited several times, always passing within a few feet without alarming the bird."

Incubation, I think, lasts about twenty-two days. The males take no part whatever in this duty, and keep to themselves till the young are grown. Many attain their full growth by the 1st of August. Young Sage Fowls are excellently flavored, superior in my opinion to either the Sooty Grouse or the Columbian Sharp-tailed Grouse, always provided that they are drawn at once after being shot. The female is devoted to her young, and will protect them at the risk of her life. They are very expert at hiding themselves, in the manner of young Ruffed Grouse.

Mr. William G. Smith writes me as follows: "While collecting in Carbon County, Wyoming, I caught six young Sage Chickens, probably about four days old, on June 10. The female flew at my legs, and followed me 200 yards to where my wagon was standing, constantly making hostile demonstrations, while the young kept calling."

Their ordinary alarm note, uttered usually only when about to take wing, is a sort of cackle "käk käk." Frequently, when they believe that they have not been observed or noticed, they will quietly sneak away, crouching low and running fairly fast. Their flight, after they are once started, is quick and often quite protracted, sailing long distances without any movement of the wings, in the manner of the Sharp-tailed Grouse.

These birds always roost on the ground, and usually in the same place, as can be seen from the amount of droppings met with in certain favorite localities. On only a single occasion have I seen three of these birds sitting on a horizontal limb of a juniper, about 2 feet from the ground, which was then covered with about a foot of snow.

Mr. George Bird Grinnell writes me about the habits of these birds as follows: "On a very few occasions I have seen the Sage Grouse standing on the branches of a sage bush, sometimes 2 or 3 feet from the ground, but I imagine that this is quite an unusual position for the bird. This species, commonly, I think, goes to water twice a day, flying down to the springs and creek bottoms to drink in the evening, then feeding away a short distance, but roosting near at hand. In the morning they drink again, and spend the middle of the day on the upland. The young birds, when feeding together, constantly call to one another with a low peeping cry, which is audible only for a short distance. This habit I have noticed in several other species of our Grouse, notably in the Dusky Grouse and the Sharp-tail.

"In western Wyoming the Sage Grouse packs in September and October. In October, 1886, when camped just below a high bluff on the border of Bates Hole, in Wyoming, I saw great numbers of these birds, just after sunrise, flying over my camp to the little spring which oozed out of the bluff 200 yards away. Looking up from the tent at the edge of the bluff above us, we could see projecting over it the heads of hundreds of the birds, and, as those standing there took flight, others stepped forward to occupy their places. The number of Grouse which flew over the camp reminded me of the oldtime flights of Passenger Pigeons that I used to see when I was a boy. Before long the narrow valley where the water was, was a moving mass of gray. I have no means whatever of estimating the number of birds which I saw, but there must have been thousands of them."

According to my own observations, confirmed by that of several other observers in widely different localities, the number of eggs laid by this species usually varies from seven to nine, and I consider eight a fair average number. In sixteen nests examined by me, but one contained ten eggs, three contained nine each, six contained eight eggs, four but seven each, and two less, unquestionably incomplete sets. Mr. W. S. Rougis, of Wyoming, gives the number as from ten to fourteen. Mr. W. M. Wolfe, writes me that the average number is fifteen, and that he has found seventeen eggs in a nest of this species. I have no reason to doubt the correctness of these statements. It only shows that no fixed rule can be laid down in such matters, and what will be found a usual set in one section, or in one season even, will not hold good in another.

The eggs of the Sage Grouse vary in shape from ovate to elliptical ovate and elongate ovate. The coloring matter is all superficial, and easily wiped off on a freshly laid egg, leaving the shell greenish white in some of the specimens and a pale pea green in others. The ground color varies from an olive buff with a greenish tinge, in a few specimens, to écru drab and greenish brown. They are more or less heavily spotted with well rounded and sharply defined spots of chocolate brown, ranging in size from a No. 2 shot to that of mustard seed. The markings vary considerably in amount, some eggs being profusely spotted, while others are but faintly so.

The average measurement of one hundred and nine specimens in the U. S. National Museum collection is 55 by 38 millimetres. The largest egg of the series measures 59.5 by 39.5, the smallest 52 by 36 millimetres.

The type specimens selected to show the differences in size, as well as in the ground color and markings, were obtained as follows: No. 17385 (Pl. 3, Fig. 11), selected from a set of nine eggs, was collected by Mr. F. A. Hirst, near Fort Bridger, Wyoming, May 10, 1877; No. 21095 (Pl. 3, Fig. 12), from a set of eight eggs, was taken near Camp Harney, Oregon, May 28, 1876; and No. 21096, (Pl. 3, Fig. 13), from an incomplete set of five eggs, was taken near Malheur Lake, Oregon, on April 4, 1877. The last two are from the Bendire collection.

Family PHASIANIDÆ.   PHEASANTS, ETC.

39.   Meleagris gallopavo LINNÆUS.

WILD TURKEY.

Meleagris gallopavo LINNÆUS, Systema Naturæ, ed. 10, I, 1758, 156.
(B 457, C 379a, R 470a, C 554, U 310.)

GEOGRAPHICAL RANGE: Eastern United States; north to Southern Canada; south to Florida and middle Texas; west to the edge of the Great Plains.

The breeding range of the Wild Turkey, the largest and finest of our game birds, is yearly becoming more and more restricted, and at the present rate of decrease its total extinction east of the Mississippi and north of the Ohio River is only a question of a few years. The northern range of this species may be defined as follows: It is still to be found in small numbers in the thinly populated and wooded portions of Pennsylvania, westward in similar localities to Ohio, southern Michigan, southern Wisconsin, and possibly southern Minnesota, as well as in some portions of Iowa, and a few probably remain in southern and western Ontario, Canada; south through Missouri and Kansas, where they are nearly exterminated, thence through the Indian Territory, eastern and central Texas, where they are still abundant, thence east through all the Gulf States to Florida and north to the first-named State. In the southwestern portions of Indiana and Illinois the Wild Turkey is not uncommon along the extensive river bottoms, and throughout nearly all the Southern States, including Maryland, it is still moderately common in suitable localities, especially in the mountainous districts. It was not uncommon in southern South Dakota and Nebraska within the last ten years, but these birds are no longer found there. At the present time it is most common in the Indian Territory and the thinly-settled portions of eastern and central Texas and Florida.

The Wild Turkey is a resident wherever found. Numerous records attest the abundance of this magnificent game bird throughout the southern New England States in former years, and evidences of its existence have been found in southern Maine.

The Wild Turkey is essentially a woodland bird, and inhabits the damp and often swampy bottom lands along the borders of the larger streams, as well as the drier mountainous districts found within its range, spending the greater part of the day on the ground in search of food, and roosting by night in the tallest trees to be found. From constant persecution in the more settled portions of its range, it has become by far the most cunning, suspicious, and wary of all our game birds, while in sections of the Indian Territory and Texas, where it has till recently been but little molested, it is still by no means a shy bird.

Capt. William L. Carpenter, Ninth Infantry, U. S. Army, writes me as follows: "March, 1880, found me encamped on the Niobrara River, Nebraska, at the mouth of Minnecadusa Creek, the advance guard to establish the new post of Fort Niobrara. This was then a wild, uninhabited country, covered with oak timber and full of large game. Turkeys were quite numerous, and several were killed near camp. It is generally conceded that this is the most wary of all game birds, yet an incident which occurred in this camp would seem to prove that the all-controlling power of animal appetite was sufficient to overcome natural caution.

"When the camp was first established, some corn was scattered at one end, in the brush. My tent was pitched about 30 yards from this spot, and it was reported that every morning and evening a Turkey came here to feed. It was certainly the first grain the bird had seen, and the new food proved an irresistible attraction. Just before sundown, while I was sitting in the tent, it came from the brush and began feeding in plain sight of several persons, amidst all the noise of a large camp. I watched it for some time, and then, without leaving the tent, 'collected' it, not, however this time, for the 'Smithsonian' as the mess clamored loudly for roast turkey without arsenic."

Dr. T. E. Wilcox, U. S. Army, who is now stationed at Fort Niobrara, Nebraska, informs me that he finds abundant and reliable testimony that the Wild Turkey was found there a few years ago and that this point marked the most western limit of its range. None occur there now.

Dr. William L. Ralph, of Utica, New York, writes me: "Fifteen years ago I found the Wild Turkey abundant in most parts of Florida, north of Lake Okeechobee, with perhaps the exception of the Indian River region, but they have gradually decreased in numbers since then, and though still common in places where the country is wild and unsettled, they are rapidly disappearing from those parts, in the vicinity of villages and navigable waters.

"At that time they would frequent the vicinity of dwellings, and where they were not molested would become quite tame. I have often seen them in company with domestic fowls, and on one occasion, where a wild gobbler associated with some tame hen Turkeys, the chicks that were hatched from their eggs showed their wild blood plainly, both in color and actions. I watched this gobbler with a great deal of interest at the times when the fowls were being fed. At first, when any one was near, he would only come to the edge of some bushes, about 40 yards away, but each succeeding day he came a little nearer, until at last he became almost as tame as the domestic birds, and would feed unconcernedly within a few feet of a person.

"One can hardly believe that the Wild Turkeys of to-day are of the same species as those of fifteen or twenty years ago. Then they were rather stupid birds, which it did not require much skill to shoot, but now I do not know of a game bird or mammal more alert or more difficult to approach. Formerly, I have often, as they were sitting in trees on the banks of some stream, passed very near them, both in rowboats and in steamers, without causing them to fly, and I once, with a party of friends, ran a small steamer within 20 yards of a flock, which did not take wing until several shots had been fired at them.

26957—Bull. 1——8

"Turkeys are still to be found quite commonly in suitable places in the localities where the following observations were made, viz: The southern half of St. Johns County and that part of Putnam County east of the St. Johns River. These birds, though resident, are given to wandering a great deal, and do not, like the Bob Whites, become attached to any particular locality. At times they will remain in a favorable place for weeks, but they are very uncertain, and will often leave such a spot for no apparent reason. When they are molested, or when there is a scarcity of food, they will keep in motion most of the time during the day, and will often travel many miles in a few hours.

"Wild Turkeys usually go in flocks, consisting of from two or three to fifteen or twenty birds, and are also occasionally found singly. Small flocks and single birds are more apt to be found now than formerly, and the large droves, consisting of several flocks associating together, are seldom if ever to be seen of late. Their favorite places of resort are woods with swamps in them or in their vicinity, and they always go to these swamps to roost or when molested.

"These birds are polygamous, and the female takes all the cares and duties of incubation upon herself. The gobblers are very pugnacious, and will often fight fiercely for the favors of the hens. The love season begins in Florida about the middle of February and lasts for about three months, and during this period the gobblers frequently utter their call and are then easily decoyed within gunshot. Native hunters have informed me that the hens roost by themselves at this season of the year.

"The nest is a slight depression in the ground, either at the foot of a tree or under a thick bush or saw palmetto. It is lined sparingly with dead leaves and grass, etc., but I could never find out whether this material was placed there by the birds or was there originally. I think these birds raise but one brood a season, though I have found fresh eggs as early as the middle of March and as late as the 1st of May. I have never found more than thirteen eggs in one nest, nor less than eight, unless they were fresh, the usual number being ten. The chicks of this species are very tender, and as they follow their mothers as soon as hatched I have often wondered how the latter could raise so many as they do. The natives of Florida say that a hen Turkey will desert her nest if the eggs are handled. Whether this be true or not I do not know, for I never tried to find out but once, and then, though the bird was gone on my second visit to the nest, I always had a strong suspicion that she was shot, for its whereabouts was known to several persons besides myself."

These birds feed on beechnuts, acorns (especially those of the white and chinquapin oaks), chestnuts, pecan-nuts, black persimmons, tuñas (the fruit of the prickly pear), leguminous seeds of various kinds, all the cultivated grains, different wild berries and grapes, and the tender tops of plants; also grasshoppers, crickets, and other insects. The actions of the gobbler during the mating season, while paying court to the female, are similar to those of the domestic Turkey, and well enough known to need no description.

The nesting season, in the southern portion of their range, begins sometimes as early as the middle of February, and later northward, where fresh eggs are occasionally found up to the middle of June. It is more than probable that in such late sets the first eggs were destroyed. The number of eggs to a set varies from ten to fourteen, eleven and twelve being most often found. The nests are usually well concealed, a favorite site being near old stumps surrounded by a dense mass of vines and bushes in bottom lands. Occasionally an exposed situation is also selected.

Capt. William L. Carpenter, U. S. Army, found such a nest on May 1, 1880, near the present site of Fort Niobrara, Nebraska, containing at that time seven eggs. It was a simple affair, on a grassy hillside, in an exposed position, and lined with dead grass, very much like that of the Sharp-tailed Grouse. On May 20, it contained eleven eggs, and on June 1 another visit was paid, when only the vacant nest and broken shells remained.

Bvt. Maj. S. L. Woodward, Tenth Cavalry, U. S. Army, sent me a set of eggs of this species, collected near Onion Creek, Archer County, Texas. This nest was found in the high grass, well protected from observation, near some rain-water holes, several hundred yards from Onion Creek. Happening to stroll close to the nest the hen flew off, and was thus discovered. It contained six perfectly fresh eggs on May 7.

Mr. J. A. Singley writes me: "The hen leaves and approaches the nest invariably by the same route, and remembering this I trailed one to its nest. There really was no nest. A dead blackjack-oak top had fallen, the wind had drifted the leaves up against it, and the eggs were laid on the leaves. I found this nest in a thicket, inside of an inclosure, on April 20. It contained eight eggs."

Sometimes two hens lay in the same nest, and an extraordinarily large set of eggs is the result. Mr. George E. Beyer, of New Orleans, Louisiana, writes me as follows: "On May 25, 1888, I found a nest with twenty-six eggs; one hen sitting on the nest and one standing by. I think both hens kept the same nest. In piney woods they like to build near old pine stumps, as the latter are generally surrounded by a growth of gallberry bushes. The nest is composed of leaves and pine straw. Our swamp Turkeys like to build in hammock lands, near old and disused fields, in blackberry or scrubby post-oak thickets."

The call notes of the Wild Turkey resemble those of the domesticated bird very much; still they differ somewhat. In feeding, their usual note is "quitt, quitt" or "pit, pit." When calling each other it is "keow, keow, kee, kee, keow, keow," and a note uttered when alarmed suddenly sounds somewhat like "cut-cut."

Usually but one brood is raised a season, but Mr. J. S. Cairns, of Weaverville, North Carolina, suspects that a second brood is occasionally raised.

having found eggs of this species as late as July 10. Incubation lasts about four weeks, and this duty is entirely performed by the female.

In shape, the eggs of the Wild Turkey are usually ovate, occasionally they are elongate ovate. The ground color varies from pale creamy white to creamy buff. They are more or less heavily marked with well-defined spots and dots of pale chocolate and reddish brown. In an occasional set these spots are pale lavender. Generally the markings are all small, ranging in size from a No. 6 shot to that of dust shot, but an exceptional set is sometimes heavily covered with both spots and blotches of the size of buckshot, and even larger. The majority of eggs of this species in the U. S. National Museum collection, and such as I have examined elsewhere, resemble in coloration the figured type of *M. gallopavo mexicanus*, but average, as a rule, somewhat smaller in size.

The average measurement of thirty-eight eggs in the U. S. National Museum collection is 61.5 by 46.5 millimetres. The largest egg measures 68.5 by 46, the smallest 59 by 45 millimetres.

The type specimen, No. 21069 (Pl. 3, Fig. 14), selected from an incomplete set of six eggs from the Bendire collection, was obtained by him through the kindness of Bvt. Maj. S. L. Woodward, Tenth Cavalry, U. S. Army, near Onion Creek, Archer County, Texas, May 7, 1874. This specimen shows the lavender markings referred to above.[1]

40. Meleagris gallopavo mexicana (GOULD).

MEXICAN TURKEY.

Meleagris mexicana GOULD, Proceedings Zoölogical Society, 1856, 61.
Meleagris gallopavo var. mexicana BAIRD, History of North American Birds, III, 1874. 410.

(B 458, C 379, R 470, C 553, U 310a.)

GEOGRAPHICAL RANGE: Tableau lands of Mexico, and north to southern border of United States from western Texas to Arizona; south to Vera Cruz (temperate region).

The breeding-range of the Mexican Turkey extends from the mountainous portions of southern and western Colorado, through similar regions in New Mexico, Arizona, western and southwestern Texas, into Mexico. In southern Colorado the bird is rare, if still found there, while specimens from the Lower Rio Grande region in Texas seem to be intermediate between true *M. gallopavo* and the *M. gallopavo mexicana*, leaning more toward the latter.

This magnificent game bird, the progenitor of our domestic Turkey, is more of a mountain-loving species than the eastern bird, and is still reasonably abundant in the wilder portions of western Texas, the Territories of New Mexico and Arizona, and very common in portions of Mexico. I believe this subspecies attains a greater size than *M. gallopavo*. I shot a specimen weighing

---

[1] Since this article was written, Mr. W. E. D. Scott has separated the Florida Turkey from the common Wild Turkey and described it under the name of *Meleagris gallopavo osceola*, in Auk, Vol. VII, 1890, p. 376.

28 pounds, after being drawn, and I have been informed that birds much heavier than this one are killed occasionally; which I can readily believe, as I have seen tracks of this subspecies along the banks of the San Pedro River, in Arizona, measuring between 5 and 6 inches in length, and unquestionably made by a much larger bird than the one killed by me.

Mr. Herbert Brown, of Tucson, Arizona, writes me as follows: "I have seen many of these Turkeys, a few in Arizona, and I believe thousands in Mexico. Without knowing it positively, I am of the belief that they raise two broods of young a season, as I have seen almost all sizes in the masting season (October) when they congregate in large numbers in the cañons to feed on *bellotes*, a small bitter acorn, common to the cañons and parks of southern Arizona and southward. I have seen their roosting places at night, in sycamore (*Alisa*) trees; I also saw one in an oak grove on the side of a hill, but they appear more to favor the cañons. On the headwaters of the Santa Domingo I have seen not less than fifty or sixty in a bunch, and Turkey, in those days, was a common camp fare. I have been told by Mexicans that coyotes catch Turkeys by running in circles under their roosting trees, till the birds get dizzy with watching them, and fall down. I never saw it done, but have been assured that it is a fact.

"This Turkey is quite abundant in the Sierra Ancha, about 130 miles northeast of Tucson, and can be found to a greater or less extent in every timbered range between here and there. One morning while camped in a box cañon in Sonora, southeast of the Sierra Azul, I counted thirty-four Turkeys flying over, at another time six, and quite frequently three and four. Times were not so peaceful then as now, and whenever the Turkeys appeared to fly wild, we would be on the lookout for Apache Indians."

Capt. William L. Carpenter, Ninth Infantry, U. S. Army, writes me as follows: "This bird is quite numerous in the White Mountains, Arizona. I never succeeded in finding the nest, but collected young, able to fly a few feet, on Black River, Arizona, July 1. Their crops contained grasshoppers and leaves. Later in the season acorns, juniper berries, and pine nuts furnish food. It appears to be stronger on the wing than the eastern Turkey and more ready to resort to long flights for safety, frequently alighting in trees for concealment; also, not so readily decoyed by calling."

The mating season commences according to latitude, from March 1 to the middle of April, by which time some of the birds commence nesting.

Mr. William Lloyd states: "In 1881, I found many of their roosts in western Texas; these birds could be seen by hundreds in the Nueces Cañon near Uvalde; they were equally common in the Frio Cañon and the valleys of the Llano and Concho, east to the Colorado River. They mated from the first to the end of March, according to the weather, and would then seek dense brush.

"Near a river their nests would be made on small islets surrounded by reeds; on the hills in shin oak clumps. Their principal food being acorns, they were most abundant where there was much mast, migrating considerable distances, but in this case only governed by food supply. I have seen them freely eating mesquite beans when they fall in the late summer, and pecan nuts are also a favorite food. They also feed on grasshoppers and other insects during April and May, running after them in the same manner as the Chaparral Cock."

They are summer residents in the higher mountain ranges, reaching an altitude from 8,000 to 10,000 feet, and retiring to the more sheltered cañons and the timbered river valleys in the late fall, congregating at such times in large flocks.

The number of eggs to a set varies from eight to fourteen, rarely more; eleven or twelve are about the average number.

That well-known ornithologist and collector, Mr. F. Stephens, took a probably incomplete set of nine fresh eggs of this species, on June 15, 1884. He writes me: "I was encamped about 5 miles south of Craterville, on the east side of the Santa Rita Mountains in Arizona; the nest was shown to my assistant by a charcoal burner. On his approach to it the bird ran off or flew before he got within good range. He did not disturb it but came to camp, and in the afternoon we both went, and I took my little camera along and photographed it. The bird did not show up again. The locality was on the east slope of the Santa Rita Mountains, in the oak timber, just where the first scattering pines commenced, at an altitude of perhaps 5,000 feet."

A good photograph, kindly sent me by Mr. Stephens, shows the nest and eggs plainly. It was placed close to the trunk of an oak tree on a hillside, near which a good-sized yucca grew, covering, apparently, a part of the nest; the hollow in which the eggs were placed was about 12 inches across and 3 inches deep. Judging from the photograph the nest was fairly well lined.

Capt. B. F. Goss found a nest of this subspecies, on May 9, 1882, in southern Texas, containing eleven eggs. He states: "It was as usual on the ground, in open, bushy country. A coarse structure, not very deeply excavated, lined with grass, weeds, and leaves, placed in quite an open situation, but well concealed by a few small bushes and bunches of growing grass.

"We were encamped quite near the nest; one morning I noticed a hen Turkey stealing through the bushes, and suspected she was going to her nest. We watched her carefully for three mornings, and having pretty nearly located the nest, commenced a close search, and examined, as we thought, every inch of ground. I was about giving up, when looking down almost at my feet, I saw the bird sitting on the nest. She at once ran; she had allowed me to pass several times within a foot of her without moving, and seemed to know at once when she was seen. I have often noticed this trait in birds of this genus; as long as unseen you can tramp all around them, but they seem to know at once when they are seen, and lose no time in getting away."

The Mexican Turkey, like the eastern subspecies, is polygamous, and the female attends exclusively to the duties of incubation, which lasts about four weeks, the male not only not assisting, but, according to observations made by Lieut. J. M. F. Partello, Fifth Infantry, U. S. Army, they often destroy the eggs and the tender young.

In addition to the different foods already mentioned, I know from personal observation that the Mexican Turkey is exceedingly fond of the wild mulberry, as well as the fruit of the prickly pear, which in southern Arizona attains a very respectable size; and again of the somewhat smaller but still more palatable fruit of the giant cactus, the pitayah or the sahuara of the natives, which is alike a favorite article of food with man, bird, and beast.

The only eggs of this species in the U. S. National Museum collection, about whose identity there can be no possible doubt, were collected on Upper Lynx Creek, Arizona, in the spring of 1870, by Dr. E. Palmer, whose name is well known as one of the pioneer naturalists of that Territory.

The eggs are ovate in shape, their ground color is creamy white, and they are profusely dotted with fine spots of reddish brown, pretty evenly distributed over the entire egg. The average measurement of these eggs is 69 by 49 millimetres. The largest measures 70.5 by 49, the smallest 67 by 48 millimetres.

The type specimen (No. 15573, U. S. National Museum collection, Pl. 3, Fig. 15) is one of the set referred to above.

Family CRACIDÆ. CURASSOWS AND GUANS.

41. Ortalis vetula maccalli BAIRD.

CHACHALACA.

Ortalida maccalli BAIRD, Birds of North America, 1858, 611.
Ortalida vetula var. *maccalli* BAIRD, History North American Birds, III. 1874, 398.
(B 456, C 378, R 469, C 552, U 311.)

GEOGRAPHICAL RANGE: Northeastern Mexico from Vera Cruz, north to lower Rio Grande Valley (both sides).

The Chachalaca occupies but a very restricted area within the borders of the United States, being found only along the thickly-timbered river bottoms of the Rio Grande, from its mouth to a short distance above Fort Ringgold, Texas, a distance of about 100 miles. It is common in suitable localities in this region, and breeds wherever found.

Asst. Surg. James C. Merrill, U. S. Army, in his notes on the "Ornithology of southern Texas," writes as follows: "The 'Chachalac,' as the present species is called on the Lower Rio Grande, is one of the most characteristic birds of that region. Rarely seen any distance from woods or dense chaparral, they are abundant in those places, and their hoarse cries are the first thing heard by the traveler on awaking in the morning. During the day, unless rainy or cloudy, the birds are rarely seen or heard; but shortly before sunrise and sunset they mount the topmost branch of a dead tree, and make the woods ring with their discordant notes. Contrary to almost every description of their

cry which I have seen, it consists of three syllables, though occasionally a
fourth is added. When one bird begins to cry, the nearest bird joins in at the
second note, and in this way the fourth syllable is made; but they keep such
good time that it is often very difficult to satisfy oneself that this is the fact.
I cannot say certainly whether the female utters this cry as well as the male,
but there is a well-marked anatomical distinction in the sexes in regard to the
development of the trachea. In the male this passes down the outside of the
pectoral muscles, beneath the skin, to within about 1 inch of the end of the
sternum; it then doubles on itself and passes up, still on the right of the keel,
to descend within the thorax in the usual manner. This duplicature is wanting
in the female. These birds are much hunted for the Brownsville market,
though their flesh is not particularly good, and the body very small for the
apparent size of the bird. Easily domesticated, they become troublesomely
familiar, and decided nuisances when kept about the house. Beyond Ringgold
Barracks this species is said to become rare, and soon to disappear; and it
probably does not pass more than 50 miles to the north of the Rio Grande.
The nests are shallow structures, often made entirely of Spanish moss, and are
placed on horizontal limbs a few feet from the ground."[1]

Mr. George B. Sennett, in his "Further Notes on the Ornithology of the
Lower Rio Grande of Texas," made during the spring of 1878, makes the following statement about this species: "Preëminently a bird of the woods; and
Lomita, without doubt, is the heart of its very limited habitat in the United
States. Here it is resident; and among the heavy timber and dense undergrowth it breeds in seclusion, secure from its enemies. A more intimate
acquaintance with this bird enables me to give a better description of its notes
than the attempt in my former memoir. The notes are loud, and uttered in
very rapid succession, and those of the female follow those of the male so
closely, while so well do they harmonize, although in different keys, that I mistook the first note of one for the last note of the other. It really utters but
three syllables, thus: 'cha-cha-lac,' instead of four, 'cha-cha-lac-ca,' as given
before. It also has a hoarse grating call or alarm note, uttered in one continuous strain and without modulation, something like 'kak-kak-kak.' Generally
this bird is seen in trees, but on one occasion four or five were seen running
about on the ground, after the manner of chickens when freed from a coop. It
does not breed in communities, but in isolated pairs, and from all accounts
raises but one brood in a season, unless the nest is despoiled, when it will lay
another clutch. The clutch almost invariably consists of three, rarely less. On
the 10th of April we obtained fresh eggs, and on the 20th sets were generally
full and fresh, after which time they contained embryos. The eggs were exceedingly hard to drill. The chicks are hatched well coated with down, and they
leave the nest as soon as hatched, the old ones leading them into the thickets,
where they are very hard to capture. I had the pleasure, at the ranch, of
seeing six hatch under a hen. The little ones looked and acted exactly like

chickens, picking up the corn-batter thrown to them, running in and out from under the hen's wings, and jumping upon her back. Four of the six died within the first two weeks, but the others lived and thrived. A few are domesticated every year at almost every ranch, and they become inconveniently familiar, getting about under foot, jumping upon tables, beds, etc.

"The young from the eggs are thickly covered with down. Upper parts mixed ash, fulvous, and brown, with a black line from the crown to the tail and a black patch on the forehead; under parts white, with the exception of the jugulum, which is fulvous ash, meeting the same colors above. When three or four weeks old, the fulvous ash and white become tawny, and the black only shows on the crown and the forehead."[1]

The breeding season begins about April 10, and lasts till the beginning of July. Mr. J. A. Singley writes me that he took eggs on June 2—probably a second laying, the first eggs having been taken or destroyed. He says: "I found this bird abundant on the Lower Rio Grande. It is a noisy fellow, and two or three can make noise enough to make the listener think that there are a dozen or more birds at hand. Noisy as they are, it is a hard matter to get sight of one, and I found the most successful plan was to hide in the thicket and wait for the birds to put in an appearance. No matter how quickly and stealthily you approached one when singing (?), he would notice you and quickly take his departure. I never saw one on the ground. They are noisiest in the morning and just before a rain.

"All the nests I found were in mesquite stubs, where the limbs had been cut off to make brush fences. These limbs are never cut close to the tree, and being close together form a cavity; leaves and twigs will fall in this and accumulate, and the bird occupies it as a nesting site. I did not find a nest that I could say was built by the bird. When the nest is approached the bird quietly flies off, rarely remaining in sight, and soon calls up its mate."

According to most observers, three eggs are said to usually constitute a full set, rarely more. Mr. Thomas H. Jackson, of West Chester, Pennsylvania, informs me, however, that of forty-four sets received by him this season (1890), all but one contained four eggs, the other being a set of three. The first of these sets was found May 13, the second May 29, and the balance between that date and July 1. In 1889, he received a single set of five, which must be considered an unusually large one. These eggs were all collected near Rio Grande City, Texas.

The eggs of the Chachalaca are a pale creamy white in color, varying in shape from ovate to short ovate, as well as elongate ovate. The shell is extremely thick, rough to the touch, and strongly granulated. The average measurement of twenty-five specimens in the U. S. National Museum collection is 58.5 by 41.5 millimetres. The largest egg measures 65.5 by 47, the smallest 53.5 by 40 millimetres.

[1] U. S. Geological and Geographical Survey, Vol. v, No. 3, pp. 426, 427.

The type specimen, No. 21068 (Pl. 3, Fig. 16), from a set of three eggs, was collected by Mr. George B. Sennett, near Lomita Ranch, Hidalgo, Texas, May 12, 1877.

Family COLUMBIDÆ.   Pigeons.

42.   Columba fasciata SAY.

BAND-TAILED PIGEON.

Columba fasciata SAY, Long's Expedition, II. 1823, 10.
(B 445, C 367, R 456, C 539, U 312.)

GEOGRAPHICAL RANGE: Western United States, from Rocky Mountains to the Pacific coast; south, through Mexico, to highlands of Guatemala.

The Band-tailed Pigeon is an irregular inhabitant of western North America, ranging from British Columbia south through Washington, Oregon, California, Arizona, New Mexico, and northwestern Texas into Mexico and Guatemala. In British Columbia, Washington, and Oregon, it is a summer resident, and only occurs regularly west of the Cascade Range, near the coast, while east of these mountains it can only be considered as a straggler, due, no doubt, to the absence of oak forests in these regions. It has also been observed in Colorado, Nevada, and Idaho, but it is doubtful if it can be considered as of regular occurrence in these localities. It probably breeds throughout the range thus indicated. In Arizona, southern New Mexico, and northwestern Texas it is found throughout the year.

Capt. William L. Carpenter writes me as follows: "The Band-tailed Pigeon occurs in the mountain regions of northern Arizona. It is sometimes quite numerous in the vicinity of Prescott, in August and September; at other seasons, without any apparent reason, very rare. It is always to be found in summer in the foothills of the White Mountains, at about 5,000 feet elevation, wherever the oak grows, although not plentiful, and always quite shy. I saw them in this region in small flocks during May, June, July, August, and September, and although not successful in finding the nest, I feel confident that they breed here.

"As far as my observation extends, this species is most numerous near the mouth of the Columbia River, where immense flocks were to be seen from May to October in 1865, which fairly rivaled those of the Passenger Pigeon, once so common over the Atlantic watershed. Their favorite food in this region appeared to be salmon berries, *Rubus nutkanus*."

Mr. H. W. Henshaw, in his article on the "Birds of the Upper Pecos River, New Mexico," makes the following statement regarding this species: "None of these birds nested near our camp, though they probably did so not far away. The latter part of August they were found feeding upon the berries of the *Sambucus racemosa*, a small shrubby plant from 2 to 4 feet high. Subsequently, when the acorns began to grow large, long before they began to ripen, they appeared to devote themselves exclusively to them, and between the Pigeons and the squirrels not an acorn was allowed to ripen. The acorns were of the scrub oak, *Quercus undulata* (two varieties), and are extremely palatable. Pigeons were shot not only with their crops full, but with the gullet crammed up to the very bill."[1]

Mr. L. Belding states: "I have seen but few of these birds in the mountains of California in summer, though it probably breeds there, as I have occasionally shot young birds at Big Trees, apparently about a month old. They were, at that age, excellent food, which cannot be said of it at any other time, its flesh being very bitter from eating acorns in winter and oak buds in spring. "It is sometimes common in the foothills in winter, but never abundant, as compared with the abundance of the Passenger Pigeon of the Atlantic States."[2]

Mr. Charles H. Townsend, in his "Field Notes on the Birds of northern California," says: "The Band-tailed Pigeon is very abundant in the foothills of the Lower McCloud River, in the fall and winter, gathering in the pine trees on the higher ridges in immense flocks. It was very seldom seen in the high mountains in summer, and did not appear to descend at all to the valleys in winter. I do not know where it breeds."[3]

Mr. O. B. Johnson, in his "List of the Birds of the Willamette Valley, Oregon," states: "An abundant summer resident, feeding chiefly on berries. They nest in various situations, much like the common Dove, *Z. carolinensis*. I found one of leaves and moss beside a tree, placed on the ground between two roots; another one upon an old stump that had been split and broken about 8 feet from the ground; another was in the top of a fir (*A. grandis*), and was built of twigs laid upon the dense flat limb of the tree, about 180 feet from the ground. These each had two eggs, pure white, and elliptical, differing from those of *Z. carolinensis* only in size; a set before me measuring 1.60 by 1.20, and 1.55 by 1.19 inches" (equaling 40.6 by 30.5, and 39.4 by 30.2 millimetres.)[4]

Mr. A. W. Anthony states: "This Pigeon is a common summer resident in Washington County, Oregon. South of Beaverton is a large spring, the waters of which contain some mineral which has great attraction for these Pigeons, and here they are always found in large numbers."[5]

Mr. Henry E. Ankeny writes me from Jacksonville, Oregon, as follows: "The Band-tailed Pigeon is not very common in the Rogue River Valley, Oregon, excepting when there is a good acorn crop. In such years they are occasionally quite plentiful in the fall months. Only a few scattering pairs breed here; all the nests I have seen have been placed on limbs of small firs, generally in thickets of these trees. They lay two eggs, and I believe rear but a single brood a season."

Mr. William Lloyd informs me as follows: "The Band-tailed Pigeon breeds in the mountains in Presidio County, and occasionally in the neigh-

[1] Auk, Vol. III, 1886, p. 80.
[2] Proceedings U. S. National Museum, 1878, Vol. I, p. 437.
[3] Proceedings U. S. National Museum, 1887, Vol. X, p. 200.
[4] American Naturalist, July, 1880, pp. 638, 639.
[5] Auk, 1886, Vol. III, p. 164.

borhood of Fort Davis, Texas. They are exceptionally abundant in the aforesaid mountains, in July, where they collect in large flocks to feed on the wild grapes, and afterward disperse in search of acorns. Each flock I saw was under the guidance of a leader who would fly around a tree several times alone, and when satisfied there was no danger it would alight, and the rest of the band followed its example. These flocks ranged from twenty to fifty. I only found eggs of this species in Mexico. They are abundant all down the Sierra Madre to at least the Sierra Nevada de Colima. It lays two eggs to a set, and I found them nesting on June 2. The nests are placed on limbs of trees, oaks preferred."

The Arizona records of the nesting of this species differ as far as the number of eggs are concerned. Mr. F. Stephens writes me that he found three nests of this species at an altitude of from 5,000 to 8,000 feet in Arizona. They prefer the open forest, where the oaks and pines mix, nesting on both kinds of trees. The nest consists of a very slight platform of sticks. In one case it was composed of a few pine twigs laid across a horizontal fork of a small branch of pine, about 20 feet from the ground. Another nest was but 8 feet, and the third 12 feet, from the ground. The earliest date of nesting was March 6, 1877, the latest July 18. Each nest contained but a single egg."

In my notes on a collection of eggs from southern Arizona, made by Lieut. Harry C. Benson, Fourth Cavalry, U. S. Army, at Fort Huachuca, I make the following statement based upon his observations: "This Pigeon is fairly common in the vicinity of Fort Huachuca during the summer months, arriving about June 1 to 10 in large flocks, frequenting the oak groves along the foothills and mountain sides. It feeds on a berry about the size of a large pea, growing on a hardwood tree not known to Lieutenant Benson, till the acorns are of suitable size, about July 15, when it feeds almost exclusively on them.

"It commences nesting about the beginning of July, and continues to lay till late in October; it does not breed in communities, however, there being but one or two nests to the acre. The nests are placed in live-oak trees (*Quercus undulata?*), from 15 to 30 feet from the ground. The nest is simply a slight platform of twigs on which the egg is laid. Eggs were taken from July 13 to September 25, 1885, inclusive.

"But a single egg is laid in a clutch, in that vicinity at least. This is elliptical ovate in shape, abruptly pointed at the smaller end; pure white in color, slightly glossy, and the five specimens sent measure 1.58 by 1.10, 1.62 by 1.10, 1.62 by 1.13, 1.68 by 1.04, 1.69 by 1.09 inches" (equal to 40.1 by 27.9, 41.1 by 27.9, 41.1 by 28.7, 42.7 by 26.4, 42.9 by 27.7 millimetres).[1]

Mr. Otho C. Poling has also made some further observations regarding the Band-tailed Pigeon during the season of 1890, in close proximity to Fort Huachuca, Arizona, and I make such extracts from his notes as are of

---

[1] Proceedings U. S. National Museum, 1887, Vol. x, p. 551.

especial interest. He writes me under date of May 5: "I have camped for the past month in Ramsey's Cañon, which is the most heavily wooded one in this locality, situated in about the center of the Huachuca Mountains. About a quarter of the way up, close to the main trail, at an elevation of about 6,000 feet, is a swampy spring, around which several acres of wild mulberry trees and bushes grow. I camped close to this place, and Pigeons were always about in abundance. Probably sixty pairs visited or staid about this spring at all times.

"I shot a number of these birds for food, and one day on approaching one sitting on a spruce bough, about 10 feet up, and not being able to start it from its perch for a wing shot, I put a 'hummer load' in my gun, and brought it down in rather a mutilated condition, as I was not over 6 feet off. It proved to be a female, and much to my surprise I found that the dust shot had smashed an egg, which I found embedded in the feathers of her belly. As I could find no sign of a nest on the limb on which she had been sitting, and the egg was entirely surrounded with feathers, I concluded she must carry it around with her. This was on March 30, and the egg was fresh.

"During the month of April, being very busy with other matters, I gave only a day or two to searching for the nests, and was not successful in finding any. On April 14, I killed a few young, about a month old, and an adult female with an egg in her ovary, about one-third size. On May 3, I made another trial, and during a hard day's hunt I found five nests, each containing a single well-incubated egg. These nests were all found in spruce pines, from 15 to 70 feet up, and were constructed simply of a few small twigs laid across a limb. Next day I found four more nests; one in an oak, 12 feet up, containing a squab about a week old; another in a mulberry, about 8 feet from the ground, with a fresh egg, and two others in spruce pines. On approaching one of these trees the female flew off heavily, and seemed to be trying to balance herself on a limb of a tree, far down below. When I climbed to the nest I found no egg in it, and am almost certain she carried it off with her. I also allowed the female to fly off from the next nest, and likewise found it without an egg. Wishing to test the matter, I afterward shot several females on the nest. In one case one dropped to the ground, and with her came the egg, breaking and spreading egg and shell among the feathers of the belly."

In some later notes sent, Mr. Poling says: "Since writing to you in regard to the Band-tailed Pigeon, on May 5, I have taken over a dozen eggs, which show my previous experiences to be *exceptions* to the rule, or else are due to difference in season or locality. As far as my knowledge now extends, I am somewhat puzzled in regard to these birds. The fact of their nesting at all seasons, without any regularity, seems to be well established, as I have taken young, two or three months old, in February; and since that time young and eggs enough to show that they lay and nest from December to August, my last fresh egg having been taken only a few days later; and the females examined show that they will lay on for two months more at least.

"In regard to their carrying the egg about I have, in addition to the cases noted, shot two other females having the egg embedded in the feathers of the belly, and further, held by the legs while flying; but in such cases they seem simply to alight on the limb of a spruce and incubate there without any nest. This accounts for the shooting of Pigeons having a broken egg smeared over the feathers, as I have done when no nest was to be seen. I have found them nesting on oaks, maple, and spruce trees, often in old nests of some other bird, or on a slight platform of twigs laid loosely on a flat limb.

"These birds sit closely, and allow one to go much nearer them than when merely feeding or resting. In fact, this is the only way to know when you have found a nest. The entire bird may be in plain sight, but no nest can be seen from below, even when within 10 feet of her; and I had generally to note the spot from whence she flew, and climb above this, when on looking down the egg might be seen, which usually laid directly over the limb, and hence was invisible from below. The average distance of the nests from the ground is about 10 feet, the extremes being 6 and 70 feet.

"Pigeons are most plentiful in these mountains during the months of June and July. They are then scattered over the entire range, from the oak groves at the bases of the foothills to the pines at the highest point. They, however, have certain points where they congregate, favorite feeding grounds, which are generally near the mouth of a cañon, and to which they repair regularly, coming and going singly or in pairs and flocks of all sizes. In coming down from the higher points to these feeding places their flight is exceedingly swift, and the noise made by their wings is wonderful. It appeared to me similar to the rush of steam from an engine, and can be heard when the birds are scarcely visible and a thousand yards overhead. The crop of a good-sized male after feeding contains from twenty-five to fifty acorns."

I have quoted, without further comment, the remarkable statement of Mr. Poling, in regard to the alleged removal of eggs by this Pigeon.

Undoubtedly the irregular distribution of the Band-tailed Pigeon at certain times is due, to a large extent, to the comparative abundance of its favorite food, acorns, which are found only in certain localities. On the eastern slopes of the Sierra Nevada in California, and the Cascade Range in Oregon and Washington, oaks are but seldom seen; hence the scarcity of these birds. I have seen flocks of the Band-tailed Pigeon repeatedly while stationed in southern Arizona, but only along the foothills of the mountains where oaks were abundant. The fruit of some of these trees is exceedingly sweet and palatable, equal to the best of nuts. There is, however, a great deal of difference in the taste of these acorns. While on one tree sweet ones may be found, those of the next, although of apparently the same species, may be bitter and unfit to be eaten by man, though quite acceptable to these birds.

In southern Arizona, at least, it would appear that the breeding season of the Band-tailed Pigeon covered nearly every month in the year, and several broods must be reared by each pair during the season. In Oregon and Washington two broods are probably the usual number.

In Arizona the nests of this species seem to be always placed in trees, while in Oregon, according to Prof. O. B. Johnson, they breed occasionally on the ground. This statement is confirmed by Dr. Cooper, who says: "In June they lay two white eggs, about the size of those of the House Pigeon, on the ground, near streams or openings, and without constructing any nests."[1]

In southern Arizona, according to the observations of several well-known and reliable collectors, but a single egg is laid at a sitting, and, as far as I am aware, at no time have two been found. Their cooing resembles that of the domestic Pigeon, and the young are fed in a similar manner. Incubation lasts from eighteen to twenty days, both sexes assisting. The young grow rapidly, and are able to leave the nest when about a month old.

The egg of the Band-tailed Pigeon is large for the size of the bird, and is somewhat peculiar in shape. This may be called a pointed elliptical ovate. All the eggs I have seen show this feature. They are pure white in color, the shell is close grained, smooth, and slightly glossy.

The average measurement of thirteen specimens is 40 by 28 millimetres; The largest of these eggs measuring 43.5 by 30, the smallest 38 by 27 millimetres.

The type specimen (No. 23240, U. S. National Museum collection, Pl. 3, Fig. 17) was obtained by Lieut. H. C. Benson, U. S. Army, near Fort Huachuca, Arizona, on September 25, 1885.

43. Columba fasciata vioscæ BREWSTER.

VIOSCA'S PIGEON.

Columba fasciata vioscæ BREWSTER, Auk, v, 1888, 86.
(B –, C –, R –, C –, U 312a.)

GEOGRAPHICAL RANGE: Lower California.

According to Mr. Brewster, this newly-described subspecies differs from the Band-tailed Pigeon in being a trifle smaller; the tail band is wanting, or only faintly indicated; the ground coloring lighter and more uniform; the vinaceous tints, especially on the head, neck, and breast, much fainter and more or less replaced by bluish ash. It has been named after Mr. Viosca, the U. S. consul at La Paz, Lower California.

Little is known regarding the habits of this bird, which are probably very similar to those of the preceding subspecies.

[1] Birds of North America, Vol. III, p. 363.

Mr. M. A. Frazar, while collecting near Pearce's Ranch, in Lower California, obtained two nests, each containing a single egg. One of these is now before me. This egg was taken July 19, 1887, and the nest in which it was found was composed of a few sticks, placed on a broken upright branch in the center of a giant cactus, about 18 feet from the ground. The egg was fresh; it is pure white, slightly glossy, elliptical ovate in shape, and not quite as notably pointed at the small end as the eggs of *Columba fasciata*; it measures 38 by 26.5 millimetres.

The type specimen (No. 23946, U. S. National Museum collection, Pl. 3, Fig. 18) above described was kindly donated by Mr. William Brewster.

44. Columba flavirostris WAGLER.

RED-BILLED PIGEON.

Columba flavirostris WAGLER. Isis, 1831, 519.
(B 446, C 368, R 457, C 540, U 313.)

GEOGRAPHICAL RANGE: Mexico and Central America; south to Costa Rica; north to southern Texas, southern Arizona, and Lower California.

The Red-billed Pigeon can only be considered as a summer visitor over the greater portion of its range within the borders of the United States. It is most commonly found along the heavily-timbered bottom lands of the valley of the Rio Grande, in Texas, where it breeds abundantly, and a few may possibly remain throughout the year near the mouth of that river. It extends thence into southern Arizona, but is of rare occurrence in that Territory.

Mr. George B. Sennett, in his "Birds of the Rio Grande of Texas," gives the following account of this interesting species: "I found this fine large Pigeon common in heavy timber, more especially in the tall scattered clumps near the larger tracts. Its appearance is so marked that it can be recognized at all times from other members of the family. Like all Pigeons, it is fond of the water. Any morning will find numbers of all the different species going to and coming from the sand-bars in the river, where they are in the habit of drinking and bathing.

"The cooing of this bird is clear, short, and rather high pitched. It is more secluded in its habits than any of the others, except the one I have lately found new to our fauna, *Æchmoptila albifrons*. In point of numbers it is much less numerous than the Carolina and the White-winged Doves; still it is shot quite extensively for the market. I found it breeding, and secured several sets of nests and eggs. * * *

"On April 30, I found my first nest of this bird in the vicinity of Hidalgo. The locality was a grove of large trees, with undergrowth, and clumps of bushes matted with vines. While prying about the thick vines, I flushed the bird off its nest, and it alighted in one of the tall trees near by. It took me but a moment or two to examine the nest and shoot the bird. In less than ten minutes' time I had also its mate. The nest was only 8 or 9 feet from the ground, and set upon the horizontal branches of a sapling, in the midst of the vines. It was composed of sticks, lined with fine stems and grasses, had a depression of an inch or more, and was about 8 inches in outside diameter by 2½ inches deep. It contained one egg with embryo just formed. Dissection of the bird showed that she would have laid no more.

"On May 8, at Lomita Ranch, a few miles above Hidalgo, in the fine grove of ebonies in the rear of the buildings of the ranch, I found two nests. Both were well up in trees, one about 25 feet and the other about 30. The nests were situated close to the body of the trees, on large branches, and were composed of sticks and grasses, with an inside depth of about 2 inches. One contained a single egg far advanced, in the other also lay a solitary egg, from which a young chick was just emerging. The parents persisted in staying about, notwithstanding we were making a great disturbance, even shooting into the same trees. Whenever we would go off some distance, they would immediately go on their nests, and seemed loath to leave on our return. These were the only ones seen breeding so near habitations. The grove was a common resort for man and beast, besides being the place where wagons, tools, etc., were kept and repaired. * * *

"From my observations I conclude that the Red-billed Pigeon breeds on our extreme southern border during April and May, that it builds a nest differing from those of other Pigeons, and lays but one egg. * * *

"The shape of my five eggs of the bird under consideration is oblong oval, with the greatest diameter in the center. Some vary slightly, tending sometimes to double pointed, and again to double rounded. The color is pure white. They measure 1.60 by 1.10, 1.55 by 1.12, 1.60 by 1.08, 1.48 by 1.08, and 1.46 by 1.07 inches, averaging 1.54 by 1.09 inches" (equaling 40.6 by 27.9, 39.4 by 28.4, 40.6 by 27.4, 37.6 by 27.4, and 37.1 by 27.7 millimetres, averaging 39.1 by 27.9 millimetres).[1]

In his "Further Notes on the Ornithology of the Lower Rio Grande of Texas," Mr. George B. Sennett gives the following information regarding this species: "Through the kindness of Dr. S. M. Finley, U. S. Army, who was stationed at camp near Hidalgo, and a good observer of animal life, I obtained valuable information concerning the arrival and departure of the Pigeons. In answer to my inquiries in regard to this species, he gave the following from his note book for 1878: 'First noticed on January 24, in flocks; about the middle of February they were seen in the woods in pairs, and cooing. The last seen of them, in 1887, was the latter part of November. These Pigeons were seen several times consorting with tame Pigeons in the ebony trees in the neighborhood of the village of Hidalgo.' This bird is resident, therefore, on the Lower Rio Grande about ten months in the year. The remaining two months it is probably in the more central part of our continent, wandering in flocks from place to place in search of food.

---

[1] Geological and Geographical Survey, Hayden, 1878, Vol IV, No. 1, pp. 45, 46.

"On April 9, the day after my arrival at Lomita, I went to Hidalgo to make arrangements about mail, supplies, etc. About a mile above the village, on familiar collecting ground of the season before, I discovered a Red-billed Pigeon on her nest in a thicket, and about 8 feet from the ground. It was not until I had approached to within arm's length that she arose and, tumbling heavily into the bushes, fluttered away over the ground in capital feint of injury, in order to attract attention away from the nest. The nest, made of twigs, was frail, saucer-shaped, and contained a single nearly fledged young. This bird breeds irregularly, and lays several times in a season. I found nests, during the whole time of my stay, containing eggs and young in all stages of development, but in no case did a nest contain more eggs or young than a single one. The parents are fond and affectionate, and both assist in incubation. Their food, when I saw them, was chiefly the hackberry fruit.

"The young from the egg have the upper parts plumbeous and sparsely covered with dark hair-like feathers. Under parts are pale and naked. The half-grown young have plumage on the body like the adult. Head and flanks do not become feathered until bird is nearly fledged, and in the half-grown young, just commences to show. From a large series of eggs I find them to average 1.55 by 1.10 inches (equal to 39.4 by 27.9 millimetres), the length varying from 1.60 to 1.45 inches (equal to 40.6 to 36.8 millimetres), and the breadth from 1.18 to 1.03 inches (equal to 45.7 to 26.2 millimetres)."[1]

Asst. Surg. James C. Merrill, U. S. Army, writes me as follows: "This handsome and large Pigeon is found in abundance during the summer months, arriving in flocks of fifteen or twenty about the last week in February. Though not very uncommon about Fort Brown, it is much more plentiful a few miles higher up the river, where the dense woods offer it the shade and retirement it seeks. Three nests, found in a grove of ash trees on the banks of the Rio Grande, near camp at Hidalgo, were frail platforms of twigs, such as are usually built by other Pigeons. Each contained one egg. These are of a pearly whiteness, and average 1.50 by 1.08 inches (equal to 38.1 by 27.4 millimetres). Both sexes incubate.[2] * * *

Capt. William L. Carpenter, Ninth Infantry, U. S. Army, writes me: "While at Fort Grant, Arizona, July 25, 1886, three specimens of this species were shown me, which had been shot that day near the post, in the foothills of the Graham Mountains, which would indicate that these birds breed in Arizona. Graham Mountain is distant about 90 miles from the Mexican border."

I can add nothing new to these accounts; the length of incubation is probably the same as that of the Band-tailed Pigeon and a single egg seems to constitute a set.

There are only four eggs of this species in the U. S. National Museum collection; the shape of two of these may be called elliptical oval, the remaining two, including the type specimen, are slightly pointed at the smaller end. Their color is pure white with but little gloss.

The type specimen (No. 20831, U. S. National Museum collection, Pl. 4, Fig. 2), was collected near Hidalgo, Texas, April 26, 1878, by Asst. Surg. James C. Merrill, U. S. Army. It measures 40 by 27.5 millimetres, another 39 by 27.4 millimetres, and the remaining two specimens, each measure 37 by 27.5 millimetres.

45. Columba leucocephala LINNÆUS.

WHITE-CROWNED PIGEON.

Columba leucocephala LINNÆUS, Systema Naturæ, ed. 10, I. 1758, 164.
(B 447, C 369, R 458, C 541, U 314.)

GEOGRAPHICAL RANGE: Southern Keys of Florida, Greater Antilles, Bahamas, also Santa Cruz, St. Bartholomew, and probably other islands of the Virgin Group and Little Antilles, also along the coast of Honduras.

The breeding range of the White-crowned Pigeon within the borders of the United States is confined to the southern Florida Keys.

Mr. W. E. D. Scott, in his paper on the "Birds of the Gulf Coast of Florida," refers to this species as follows: "A regular summer resident at Key West and vicinity, and Mr. Atkins also took it on one occasion at Punta Rassa (see Auk, Vol. v, p. 185). Though a regular summer resident, it seems not at all common at Key West, though quite abundant on neighboring keys in July, August, and September. Mr. Atkins says that they arrive at Key West from May 1 to 15, and remain till November."[1]

Dr. Henry Bryant, in his "List of Birds seen at the Bahamas in 1859," makes the following statement about this species: "This bird is a constant resident, though not frequently seen in winter, at which time it is much less gregarious in its habits than in spring and summer. The number is probably augmented, during and after the breeding season, by birds that have passed the winter further south. It breeds in communities, in some places, as at Grassy Kays, Andros Island, in vast numbers; here the nests were made on the tops of the prickly pear, which cover the whole kay; at the Biminis and Buena Vista Kay, Ragged Island, on the mangroves; and at Long Rock, near Exuma, on the stunted bushes. I do not think they ever select a large kay for their breeding place. The eggs are laid by the middle of May, and the young leave the nest about the 1st of July, previous to which great numbers are killed by the negroes. It is a shy bird when not breeding, even in the most uninhabited localities. Its food consists entirely of berries and fruits."[2]

According to Audubon, the White-headed Pigeon arrives on the southern keys of Florida about the 20th of April, sometimes not till the 1st of May. They are at all times exceedingly wary, probably on account of the war that

[1] Auk, VI, 1889, p. 246.
[2] Proceedings Boston Society Natural History, 1859, pp. 121, 122.

is incessantly waged against them, their flesh being very juicy and finely flavored. This shyness is only partially abated, even during the breeding season, as they will silently slide from their nest when sitting, if it is approached, and retreat to the dark shade of the mangroves, and do not return for an interval to their charge. They were more abundant in the more southern keys, except the sterile Tortugas.

Audubon found the nests placed high or low, according to circumstances, but never saw two on the same tree. He has met with them on the top of a cactus only a few feet from the ground, or on a low branch of a mangrove almost touching the water. They are said to resemble those of the common Passenger Pigeon, but are more compact and better lined, the outer part being composed of small dry twigs, the inner of fibrous roots and grasses. The eggs are two, of an opaque white, roundish, and as large as those of the common Pigeon. Audubon thinks that these birds may have several broods in a season.[1]

I am unable to add anything new to these accounts. The eggs of this species in the U. S. National Museum collection, obtained mostly from Cuba and the Bahamas, are more glossy, and the shells are much smoother, than in the eggs of other species of this family. They are pure white in color and elliptical oval in shape; some are nearly oval.

The average measurement of eleven specimens in the U. S. National Museum collection is 37 by 26.5 millimetres. The largest egg measures 39.5 by 26, the smallest 34.5 by 26 millimetres.

The type specimen (No. 556, Pl. 4, Fig. 4), from a set of two, was collected on Indian Key, Florida, in the spring of 1859, by Mr. G. Würdemann.

46. Ectopistes migratorius (LINNÆUS).

PASSENGER PIGEON.

Columba migratoria LINNÆUS, Systema Naturæ, ed. 12, I, 1766, 285 (♂).
Ectopistes migratoria SWAINSON, Zoölogical Journal, III, 1827, 362.
(B 448, C 370, R 459, C 543, U 315.)

GEOGRAPHICAL RANGE: Deciduous forest regions of eastern North America; west, casually, to Washington and Nevada; Cuba.

The breeding range of the Passenger Pigeon to-day is to be looked for principally in the thinly settled and wooded region along our northern border, from northern Maine westward to northern Minnesota; in the Dakotas, as well as in similar localities in the eastern and middle portions of the Dominion of Canada, and north at least to Hudson Bay. Isolated and scattering pairs probably still breed in the New England States, northern New York, Pennsylvania, Michigan, Wisconsin, Minnesota, and a few other localities further south, but the enormous breeding colonies, or pigeon roosts, as they were formerly called, frequently covering the forest for miles, and so often mentioned by naturalists

[1] Birds of North America, 1874, Vol. III, pp. 364, 365.

and hunters in former years, are, like the immense herds of the American bison which roamed over the great plains of the West in countless thousands but a couple of decades ago, things of the past, probably never to be seen again.

In fact, the extermination of the Passenger Pigeon has progressed so rapidly during the past twenty years that it looks now as if their total extermination might be accomplished within the present century. The only thing which retards their complete extinction is that it no longer pays to net these birds, they being too scarce for this now, at least in the more settled portions of the country, and also, perhaps, that from constant and unremitting persecution on their breeding grounds they have changed their habits somewhat, the majority no longer breeding in colonies, but scattering over the country and breeding in isolated pairs.

Mr. William Brewster, in his article "On the Present Status of the Wild Pigeon," etc., writes as follows: "In the spring of 1888 my friend, Captain Bendire, wrote me that he had received news from a correspondent in central Michigan to the effect that Wild Pigeons had arrived there in great numbers and were preparing to nest. Acting on this information, I started at once, in company with Mr. Jonathan Dwight, jr., to visit the expected 'nesting' and learn as much as possible about the habits of the breeding birds, as well as to secure specimens of their skins and eggs.

"On reaching Cadillac, Michigan, May 8, we found that large flocks of Pigeons had passed there late in April, while there were reports of similar flights from almost every county in the southern part of the State. Although most of the birds had passed on before our arrival, the professional Pigeon netters, confident that they would finally breed somewhere in the southern peninsula, were busily engaged getting their nets and other apparatus in order for an extensive campaign against the poor birds.

"We were assured that as soon as the breeding colony became established the fact would be known all over the State, and there would be no difficulty in ascertaining its precise location. Accordingly, we waited at Cadillac about two weeks, during which time we were in correspondence with netters in different parts of the region. No news came, however, and one by one the netters lost heart, until finally most of them agreed that the Pigeons had gone to the far north, beyond the reach of mail and telegraphic communication. As a last hope, we went, on May 15, to Oden, in the northern part of the southern peninsula, about 20 miles south of the Straits of Mackinac. Here we found that there had been, as elsewhere in Michigan, a heavy flight of birds in the latter part of April, but that all passed on. Thus our trip proved a failure as far as actually seeing a Pigeon 'nesting' was concerned; but partly by observation, partly by talking with the netters, farmers, sportsmen, and lumbermen, we obtained much information regarding the flight of 1888, and the larger nestings that have occurred in Michigan within the past decade, as well as many interesting details, some of which appear to be new about the habits of the birds.

"Our principal informant was Mr. S. S. Stevens, of Cadillac, a veteran Pigeon netter of large experience, and, as we were assured by everyone whom we asked concerning him, a man of high reputation for veracity and carefulness of statement. His testimony was as follows: 'Pigeons appeared that year in numbers near Cadillac, about the 20th of April. He saw fully sixty in one day, scattered about in beech woods near the head of Clam Lake, and on another occasion about one hundred drinking at the mouth of the brook, while a flock that covered at least 8 acres was observed by a friend, a perfectly reliable man, flying in a northeastly direction. Many other smaller flocks were reported.'

"The last nesting of any importance in Michigan was in 1881, a few miles west of Grand Traverse. It was only of moderate size, perhaps 8 miles long. Subsequently, in 1886, Mr. Stevens found about fifty dozen pairs nesting in a swamp near Lake City. He does not doubt that similar small colonies occur every year, besides scattered pairs. In fact, he sees a few Pigeons about Cadillac every summer, and in the early autumn young birds, barely able to fly, are often met with singly or in small parties in the woods. Such stragglers attract little attention, and no one attempts to net them, although many are shot.

"The largest nesting he ever visited was in 1876 or 1877. It began near Petosky, and extended northeast past Crooked Lake for 28 miles, averaging 3 or 4 miles wide. The birds arrived in two separate bodies, one directly from the south by land, the other following the east coast of Wisconsin, and crossing at Manitou Island. He saw the latter body come in from the lake at about 3 o'clock in the afternoon. It was a compact mass of Pigeons, at least 5 miles long by 1 mile wide. The birds began building when the snow was 12 inches deep in the woods, although the fields were bare at the time. So rapidly did the colony extend its boundaries that it soon passed literally over and around the place where he was netting, although when he began, this point was several miles from the nearest nest. Nestings usually start in deciduous woods, but during their progress the Pigeons do not skip any kind of trees they encounter. The Petosky nesting extended 8 miles through hardwood timber, then crossed a river bottom wooded with arborvitæ, and thence stretched through white pine woods about 20 miles. For the entire distance of 28 miles every tree of any size had more or less nests, and many trees were filled with them. None were lower than about 15 feet above the ground.

"Pigeons are very noisy when building. They make a sound resembling the croaking of wood frogs. Their combined clamor can be heard 4 or 5 miles away when the atmospheric conditions are favorable. Two eggs are usually laid, but many nests contain only one. Both birds incubate, the females between 2 o'clock p. m. and 9 or 10 o'clock the next morning; the males from 9 or 10 o'clock a. m. to 2 o'clock p. m. The males feed twice each day, namely, from daylight to about 8 o'clock a. m., and again late in the afternoon. The females feed only during the forenoon. The change is made with great regularity as to time, all the males being on the nest by 10 o'clock a. m.

"During the morning and evening no females are ever caught by the netters; during the forenoon no males. The sitting bird does not leave the nest until the bill of its incoming mate nearly touches its tail, the former slipping off as the latter takes its place.

"Thus the eggs are constantly covered, and but few are ever thrown out despite the fragile character of the nests and the swaying of the trees in high winds. The old birds never feed in or near the nesting, leaving all the beechmast, etc., there for their young. Many of them go 100 miles each day for food. Mr. Stevens is satisfied that Pigeons continue laying and hatching during the entire summer. They do not, however, use the same nesting place a second time in one season, the entire colony always moving from 20 to 100 miles after the appearance of each brood of young. Mr. Stevens, as well as many of the other netters with whom we talked believes that they breed during their absence in the South in the winter, asserting as proof of this that young birds in considerable numbers often accompany the earlier spring flights. * * *

"Five weeks are consumed by a single nesting. Then the young are forced out of their nests by the old birds. Mr. Stevens has twice seen this done. One of the Pigeons, usually the male, pushes the young off the nest by force. The latter struggles and squeals precisely like a tame squab, but is finally crowded out along the branch, and after further feeble resistance flutters down to the ground. Three or four days elapse before it is able to fly well. Upon leaving the nest it is often fatter and heavier than the old birds; but it quickly becomes much thinner and lighter, despite the enormous quantity of food it consumes.

"On one occasion an immense flock of young birds became bewildered in a fog while crossing Crooked Lake, and descending struck the water and perished by thousands. The shore for miles was covered a foot or more deep with them. The old birds rose above the fog, and none were killed.

"At least five hundred men were engaged in netting Pigeons during the great Petosky nesting of 1881. Mr. Stevens thought that they may have captured on the average 20,000 birds apiece during the season. Sometimes two carloads were shipped south on the railroad each day. Nevertheless he believed that not one bird in a thousand was taken. Hawks and Owls often abound near the nesting. Owls can be heard hooting there all night long. The Cooper's Hawk often catches the stool pigeon. During the Petosky season Mr. Stevens lost twelve stool birds in this way.

"There has been much dispute among writers and observers, beginning with Audubon and Wilson, and extending down to the present day, as to whether the Wild Pigeon lays two eggs or one. I questioned Mr. Stevens closely on this point. He assured me that he had frequently found two eggs or two young in the same nest, but that fully half the nests which he had examined contained only one.

"Our personal experience with the Pigeon in Michigan was as follows:

"During our stay at Cadillac we saw them daily, sometimes singly, usually in pairs, never more than two together. Nearly every large tract of old growth mixed woods seemed to contain at least one pair. They appeared to be settled for the season, and we were convinced that they were preparing to breed. In fact, the oviduct of a female, killed May 10, contained an egg nearly ready for the shell.

"At Oden we had a similar experience, although there were perhaps fewer Pigeons there than about Cadillac.

"On May 24, Mr. Dwight settled any possible question as to their breeding in scattered pairs, by finding a nest on which he distinctly saw a bird sitting. The following day I accompanied him to this nest, which was at least 50 feet above the ground, on the horizontal branch of a large hemlock, about 20 feet out from the trunk. As we approached the spot an adult male Pigeon started from a tree near that on which the nest was placed, and a moment later a young bird, with stub tail and barely able to fly, fluttered feebly after it. This young Pigeon was probably the bird seen the previous day on the nest, for on climbing to the latter, Mr. Dwight found it empty, but fouled with excrement, some of which was perfectly fresh. A thorough investigation of the surrounding woods, which were a hundred acres or more in extent, and composed chiefly of beeches, with a mixture of white pines and hemlocks of the largest size, convinced us that no other Pigeons were nesting in them.

"All the netters with whom we talked believe firmly that there are just as many Pigeons in the West as there ever were. They say the birds have been driven from Michigan and the adjoining States, partly by persecution, and partly by the destruction of the forests, and have retreated to uninhabited regions, perhaps north of the Great Lakes in British North America. Doubtless there is some truth in this theory; for, that the Pigeon is not, as has been asserted so often recently, on the verge of extinction, is shown by the flight which passed through Michigan in the spring of 1888. This flight, according to the testimony of many reliable observers, was a large one, and the birds must have formed a nesting of considerable extent in some region so remote that no news of its presence reached the ears of the vigilant netters. Thus it is probable that enough Pigeons are left to restock the West, provided that laws sufficiently stringent to give them fair protection be at once enacted. The present laws of Michigan and Wisconsin are simply worse than useless, for, while they prohibit disturbing the birds *within* the nesting, they allow unlimited netting only a few miles beyond its outskirts *during the entire breeding season*. The theory is, that they are so infinitely numerous that their ranks are not seriously thinned by catching a few millions of breeding birds in a summer, and that the only danger to be guarded against is that of frightening them away by the use of guns or nets in the woods where their nests are placed. The absurdity of such reasoning is self-evident, but, singularly enough, the netters, many of whom struck me as intelligent and honest men, seem really to believe in it. As they have more or less local influence, and, in addition, the powerful backing of the large game dealers in the cities, it is not likely that any really effectual laws can be passed until the last of our Passenger Pigeons are preparing to follow the Great Auk and the American bison."[1]

In order to show a little more clearly the immense destruction of the Passenger Pigeon *in a single year and at one roost* only, I quote the following extract from an interesting article "On the habits, methods of capture, and nesting of the Wild Pigeon," with an account of the Michigan nesting of 1878, by Prof. H. B. Roney, in the Chicago Field (Vol. x, pp. 345–347):

"The nesting area, situated near Petosky, covered something like 100,000 acres of land, and included not less than 150,000 acres within its limits, being in length about 40 miles by 3 to 10 in width. The number of dead birds sent by rail was estimated at 12,500 daily, or 1,500,000 for the summer, besides 80,352 live birds; an equal number was sent by water. We have," says the writer, "adding the thousands of dead and wounded ones not secured, and the myriads of squabs left dead in the nest, at the lowest possible estimate, a grand total of 1,000,000,000 Pigeons sacrificed to Mammon during the nesting of 1878."

The last-mentioned figure is undoubtedly far above the actual number killed during that or any other year, but even granting that but a million were killed at this roost, the slaughter is enormous enough, and it is not strange that the number of these Pigeons are now few compared with former years.

Capt. B. F. Goss, of Peewaukee, Wisconsin, writes me: "Ten years ago the Wild Pigeon bred in great roosts in the northern parts of Wisconsin, and it also bred singly in this vicinity; up to six or eight years ago they were plenty. The nest was a small rough platform of twigs, from 10 to 15 feet from the ground. I have often found two eggs in a nest, but one is by far the most common. These single nests have been thought by some accidental, but for years they bred in this manner all over the county, as plentifully as any of our birds. I also found them breeding singly in Iowa. These single nests have not attracted attention like the great roosts, but I think it is a common manner of building with this species."

Mr. Frank J. Thompson, in charge of the Zoölogical Gardens at Cincinnati, Ohio, gives the following account of the breeding of the Wild Pigeon in confinement: "During the spring of 1877, the society purchased three pairs of trapped birds, which were placed in one of the outer aviaries. Early in March, 1878, I noticed that they were mating, and procuring some twigs, I wove three rough platforms, and fastened them up in convenient places, at the same time throwing a further supply of building material on the floor. Within twenty-four hours two of the platforms were selected; the male carrying the material, whilst the female busied herself in placing it. A single egg was soon laid in each nest and incubation commenced. On March 16, there was quite a heavy fall of snow, and on the next morning I was unable to see the birds on their nests on account of the accumulation of the snow piled on the platforms around them. Within a couple of days it had all disappeared, and for the next four or five nights a self-registering thermometer, hanging in the aviary, marked from 14° to 19°. In spite of these drawbacks both of the eggs were hatched and the young ones reared. They have since continued to breed regularly, and now I have twenty birds, having lost several eggs from falling through their illy-contrived nests and one old male."[1]

The Passenger Pigeon has been found nesting in Wisconsin and Iowa during the first week in April, and as late as June 5 and 12 in Connecticut and Minnesota. Their food consists of beechnuts, acorns, wild cherries, and berries of various kinds, as well as different kinds of grain. They are said to be very fond of, and feed extensively on, angle worms, vast numbers of which frequently come to the surface after heavy rains, also on hairless caterpillars.

Their movements, at all seasons, seem to be very irregular, and are greatly affected by the food supply. They may be exceedingly common at one point one year, and almost entirely wanting the next. They generally winter south of latitude 36°.

Their notes during the mating season are said to be a short "coo-coo," and the ordinary call note is a "kee-kee-kee," the first syllable being louder and the last fainter than the middle one.

Opinions differ as to the number of broods in a season; while the majority of observers assert that but one, a few others say that two, are usually raised. The eggs vary in number from one to two in a set, and incubation lasts from eighteen to twenty days, both sexes assisting. These eggs are pure white in color, slightly glossy, and usually elliptical oval in shape; some may be called broad elliptical oval.

The average measurements of twenty specimens in the U. S. National Museum collection is 37.5 by 26.5 millimetres. The largest egg measures 39.5 by 28.5, the smallest 33.5 by 26 millimetres.

The type specimen figured (No. 18544, Pl. 4, Fig. 6), was taken from the oviduct of a bird, in May, 1882, by Messrs. H. T. Phillips & Co., Detroit, Michigan.

[1] Bulletin Nuttall Ornithological Club, Vol. VI, 1881, p. 122.

47. Zenaidura macroura (Linnæus).

MOURNING DOVE.

Columba macroura Linnæus, Systema Naturæ, ed. 10, 1758, 164 (part).
Zenaidura macroura, Ridgway, Proceedings U. S. National Museum, viii, 1885, 355.
(B 451, C 371, R 460, C 544, U 316).

Geographical range: Whole of temperate North America; north to Canada, southern Maine and British Columbia; south to Panama and West Indies.

The breeding range of the Mourning Dove, also called the "Carolina" and "Turtle" Dove, extends over the entire United States, from Florida and the Gulf of Mexico to the Canadian border, and from the Atlantic seaboard to the Pacific Ocean.

It is only a summer visitor in the northern portions of its range, and occurs sparingly during this season in southern Canada, and more frequently in Manitoba. On the Pacific coast it is not uncommon on Vancouver Island, and it extends well into the interior of British Columbia. I have personally met with it in considerable numbers, both in northern Washington and northern Idaho.

It winters, and is a constant resident throughout the year, in nearly all the Southern States. Mr. W. W. Cooke states: "From latitude 36° southward, this Dove can be found regularly and abundantly throughout the year. Between latitude 36° and 38° it is a regular but not an abundant winter resident, occurring in flocks. North of latitude 38°, although many are found each winter, they are merely single birds that have found exceptionally favorable quarters."[1]

The Mourning Dove is a well-known bird, tame and gentle when not molested, frequently breeding in the gardens and shrubbery near dwelling houses, and often feeding in the barnyard amongst domestic fowls. They never occur in such large flocks as the Passenger Pigeon, but are usually found in small parties of from six to a dozen or more, and in the late fall, previous to their migration, I have often flushed fifty and upward from a favorite feeding ground. They like to alight in roads, where they may often be seen traveling along in search of suitable food or gravel, or for the purpose of taking a dust bath, of which they are very fond.

In the more arid portions of the West, especially in southern Arizona, I have often noticed this Dove a long way from water, but as they are exceedingly strong and rapid flyers, distance is but a trifling matter to them. They usually visit their regular watering places in the morning, and in the evening just before sundown, and, where water holes or springs are scarce, they can be seen coming from all directions in search of such localities, usually in pairs or little parties of from four to six. At this time, if closely watched, they are a sure guide to water; old mountaineers are well aware

---

[1] Bulletin 11, Department of Agriculture, Report on Bird Migration in the Mississippi Valley, p. 109.

of this fact, and, if not familiar with the country, they shape their course after the line of travel of these Doves, which is always a direct and straight route to the objective point.

The mating season begins early in March in the southern portions of their range, and later northward. Fresh eggs have been found by Mr. George E. Beyer, near New Orleans, Louisiana, on March 30; and I have taken perfectly fresh specimens as late as September 14, and might probably have found them still later had I looked for them at my camp on Rillitto Creek, near Tucson, Arizona.

Their peculiar love notes are well known to most observers, and can be frequently heard in the early spring wherever these birds are found, being a sound of three syllables in the beginning, like "cŏŏ, cŏŏ, rŏŏ," and ending with four and five, with slight variations like "Mŏŏ, ŏă, ŏŏŏ, ŏŏŏ, ŏŏŏ." These notes are guttural and difficult to describe exactly. It is a low, mournful, but nevertheless far-penetrating note, though the actions of the male while uttering them seem cheerful and lively enough. He pays devoted court to his mate at all times, and I am inclined to the belief that many remain paired throughout the year, as single pairs may be seen in winter as well as summer.

The nesting season begins about the middle of March in Florida and other Southern States, and from three to four weeks later in the more northern portions of its range.

The nesting sites of this species are exceedingly variable. Usually the slender and frail platform of twigs, which answers the purpose of a nest, is placed on some flat spreading limb of a tree, at a height of from 10 to 20 feet from the ground, no especial preference being shown for any particular kind, unless perhaps, the evergreens, such as small cedars, junipers, and pines. Various sorts of bushes are also used; in fact, nests may be found at a height of only a few inches from the ground and again 50 and more feet up. In the Carolinas, Dr. Elliott Coues states, in his "Birds of the Northwest," that they nest chiefly on the ground. Personally I have more than once found the eggs of this species lying on the bare ground under the shelter of some little bush, and usually close to a creek, both in Washington and Idaho, but the majority of these birds nested there in willow thickets. At Fort Custer, Montana, I found their nests on flat sticks of cordwood in woodpiles; and in Arizona they often lay on a cottonwood stump. I found a nest of the Mourning Dove on top of an abandoned and broken down Magpie's nest, near Camp Harney, Oregon; in fact, they use all sorts of strange sites, flat tops of bowlders, ledges of cliffs, old nests of other species on top of which a slight platform of sticks is built, etc. In the more open prairie regions of the Western States they probably breed as frequently on the ground as above it, but in Arizona, although I noticed many of their nests, none were found on the ground, no doubt due to the many reptiles inhabiting that country. Here they usually selected the thick shrubbery found along water courses, and again, in some localities they were quite partial to the thorny mesquite and catsclaw bushes, and even to the ever-present cholla cactus.

Strange and unusual nesting sites are rather common with this species, and I will mention a few of them:

Mr. Lynds Jones found a Mourning Dove's nest on top of a straw shed, and another beneath a shed roof. Mr. R. B. McLaughlin reports one found in an old nest of a Green Heron, and another in a cavity of a tree. He writes me as follows: "The excavation had been made by Woodpeckers, and did not extend downward for more than an inch or so, but had been dug almost straight in. This made a nice foundation upon which the nest was built. The site was all that was peculiar about it."

This Dove will occasionally lay its eggs in nests occupied by other species, and one of the strangest combinations of this kind is thus described in Forest and Stream, September, 1889, by Mr. J. L. Davidson, in his "Birds of Niagara County, New York;" he says: "June 17, 1882, I found a Black-billed Cuckoo and a Mourning Dove sitting together on a Robin's nest. The Cuckoo was the first to leave the nest. On securing the nest I found it contained two eggs of the Cuckoo, two of the Mourning Dove, and one Robin's egg. The Robin had not quite finished the nest when the Cuckoo took possession of it and filled it nearly full of rootlets, but the Robin got in and laid one egg. Incubation had commenced in the Robin's and Cuckoo's eggs, but not in those of the Mourning Dove. I have the nest and eggs in my collection. This was first published under the head of 'A strange story' (Forest and Stream, August 24, 1882, p. 65)."

Mr. Whitmer Stone writes me: "In Lancaster County, Pennsylvania, where this species breeds extensively in orchards, I have found them building a thin platform of sticks on top of old nests of the American Robin, *Merula migratoria*, and the King Bird, *Tyrannus tyrannus*, and once in August I found a Dove sitting on two eggs, which were deposited in a Robin's nest from which the young birds had departed a month before. Fragments of the shells of the Robin's eggs were still to be seen in the bottom of the nest, and the Doves did not seem to have done anything at all in the way of building."

That they are hardy birds is well attested by Mr. J. W. Preston, of Baxter, Iowa, who writes me as follows: "A few remain here during winter, and are seen about stockyards, where they come from the woods to feed. Five are with me this winter, 1890, and are doing well, though they were a pitiful sight when the snow was 16 inches deep, and the mercury fell 20° below zero. Our Doves have a habit of repairing to some dry ditch to roost. Hundreds will flock to some chosen rendezvous, and they are very regular in their movements. This seems to occur in the more sparsely timbered regions, in the autumn."

Their food consists of small seeds, the various kinds of grain, berries, beechnuts, small acorns, wild pease, and the tender tops of plants, worms and insects of different kinds. Mr. C. S. Brimley, of Raleigh, North Carolina, accuses this species of pulling up the sprouting corn; and he says that it also does great damage to the early pease in the trucking districts. In Florida, according to Dr. William L. Ralph, they are very partial to feeding on grounds recently burnt over.

Incubation, in which both sexes assist, is said to last about two weeks. Two, and possibly three, broods are reared in a season. The nests are frail affairs, simply slight platforms of small sticks or twigs, very shallow, sometimes lined with bits of dry grass, a few leaves, Spanish moss, or pine needles, whichever is most convenient to the nesting site, and sometimes they are without any lining whatever, so that the eggs can readily be seen through the bottom of the nest. The nest, frail as it may often be, requires, nevertheless, considerable labor to construct, as this Dove, like most of the members of this family, is a poor nest builder; and, as far as my observations go, this seems to be done entirely by the female, the male looking on and cooing most of the time, but not assisting its mate in any way. On the ground there is even less of an attempt at making a nest, and I have seen the eggs lying in a slight hollow on the bare ground, not even surrounded by a few sticks or grass.

Mr. R. B. McLaughlin says: "Like other species which never lay more than two eggs, the Mourning Dove skips one day in the laying of its eggs; that is, if it lays the first one to-day it will not lay to-morrow, but the day after the second egg will be deposited."

The eggs are usually two in number, sometimes only one. Occasionally three, and even four, have been found in a nest, presumably laid by different birds. Mr. F. Stephens found a nest containing three eggs, on April 13, 1879. In two of these eggs incubation had commenced, the third was fresh.

Mr. Lynds Jones found a set of four near Grinnell, Iowa, and thinks they were all laid by one bird. He further states: "These Doves are very loving all through the year, and both parents are very attentive to their young, even long after they leave the nest. I have often found the female covering fully fledged young, always sitting crosswise of them. The young nestlings are fed on cutworms and other worms, as well as bugs; later they feed on grain and small seeds. If the sitting bird be flushed, she will tumble from the nest with piteous cries and in a very dilapidated condition. The male is always near by. The young grow very rapidly, and leave the nest early. In this locality nests on the prairie contain eggs fully two weeks earlier than nests in the woods."

The eggs of the Mourning Dove vary considerably in shape as well as in size. The majority may be called elliptical oval, others are elliptical ovate, and a few oval. They are pure white in color; the shell is smooth and moderately glossy. The average measurement of seventy-nine specimens in the U. S. National Museum collection, is 28 by 21 millimetres. The largest egg in this series measures 30.5 by 22.5, the smallest 25 by 19.5 millimetres.

Of the type specimens, No. 20818 (Pl. 4, Fig. 8), from the Merrill collection, was taken by Asst. Surg. James C. Merrill, U. S. Army, near Fort Brown, Texas, on June 8, 1876; and No. 20796 (Pl. 4, Fig. 9), from the Bendire collection, was taken May 26, 1872, by the writer, near Tucson, Arizona.

48. Zenaida zenaida (Bonaparte).

ZENAIDA DOVE.

Columba zenaida Bonaparte, Journal Academy Natural Sciences, Phila., v, 1825, 30.
Zenaida zenaida Ridgway, Proceedings U. S. National Museum, viii, 1885, 355.
(B 449, C 372, R 462, C 545, U 317.)

GEOGRAPHICAL RANGE: Florida Keys, Bahamas, Cuba, Jamaica, Porto Rico, Santa Cruz, Sombrero, and coast of Yucatan.

The breeding range of the Zenaida Dove within our borders is restricted to the southern Florida Keys, where it seems to be an irregular visitor. According to Audubon "they made their appearance among the islands at Indian Key about the 15th of April, where they remained until October, returning then to the West India Islands whence they came and where they are most numerous.

"They breed on the few keys that are covered with grass and low shrubs. They always place their nest on the ground, often with so little concealment that it may be easily discovered by any one searching for it. Occasionally it is placed between tufts of grass, the tops of which bend over and conceal it. A small hole is scooped in the sand, in which a slight nest, composed of matted blades of dry grasses, is placed, circular in form, and embedded in an outer collection of dry leaves and twigs. The whole fabric is said to be more compact than the nest of any other Pigeon."[1]

Mr. Charles B. Cory, in "Birds of the Bahama Islands" (p. 139), says: "This beautiful Dove is found throughout the Bahamas, but does not appear to be very abundant. It seems to be rather solitary in its habits, and is never met with in flocks. The nest is composed of small sticks loosely put together. On May 27 I procured a nest which was placed in the crotch of a fallen tree about 3 feet from the ground. It contained two white eggs."

Dr. Henry Bryant, in the "Proceedings of the Boston Society of Natural History, for 1859," (p. 120), makes the following statement about this species: "The Zenaida Dove, though more seldom seen in the Bahama Islands than the White-crowned Pigeon, *Columba leucocephala*, is still by no means rare. It never collects in flocks, and does not breed in communities like the former. In its habits it is intermediate between the *Z. carolinensis* and the *C. passerina*. It feeds and passes the principal part of its time on the ground, and when flushed, flies off in a straight line, very much as the common Quail. The crops of those killed by me were filled with small seeds, about the size of a mustard seed, apparently all of the same kind. All the nests I saw were made in holes in the rocks, and consisted, as is always the case in this family, of but a few sticks."

From these accounts it will be seen that the Zenaida Dove differs considerably in its nesting habits in certain localities. According to Mr. W. T.

---

[1] Birds of North America, 1874, Vol. iii, pp. 379, 380.

March, who found this species breeding near the Salt Ponds, Spanishtown, Jamaica, in May, 1863, it nests indiscriminately on the ground or in trees, making a slight platform of sticks and twigs loosely put together. It breeds from April to the end of July, and probably raises two broods a season.

None of the more recent explorers seem to have met with this species on the Florida Keys, and if found there at all now, it must be considered as rather rare.

I can add nothing new to these accounts. The eggs are two in number, pure white in color, oval in shape, and more rounded than Pigeon eggs generally are.

The average measurement of eleven eggs of this species in the U. S. National Museum collection, is 31 by 23.5 millimetres. The largest measures 34 by 23, the smallest 29 by 22.5 millimetres. None of these eggs were collected within the limits of the United States.

The type specimen (No. 6140, Pl. 2, Fig. 21) was collected near Spanishtown, Jamaica, by Mr. W. T. March, in May, 1863.

49. Engyptila albifrons (BONAPARTE).

WHITE-FRONTED DOVE.

Leptoptila albifrons BONAPARTE, Conspectus Avium, II, December, 1854, 74.
Engyptila albifrons COUES, Bulletin Nuttall Ornithological Club, v, April, 1880, 100.
(B —, C —, R 463, C 542, U 318.)

GEOGRAPHICAL RANGE: Valley of the Lower Rio Grande in Texas, and southward to Mexico and Guatemala.

The breeding range of the White-fronted Dove within the limits of the United States, as at present known, is confined to the Lower Rio Grande Valley, Texas, where Mr. George C. Sennett obtained the first specimen found within our border, on April 18, 1877, in a tract of timber near the bank of the Rio Grande, a mile below Hidalgo, Texas. In his "Further Notes on the Ornithology of the Lower Rio Grande of Texas," he writes as follows about this species: "Dr. Finley reports the arrival of this Pigeon at the vicinity of Hidalgo and Lomita about the middle of February, its departure having taken place in November. Although it is less numerous than the Red-billed Pigeon, yet by its peculiar note is is easily distinguished from all other species, and can thus be readily obtained. We heard it daily. It is so much more retiring in its habits than other Pigeons, that, were it not for the peculiarity we mention, it would be met with very seldom. It frequents the dense and heavy growth of timber, and long and frequent were our endeavors to find its nest."[1]

A nest was finally found by a Mexican assistant of Mr. Sennett's, in some dense shrubbery in the river bottom, containing two eggs. The parent was also shot at the same time. The two eggs obtained enabled him to identify two

---

[1] U. S. Geological and Geographical Survey, Hayden, Vol. v, No. 3, p. 424.

others that were brought in with a lot of White-winged Dove's eggs. The nest was situated in the forks of a bush about 5 feet from the ground, was flat and quite large for a Pigeon's nest, and composed of the dead branches, twigs, and bark of pithy weeds.

Dr. J. C. Merrill, assistant surgeon U. S. Army, while stationed at Fort Brown, Texas, met with this Pigeon there. He says: "It is not rare in the vicinity of Fort Brown, but is shy and not often seen. The only nest I have found was taken June 8, 1878, on the Government reservation. It was about 7 feet from the ground, supported by the dense interlacing tendrils of a hanging vine growing on the edge of a thicket." [1]

Mr. William Lloyd writes me that this Pigeon breeds abundantly in the Sierra Madre, from southern Chihuahua to Beltran, Jalisco, Mexico, at an altitude of from 1,100 to 2,200 feet. The nests, usually placed in thorny shrubs. *Huisache, Acacia farnesiana* 10 to 12 feet from the ground, are substantially made of straw. He found eggs as early as May 10, and up to June 13, when they were much incubated. It frequents deep arroyas mostly during the breeding season. They are only summer residents in northern Mexico, migrate in pairs, and feed principally on fruits. The eggs of this species are elliptical oval in shape; their color is cream-buff, and the shell is smooth and glossy.

The only specimens in the U. S. National Museum collection, are the two taken by Dr. James C. Merrill, U. S. Army, near Fort Brown, Texas, on June 8, 1878, referred to previously. These measure 31 by 23, and 30 by 22.5 millimetres. One of these, No. 20830, is figured on Pl. 2, Fig. 22.

50. Melopelia leucoptera (LINNÆUS).

WHITE-WINGED DOVE.

Columba leucoptera LINNÆUS, Systema Naturæ, ed. 10, I. 1758, 164.
Melopelia leucoptera BONAPARTE, Conspectus Avium, II, December, 1854, 81.
(B 450, C 373, R 464, C 546, U 319.)

GEOGRAPHICAL RANGE: Mexico, south to Costa Rica; north to southern border of United States (Florida and Texas to Arizona); Lower California, Cuba, Jamaica.

Within the borders of the United States the breeding range of the White-winged Dove is confined to the southwestern parts of our domain, including Arizona and New Mexico, as well as western and southwestern Texas, as far as known at present. Stragglers have also been observed in southern Colorado and at Key West, Florida, and it probably occurs at other points of the Gulf coast, and may possibly breed there also. It is one of the most common birds in Southern Arizona, and found there at altitudes up to about 4,000 feet, seldom higher.

According to my observations, it is equally as abundant in the foothills of the mountains as in the lower and hot valleys of the San Pedro, the Santa Cruz, the Gila, Verde, and Salt Rivers. It is partial to certain localities,

---

[1] Proceedings U. S. National Museum, Vol. 1, 1878, p. 158.

and in such they may be quite common, while in others, apparently as well suited, they are, for some reason comparatively rare, or entirely wanting. I found the White-winged Dove fairly common in 1872 in the foothills of the Santa Catalina Mountains and among the undergrowth bordering the Rillitto Creek, the present site of Fort Lowell, Arizona.

Both Mr. George B. Sennett, and Asst. Surg. James C. Merrill, U. S. Army, report this species as exceedingly abundant in the Lower Rio Grande Valley in Texas.

In the vicinity of Tucson, Arizona, it is partially resident throughout the year. I have observed specimens during every month of the winter of 1872 and 1873, but they are not as abundant then as in the summer. The conspicuous white wing-patch easily distinguishes this bird from any of the other Doves found in that region.

The mating season begins early, usually about the middle of March, and the cooing and love-making of the male can be heard and seen almost any day during a stroll among the shrubbery found along the borders of the water courses, which seem to be favorite resorts for many of these birds.

Their call notes are varied, much more so than those of any other species of this family found with us; they are sonorous, pleasing, and rather musical. On this account the natives keep many of them as cage birds, calling them *Paloma cantador*, Singing Dove. They soon become very gentle and reconciled to captivity, feeding readily out of one's hand and allowing themselves to be handled without fear.

One of their most characteristic call notes bears a close resemblance to the first efforts of a young Cockerel when attempting to crow, and this call is frequently uttered and in various keys. While thus engaged the performer usually throws his wings upward and forward above the head and also spreads his tail slightly. Some other notes may be translated into "cook for you," or "cook for two," "cook-kara-coo," besides a variety of calls, one of these a querulous harsh one, resembles somewhat the syllables "chüä-hüä."

Mr. Herbert Brown, of Tucson, Arizona, writes me: "There are but few of these Doves found in the immediate vicinity of Tucson, but they are numerous all over the country generally. They are not particularly partial to open mesas or bottom lands as a rule, but confine themselves largely to the rough and rocky foothills, covered with the sahuara cactus and palo verde bushes. I first met with this species at New River, on the Black Cañon road, about 135 miles north of this place, and I consider this point their northernmost limit in this Territory. There I met with them for the first time, fourteen years ago. What particularly drew my attention to them at the time was their call, which so much resembled the crow of a young rooster, that I remarked to my companion, 'We must be in the neighborhood of a ranch;' and it was only when I saw the bird in the act of doing the barnyard honors that I learned my mistake. I have seen many of their nests and eggs but never disturbed them. A few years ago I found about a dozen of their nests in almost as many minutes.

They were all placed in a little mesquite grove between a couple of volcanic hills west of town."

Nidification usually begins in the latter part of April in southern Arizona, and sometimes later. The nests of the White-winged Dove, like those of most of the members of this family, are as a rule rather frail structures, consisting of a slight platform of small sticks and twigs interlaced with each other, and lined more or less with bits of old weeds, stems of mesquite leaves, and dry grasses. Mr. G. B. Sennett found a nest of this species made of Spanish moss. No particular preference seems to be shown for certain trees. I have found many of my nests in mesquite trees, the most abundant in that locality; others were placed in walnut trees, willows, and cholla cactus, at various heights from the ground, from 2½ up to 30 feet.

I found my first nest on May 15, 1872. This was placed on a live mesquite stump from the top of which a number of green sprouts had grown out, the nest being only 2½ feet from the ground, the lowest situation in which I found any. It contained two fresh eggs. A second nest, found May 24, was placed on a horizontal limb of a mesquite tree, 6 feet from the ground. A third I found on a walnut tree close to the trunk, among some young sprouts. It was fully 20 feet from the ground. An occasional nest is placed in cacti, but the majority in trees.

As stated above I found, my first nest of this species on May 15, and my last, the fourteenth, on June 24. Although I searched carefully for others during the next three months, when both *Zenaidura macroura* and *Columbigallina passerina pallescens* were still laying, I failed to find a single nest of the White-winged Dove with eggs after that date.

Dr. J. C. Merrill, U. S. Army, took eggs of this species near Fort Brown, Texas, up to July 3.

I believe but one brood is usually raised in a season. As near as I can judge, incubation lasts about eighteen days. The male relieves the female somewhat in these duties, but does not assist to any great extent; he, however, assiduously helps to care for the young.

Their food consists of insects, small seeds, grain, if procurable, berries, mesquite beans, and the fruit of the sahuara cactus, *Cereus giganteus*, which seems to be a favorite article of food with many birds in Arizona.

In the late summer these birds collect in small parties, and I found them not at all shy. I have frequently seen as many as a dozen feeding among the cavalry horses along the picket line in my camp, allowing the men to walk within 10 feet of them without flying off.

The eggs are generally two in number, seldom one. They are mostly elliptical oval in shape, that is, equally rounded at each end; a few may be called oval. Their color is a rich creamy tint when fresh, and readily perceptible then, but in many specimens, especially in such as were considerably advanced in incubation when taken, this delicate tint fades, in time leaving the egg a dull dead white.

The average measurement of seventy-four specimens in the U. S. National Museum collection, is 30.5 by 23 millimetres. The largest egg measures 34 by 24, the smallest 28 by 19.5 millimetres. The type specimen (No. 20822, of the Merrill collection, from a set of two, Pl. 2, Fig. 23) was collected, near Edinburgh, Texas, by Asst. Surg. J. C. Merrill, U. S. Army, May 16, 1876.

51. Columbigallina passerina (LINNÆUS).

GROUND DOVE.

Columba passerina LINNÆUS, Systema Naturæ, ed. 10, I, 1758, 165.
Columbigallina passerina ZELEDON, Proceedings U. S. National Museum, VIII, 1885, 112.
(B 453, C 374, R 465, C 547. U 320.)

GEOGRAPHICAL RANGE: West India Islands; north to southern Atlantic and Gulf States excepting Texas (casually to District of Columbia).

The little Ground Dove breeds abundantly throughout the southern Atlantic and Gulf States, from South Carolina, south, through Georgia, Alabama, Florida, and southern Louisiana, being much more abundant in the immediate vicinity of the coast than in the interior.

Dr. William L. Ralph, who is familiar with this species, writes me as follows about it: "Few of the birds I am acquainted with have so much interested me as these little Doves. They are very tame, and seem to delight in frequenting the vicinity of houses, often building their nests within a few feet of the ground, near places where people are constantly passing. In Florida they are particularly fond of resorting to orange groves, both for nesting and feeding. In other localities they prefer fields in the vicinity of woods, with very little undergrowth.

"Their nests, which are usually larger in proportion and more compactly built than is common with birds of this family, are composed of twigs, weeds, dead leaves, pine needles, Spanish moss, etc., and lined with fine grass or some similar material. They are generally placed on a horizontal branch of a tree, preferably an orange tree, from 5 to 25 feet above the ground, most of them being under 10 feet. I have not infrequently found them in bushes or vines, and occasionally on the ground, but I believe the latter situation is not at all common.

"I have found these birds nesting at all times between the 1st of March and the 1st of June, but they do not begin breeding in any numbers before the middle of April. The earliest and latest dates of sets of my own collecting, now at hand, are March 21 and May 7. In all probability these birds nest several times during the season, but I have no personal knowledge of their breeding more than twice, as I never remained in Florida after the 1st of June.

"Both parents assist in incubation, and show a great deal of concern in the care of their homes. When one is driven from a nest containing eggs, it will drop to the ground as if shot, and will then flutter around as if wounded, to try to draw the person disturbing it away from the nest, but, whether it succeeds or not, it will soon fly off. When a nest contains young, however, the bird will become almost frantic with anxiety, and will tumble around until it appears to be nearly exhausted. I have often refrained from taking nests that I have wanted, on account of the evident distress of the parent birds.

"These Doves seem to dislike being alone, and go about in small flocks even during the breeding season. They are constant residents of Florida, and ten or fifteen years ago I found them abundant throughout the northern and central parts of the State. In the locality where these notes were taken they are still common, though fast decreasing in numbers, owing principally to the causes that are fast exterminating most Florida birds, viz, plume hunters and tourists.

"Their notes are very much like those of the Mourning Dove, and during the mating and breeding seasons, they will frequently sit for an hour or more at a time on the roof of some building, or in a tree, uttering their mournful calls. For birds so small they make a great deal of noise when they fly, but their wings do not produce a whistling sound like those of the Mourning Dove. Their food consists of seeds, grain, etc., and where they can get them, they will eat green pease."

Mr. T. D. Perry, of Savannah, Georgia, states: "The breeding season of the Ground Dove covers a long period, commencing, as it does, early in April and continuing through June and sometimes July. I have found their nests as early as April 9, with eggs slightly incubated, and as late as July 9, with fresh eggs. The nest is a very slight affair, consisting of a few twigs and pine needles, and very often grass and pine needles. The birds seem to have no regular place for their location. In one instance I found a nest on the ground, and in another case one on a stump. Generally, however, it is located in vines, or in a pine sapling or a myrtle bush. The height from the ground varies from 2 to 6 feet in most instances, though I once found a nest that was 10 feet up, and another that was 20 feet from the ground. They seem to prefer the pine lands, where the undergrowth of young saplings make the woods dense, and here their nests are generally found."[1]

Since Mr. Perry wrote the above he informs me that sets of eggs of this species have been found near Savannah, Georgia, as early as March 13 and 17, and Mr. Arthur T. Wayne records taking a set of eggs of the Ground Dove near Charleston, South Carolina, containing small embryos, on October 19, 1886, securing the male while incubating.[2]

Incubation is said to last two weeks, both parents assisting. Two broods, occasionally three, and, in exceptional cases, even four are raised in a season.

The eggs of the Ground Dove are usually two in number, pure white in color, and most commonly elliptical oval in shape; a few are oval, and some elliptical ovate (slightly pointed at one end).

---

[1] Ornithologist and Oölogist, Vol. XII, 1887, p. 102.   [2] Ibid., p. 7.

The average measurement of forty-three eggs of this species in the U. S National Museum collection, is 21.5 by 16.5 millimetres. The largest egg in the series measures 24 by 17.5, the smallest 19 by 14.5 millimetres. No specimen is figured, as they are indistinguishable from the egg of the subspecies following, of which one is represented.

52. Columbigallina passerina pallescens (BAIRD).

MEXICAN GROUND DOVE.

Columbigallina passerina pallescens RIDGWAY, Manual North American Birds, 1887, p. 586, Appendix.

(B 453, part; C 374, part; R 465, part; C 547, part; U 320a.)

GEOGRAPHICAL RANGE: Southwestern United States; Texas to Arizona and Lower California; south through Mexico (both coasts) to Central America.

The breeding range of the Mexican Ground Dove within our borders is confined to southwestern Texas and southern Arizona, and probably to southern New Mexico, although there are as yet no records of its breeding in the latter Territory, as far as I am aware. A few stragglers breed probably also in southern California, where it has been taken on several occasions. It is quite common in Lower California, where Mr. J. Xantus took its eggs near Cape St. Lucas, and Mr. L. Belding at San José del Carbo.

Dr. James C. Merrill, U. S. Army, found this subspecies abundant in the vicinity of Fort Brown, Texas, particularly during the summer. Several sets of eggs collected by him while stationed at this post are in the U. S. National Museum collection. He says: "The small and rather compact nests are placed on the horizontal branch of a stout bush or tree, and are lined with a few straws. On one occasion I found the eggs in a roughly made nest on the ground on the edge of a prairie."[1]

It was also noticed by Mr. G. B. Sennett at Hidalgo, Texas, and they breed all along the Lower Rio Grande.

Mr. William Lloyd writes me that he found the Mexican Ground Dove nesting between April 8 and June 1, sometimes under a bush or on an open hillside, and again on a limb or in a fork of a tree from 3 to 8 feet from the ground. These birds are often found in the streets of towns, and especially in corrals where horses and cattle are kept, ranging from sea level to an altitude of 9,000 feet, in Mexico. They are usually seen in flocks from ten to twelve, and feed on small grain and seeds. He gives their call note as "pas-cual, pas-cual, pas-cual."

According to Mr. Herbert Brown, the Mexican Ground Dove is common about Tucson. He says: "In the fall, when the weeds and grass seeds are ripe, I have seen as many as fifty of these birds in a flock, during the winter months but few are seen, and then only in pairs.

Proceedings U. S. National Museum, Vol. 1, 1878, p. 158.

"They lay two eggs, and nest in trees or bushes. A nest found June 11, 1887, was constructed of a few dead twigs and grass placed on a limb of a willow near the ground; the female was on the nest. One found June 19 was also in a willow tree 20 feet from the ground and out on a limb 15 feet from the body of the tree, and made of a few dried stalks of alfalfa. It contained two eggs and the female was on the nest. A third nest, found June 26, containing two eggs, was made of long stems of dry grass and placed about 10 feet from the ground. Whether this was a first laying, I cannot say. The nests are almost flat. I do not think I ever saw a cavity more than half an inch deep."

I found this subspecies a resident, during the entire year, at my camp on Rillitto Creek, near Tucson, in 1872 and 1873. In the winter they became exceedingly tame, and I have seen them frequently alight within a few feet of my tent, picking up bread crumbs which I threw to them. They are lovely and affectionate little birds, their call notes and cooing, however, have rather a doleful sound. My picket line was always a favorite resort for them, as well as for other species, particularly during the winter, when from six to ten Doves might be seen almost any time of the day feeding on the scattered grain left by the horses. I am inclined to believe that they remain paired for life.

Love-making begins about the first week in April, but although I searched carefully for their nests I failed to find any till May 30. From this date until the middle of September I examined a number. All of the nests seen by me were placed on bushes or on trees, from 3 to 21 feet off the ground, not a single one was found on the ground.

The first one found, on May 30, was placed in a syringa bush, about 3 feet from the ground. The little platform of small twigs and grass stems was very slight, about 4½ inches in diameter, and almost perfectly flat. The eggs were fresh.

Other nests, subsequently noticed, were placed in various trees and bushes, mostly in mesquite thickets, a few in willows, and two in walnut trees. A nest found July 28 was placed in a tree of this kind, about 20 feet from the ground. The tree was leaning, and some young sprouts had grown out from the main trunk, among which the nest was placed. The eggs were fresh, probably a second laying. All the nests examined by me were found in the creek bottoms or else close by, generally in clumps of mesquite bushes.

I found a set of eggs, slightly incubated, on September 14, and might have found them breeding still later, I presume, if I had looked for their nests.

Although the Mexican Ground Dove begins breeding rather late, I believe they usually rear two, if not three, broods a season. The young are fed on small seeds and berries of different kinds, as well as grain, whenever it is procurable. Gravel, in considerable quantities, is used by them to help in grinding up the various seeds they feed on.

The eggs are usually two in number, seldom but one. Incubation lasts about fourteen days, the male assisting in these duties. The eggs are pure white in color, elliptical oval in shape, with now and then one which may be slightly pointed, and a few that may be called oval.

The average measurement of fifty-four specimens in the U. S. National Museum collection, is 21.5 by 16.5 millimetres. The largest egg in the series measures 23 by 17.5, the smallest 20 by 16 millimetres.

The type specimen (No. 20804, Pl. 2, Fig. 24), from a set of two, from the Bendire collection, was taken by myself, near Tucson, Arizona, June 28, 1872

53. Scardafella inca (LESSON).

INCA DOVE.

Chamœpelia inca LESSON, Descriptive Quadrupeds, etc., Buffon, 1850, 211.
Scardafella inca BONAPARTE, Conspectus Avium, II, December, 1854, 85.
(B 452, C 375, R 466, C 549, U 321).

GEOGRAPHICAL RANGE: Southern border of United States from Rio Grande Valley of Texas to southern Arizona and Lower California; south to Mexico and Guatemala.

The breeding range of the Inca Dove within our border includes the Rio Grande Valley of Texas, from the vicinity of Laredo to southern Arizona, and its distribution seems to be somewhat irregular. It is also an occasional visitor in the interior of Texas (Austin).

None of the ornithologiststs who have explored the Lower Rio Grande Valley within the past few years, as well as other portions of Texas, appear to have met with this species. Dr. H. B. Butcher, however, seems to have found it abundant near Laredo, Texas, in 1866, and took its nest and eggs there on July 1 of that year. It was placed in the forks of a small mesquite tree.[1]

In southern Arizona it has been met with by different parties. I found it rather a rare resident near Tucson in 1872.

Mr. W. E. D. Scott writes as follows: "The only points where I have seen this species are Tucson and Florence, where it is, especially in the latter place, of common occurrence during the warmer portion of the year. The birds are very tame, and seem to affect particularly the streets, corrals, and gardens in the heart of the town."[2]

Mr. Herbert Brown informs me: "I have examined a number of the nests of the Inca Dove at different times. They are as a rule much better constructed than those of the Mexican Ground Dove. The cavity is about half an inch deep, and the materials used, fine dead twigs, are much more compactly put together than in the nests of the latter. On September 20, 1884, I found a nest of this species in an apricot tree about 10 feet from the ground. The nest contained two eggs, which I did not disturb."

Mr. Xanthus found this Dove breeding abundantly at Cape St. Lucas. One

---

[1] History of North American Birds, 1874, Vol. III, p. 588.   [2] Auk, Vol. III, 1886, p. 421.

nest was found in a leafless acacia about 6 feet from the ground; another, found
May 26, was about 5 feet high, in a small thorn bush; a third was at the height
of 8 feet and also placed in a bush; others were placed in small oaks, cacti,
opuntias, and in other situations, all above the ground, at heights varying from
5 to 8 feet.[1]

I first met with this species in February, 1872, and saw others from
time to time, but never abundantly. I am sure a few pairs bred within a
short distance of my camp near Tucson, but I did not find their nests. The
only one I obtained was taken in the valley of the Santa Cruz River, near
Tubac, Arizona, on June 6, 1872. It was placed in a thick mesquite bush,
about 4 feet from the ground, and contained two fresh eggs. The nest was
a slight platform of twigs and grasses about 5 inches in diameter. The
female was on the nest. The eggs show a slight creamy tint, due perhaps
to not being thoroughly cleaned at the time. A set collected by Colonel
Greyson, near Mazatlan, Mexico, are pure white in color.

This bird is easily distinguished from the Mexican Ground Dove by its
much longer tail; its habits are very similar, and it probably raises two
broods in a season.

The eggs, two in number, are scarcely distinguishable from those of the
Ground Dove; they are white in color, and elliptical oval in shape.

The average measurement of the four specimens in the U. S. National
Museum collection is 21 by 16.5 millimetres. The largest egg measures 22
by 17.5, the smallest 20 by 16 millimetres.

The type specimen (No. 20802 Pl. 2. Fig. 25), from the Bendire collection, was taken by the writer near Tubac, Arizona, June 6, 1872.

54. Geotrygon martinica (GMELIN).

KEY WEST QUAIL-DOVE.

Columba martinica GMELIN, Systema Naturæ, I, ii, 1788, 781.
Geotrygon martinica BONAPARTE, Conspectus Avium, II, December, 1854, 74.
(B 454, C 376, R 467, C 550, U 322.)

GEOGRAPHICAL RANGE: Haiti, Cuba, Bahamas, and Florida Keys.

Within the limits of the United States the breeding range of the Key
West Quail-Dove is confined to the island of Key West and the extreme
southern Florida Keys; it appears exceedingly rare now in places where
Audubon found them fairly abundant.

Dr. J. Gundlach, in his "Beiträge zur Ornithologie Cuba's" states: "It is
not rare. Its general habits resemble those of the Blue-headed Quail-Dove,
preferring rocky and wooded regions. I scarcely ever met them outside of the
forest. In the densely timbered portions it may be seen scratching among
the leaves for food. It likes to perch on horizontal limbs, especially on
limbs of the zarsas.

[1] History of North American Birds, 1874, Vol. III, p. 388.

"Its rather doleful call note resembles the syllables 'hu-up.' The nest, consisting of a slight platform of sticks, is usually placed cn the top crown of certain parasitic creepers, found in the more open but shady primitive forests. The eggs are two in number, of a pale ochre yellow color, and measure 31.5 by 24 millimetres. I found nests between the months of February and July."

Dr. Juan Vilaró, of the University of Habana, writes me: "The Key West Quail-Dove is quite common and a constant resident on the Island of Cuba. Its food consists of fruits, seeds, small snails, etc. It walks with neck contracted and the tail slightly raised. It perches on trees. It is commonly called *Boyero* (Ox Driver), on account of the resemblance of its call to the signal to stop, 'huup,' given by drivers to their oxen. *Barbequejo* (a kind of bridle) is another of its names, on account of the white stripe extending from the antero-inferior region of the eye almost to the occiput, and resembling the ornamental part of a bridle. In the central and eastern parts of the island it is called 'Torito' (Little Bull), its call notes sometimes resembling a roaring. The nest is placed in high trees as a rule, usually *Curujeyes*. It commences breeding in February and lays until July. The eggs are two in number, ochraceous white in color, and measure 31 by 24 millimetres."

Mr. J. W. Atkins, of Key West, informs me, under date of December 29, 1889: "I have taken but a single Key West Quail-Dove; I have seen several others, but have been unable to find their nests and eggs; and you may infer the rarity of this species here, when I tell you that many local sportsmen to whom I have showed the bird I killed had never seen one or knew that it occurred here."

There are no eggs of this species in the U. S. National Museum collection, and I have been unable to secure a specimen to figure.

55. Geotrygon montana (LINNÆUS).

RUDDY QUAIL-DOVE.

Columba montana LINNÆUS, Systema Naturæ, ed. 10, I, 1758, 163.
Geotrygon montana BONAPARTE, Conspectus Avium, I, 1850, 72.

(B —, C —, R —, C —, U 322.1.)

GEOGRAPHICAL RANGE: Tropical America in general (including the West Indies); north to Cuba, and eastern Mexico (Mirador); accidental at Key West, Florida.

The Ruddy Quail-Dove is entitled to a place in our avifauna, a single specimen having been taken at Key West, December 10, 1888 (Auk, VI, April, 1889, 160, 161, and July, 1889, 246). It is very doubtful if it breeds within our limits.

Dr. Jean Gundlach, in his "Neue Beiträge zur Ornithologie Cuba's," says: "The habits of this species are similar to those of *G. martinica*. It is not found on the cayos or small islands of the coast of Florida, but probably extends to the South American mainland. Nidification and the color of the eggs are the same as with those of the preceding species, *G. martinica*, only the latter are a trifle smaller, measuring 28.5 by 21.5 millimetres. I found a nest in August, containing two buff-colored eggs which were broken before I measured them. Its call notes are rather misleading, seeming near when distant, and again the reverse. They resemble the syllable 'hup' repeatedly and quickly given. It is not rare; breeds and inhabits the forests, and remains a good deal of the time on the ground searching for food among the dry leaves."

Mr. Philip Henry Gosse, in his "Birds of Jamaica," says: "The Mountain Partridge affects a well-wooded country, and is found in such woods as are more choked with bushes. * * * It is essentially a ground Pigeon, walking in couples or singly, seeking for seeds and gravel on the earth. * * * It is often seen beneath a pimento picking up the fallen berries; the physic nut, also, and other oily seeds afford it sustenance. I once observed a pair of these Doves eating the large seeds of a mango that had been crushed. With seeds, I have occasionally found small slugs, a species of *Vaginulus*, common in damp places, in its gizzard. * * *

"In the Short Cut of Paradise, where the sweet wood abounds, the Partridge is also numerous; in March and April, when these berries are ripe, their stomachs are filled with them. Here, at the same season, their cooing resounds, which is simply a very sad moan usually uttered on the ground, but on one occasion we heard it from the limb of a cotton tree at Cave, on which the bird was sitting with its head drawn in; it was shot in the very act. * * *

"One day in June I went down with a young friend into a wooded valley at Content to look at a Partridge nest. As we crept cautiously toward the spot, the male bird flew from it. I was surprised at its rudeness; it was nothing but a half dozen decayed leaves laid on one another, and on two or three dry twigs, but from the sitting of the birds it had acquired a slight hollowness, about as much as a skimmer. It was placed on the top (slightly sunk among the leaves) of a small bush not more than 3 feet high, whose glossy foliage and small white blossoms reminded me of a myrtle. There were two young recently hatched, callow and peculiarly helpless, their eyes closed, their bills large and misshapen; they bore little resemblance to birds. On another occasion I saw the male shot while sitting; the nest was then placed on a slender bush, about 5 feet from the ground. There were but two eggs, of a very pale buff color; sometimes, however, they are considerably darker."

Seven eggs of this species are in the U. S. National Museum collection, all taken in the West India Islands. They are oval in shape, and rather more rounded than is usually the case with Pigeon eggs. In color they vary considerably, from a pale cream to a salmon buff. The average measurement is 27 by 21 millimetres. The largest egg measures 29.5 by 22, the smallest 24 by 19.5 millimetres.

The type specimen (No. 17886, Pl. 2, Fig. 26), from a set of two, was collected at Sainte Marie, West Indies, April, 1877, by Mr. F. A. Ober. One of these eggs is still darker colored, and three are somewhat lighter tinted. Their shells are very smooth and glossy.

56. Starnœnas cyanocephala (LINNÆUS).

BLUE-HEADED QUAIL-DOVE.

Columba cyanocephala LINNÆUS, Systema Naturæ, ed. 10, 1, 1758, 163.
Starnœnas cyanocephala BONAPARTE, Geographical and Comprehensive List, 1838, 41.
(B 455, C 377, R 468, C 551, U 323.)

GEOGRAPHICAL RANGE: Florida Keys and Cuba.

The Blue-headed Quail-Dove is another rare visitor within our borders, and it is doubtful if it breeds with us. No specimens have been taken of late years on any of the Florida islands and smaller keys.

Dr. Jean Gundlach, in his "Beiträge zur Ornithologie Cuba's," writes about the Blue-headed Quail-Dove as follows: "This species is a constant resident of the Island of Cuba, and is likewise found in the extreme southerly portions of the United States, as well as on the Island of Jamaica, but does not seems to occur on the remaining West India Islands.

"It is not uncommon in the extensive forest, especially in such in which the ground is rocky, but is scarcely ever found in cultivated fields or open prairie country. It moves slowly, with the neck contracted and tail erected, while searching for food among the dead leaves on the ground. This consists of seeds of various kinds, berries, and occasionally small snails. After feeding, it usually flies into a tree and perches on a leafless horizontal limb, or on one of the numerous parasitic vines, to rest. In the early mornings, should its plumage, perchance, have become wet while traveling through the dew-laden shrubbery, it selects a sunny spot to dry itself. From time to time this Dove utters her call note, consisting of two hollow-sounding notes, 'hu-up,' the first syllable long drawn out, the second short and uttered very quickly. Besides this note a low muttering is occasionally heard. Their call notes are deceptive, appearing near when distant, and distant when close by. Its flight is noisy when starting, similar to that of the European Partridge, from which it receives its misleading name 'Perdiz.'"

It nests in April and May; the nest is a simple affair, consisting of a few twigs. It is usually placed in the tops of parasitic vines, *Tillandsia*. It lays two white eggs, measuring 35 by 25.5 millimetres.

Dr. Juan Vilaró, professor of the University of Habana, Cuba, writes me: "This Dove is constantly decreasing in numbers, being continually persecuted, notwithstanding it is protected at certain times by the hunting laws."

According to Dr. Bachman, of Charleston, South Carolina, the egg of this species is of a uniform, creamy-white color, and measures 1.43 inches in length by 1.10 in breadth (about 36.3 by 27.9 millimetres). The egg referred to was laid in confinement in his aviary.[1]

I consider the latter description as more probably the correct one. There are no specimens of the eggs in the U. S. National Museum collection.

[1] Birds of North America, 1874, Vol. III, p. 386.

BIRDS OF PREY.

Family CATHARTIDÆ. AMERICAN VULTURES.

57. Pseudogryphus californianus (SHAW).

CALIFORNIA VULTURE.

Vultur californianus SHAW, Naturalists' Miscellany, IV, 1797, Pl. ccci.
Pseudogryphus californianus RIDGWAY, History North American Birds, III, 1874, 338.
(B 2, C 364, R 453, C 536, U 334.)

GEOGRAPHICAL RANGE: Pacific coast region of the United States, from Oregon southward, to northern Lower California; (southern Utah?)

The breeding range of the California Vulture, as far as known, is restricted to the State of California and to a comparatively small portion thereof. To define it more definitely it may be located in that part of the State situated between the Sierra Nevada and the Pacific Ocean, between latitude 32° 30' and 38°. It is possible that a few of these birds breed in the mountains of Lower California also.

Mr. Walter E. Bryant in his "Catalogue of the Birds of Lower California," makes the following statement in reference to this species: "Mr. Anthony is the only one who has reported this species from the peninsula; he has observed them at several places from the sea level to an altitude of 11,000 feet. From the fact of their primary and secondary quills being prized by Mexican and Indian gold miners for use in carrying gold dust, an opportunity to kill a vulture is never allowed to pass unimproved."[1]

Col. N. S. Goss, of Topeka, Kansas, tells me that this Vulture is occasionally found on the Mexican side of our border, in Lower California, and is said to breed there in small numbers.

No other of our larger Raptores has such a restricted range. It has been reported as being seen as far north as the Columbia River, and even on Vancouver Island, by some of our earlier ornithologists, but I believe has not been met with there within recent years. Stragglers have been reported from southern Utah and the vicinity of Fort Yuma on the Colorado River.

This Vulture is readily recognized by its superior size and the conspicuous white area visible on the lower parts of its long and powerful wings when soaring through space. Its flight is graceful beyond comparison as it sails majestically overhead in gradually contracting or expanding circles, now gently falling with the wind and again rising easily against it, without

---

[1] Proceedings California Academy Sciences, 2d series, Vol. II, 1889, p. 278.

a perceptible motion of its pinions. While on the wing it looks more than the peer of any of our birds, the Golden Eagle not excepted.

Even when comparatively common during the years of 1866–1868, I do not remember ever noticing a single bird of this species on the eastern slopes of the Sierra Nevada. At that time I was stationed at Camp Independence, in Inyo County, California, while directly west of the mountains, not more than a hundred miles in a bee line, these birds were then moderately abundant on the great plains of the Tulare Valley; and I have seen assemblies of them numbering from six to fifteen on several occasions. However, they were never so plentiful at any time as its smaller relative, the Turkey Vulture. Why the range of this Vulture should be so restricted is hard to explain, but to this its rapid decrease is undoubtedly due, and poison has so far been the principal agent.

The home of the California Vulture is among the almost inaccessible cliffs of the minor mountain ranges running parallel to the Sierra Nevada. Stockraising has increased enormously in southern California during the past twenty years, and these fastnesses have been completely overrun by stockmen to find pasture for their flocks during the hot summers when everything is dried up in the valleys. Necessity compelled this invasion of the retreats of numerous predatory carnivora, like the grizzly bear, the panther, lynx, and the prairie wolf. These, as a matter of course, preyed on the calves and flocks of sheep that were to be found almost everywhere in the mountains at that time, to be had for the taking, and they naturally enough committed a great deal of damage.

The simplest and certainly the safest way for the stockmen to get rid of such undesirable neighbors was to bait them with poisoned carcasses. This means was resorted to almost everywhere, and generally with considerable success.

The Vultures, too, with their keen sight and scent, found many of these, to them, tempting baits, and being sociable in disposition many of these birds were destroyed by this means, so that by this time comparatively few are said to be left.

From recent information it appears, however, that within the past few years these birds have again commenced to hold their own, and in a few of the more barren and inaccessible mountain ranges in the vicinity of Santa Barbara they do not seem to be decreasing, and may in time regain their former numbers.

Mr. A. L. Parkhurst writes me to the same effect from Monterey, California. He says: "They are still abundant in the wild and rugged districts of this county, are extremely shy, perfectly capable of taking care of themselves, and breed mostly in inaccessible cliffs. One is rarely shot.

"They are retiring from the presence of civilization, as it demands the conversion of the large cattle ranges into small farms. I doubt if they are decreasing in numbers. I have seen over a dozen a day many times."

If, as Mr. Parkhurst states, these birds are still common in Monterey County, it is undoubtedly due to the breaking up of the large cattle ranches and their conversion into small farms. Poison, which has been resorted to on most of the larger stock ranches to kill the carnivora, has certainly almost exterminated the California Vulture as well, and in more than one locality, where they were formerly abundant, their very perceptible decrease is, in my opinion, mainly due to this cause.

Mr. William R. Flint writes me as follows about this species: "I am sorry that I cannot give you any reliable information as to the nesting habits of these birds. I feel confident, however, that they nested not far from my home in San Benito County. I have seen many of these birds there, and my brother claims to have seen some young ones among a group of six or seven old ones. The last time I was at home, in the spring of 1889, I searched for their nests, but failed to find any. I saw several of these birds about, and as a rule they associate with the Turkey Buzzards, and generally fly much higher, so that to an ordinary observer they appear nearly of the same size. The largest number I have ever seen at one time during late years was in the summer of 1884, when I saw fourteen together, and these allowed me to approach within 30 yards of them.

"I remember one Vulture which had its roosting place in an old, half-dead white oak, thickly surrounded by brush and smaller trees. Here he roosted regularly every evening, always appearing by sundown, and usually just before. He was a late riser, as I have seen him there more than once when the sun was two hours high. In this respect they are very much like the Turkey Vulture. I never saw any others roosting in or near this tree. They will often so gorge themselves with food as to be unable to rise from the ground and fly, and I have passed within a few yards of them in a buggy and they would make no effort to get away.

"A vaquero on the ranch lassoed one a few years ago with his riata when it was overgorged with food, and brought it to San Juan, where it was kept chained to one of the adobe pillars in the old mission church. It would show fight on slight provocation, and utter a hoarse, hissing sound when disturbed, flapping its wings violently at the same time.

"I have talked with two parties who claim to have seen nests of these birds. In one instance the nest was placed on a high rugged rock on a cliff, along one of the forks of the San Joaquin River, the other was discovered in a timber tract of the Santa Cruz Mountains, by an old wood-chopper. This nest was said to have been a huge affair, and was placed on the first large limb of a redwood tree, about 75 feet from the ground and close to the trunk. His camp being near this tree, he had a good chance to observe the birds, and says the young, of which there were two, were nearly three weeks learning to fly, after being large enough to leave the nest and crawl out on the limbs. When they finally left the nest they flew over the cañon to a rocky cliff on the opposite side, about a quarter of a mile distant.

"I think there is no foundation for the belief that this Vulture kills lambs and sickly calves. I have never known such a thing to happen, and have seen many Vultures during lambing time, when there were thousands of young lambs to eat, if they had felt disposed to kill any; neither have I noticed them eating dead flesh unless in an advanced state of decomposition."

Mr. Walter E. Bryant writes me as follows: "My experience has been that the California Vulture is extremely rare now. I doubt if I have ever seen it in its wild state. Many reports of its being common, its breeding, etc., which I have investigated at considerable trouble, time, and expense, have been either the Turkey Buzzard or the Golden Eagle seen on the wing."

About their nesting habits, nothing but what has already been published is known, and the egg remains one of the rarest in collections, and is likely so to continue. I have not been able to learn whether any have been taken during the past twelve years.

Dr. Heermann states that a nest of this bird, with young, was discovered in a thicket on the Tuolumne River. It was about 8 feet back from the entrance of a crevice in the rocks, completely surrounded and masked by thick underbrush and trees, and composed of a few loose sticks thrown negligently together He found two other nests of like construction and similarly situated, at the head of Merced River, and in the mountains. From the latter the Indians were in the habit of yearly robbing the young to kill at one of their festivals.

Mr. Alexander S. Taylor, of Monterey, published a series of papers in a California journal relative to this Vulture. In one of these he mentions that a Mexican ranchero, in hunting among the highest peaks of the Santa Lucia Range, disturbed two pairs from their nesting places, and brought away from one a young bird a few days old, and from the other an egg. There was no nest, the eggs having been laid in the hollow of a tall old robles oak, in a steep barranca, near the summit of one of the highest peaks. These birds are said by some hunters to make no nest, but simply to lay their eggs on the ground at the foot of old trees, or on the bare rocks of solitary peaks. Others affirm that they sometimes lay their eggs in old nests of Eagles and Buzzards. Mr. Taylor states that the egg was of a dead, dull white color, and that the surface of the shell was slightly roughened. It was nearly a perfect ellipse in shape, and measured 4.50 inches in length by 2.38 inches in diameter (114.3 by 60.45 millimetres). The eggshell held 9 fluid ounces of water. The young Vulture weighed 10 ounces; its skin was of an ochreous yellow, covered with a fine down of dull white.[1]

I have a sketch of this egg before me, drawn by Mr. W. M. Ord, at Monterey, California, in April, 1859. The exact measurements of this drawing are 116 by 72 millimetres. They do not correspond with those given in the "History of North American Birds," previously mentioned. On the drawing it is stated "color dead white."

I believe that the mode of nidification of the California Vulture is sim-

---

[1] History of North American Birds, 1874, Vol. III, p. 342.

THE CALIFORNIA VULTURE. 161

ilar to that of the common Turkey Vulture, and that as a rule they make but little of a nest, usually laying their eggs on rubbish on the ground found in the immediate vicinity of the nesting site, alongside or in a hollow log, or in crevices of rocky cliffs. It is possible that at times they make use of the abandoned nests of the Golden Eagles, which are common in that part of California, and the nest described to Mr. Flint as being placed in a large redwood tree in the Santa Cruz Mountains was probably such an one, and was made use of by the Vultures after being abandoned by the Eagles.

I have only seen two eggs of this species, both taken by Dr. C. S. Canfield, near San Rafael, California. One of these, No. 9983, is now in the U. S. National Museum collection; the other I saw in the collection of the Academy of Natural Sciences of Philadelphia, Pennsylvania, in 1880, but it has since disappeared. Both of these eggs were of a uniform light grayish-green color and unspotted. The shell of the specimen now before me is close grained and deeply pitted, differing in this respect from the eggs of other Vultures, and is, like those, slightly glossy. It is elongate ovate in shape, and is figured on Pl. 4, Fig. 5; it measures 114 by 65 millimetres. The late Dr. T. M. Brewer described this specimen as of a uniform pale greenish blue, almost an ashy greenish white.

58. Cathartes aura (LINNÆUS).

TURKEY VULTURE.

Vultur aura LINNÆUS. Systema Naturæ, ed. 10, 1, 1758, 86.
Cathartes aura SPIX, Avium Brasiliam, 1, 1825, 2.
(B 1, C 365, R 454, C 537, U 325.)

GEOGRAPHICAL RANGE: Nearly the whole of temperate and tropical America, including West Indies; south to Falkland Islands and Patagonia; north, more or less regularly, to southern New England, New York, Saskatchewan, and British Columbia.

The breeding range of the Turkey Vulture, more commonly known as the Turkey Buzzard, includes the greater portion of the United States, with the exception of the higher mountain regions of the interior, and of New York and the New England States, where it only occurs as a straggler at rare intervals. An occasional pair may breed in the extreme southern part of Long Island, specimens having been observed there on repeated occasions within recent years during the summer months. It is quite common throughout the South, gradually becoming rarer as it advances northward. East of the Rocky Mountains it is resident throughout the year from about latitude 39° southward; while on the Pacific coast it winters as far north as latitude 46°, near the mouth of the Columbia River.

North of our boundary, Mr. Ernest E. Thompson reports it as common in the Assiniboine Valley, Manitoba; and Dr. Richardson found them late in

26957—Bull. 1——11

the month of June, on the banks of the Saskatchewan, in latitude 55°, which probably marks the most northern point of their breeding range.

The Turkey Vulture is a well known resident throughout the Middle and Southern States, as well as on the Pacific coast. I have seen these birds everywhere in Arizona, California, Nevada, Oregon, Washington, and Idaho. Let an animal be killed accidentally, or die on the march, and you will not have long to wait before some of the Vultures are about.

They look their best aloft, as their flight is exceedingly easy and graceful; while the apparent absence of all effort as they sail in stately manner overhead, in ever changing circles, and without any apparent movement of their well shaped wings, makes them really attractive objects to watch; but let them once descend to the ground or alight on a tree, and their attractiveness ceases; now they are anything but prepossessing, and it requires no effort to place them where they properly belong, among the "scavengers of the soil."

Dr. William L. Ralph writes me: "In Florida they are abundant and appear to decrease but little in numbers. When not molested they become very tame, and in many of the Southern cities and villages they can be seen walking around the streets or roosting on the house tops with as little concern as domestic animals.

"Although they eat carrion, these birds prefer fresh meat, and the reason of their eating it when decayed is that they cannot kill game themselves and their bills are not strong enough to tear the tough skin of many animals until it becomes soft from decomposition. I have often had Ducks and other game, which I had hung in trees to keep from carnivorous animals, eaten by them. When they find a dead animal they will not leave it until all, but the bones and other hard parts, has been consumed, and if it be a large one, or if it have a tough skin, they will often remain near it for days, roosting by night in the trees near by. After they have eaten—and sometimes they will gorge themselves until the food will run out of their mouths when they move—they will, if they are not too full to fly, roost in the nearest trees until their meal is partly digested, and then commence eating again.

"Many times I have seen these birds in company with the Black Vulture floating down a stream on a dead alligator, cow, or other large animal, crowded so closely together that they could hardly keep their balance, and followed by a number on the wing. I never have seen them fight very much when feeding, but they will scold and peck at one another, and sometimes two birds will get hold of the same piece of meat and pull against each other until it breaks or until the weaker one has to give it up."

A specimen shot by me on September 22, 1872, at my camp on the Rillitto Creek, near Tucson, Arizona, was so completely gorged with small minnows, each about 1½ inches long, that they filled its mouth. How and where it got these fish has always been a puzzle to me; they were not decayed, and must have been caught alive, or found very shortly after death. I noticed that in Arizona, during the hot summers, these birds generally kept their wings partly open, both when perching on trees or alighting on a carcass, giving them an extremely drowsy appearance.

The young when first hatched are covered with soft, white down, and the nesting sites are not always the filthy places described by many observers.

On July 12, 1883, I found, near the Klamath Indian Agency, a nest containing two young, perhaps a week old, in a cavity of rocks that was quite clean, and there was scarcely any disagreeable odor perceptible about it. The young were sitting on the bare ground and made a slight hissing sound when touched. They are fed in the same manner as young Pigeons, the parents disgorging food into the mouths of the young.

In most of the Southern States nidification begins usually about the latter part of March, occasionally even in February; in the Middle States generally about the last week in April or the beginning of May, and in the more northern portions of its range it may be protracted till June, according to the season.

Capt. B. F. Goss writes me: "I have found this species nesting in Kansas, in caverns and crevices of rock, in hollow trees, and on the ground in hollows of old logs. In Texas I found a large number of nests, one in an old Heron's nest, one on a cactus about 3 feet from the ground, the rest all on the ground under thorny shrubs. They make little or no nest; I have found eggs lying on the bare rock. They sit pretty close, but on near approach will leave the nest if the way is open, but if confined in a cavern or hollow log, will often refuse to move, and if disturbed will disgorge the foul contents of their stomachs, when the intruder will be glad enough to beat a retreat. While rearing their young the nest is always foul and sometimes the stench is unbearable. In southern Texas I found eggs as early as February 15, and up to May 2. In a single instance I found three eggs in a nest; this was placed on the ground under a bunch of thick bushes."

Mr. P. W. Smith, jr., found a nest of this species, which also contained three eggs, near Greenville, Illinois, and four young are recorded to have been found in the same nest by W. W. Edwards, near Abbeville, Louisiana.[1]

Mr. Lynds Jones writes me from Grinnell, Iowa; "I once started a Turkey Buzzard from her nest, and found among the matter thrown up mice and pieces of a skunk, evidently very recently killed.

"The nesting site was a hollow stump, resorted to year after year, until it fell to pieces from old age. At the time I flushed the bird no others were anywhere in sight, but very soon six others came at the summons of the female. When they alighted, their wings remained spread. The only sound they made was a peculiar choking, hissing noise, as they sailed past me about a rod away. The eggs were deposited on the punk at the bottom of the stump, with no attempt at a nest."

[1] Ornithologist and Oölogist, Vol. X, 1885, No. 5, p. 80.

In the West, they breed most frequently in rocky cañons, in the foot-hills of the mountains, up to an altitude of 7,500 feet, depositing their eggs in small caves or crevices of the many cliffs. Mr. W. E. D. Scott found a nest on May 2, 1885, on the Santa Catalina Mountains at an altitude of 5,000 feet. It was placed between two large bowlders.

In the Middle and Southern States the Turkey Vulture breeds more often in the heavily timbered bottom lands, adjacent to the larger streams, as well as in swampy regions. The nest may be found on the ground, under an old log, or in a hollow tree; and even an old Hawk's nest, as high as 40 feet, is occasionally used.

Mr. J. W. Preston states: "The Turkey Vulture will sometimes go into an aperture far up a tree and follow the hollow clear to the ground. At Spirit Lake, Iowa, I took a set of two eggs from an old elm tree, which leaned in the form of an arch; the bird made its way into the tree at the broken off top, and deposited the eggs near the roots of the tree, where I relieved her of them, by the help of an axe. I have seen the female leave the nest to feed, and the male has been seen visiting the nest while the former incubated, but I do not know that he carried food to her; he is very solicitous, however."

The only note they have is of a hissing wheezy sound when disturbed, and this is generally only uttered on such occasions. In southwestern and southern Texas, as well as in other regions where this species is abundant, they breed as often in communities as singly; the nests, if they can be called such, being generally placed on the ground, under the shelter of small bushes on a side hill, and again in low places on salt marshes of the seashore, or in thickets in river bottoms. They are not at all particular in the selection of a nesting site. Two eggs are usually laid, occasionally but one, and very rarely three. These are among the handsomest of the eggs of the Raptores. Their ground color is generally a light creamy tint, occasionally a dull dead white, with a very faint trace of green in some few instances. They are blotched, smeared, and spotted with various shades of reddish brown, choco-late, and lavender, the markings usually predominating about the larger end of the egg, and are very irregular in outline. In eggs belonging to the same set, the markings frequently differ greatly in size and intensity, one being heavily marked and the other but slightly. Occasionally an egg is found which is entirely unspotted. These eggs also vary greatly in shape: the majority are elongate ovate, a few are ovate, others elliptical ovate, and now and then one is perfectly cylindrical ovate.

The average measurement of thirty-four specimens in the U. S. National Museum collection is 72 by 49 millimetres. The largest egg of the series meas-ures 83.5 by 50, the smallest 68 by 46 millimetres.

The two type specimens show about the average types of the heavier and lighter colored eggs of this species.

No. 17608, from a set of two (Pl. 4, Fig. 1), was collected by Mr. Robert

Ridgway, near Momit Carmel, Illinois, May 6, 1878. No. 21574, from a set of two (Pl. 4, Fig. 3), was taken near Ilchester, Maryland, on May 30, 1884, by Mr. C. W. Beckham.

59. Catharista atrata (Bartram).

BLACK VULTURE.

Vultur atratus Bartram, Travels in Carolina, 1792, 285.
Catharista atrata Gray, Handlung 1, 1869, 3.
(B 3, C 366, R 455, C 536, U 326).

Geographical range: Whole of tropical and warm temperate America, south to Argentine Republic and Chile, north regularly to North Carolina and Lower Mississippi Valley, western Texas; irregularly or casually to Maine, New York, Ohio, Indiana, Illinois, Kansas, and South Dakota.

The Black Vulture or Carrion Crow has a much more restricted distribution than the preceding species. It is less roving than the former and is generally a constant resident wherever found, except in the extreme northern portion of its range. It is also far more abundant on the seaboard than in the interior.

The breeding range of this Vulture may be defined as follows: On the Atlantic coast, from southern North Carolina, southward through the South Atlantic and Gulf States. In the interior it has been found breeding in both southern Indiana and southern Illinois, but seems to be very irregularly distributed. It occasionally straggles into southern Ohio, and specimens have been taken in New York, and even in Maine. It has also been found breeding in Kansas, but on a single occasion only, as far as known. In Texas it is a summer resident in the central and western parts of the State, and in the southern portions it is found throughout the year.

Mr. William Lloyd found it breeding near San Angelo, Tom Green County, in June, 1884, and he writes me in this connection as follows: "A strange habit of both the Black and the Turkey Vulture in western Texas is the fact that though abundant in summer, in winter we are left without a single one. This is not due to lack of carrion, for every winter the place is full of dead cattle, nor to temperature, for here in Presidio County it rarely freezes. The altitude, too, is rather less than in Concho County and the Staked Plains, where the same state of affairs obtains. Then why is it that this bird, and especially the Turkey Vulture, which is well known to winter much farther north in other regions, is entirely absent from the bend of the Rio Grande and the Concho Valley during this season?"

The Black Vulture is only common in the southern parts of the United States, usually outnumbering the Turkey Vulture near the seashore and being outnumbered by the latter in the interior. It rarely breeds north of latitude 36°.

In its habits it differs but little from the Turkey Vulture. It is not nearly so graceful a bird on the wing as the latter, its flight being much heavier and apparently laborious, and is accompanied by considerable flapping of the wings.

Mr. Clement S. Brimley, of Raleigh, North Carolina, found a pair of these birds breeding on April 21, 1890. The nest was a slight affair and contained two slightly incubated eggs. It was placed under the elevated end of a prostrated log near a swamp. Another, found on May 5, containing also two eggs, was placed under the shelter of two huge bowlders on a sloping hillside near a stream; incubation had begun.

Mr. Walter Hoxie describes the breeding habits of this species as follows: "Buzzard Island lies in a bend between Ladies Island and Wassa Island, and is about 3 miles in a direct line from Beaufort, South Carolina. It is about 1½ acres in extent, and surrounded by boggy marsh, beyond which, at low tide, stretch wide flats of gray mud, liberally dotted with banks of coon oysters. * * *

"Perhaps a dozen or twenty pairs breed here regularly, the most of them being the black species (*Catharista atrata*), though a pair or two of the Turkey Buzzards may be observed every year. The portion of the island most frequented by them is the west end. Here, under a dense growth of yucca, I have taken nineteen eggs in one afternoon, and seen at the same time five or six pairs of newly hatched young. There is never the slightest attempt at forming a nest, or even excavating a hollow. The eggs are laid far in under the intertwining stems of the yuccas, and, in the semishadows, are quite hard to be seen. The parent birds, however, have the habit of always following the same path in leaving and approaching their precious charge, and after a little experience I learned to distinguish these traces so well that I seldom failed to follow them up and secure the coveted specimens. This track is seldom if ever straight. It winds under and around the armed stems, and the difference in bulk between a man and a Buzzard being considerable, the pointed leaves find a good many of a fellow's weak points before he reaches his prize. * * *

"Quite rarely I have found eggs on the other parts of the island, and once or twice in completely exposed situations, with not even an attempt to get under the protection of an overhanging bush. Possibly these belonged to young birds which had still much to learn in regard to the ways of housekeeping. I have also occasionally found isolated nests upon the outer Hunting Islands. In these latter cases the eggs have always been easy to find, for being among clean sylvan surroundings the collector need only follow his nose—if it is a good one success is certain.

"Both sexes assist in the work of incubation. A week or ten days often elapse between the deposition of the two eggs, but I have never observed over a day's difference in the time of hatching. Indeed, I have never found a bird sitting on a single egg. The period of incubation is very nearly thirty days, but I have not yet decided this quite to my satisfaction. I have never taken more than two eggs in a set, but my friend, Mr. Alfred Cuthbert, of this place, took a set of three in 1884. I am not certain that two broods are not sometimes raised. I have myself taken eggs only from April 2 to May 26, but I have heard of young observed as late as August."[1]

[1] Auk, Vol. III, 1886, pp. 245-247.

Dr. William L. Ralph writes me: "In Florida the Black Vulture resorts to the cypress swamps to nest, and its eggs are laid in slight depressions, on small hillocks that rise above the water, and lined only with the dead leaves and other vegetation that were there originally. Although I have never found many of their nests, I believe that they usually breed in these swamps, for I have often found them, sometimes in considerable numbers, in such situations during the breeding season. As a rule their eggs are very hard to find, for the birds leave them before one can get within sight of their nests, and will not return as long as anyone remains in the vicinity. The earliest set that I have collected was taken March 29, and I have found fresh eggs from that time until the middle of April."

Mr. W. W. Worthington found a small colony of the Black Vultures breeding near Beaufort, South Carolina, March 15, 1886. The nests, of which he secured four, were all placed under a thick growth of yucca, in a small hammock surrounded by marsh. Three nests contained two eggs each, the fourth but a single one, and in this incubation had begun.

Mr. A. W. Butler, of Brookville, Indiana, informs me that a pair of these birds nested 4 miles west of Brookville, in May, 1889. The nesting site was a hollow sycamore about 20 feet high; the two eggs were placed upon the ground, inside.

Both Mr. G. B. Sennett and Dr. James C. Merrill, U. S. Army, report these birds as abundant on the Lower Rio Grande in Texas, nesting on the ground or under hollow logs in the woods. Capt. B. F. Goss also found the Black Vulture abundant near Corpus Christi, Texas, nesting on the ground, under thick bushes or under logs. Eggs were taken by him between April 4 and May 13.

From the foregoing accounts it will be seen that the Black Vulture is more or less gregarious in its habits at all times, breeding frequently in small communities, making little or no nest, and the eggs, usually two in number, are, perhaps with exceedingly rare exceptions, always placed on the ground, in canebrakes, under bushes, old logs, on rocks, and again in perfectly open and unsheltered situations. Occasionally but one egg will be laid, and very rarely three. In the more Southern States nidification begins about the 1st of March, and later northward.

Probably but one brood is raised in a season. The young when first hatched are covered with light buff colored down, and they are fed in the same manner as the young of the preceding species.

The eggs of the Black Vulture are readily distinguished from those of the Turkey Vulture by their different ground color, somewhat larger size, and fewer markings as a rule. By far the greater number of eggs are elongate ovate, a few are short ovate, others elliptical ovate. Their ground color is a pale gray green; in none of the specimens before me can it be called a creamy white; the tint is perceptibly different. In an occasional specimen it may be called pale bluish white, like well watered milk, but the first mentioned color predominates.

The markings vary from chocolate to reddish brown of different tints, and mixed among these, in about one-half the specimens, are found shell markings of lilac and lavender; in an occasional specimen these predominate over the first mentioned tints. In the series before me all the markings are rather irregular in shape and are clustered about the larger end of the egg. They are usually large, and seldom confluent. A few eggs are but slightly marked, and the spots are small and fine, but none are entirely unspotted.

The average measurement of seventeen specimens in the U. S. National Museum collection is 76.5 by 52 millimetres. The largest egg measures 84 by 55, the smallest 68.5 by 47 millimetres.

Of the type specimens, No. 20704 (Pl. 4, Fig. 7) was collected March 12, 1875, in Comal County, Texas, and No. 20705 (Pl. 4, Fig. 10) was collected April 12, 1885, in the same county and place. Both types are from the Bendire collection, and were obtained originally from the late Mr. E. Ricksecker, of Nazareth, Pennsylvania. They show the heavier and lighter marked types found in the eggs of this species.

Family FALCONIDÆ. VULTURES, FALCONS, HAWKS, EAGLES, ETC.

6o.   Elanoides forficatus (LINNÆUS).

SWALLOW-TAILED KITE.

Falco forficatus LINNÆUS, Systema Naturæ, ed. 10, I, 1758, 89.
Elanoides forficatus COUES, Proceedings Academy Natural Sciences, Philadelphia, 1875, 345.

(B 34, C 337, R 426, C 493, U 327.)

GEOGRAPHICAL RANGE: Tropical and warm temperate portions of continental America; north in the interior regularly to Illinois, Iowa, Minnesota, casually to Manitoba, and Assiniboia, etc.; along the Atlantic coast casually to Pennsylvania, New York, and southern New England; accidental in England.

Although the breeding range of the Swallow-tailed Kite within the limits of the United States must be considered as quite extensive, it is a very irregular one, and the birds are only summer residents over the greater part of their range. It breeds regularly in Florida and South Carolina, and probably farther north along the Atlantic seaboard, in the States of North Carolina and Virginia, and apparently even in New York State, where several Kites were observed on different occasions in Rensselaer County, in the latter part of July and the beginning of August, 1886, strongly suggesting their breeding in that vicinity during the season in question.

From Florida westward it is irregularly distributed through the Gulf States, including the greater portion of Texas. It also breeds in the interior in suitable localities throughout the entire length of the Mississippi Valley, and in the States adjacent to our northern boundary, and a few pass this, as it has been observed on different occasions in the British provinces of Assiniboia and Manitoba, in latitude 50°.

Stragglers have been taken during the summer months in many of our Northern States, as well as in England, but few winter within our borders, and, excepting in portions of Florida and Texas, it must be considered as rather a rare species throughout its summer range. It is also found throughout the greater part of the South American continent as far south as the Argentine Republic and Chile.

Dr. William L. Ralph, who has had excellent opportunities to observe the Swallow-tailed Kite in Florida during several seasons, writes me as follows about it: "Excepting, perhaps, the Turkey Vulture, I think this bird is the most graceful of any while on the wing. It has the same easy floating motion, but at times it flies very rapidly and turns very quickly, which is something I have never seen the former bird do. Their motions are very 'swallow-like,' and that, with their forked tails, makes them look like gigantic Barn Swallows; and like the Chimney Swifts they have a habit of traveling together in small companies, usually consisting of three individuals, especially when they first return from the south. During the breeding season flocks consisting of from two or three to ten or twelve birds, but oftener of three, may be seen following one another around, frequently uttering their calls and circling in and out among the tree tops so fast as to make one dizzy to look at them. Except during this season one seldom sees one of these birds unless it is flying, and I have often wondered if they did not at times sleep while on the wing. At least I know that they usually if not always eat while flying, for I have many times seen one sailing leisurely along, occasionally bending its head to tear a piece from a small snake that it held in its talons, and I have never seen one alight to eat its food, as other birds of prey do.

"When hunting they fly quite close to the ground, like Marsh Hawks, but at other times they sail above the tree tops, and sometimes so far above that it takes a good eye to see them. Their food consists almost entirely of reptiles. Small snakes seem to be a favorite article of food with them. I never have seen one catch a bird, and believe they do not. This habit of eating snakes has given them the name of 'Snake Hawk' among the natives of Florida.

"Swallow-tailed Kites begin to arrive in this State from the south about the middle of March, but do not become common until two or three weeks later. They appear to be as abundant now as formerly, probably because most of the tourists have left Florida before they arrive in any numbers. Although these birds are common in the southern half of St. Johns County, and that part of Putnam County east of the St. Johns River, and though I have found quite a number of their nests, I have never been able to get but two sets of their eggs, owing to a habit they have of building in places that are very hard to reach.

"The first nest was taken April 22, 1887, 11 miles northeast of Palatka, Florida, and contained two eggs, so nearly hatched that the embryos in them were feathered. It was situated 90 feet above the ground in, or rather on, the top of a very slender pine tree growing on the edge of a cypress swamp. The trunk of this tree at a height of 5 feet above ground was not more than 15 inches in diameter, and at the place where my climber stood, as he took the eggs, it was less than 3 inches, while the limbs he stood on were only about an inch thick. The nest was composed of large twigs thickly covered with Spanish moss (*Tillandsia usneoides*) and long moss (*Usnea barbata*), lined with the same materials, with the addition of a few feathers from the birds. It measured 20 inches in length, 15 inches in width, and 12 inches in depth on the outside, and 6 inches in diameter by 4 inches deep on the inside.

"The nests of this species are usually so very irregular, that I should think they simply hollowed out bunches of mossy twigs that they found lodged in the tops of trees, had I not often seen them carrying this material to nests that they were building.

"Both birds were present when the eggs were taken and made much ado They would dive at the head of my climber, uttering their shrill but rather feeble cry, and at times were so fierce that he had to stop and strike at them with his hat to prevent them from striking him. This set of eggs is the earliest I have record of, and if it takes these birds four weeks to hatch their eggs, as it usually does the larger Hawks in the north, they must have been laid before April 1. They usually commence laying about the middle of April, and I have found them sitting on their nests from that time until the 1st of June, the latter being the latest date I have ever remained in Florida. Most of them have their eggs laid by the middle of May. One nest which I saw these birds building was deserted for three or four weeks and then reoccupied, but whether or not by the original pair, I do not know.

"The second nest was taken 7 miles northeast of San Mateo, Florida, April 14, 1888, and contained two fresh eggs. It was also situated in the extreme top of a slender pine, in every respect an exact counterpart of the one that held the first nest. The difference in the height of the two was less than a foot. As nearly as I could judge, about three-fourths of the nests of this species found by me were about the same distance above the ground, i. e., they were 90 feet, and the remainder from a little above that height to 125 or 130 feet. The birds to which the second set belonged were not so pugnacious as the owners of the first, but they made a great fuss, and soon had four others of the same species with them to see what was going on, and these seemed as much concerned at the disturbance of the nest as the owners themselves. They were less fierce than the first on account of their eggs being fresher, for, like all birds, they exhibit more anxiety just before and just after the eggs are hatched than at any other time. The nest of this pair was composed of large twigs, Spanish moss, and pine needles, lined with green moss and small twigs. The earliest date on which I found this species breeding was April 5, 1891, when I took a set of two eggs, 8 miles southeast of San Mateo, Florida. Both eggs were rather smaller than the usual size and also lighter colored.

One had been incubated for about a week, the other was fresh. The nest was situated in the extreme top of a slender pine, 86 feet from the ground. Both parents made much ado, flying down at my climber from above at an angle of about 45 degrees. The call note of this species sometimes sounds very much like the 'peet, peet,' of the Spotted Sandpiper.

"I think both parents assist in incubation, and that but one brood is raised in a year. In Florida, like the Bald Eagles, they nearly always nest in pine trees and in the tallest they can find, but, unlike the latter, which always select trees of the greatest diameter, they choose the very slimmest.

"They usually breed in wild uninhabited localities, but, except in regard to their nests, they appear to have but little fear of man, and are often to be seen flying around among the houses of the small villages in this vicinity. The places resorted to for breeding are the low-lying pine woods, and the nests are usually built in trees that grow in or near the cypress swamps, so common in these situations.

"The Swallow-tailed Kite has a peculiar way of leaving its nest, for instead of flying directly from one side, as other birds do, it nearly always rises straight up for a short distance first, as if it were pushed up with a spring, and, when about to alight on its nest, it will poise itself a short distance above its eggs and then gradually lower itself down on to them. When they are thus poised above their nests there is scarcely a perceptible movement of their wings, and they often lower themselves so gradually that one can hardly tell when they have reached their eggs."

According to Mr. J. W. Preston, these birds nest in Becker County, Minnesota, from about May 15 through June, the nests being usually placed in basswood trees and in the extreme tops, from 50 to 80 feet high. An unusual nesting site found by him was in a slender canoe birch about 40 feet from the ground. Their call notes are a shrill keen "e-e-e," or "we-we-we," uttered in a high key, which is very piercing and may be heard at a great distance. In Minnesota their nests are usually found in dense woods not far from lakes. They are highly gregarious; disturb a nest and every Kite in the neighborhood is soon on hand. He has found them not apt to attack a person, but they do attack other birds coming in the vicinity of their homes. He says: "Of all aërial performances I have ever witnessed, the mating of the Swallow-tailed Kite excels. Ever charming and elegant, they outdo themselves at this season. In the spring of 1886 they chose as their mating ground an open space over the mouth of an ice-cold brook that made its way out from a dark tangled larch swamp. From my boat on the lake I had an excellent view of them. All the afternoon seven of these matchless objects sported, chasing each other here and there, far and near, sailing along in easy curves, floating, falling, and rising, then darting with meteor-like swiftness, commingling and separating with an abandon and airy ease that is difficult to imagine.

"Next day three pairs were selecting nesting sites. They are extremely particular in regard to the matter of a nest, and may not be disturbed in the least, or they choose another home. Nesting materials (twigs and moss) are carried by the female in her talons, the male following close, and going on the nest to arrange them. Days, and sometimes even weeks, are required to suitably complete the structure. During this time they work in the morning and fly over the lakes and woods in the afternoon. The nest is usually built on the foundation of an old one of a previous year. The female does not alight to secure nesting materials, but snatches them while in full flight. Once, while standing in a larch swamp, a Kite dashed by me and took up a small twig, heavily draped with *usnea*, and proudly soared out over the woods with it. Near their breeding ground they seem to be constantly on the wing. I have known as many as six pairs to nest within a small area. Once incubation has commenced the female sits nearly constantly and seldom leaves the nest, and the male faithfully feeds and protects her. Some days before laying the eggs the female rests, being very quiet and droopy, and I noted this in many birds. I have taken but two nests of this species, one containing two eggs, the other but a single one."

Mr. George G. Cantwell writes me as follows: "The Swallow-tailed Kite is not uncommon in the vicinity of Minneapolis, Minnesota, and especially around Lake Minnetonka. I think about a dozen pairs, and possibly more, breed every year about this lake. They are quite fearless and hunt for their prey right on the lawns in front of the summer hotels and the numerous cottages on the shores of the lake.

"I took but a single nest of this species there about May 15, 1887. It was placed in the top branches of a tall maple, about 60 feet from the ground, and which contained four eggs on the point of hatching. A farmer who showed me this nest, told me that about a week before he destroyed another close to his house containing the same number of eggs and killed the parents fearing that they might some day carry off some of his poultry, as a good many other Hawks had already done so."

Mr. George E. Beyer also found a set of four eggs on Milton Island, Lake Pontchartrain, Louisiana. This nest was placed in the top and near the trunk of a cypress tree, 60 feet from the ground. It was taken June 16, 1889; the eggs contained embryos about a week old. The nest was composed of dry twigs carelessly and loosely laid together, lined with moss and a few pieces of snakeskin, evidently *Bascanium constrictor*.

The Swallow-tailed Kite is, on the whole, a perfectly harmless and beneficial bird, feeding to great extent on reptiles of various kinds, beetles, grasshoppers, crickets, cotton worms, small frogs, and tree toads. It is doubtful if it ever kills a bird.

In Texas, the Indian Territory, and Kansas, this species builds frequently in the tops of the tallest cottonwood trees, occasionally in pin oaks or pecans, where these are found, and always as near to the tops of the trees as the nest can safely be placed.

Nidification varies according to locality, beginning about the first week in

April in the more southern portions of its breeding range, and correspondingly later farther north, sometimes not before the first or second week in June.

Two eggs are generally laid to a set, occasionally but one, and rarely three or four. The average measurement of twenty specimens from different parts of the United States is 47 by 37 millimetres. The largest egg in this series measures 50 by 39, the smallest 41.9 by 34.5 millimetres. I consider these eggs as handsome as those of any of our Raptores. They are usually oval in shape; some approach an elliptical ovate. The shell is moderately smooth and close grained. The ground color varies from a dull to an ashy white, and again it may be a delicate cream color.

The eggs are spotted and blotched with different shades of rich brown and ferruginous, usually irregular in outline, and varying considerably in amount. These markings sometimes form an irregular band running from the center to the smaller end, and frequently become confluent. Occasionally a specimen is found in which the markings are very few and small in size, scarcely any being larger than a No. 10 shot, and the majority smaller. In a few specimens light lavender colored shell markings, generally of small size, are also visible. There is a great deal of difference in the style and markings of these eggs if a number are compared, but they can readily be distinguished from the eggs of any of our Raptores.

Of the type specimens, No. 20671, U. S. National Museum collection (Pl. 5, Fig. 1), selected from a set of two, represents one of the more lightly marked types; it was originally in my collection, and taken in Black Hawk County, Iowa, June 3, 1875. Another (Pl. 5, Fig. 2), from the collection of Dr. William L. Ralph, Utica, New York, also from a set of two, was taken near San Mateo, Florida, April 14, 1888, and kindly loaned for the purpose of figuring.

61. Elanus leucurus (VIEILLOT).

WHITE-TAILED KITE.

Milrus leucurus VIEILLOT, Nouveau Dictionaire, xx, 1818, 563 (errore 556).
Elanus leucurus BONAPARTE, Geographical and Comparative List, 1838, 4.
(B 35, C 336, R 427, C 492, U 328.)

GEOGRAPHICAL RANGE: Tropical and subtropical America (except West Indies), north to South Carolina, southern Illinois (casual?), the Indian Territory and middle California. Accidental in Michigan.

The breeding range of the White-tailed Kite, so far as is actually known by the taking of their nests and eggs, seems to be confined to South Carolina (where Mr. Ward, Audubon's assistant, found it nesting on the Santee River early in March), Florida, the Indian Territory, Texas, and the middle portions of California.

It is said to occur in Georgia, Alabama, Mississippi, Louisiana, and southern Illinois, and probably breeds sparingly in all these localities, but it must be considered rare east of the Mississippi River, and is not a common species anywhere within the limits of the United States. It extends over the South American continent to the Argentine Republic. Within our borders it is perhaps most often found in California, where it is a constant resident. It probably also winters in Louisiana and southern Texas. Stragglers are reported to occasionally reach Michigan, where G. A. Stockwell is said to have met with it.

The first eggs of this species, a set of four, obtained by the Smithsonian Institution, were taken by Mr. J. H. Clark, May 9, 1861, near Fort Arbuckle, Indian Territory, and are now in the collection.

Prof. B. W. Evermann took several sets in California, in 1880 and 1881. He writes me: "I found my first set, containing three eggs, May 4, 1880. They were nearly ready to hatch. The nest was near the end of one of the topmost limbs of a cottonwood (*Populus*) near Santa Paula. It was constructed of coarse sticks, lined with shreds or strippings of the inner dead bark of the cottonwood. I took no others in 1880, but in 1881 I met with better success. On April 12, I got a set of four and another of five eggs, all fresh. These nests were placed in the extreme tops of two unusually tall live oaks (perhaps 45 or 50 feet from the ground), and, like the other, were rather flat structures of sticks, lined with the same material as the first, with the addition of a little straw (barley, I think). These two nests were also found near Santa Paula. In one of them a second set was laid early in June.

"The actions of this species when their nest is approached are interesting. I think in every case the bird would leave the nest while I was quite a distance from the tree, and quietly fly off to another near by. There it would remain until I had nearly reached the nest, when it would fly toward me, and when about 20 to 30 feet above me and the nest, it would balance itself in the air as Sparrow Hawks and Bluebirds often do, and with legs hanging down would utter its distress note a few times and then fly away, probably not to return again, but simply to watch me from some tree top several rods away. The note, as I remember it, is a broken cry or scream. When the female hovered over my head, the male made his appearance also, but came no nearer than a tree several rods away.

"On November 20, 1880, I shot a White-tailed Kite from a telegraph pole while sitting in my buggy, and as it fell to the ground its mate flew to it, and was also secured. I do not regard this bird as at all common. I have seen solitary individuals skimming over the fields near San Buenaventura, the marshes up towards Saticoy, and at various other places up the Santa Clara Valley, as far as Newhall, 50 miles from the coast."

Mr. L. Belding gives the White-tailed Kite as a constant and common resident about Stockton, California, and says: "I have seen as many as twenty at the same moment within a circle of half a mile. I have also noticed it at Marysville in winter. It is rarely seen away from the tule marshes."

Mr. A. L. Parkhurst informs me that he found this species breeding near San José, California, from March 15 to April 10—their nests being generally well hidden among the leaves near the tops of the trees, usually at a height of about 30 feet. Their usual resorts during the breeding season are the banks of streams or the fresh water marshes, especially if a few scattered live oaks or willow groves are close by, and their favorite nesting sites are the tops of live oaks, although other trees are also made use of whose foliage securely conceals the nest during incubation. These birds can be found near the tree in which they intend to breed as early as January. Sometimes several birds are anxious to secure the same site, and in such a case fights are sure to result, not alone for first choice of a nesting site, but also for a mate. As soon as a pair have mated, they proceed to drive away from the neighborhood all others of their kind, and even larger and more powerful birds of prey are taught to respect their claims to the locality selected.

"As soon as the male feels secure in his possession he takes his stand upon the top of a tree near by, whose foliage is not so thick as to obstruct the view, yet where he is practically concealed, and from this station he is ever watchful for the approach of an enemy. I have not been able to get sufficiently close to a nest to enable me to observe the manner of nidification without being detected by the male. I am satisfied, however, that it is accomplished by the unaided effort of the female, while the former is doing picket duty.

"The nest is composed of small dead twigs, placed in the upright forks of a limb, and is neatly lined with dry stubble and grasses. The cavity is usually about 3 inches deep. It is much of the same order of architecture as an ordinary Crow's nest, but is a trifle larger and quite as neat. The date of laying varies somewhat with the season. The first nest I found was on April 6, 1883. It contained four eggs that were about half hatched. I took three sets of eggs from this tree in as many different years, but each time a different nest was built. These birds breed in the same locality from year to year, often in the same tree, but according to my observations they always build a new nest each season."

Other observers describe the nest as a slight structure and shallow, and say it is placed occasionally in sycamore trees or maples, but their favorite tree for a nesting site seems to be the live oak.

Messrs. Sclater and Hudson give a good account of the habits of this species in "Argentine Ornithology" (Vol. ii, pp. 71, 72). Extracts from this are as follows: "This interesting Hawk is found throughout the Argentine Republic, but it is nowhere numerous. It is a handsome bird, with large ruby-red irides, and when seen at a distance its snow-white plumage and buoyant flight gave it a striking resemblance to a Gull. Its wing power is indeed marvelous. It delights to soar, like the Martins, during a high wind, and will spend hours in this sport, rising and falling alternately, and at times, seeming to abandon itself to the fury of the gale, is blown away like thistle down, until, suddenly recovering itself, it shoots back to its original position. Where there are tall poplar trees these birds amuse themselves by perching on the topmost slender twigs, balancing themselves with outspread wings, each bird on a separate tree, until the tree tops are swept by the wind from under them, when they often remain poised almost motionless in the air until the twigs return to their feet.

"When looking out for prey, this Kite usually maintains a height of 60 or 70 feet above the ground, and in its actions strikingly resembles a fishing Gull, frequently remaining poised in the air with body motionless and wings rapidly vibrating for fully half a minute at a stretch, after which it flies on or dashes down upon its prey.

"The nest is placed on the topmost twigs of a tall tree, and is round and neatly built of sticks, rather deep and lined with dry grass. The eggs are eight in number. An approach to the nest is always greeted by the birds with long distressed cries, and this cry is also uttered in the love season, when the males often fight and pursue each other in the air. The old and young birds sometimes live together until the following spring."

The food of the White-tailed Kite is said to consist of small rodents, snakes, and small birds. It is evidently a harmless bird in this respect, and rather beneficial than otherwise. Both parents assist in the care of the young.

But one brood is raised in a season; if the first set of eggs should be taken a second one is occasionally laid in the same nest about a month later.

In the United States the number of eggs laid to a set varies from three to five, generally four. Their ground color is creamy white, and they are heavily marked over their entire surface with irregular confluent blotches and smears of dark blood red and claret brown, of different degrees of intensity, the smaller end being often the more heavily colored. But little of the ground color is visible in the majority of the specimens. Some sets are much lighter than others, possibly a second laying; the eggs are usually oval in shape.

The average measurement of eight specimens is 42 by 33.5 millimetres. The largest egg measures 44 by 34, the smallest 41 by 32 millimetres.

The type specimen (No. 2927, U. S. National Museum collection, Pl. 5, Fig. 3), selected from a set of four, was collected by Mr. J. H. Clark, May 9, 1861, near Fort Arbuckle, Indian Territory. The second (Pl. 5, Fig. 4), selected from a set of five eggs, now in the collection of Mr. Josiah Hoopes, West Chester, Pennsylvania, was taken near Santa Barbara, California, on April 14, 1886, and kindly loaned for illustration.

62. Ictinia mississippiensis (WILSON).

MISSISSIPPI KITE.

Falco mississippiensis WILSON, American Ornithology, III, 1811, 80, Pl. 25, Fig. 1.
Ictinia mississippiensis GRAY, Genera of Birds, I, 1845, 26.
(B 36, C 335, R 428, C 491, U 329.)

GEOGRAPHICAL RANGE: More southern United States, east of the Rocky Mountains, north regularly to Georgia, southern Illinois, Kansas, etc., casually or irregularly, to Pennsylvania, Wisconsin, and Iowa; south through eastern Mexico to Guatemala.

The breeding range of the Mississippi Kite is confined to the southern portions of the United States, from Louisiana, Mississippi, and Texas northward to southern Illinois, the Indian Territory, and Kansas. It likewise occurs in South Carolina, Georgia, Florida, and Alabama, and probably breeds more or less frequently in all these States. It is a casual summer visitor in Iowa and Wisconsin, and is much more abundant west of the Mississippi than east of this stream. A few winter in the Southern States.

Mr. B. G. Gault writes me: "I found the Mississippi Kite abundant in the Red River region of Texas, especially in that portion of it included within Bowie County, where they can be seen at all times of the day and in all sorts of places; but for feeding grounds they seem to prefer the cotton fields. At King's plantation they were particularly abundant, and I have seen as many as eight or ten in the air at one time. Although they do not possess the swallow-like movements in so high a degree as their larger and handsomer relative, the *Elanoides forficatus*, still they do indeed present a pleasing sight, appearing now before us, perhaps 100 or 200 feet above ground; the next minute they are skimming just above the tops of the cotton plants or between the rows, and again they are high in the air, seeming never to tire of these wing evolutions, and it is rarely that one is seen to alight.

"Occasionally, however, I have come suddenly upon them, perched quietly on some dead or decaying tree, and have been greatly surprised at their stupidity in allowing me to approach within 40 or 50 feet of them before they would attempt to take wing. Owing to their numbers, one would readily imagine it not a difficult task to discover the nest of this bird, but in this, after repeated searches, I was badly disappointed.

"July 2, 1888, I came upon a pair of these birds that could not be induced to leave a particular locality on the edge of an old plantation. Very evidently a nest or young were close at hand, but the most careful search failed to reveal one or the other."

The well known ornithologist, Col. N. S. Goss, found this species breeding abundantly in Barber County, Kansas. He writes as follows: "While collecting in this State I found, May 9, 1887, quite a number of the Mississippi Kites sailing over and into the timber skirting the Medicine River, near Sun City, Barber County, and from their actions knew that they were mating and upon their breeding grounds—a lucky find, worth following up. On the 11th I noticed several of these birds with sticks in their bills (green twigs in leaf) flying aimlessly about as if undecided where to place them, keeping hidden within the trees as much as possible, dropping the sticks when from fright or other cause they arose high above the tree tops. * * *

"I returned to the Kites on the 16th, and remained watching the birds until the morning of the 22d, at which time the nests found, seven in number, appeared to be completed, and I saw a pair of the birds in the act of copulation. A business matter called me home, and I hired the man, with whom I stopped, to climb the trees on the 28th for the eggs, but a hailstorm on the 25th injured the nests badly, and in one case beat the nest out of the tree. On the 31st he collected four sets of two eggs each, and one with only one egg. It being a hard tree to climb, he decided to take the egg rather than wait to see if the bird would lay more. Not hearing from him, I returned to the ground June 10, and put in the day examining the nests, etc., collecting two more sets of two eggs each. * * * The old nests had a few leaves for lining, in addition to the leaves attached to the twigs used in repairing the same, but the new ones appeared to be without additional leaves. They were all built either in the forks from the main body, or in the forks of the larger limbs of the cottonwood and elm trees, and were at least from 10 to 100 rods apart; were not bulky, and when old would be taken for the nests of the common Crow. They ranged in height from 25 to 50 feet from the ground.[1]"

Mr. George E. Beyer writes me that he "took two nests of this species in Louisiana during the season of 1889." According to his observations "but one brood is raised in a season, nidification beginning late, usually about the end of May. The nests are placed in the tops of loblolly pines (*Pinus tœda*) or white oaks (*Quercus alba*), at a height of from 50 to 60 feet. Pine woods are the favorite localities.

"Though of a peaceful disposition, under ordinary circumstances, the Mississippi Kite vigorously attacks all intruders coming too close to its nesting site. A nest found June 10, 1889, was built in a loblolly pine between a fork near the top. It was composed of small dry twigs of water or pin oak and lined with Spanish moss and contained but a single egg. A second nest, found in an oak about 40 feet from the ground, on June 24, 1889, was similarly constructed and lined with the dry leaves of the loblolly pine and moss mixed. This set contained three well incubated eggs, nearly ready to hatch. Both sexes incubate, and the male also assists in building the nest. I have seen this species viciously attacking Crows and Jays when too close to its nest, uttering at the same time a sharp hissing sound composed of two syllables."

According to Dr. A. K. Fisher's experience in Louisiana it would appear

---

[1] Auk, Vol. IV, 1887, pp. 344, 345.

that this Kite does not always breed so late in that region, as he shot a fully fledged young bird on May 31.

Mr. J. A. Singely has taken several sets of the eggs of this species in Texas. A nest taken in Lee County is thus described: "The foundation and sides of the nest are built entirely of small sticks, the interior portion of small green oak twigs in leaf and leaves of the mesquite tree; the lining is of green moss, on top of which are placed green leaves of the pecan tree. The outer diameter from the ends of the longest twigs is 17 inches, the most compact portion 11 inches; interior diameter 5 inches, outer depth 7 inches, inside depth 1½ inches."[1]

Occasionally a nest is lined with willow twigs in leaf. Mr. H. Nehrling has also found this species breeding in Texas, near Houston, but it is not common there.

The food of this Kite seems to consist principally of grasshoppers, locusts, and other insects, probably varied with a diet of small rodents, lizards, and snakes. It is unquestionably a perfectly harmless species, if not actually a beneficial one, from an economic standpoint.

"The eggs of this species are usually two or three in number and are ordinarily deposited from the middle of May to the latter part of June. But one brood is raised in a season. The same nest is often used from year to year, when not disturbed, the necessary repairs required being made each season. The first eggs of this species deposited in the U. S. National Museum collection, were obtained through Mr. C. S. McCarthy, who found several of their nests on the Canadian River in the Indian Territory, between June 5 and 21, 1862.

They are rounded ovate in shape, pale bluish white in color, and unspotted; the type specimen, however, shows a few minute deeper blue shell markings, not usually found in the eggs of this species. They are frequently badly stained by contact with the decaying green leaves on which they are usually placed.

The average measurement of twenty-nine eggs of the Mississippi Kite is 41 by 34 millimetres. The largest egg in the U. S. National Museum collection, measures 44.5 by 36.5, the smallest 39 by 32 millimetres.

The type specimen, one of a set of two (No. 23160, Pl. 5, Fig. 5), was taken by Col. N. S. Goss, in Barber County, Kansas, May 31, 1887.

---

[1] Nests and Eggs of North American Birds Davie, 1889, p. 166.

63. Rostrhamus sociabilis (VIEILLOT).

EVERGLADE KITE.

Herpetotheres sociabilis VIEILLOT, Nouveau Dictionaire, XVIII. 1818, 318.
Rostrhamus sociabilis D'ORBIGNY, Voyage; Oiseaux, II, 1847, 73.
(B 37, C 334, R 429, C 490, U 330.)

GEOGRAPHICAL RANGE: Whole of tropical America, except part of West Indies; south to Argentine Republic and Ecuador; north to Florida and Atlantic coast of Mexico.

The Everglade Kite, a common South American species, has but a very restricted range within the United States, being a resident of the swamps and marshes of southern Florida, and said to be fairly abundant in the Everglades. It breeds in suitable localities throughout these little known regions.

Messrs. Sclater and Hudson say: "This Hawk in size and manner of flight resembles a Buzzard, but in its habits and the form of its slender and very sharply hooked beak it differs widely from that bird. The name of the 'Sociable Marsh Hawk,' which Azara gave to this species, is very appropriate, for they invariably live in flocks of from twenty to a hundred individuals, and migrate and even breed in company. In Buenos Ayres they appear in September and resort to marshes and streams abounding in large water snails (*Ampullaria*), on which they feed exclusively. Each bird has a favorite perch or spot of ground to which it carries every snail it captures, and after skillfully extracting the animal with its curiously modified beak, it drops the shell on the mound. When disturbed or persecuted by other birds they utter a peculiar cry, resembling the shrill neighing of a horse. In disposition they are most peaceable, and where they are abundant all other birds soon discover that they are not like other Hawks and pay no attention to them. When soaring, which is their favorite pastime, the flight is singularly slow, the bird frequently remaining motionless for long intervals in one place, but the expanded tail is all the time twisted about in the most singular manner, moved from side to side, and turned up, until its edge is nearly at a right angle with the plane of the body. * * *

"Concerning its breeding habits, Mr. Gibson writes: 'In the year 1873, I was so fortunate as to find a breeding colony in one of our largest and deepest swamps. There were probably twenty or thirty nests placed a few yards apart in the deepest and most lonely part of the whole 'cañadon.' They were slightly built platforms, supported on the rushes and 2 or 3 feet above the water, with the cup-shaped hollow lined with pieces of grass and water rush. The eggs never exceeded three in a nest; the ground color generally bluish white, blotched and clouded very irregularly with dull red brown, the rufous tint sometimes being replaced with ash gray.'"[1]

Mr. C. J. Maynard describes a nest found by him in the Everglades, which was placed in a magnolia bush about 4 feet from the water, as quite flat and

[1] Argentine Ornithology, 1887, Vol. II, pp. 72, 73.

about a foot in diameter. It was composed of sticks, carelessly arranged, lined with a few dry heads of saw grass, and contained one egg. Upon dissecting the female, which was shot at the same time, another egg was found just ready to be laid; this was unspotted, and of a blue color throughout. Several sets of these eggs were brought to him subsequently by Seminole Indians.[1]

Mr. E. W. Montreuil, in a letter published by Mr. H. B. Bailey, describes the breeding habits of this species as follows: "This bird (*Rostrhamus sociabilis*) is found in numbers in the Everglades of Florida, especially on the east side. They lay their eggs early in March, but some pairs later than others, as the set you have were taken March 16, and were fresh, while all the other nests had young in them. When they breed a male and female are by themselves, always near a small island, which they make their rendezvous, and while resting on a branch they can have an eye on their nest for enemies, especially Crows, who rob their nests whenever they can. Around some of the islands there are several pairs of *Rostrhamus*, but they always place their nests a few acres apart from each other. * * * They built their nests with dry branches and grasses, attached to saw grasses about 12 inches below the tops, just so as to be out of sight. They measure about 12 inches in diameter and 6 inches high, and the cavity is about 3 inches deep. They lay two or three eggs. The old birds usually bring their throats full of the animals of the Everglade shells, but sometimes they bring the animal in the shell, as many of the nests contained a lot of these shells. While they have young they are not wild, flying over one's head when near the nest."[2]

Mr. J. F. Menge, who found the Everglade Kite nesting near Meyers, Florida, writes me "They generally lay but two eggs, commencing to nest as early as March 1, and up to the end of April. Their favorite nesting sites are swamps, overgrown with low willow bushes, the nests usually being placed about 4 feet from the ground. They frequent the borders of open ponds and feed their young entirely on snails. According to my observations the female does not assist in the building of the nest. I have watched these birds for hours. She sits in the immediate vicinity of the nest and watches while the male builds it. The male will bring a few twigs and alternate this work at the same time by supplying his mate with snails, until the structure is completed. They feed and care for their young longer than any other birds I know of, until you can scarcely distinguish them from adults."

A nest of this species now before me, taken by Mr. Menge, and kindly forwarded, measures 16 by 13 inches in diameter, and is about 8 inches thick. It is not an artistic looking structure, but rather carelessly put together. The base consists of dry willow twigs, some of them half an inch in diameter; the greater portion are, however, smaller. The inner cavity is about 7 inches wide by 1½ inches deep. This is lined with small stems of a vine and a few willow leaves. The latter look as if the twigs, to which some of them are still

---

[1] Birds of Florida and Eastern North America, 1881, pp. 284-290.
[2] Auk, Vol. 1, 1884, p. 95.

attached, might have been broken off by the birds while green; the first mentioned material predominates in the lining. This nest was found on April 3, and had just been finished. On April 23, it was visited again, and it then contained two eggs, in which incubation was so far advanced that they could not be saved.

From the foregoing accounts it will be seen that the nesting sites selected by these birds are rather variable, and that while in some sections in South America they may be said to breed in small colonies, this does not appear to be the case in Florida, at least as far as we know at present. The food of this species in Florida, about which all observers agree, seems to consist entirely of a fresh water snail, *Pomus depressa* (Say), and according to Mr. W. E. D. Scott, the local name of the Everglade Kite in the vicinity of Panasofka Lake, where he found them very abundant during the month of March, 1876, is "Snail Hawk."

The number of eggs is two or three, and these seem to be deposited from the latter part of February till the 1st of May, usually in March or the beginning of April.

In shape they vary from a rounded ovate to oval. The ground color is pale greenish white. In some specimens this is scarcely perceptible, the whole surface of the egg being covered with rusty or brownish red blotches and smears of different degrees of intensity, and running into each other, hiding it completely. None of the markings in the majority of specimens before me are well defined, except in a single one. In this case the ground color is distinctly visible, the egg being but slightly marked, principally about the center. These markings are well defined in the shape of small irregular blotches connected with each other by lines and scrawls similar to those found on the eggs of the Grackles. They vary from fawn color to light brown.

The average measurement of six specimens is 44.5 by 37 millimetres. The largest egg measures 47 by 37.5, the smallest 42 by 37 millimetres.

The type specimen (No. 16827, U. S. National Museum collection) was taken in the spring of 1873, in Florida, by Mr. C. J. Maynard. It is figured on Pl. 5, Fig. 6. Another, from a set of three eggs taken by Mr. E. W. Montreuil, March 16, in the Everglades of Florida, was kindly loaned by the American Museum of Natural History, New York City, for the purpose of figuring. It is the lightest colored egg of this species I have seen. This is figured on Pl. 5, Fig. 7.

64. Circus hudsonius (Linnæus).

MARSH HAWK.

Falco hudsonius Linnæus, Systema Naturæ, ed. 12, t, 1766,-128.
Circus hudsonius Vieillot, Oiseaux, Amerique Septentrionale. 1, 1807, Pl. 9.
(B 38, C 333, R 430, C 489, U 331.)

Geographical range: North America in general, south in winter to Panama, Bahamas, and Cuba.

The Marsh Hawk, also called the Harrier and Mouse Hawk, is one of the best known of the Raptores found within the limits of the United States. Its breeding range extends over nearly the entire North American continent. In portions of the South Atlantic States it must, however, be considered as a rather rare summer resident, and in certain sections of this region, like the Piedmont hill country of South Carolina, according to Dr. Leverett M. Loomis, it does not breed at all, and I think the same remarks apply to the greater part of Florida, where only a few remain during the summer, principally in the northern parts of the State. It is a fairly common summer visitor in the Arctic regions, and east of the Rocky Mountains winters regularly as far north as latitude 39°, while on the Pacific coast I have found a few remaining in latitude 46°. It is likely, however, that these birds bred farther north and wintered here, and that the actual summer residents migrated south, and were, to a certain extent, replaced by these visitors from colder regions.

The Marsh Hawk is easily recognized by the conspicuous white patch at the base of its long tail, and its great stretch of wing, while flying, makes it appear a much larger bird than it really is. In the South Atlantic States it is by no means as common during the breeding season as in the more northern portions of its range.

On the extensive prairies of the West, it is a familiar sight to see a pair, and often several, of these birds skimming close to the ground, now along the borders of a meadow, or the shrubbery found close to the banks of small streams, and the tule covered borders of fresh or salt water marshes, actively engaged in search of their prey. From the fact that even in winter one frequently sees pairs of these birds hunting in company as often as singly, I am inclined to believe that many remain mated throughout the year.

Its flight is singularly easy and graceful. One moment it may be seen sailing or drifting along before a strong breeze without an apparent movement of its wings, in the next it may raise or lower itself or turn completely over, in undulating motions; dropping suddenly in the grass, or staying suspended in the air over some point which might be suited to the location of its intended quarry.

Its food consists principally of meadow mice, small ground squirrels, and other rodents, frogs, grasshoppers, locusts, and, in portions of the West, the large and destructive crickets (*Anabus simplex*); where these are abundant they feed almost exclusively on them. Now and then small birds or snakes are also eaten, probably when other food is not so easy to obtain. It is an extremely useful bird, and well deserves the fullest protection.

According to Mr. W. G. Smith, of Loveland, Colorado, the Marsh Hawk will occasionally feed on decaying animal matter, such as dead Ducks, and will kill and devour such game when found wounded.

Dr. W. L. Ralph tells me: "In Florida I have often had these birds come at the report of my gun and try to carry off the ducks that I had shot, and sometimes they would be so persistent that I could almost catch them before they would leave. I saw a couple of these birds on March 27, and another pair on March 28, 1891, a few miles from San Mateo, Florida. Still another pair was seen by one of my men on the last day of that month, and I think that is rather late for them to be so far south, unless they were going to nest. Sometimes they will fly just in front of the blaze of a forest or prairie fire to catch the small mammals driven out by the heat."

In the West, where I have principally observed these birds, they are not very shy, often sailing in search of food quite close to a person. Sometimes they seem to be so entirely absorbed in such occupation that I frequently had them pass within 20 feet without apparently seeing me, and, after noticing me, turning only a little out of the way and altering their course but very slightly.

The mating season begins about April 1, sometimes earlier. Pairs of these birds may be seen at this time playing with each other, sailing around in graceful gyrations, turning over and over, and uttering shrill screams of delight while engaged in these aërial evolutions. Soon thereafter a suitable nesting site is selected, and this is always on the ground or close to it. It is usually not far from water, in a thick bunch of grass, on a slight hillock in a marsh, in bunches of rank weeds, low bushes, among rushes, flags, or on tule drifts. The male assists faithfully in the construction of the nest, which, in most cases, is but a slight affair. Now and then one is found somewhat more elaborately built than the average, but, as a rule, there is but little material used, and the inner lining is also generally rather scant. A reasonably level surface in the high grass or thick bushes having been selected for this purpose, bits of dry grasses, stubble, or pieces of weed stalks are carried to the site, and this is arranged in a circle, leaving a rather shallow cavity in the center, lined with similar material, and the eggs are deposited on this. Occasionally, especially when placed in wet places subject to overflow, the nest proper is put on a slight platform of sticks to raise it slightly from the ground. I have never found a nest lined with either hair or feathers, excepting such as drop from the sitting bird.

In the Southern States nidification begins in the early part of April; in the Middle States usually about the second week in May, rarely earlier, and correspondingly later northward, where it is protracted till the first two weeks in June. When not disturbed the same sites are resorted to from year to year, and in such cases the nests are frequently bulky affairs. The usual number of eggs laid by this species is from four to six, rarely more.

Mr. John Swinburne writes me that in Arizona the usual number of eggs found by him is two or three. I found a nest near the Laguna, 9 miles from Tucson, Arizona, with two newly hatched young and an egg about to hatch. I also took a much incubated set of three eggs near Fort Lapwai, Idaho, June 15, 1871. As an offset to such small sets, I had the good fortune to find one near Walla Walla, Washington, on May 8, 1882. The nest was placed in a pasture, close to a small stream, alongside of and concealed by a tall bunch of rye grass. It was principally constructed of dry grasses carelessly arranged. The eggs were in different stages of incubation, five contained large embryos that required rotting out, one was quite fresh, and two nearly so.

The earliest date on which any of the eggs were taken among the series of this species in the U. S. National Museum collection is April 29, 1837, collected by Audubon, near Galveston, Texas. On the Pacific coast the Marsh Hawk lays in the beginning of April, and even at Camp Harney, Oregon, I have taken their eggs on April 29, 1877.

In Connecticut they nest usually between May 15 and 25. Mr. John N. Clark, of Saybrook, writes me: "I observe that the female sits very closely and the male procures the food with great diligence. I have located the nests often by watching him. The moment he appears in sight the female utters shrill cries from the nest, and leaves it as he approaches, rising in the air to meet him, and seizing with dexterity the morsel that he drops in her talons. She then retires to the nest, to feed or distribute it to the young. I have examined the stomachs of several specimens, and found meadow mice, their favorite food. One I examined contained not less than eleven, another nine, and nothing else. I have rarely seen them disturb domestic fowls. I have reared these birds from the nest, to study their habits, and have found them very interesting, though never becoming very gentle pets. They showed a decided fondness for frogs and mice of all kinds, and they rarely declined fish."

Mr. J. W. Preston, of Baxter, Iowa, writes me: "The nests of the Marsh Hawk are usually built of weeds and grass, but I remember seeing one placed on a tussock over water, which was bulky and well made. The foundation consisted of plum twigs and small branches. It was thick, the depression deep, and was well lined with grass and down. They nest here sometimes on the highest ridges. Their call note is a peevish scream, not unlike that of the Red-tailed Hawk, though not so strong. A peculiar clucking or cackling, in a short jerky way, is also uttered at times, while ascending in the air. This seems to be done for diversion. The male assists in incubation, and I have seen him reluctant to abandon the nest to its mate. The earliest date on which I have found these birds nesting here was April 20. In a set of eight found by me the eggs ranged from fresh to well incubated. This nest was placed in a low hazel thicket, on a high ridge."

Mr. Lynds Jones, of Grinnell, Iowa, says: "Once during the breeding season I saw a male catch a large garter snake and fly up with it several hundred feet, then drop it to the female, who just then came flying along near the ground; she caught and carried it to the nest, followed by the male."

Mr. George G. Cantwell found a nest of this species placed on a haycock.

The male assists to a certain extent in incubation, which lasts somewhat over three weeks, and seems to begin, occasionally at least, before the complete set of eggs is laid. Both parents assist in caring for the young, and the family remains together for sometime after they leave the nest. The young when first hatched are covered with a grayish buffy down. But one brood is raised in a season.

The eggs, commonly from four to six in number, and usually laid at intervals of two and three days, are pale greenish or bluish white in color, and the majority are unspotted. Quite a number, however, are more or less blotched and spotted with pale buff and brownish markings. About two-fifths of the series are more or less plainly marked. The shell is smooth and slightly glossy. In shape these eggs vary greatly, ranging through the different forms of ovate.

The average measurement of seventy-three specimens in the U. S. National Museum collection is 46 by 36 millimetres. The largest egg in the series measures 52 by 38 millimetres; this was taken on the Lower Anderson River, Arctic America, by Mr. R. MacFarlane, of the Hudson Bay Company. The smallest of the series measures 43 by 34 millimetres, and was taken by the writer near Fort Walla Walla, Washington.

Of the type specimens figured, No. 13249 (Pl. 5, Fig. 8), a single egg, was taken in Maine by Dr. B. Dixon; No. 20686 (Pl. 5, Fig. 9), from a set of five eggs, four of which are distinctly spotted, was taken near Fort Lapwai, Idaho, May 14, 1870; and No. 20690 (Pl. 5, Fig. 10), from a set of eight eggs, was collected near Fort Walla Walla, Washington, on May 8, 1882. The last two types are from the Bendire collection.

65. Accipiter velox (WILSON).

SHARP-SHINNED HAWK.

Falco velox WILSON, American Ornithology, v, 1812, 116, Pl. 45, Fig. 1.
Accipiter velox VIGORS, Zoölogical Journal, ı, 1824, 338.
(B 17, C 338, R 432, C 494, U 332.)

GEOGRAPHICAL RANGE: North America in general, south in winter to Guatemala.

The Sharp-shinned Hawk breeds throughout the entire United States, but in the more southern portions, excepting the mountain regions, it must be considered as a rather rare summer resident. The only eggs of this species from the South in the U. S. National Museum collection are two, said to have been taken near Edinburg, Texas. Mr. Würdemann has found it nesting in Florida, and Mr. G. E. Beyer informs me that it is a summer resident in southern

Louisiana, having shot it there in July. In the Middle and Northern States it is fairly common in suitable localities, and the same holds good in the Rocky Mountain region and the mountains of the Pacific coast, where it is known to breed from California to Alaska. It is not uncommon in the higher mountain regions of Arizona, and has been met with in similar localities in Lower California.

It seems to be a regular summer visitor throughout the southern portions of the Dominion of Canada, while in the interior of British North America it breeds at least as far north as latitude 62°; Mr. R. Kennicott taking its nest and eggs near Fort Resolution, on the shores of the Great Slave Lake, on June 16, 1860, and Mr. J. Lockhart, of the Hudson Bay Company, found it breeding in the same locality in June, 1863. The eggs taken by these gentlemen are in the U. S. National Museum collection. Mr. B. R. Ross noticed it also at Fort Simpson on the Mackenzie River, in about latitude 63°.

In the northern portions of their range they are only summer residents, a few wintering as far north as latitude 40°, the majority passing farther south. The southward migration takes place in the latter part of September and the beginning of October. Dr. A. K. Fisher tells me of having seen several hundred of these birds at this time in one day's tramp, the majority of them flying high. They return to their summer homes about the latter part of March or the first days in April.

The Sharp-shinned Hawk, though small in size, is full of dash and courage, frequently attacking birds as large as itself and killing them with ease. Like its larger relatives, the vicious Cooper's Hawk and Goshawk, it has very destructive instincts. It lives mostly on small birds, but occasionally on some fully as large as itself—among these, Pigeons, Bob Whites, Mourning Doves, Purple Grackles, and Robins may be mentioned. Small rodents appear to furnish but a very limited portion of its food, and insects are likewise rarely eaten.

Although the wings of the Sharp-shinned Hawk are rather short, its flight when in pursuit of its prey is unerring and exceedingly swift. No matter which way the selected victim may turn and double, his untiring pursuer is equally prompt, and only rarely will it miss capturing its quarry. Once struck, death fortunately follows quickly, as it fairly transfixes its victim's vitals with its long and sharp talons.

It is said to be rather fond of young pullets, and does not hesitate to help itself to these as long as one is left. It is one of the few species that must be considered as more harmful than beneficial, looked at from an economic point of view.

Mr. William L. Ralph writes me that "Sharp-shinned Hawks are quite common in the northern part of Oneida County, New York, and do not seem to decrease in numbers so fast as other Hawks; which is partly due to their retiring habits, and again to their quick movements, which make it difficult to shoot them. They do not, like most other Hawks, circle through the air in search of food, but skulk around in thick trees and bushes and pounce on their prey when least expected. When they seize a bird or mammal, no matter how small it may be, they always fly at once to the ground with it. When they wish to carry their prey to any distance, they do it by short flights just above the ground. They have a peculiar habit of stretching out their legs as far as they can, as soon as they seize their quarry, as if they were afraid of what they had caught. Their food, according to my observation, consists principally of mice and other small mammals, as well as of small birds, but in this region they kill very few of the latter.

"The only call note I have ever heard the Sharp-shinned Hawk utter sounds very much like the "cac, cac, cac" of the Flicker, and is exactly like that of the Cooper's Hawk, excepting perhaps, that it sounds a little shriller and not quite so loud. These birds resort to the woods, thick with evergreen trees, during the breeding season, but at other times they are often seen in the open, especially in swampy places, where they prey on mice and moles, usually common in such situations. I have often seen them in Florida during the winter, but have never found them breeding there. Southern people call this Hawk the 'The Little Blue Darter,' and Cooper's Hawk the 'Big Blue Darter.'

"The nests of the Sharp-shinned Hawk are usually very large for the size of the bird, so much so that it is impossible by looking from the ground to tell whether one is occupied or not. They measure sometimes from 25 to 30 inches in diameter and from 4 to 6 inches in depth on the outside, and from 10 to 12 inches in diameter by from 2 to 4 inches on the inside. They are placed in evergreen trees, usually against the trunk, on branches growing out from it. I have never found one in any other situation. The nest is generally made of large twigs and lined with small ones, but sometimes a few pieces of hemlock bark, or a few strips of the inner bark, or a little of both, are added to the lining. They are situated from 25 to 50 feet above the ground, usually about 40 feet. The nests are evidently made by the birds themselves, as I know of no other species that builds exactly like them.

"In this locality Sharp-shinned Hawks commence laying about the middle of May, and fresh eggs can be found from that time until the 1st of June. The male may assist in incubation, but I think not, for I have found a number of their nests and some of them I visited several times, but every time that I could make sure of the sex of the bird it proved to be the female. I have, however, seen the male bring food to its mate while she was incubating.

"I have never known of but one instance here in the North of a nest belonging to birds of prey being defended by its owners, and this was a nest owned by a pair of this species, that was found during the present season, 1890. Usually, when a nest of these birds is disturbed, the female will fly around among the trees some distance away and utter her Flicker-like cries, while the male will keep out of sight altogether. This nest was found on May 9, and even then the female was quite angry and struck at the climber several times, although she had laid but three eggs. On May 16, Mr. Egbert Bagg and I went to collect the eggs, and on our arrival found the female sitting and the male just flying to the nest with a mole in his claws. As we reached the foot of the tree the female flew to a limb near by and the male disappeared from view, and till the climber reached the nest the only demonstration made by either was the cry uttered by the female. Nevertheless she was fierce enough then, striking at him every few seconds while he was in the tree, a period of about a quarter of an hour, except in a single instance, when she hurt herself by flying against a limb, which kept her quiet for two or three minutes. During the last half of the fight the male took part and struck at the climber as determinedly as his mate. This is the only time I have ever known a male of this species to make any kind of demonstration while its nest was being robbed. The set of eggs belonging to this pair of Sharp-shinned Hawks is the earliest I have ever found. The nest was peculiar, much deeper than usual, measuring about 7 inches in depth on the inside, while the majority are generally quite shallow in comparison with the diameter. This little Hawk is not uncommon in portions of St. Johns and Putnam Counties, Florida, during the winter, and I think that a very few remain to breed. The native hunters say that they stay during the summer, and I have spoken with some that claim to have found their nests. The latest date on which I saw any of these birds during the season of 1891 was March 26. Like the Marsh Hawk, they will sometimes fly just in front of a forest or prairie fire to catch the small mammals that are driven out by the heat."

Mr. J. H. Sage found a nest of this species containing four fresh eggs in a small white pine tree, in a grove of these trees and hemlocks, near Portland, Connecticut, May 30, 1889. This nest was composed entirely of pine and hemlock sticks and twigs; it was saddled on the base of a limb and built partly around the trunk of the tree. The nest was 8 inches thick, 30 inches in longest and 23 inches in shortest diameter; the depression for the eggs was very slight; it was 48 feet from the ground. The female was on the nest and was shot as she flew from it; she was in the young plumage. The male was flying about, continually uttering the peculiar cry of this species. The Sharp-shinned is one of the most abundant Hawks in the vicinity of Portland, and is seen occasionally in the winter.

Mr. Lynds Jones, of Grinnell, Iowa, found eggs of this species on May 2, and writes me "that in this locality they breed occasionally in hollows in 'American lindens,' and in such cases the nest is made of the inner bark of this tree and of the wild grapevine, with a lining of grass and feathers. When built in a tree (an open nest) sticks are used. It generally chooses limbless trees, most frequently oaks, to nest in, from 15 to 60 feet up. I believe the female alone incubates, from the fact that at this time the male only is seen searching for food."

In the vicinity of St. Johns, New Brunswick, Mr. J. W. Banks says this species nests usually in the latter part of May and the beginning of June;

the nests being placed against the trunks of spruce and fir trees, at a height of about 30 feet. He found two fresh eggs on one occasion, May 5, 1880; and on May 8 the nest contained four eggs, but this is unusually early for this locality. Their call note is a clear "chee-up, chee-up."

In the West, in Oregon, Washington, and Idaho, these birds are not very common, and I found only two nests. One on Craig's Mountain, near Fort Lapwai, Idaho, May 24, 1870, was in a spruce grove, in a tree of this species, close to the main stem, and about 20 feet from the ground; it contained four nearly fresh eggs. The nest was composed of small twigs of the service-berry bush, rather shallow, and contained no lining of any kind.

Another nest of this species was taken by me near Fort Klamath, Oregon, May 18, 1883, containing five handsome fresh eggs. This was composed entirely of fine willow and sage bush twigs, none thicker than a lead pencil, and evidently selected with care as to size. The nest was about 20 inches in diameter outside by 7 inches deep, and placed in the top of a small bushy black pine, close to the main trunk about 25 feet from the ground. In both instances the birds themselves betrayed the location of the nests by their solicitous actions. The females were in each case noisy and demonstrative, and the owner of the last nest swooped twice pretty close to me, scream-ing fiercely at the same time. This nest was shallow and contained no lining, but was very compactly and strongly built, and well hidden by the lower branches.

According to Mr. C. F. Morrison this species breeds in abandoned Magpies' nests along the La Plata River in Colorado; and three eggs taken by him near Fort Lewis, Colorado, on June 22, 1886, now in the U. S. National Museum collection, were said to have been deposited in a dilapi-dated nest of a Magpie, the arched roof of which had fallen upon the main nest, forming a hollow, which was lined with a few feathers upon some dead leaves which had partially filled it the fall before.[1]

Occasionally the Sharp-shinned Hawk is said to nest in cliffs, and while this may possibly be of more frequent occurrence in the Arctic regions, where the habits of our birds are not as well known as one could wish, such nesting sites, as well as those in hollow trees, must, in the United States at least, be considered as decidedly rare and exceptional. The Sharp-shinned Hawk nests in evergreens from choice, and if these be absent, it may resort to a birch, an oak, a maple, chestnut, or hickory tree; if any conifers are found in the vicin-ity of their breeding grounds it invariably nests in these. It is a late breeder, in fact one of the tardiest of our Raptores. As a rule the eggs are seldom laid earlier than the first week in May, usually in the last half of this month, and not infrequently in June. A set of four was taken as late as May 20, near Washington, D. C.

The nests of this species are usually fairly well constructed, better than those of the majority of our Raptores, and it generally builds its own nest, seldom using those which have been abandoned by other species.

---

[1] Ornithologist and Oölogist, vol. XII, 1887, No. 2, p. 27.

The usual number of eggs laid by the Sharp-shinned Hawk is four or five; larger sets are rare. Mr. C. L. Rawson, of Norwich, Connecticut, took a set of seven on June 1, 1881. Aside from the unusually large number of eggs, the situation of the nest from which they were taken was also rather remarkable, being placed in a low pine, only 10 feet from the ground.

The eggs are laid at intervals of one and two days, and incubation does not commence till the set is complete; the female guarding her nest constantly, however. Incubation lasts probably about three weeks, and the male does not appear to assist in this duty, but supplies the food for his mate during the time. The young, when first hatched, are covered with white down and grow rapidly. But one brood is raised in a season.

The shape of these eggs is nearly a perfect oval; an occasional specimen may be called short ovate. The ground color is usually a pale bluish or greenish white, which fades with age into a dull gray white. In a few specimens the ground color is almost completely hidden by confluent markings of cinnamon-rufous.

The eggs are mostly heavily blotched, spotted, and marbled with various shades of brown, the darker of these tints predominating; these are again mixed with different shades of drab, fawn color, lavender, and clay color. The different patterns of markings are endless in variety, in some specimens they are heaviest on the larger, in others the smaller end; in a few they are disposed in the shape of a wreath in the center, leaving both ends of the egg nearly unspotted. In others, again, they are pretty evenly distributed over the entire egg, and a few are almost unspotted. Finally, in some specimens, the markings run into each other, giving the egg a clouded appearance.

The average size of fifty specimens in the U. S. National Museum collection is 37 by 30.5 millimetres. The largest egg measures 39 by 32, the smallest 35 by 29 millimetres.

Of the type specimens, No. 20675 (Pl. 5, Fig. 11), selected from a set of three eggs, was taken in Massachusetts, May 29, 1875; No. 20677 (Pl. 5, Fig. 12), from a set of four, taken June 1, 1878, near Eastford, Connecticut; No. 20679 (Pl. 5, Fig. 13), from a set of five, taken at Fort Klamath, Oregon, May 18, 1883; No. 20680 (Pl. 5, Fig. 14), taken May, 1879, at Blue Ridge, Pennsylvania; all from the Bendire collection. No. 21042 (Pl. 5, Fig. 15), from an incomplete set of two, taken near Edinburg, Texas, in 1878, from the Merrill collection. No. 22812 (Pl. 5, Fig. 16), from a set of four, taken near Redding, California, by Mr. L. W. Green, May, 1886; and No. 23566 (Pl. 5, Fig. 17), from a set of three taken by Mr. Charles F. Morrison, near Fort Lewis, Colorado, June 22, 1886.

66. Accipiter cooperi (Bonaparte).

COOPER'S HAWK.

Falco cooperi Bonaparte, American Ornithology, II, 1828, 1, Pl. x, Fig. 1.
Accipiter cooperi Gray, List of Birds in British Museum, Accipitres, 1844, 38.
(B 15, 16, C 339, R 431, C 495, U 333.)

Geographical range: Whole of temperate North America, including the greater part of Mexico.

With the exception of Alaska, the breeding range of the Cooper's Hawk is coextensive with the limits of the United States. It is known to breed from Maine to Florida, from Louisiana and Texas throughout the interior, and on the Pacific coast from northern Lower California (Cape Colnett, latitude 31°), to our northern border. It is also met with along the southern portions of the Dominion of Canada, from Newfoundland and New Brunswick westward to Manitoba, where it is reported as tolerably common in the vicinity of Winnipeg, up to latitude 50°, and stragglers probably reach points still farther north, as Dr. T. M. Brewer reported it from the Saskatchewan. It also occurs in British Columbia, both on the coast and in the interior, and undoubtedly breeds there as well. In the more northern portions of its range it is only a summer resident, wintering from about latitude 39°, southward, and passing into Mexico. It breeds most commonly in the Middle and Northern States, becoming rarer as the extreme limits of its northern and southern range are reached.

Cooper's Hawk must be considered as one of the few really injurious Raptores found within our limits, and as it is fairly common at all seasons throughout the greater part of the United States, it does, in the aggregate, far more harm than all other Hawks. It is well known to be the most audacious robber the farmer has to contend with in the protection of his poultry, and is the equal in every way, both in spirit and dash, as well as in bloodthirstiness, of its larger relative, the Goshawk, lacking, however, the strength of the latter, owing to its much smaller size. It is by far the worst enemy of all the smaller game birds, living to a great extent on them as well as on small birds generally.

It does not appear to be especially fond of the smaller rodents; these, as well as reptiles, batrachians, and insects seem to enter only to a limited extent into its daily bill of fare, and unfortunately it is only too often the case that many of our harmless and really beneficial Hawks have to suffer for the depredations committed by these daring thieves.

The flight of Cooper's Hawk is both easy and graceful, and ordinarily not especially swift. He may most often be seen skimming along close to the ground, in rather a desultory manner, usually skirting the edges of open woods or clearings; but once in sight and in active pursuit of its selected prey it darts in and out through the densest thickets with amazing swiftness, where it would seem impossible for it to follow successfully; especially is this the case when chasing some small bird that generally tries to take refuge in such places. It manages, however, with the assistance of its long tail, which helps it very materially, to turn suddenly and double with remarkable ease, even in dense undergrowth, arresting its flight instantly, and darting off, perhaps, at a right angle the next second to capture its selected victim.

The favorite haunts of Cooper's Hawk are moderately timbered districts, interspersed with cultivated fields and meadows, but it is also found in the more extensive and heavily wooded mountain regions, and in the West on the open and almost treeless prairies. It is more or less migratory throughout its range, excepting in the extreme southern parts, where it is perhaps a constant resident. Birds nesting in the northern half of the United States I believe migrate regularly, and are replaced by others coming from points still farther north, which would naturally lead to the belief that they were constant residents when this is not the case.

In the Middle States they usually return from their winter haunts about March 15, and by the end of the month they are located again on their breeding grounds, which are generally resorted to from year to year by the same birds, if not persistently disturbed.

As a rule I believe a new nesting site is selected each season at no great distance from that of the previous year, but occasionally the old one is resorted to for successive seasons. In the choice of these, old Crows' nests when available are given the preference, but larger Hawks' nests, as well as those of squirrels, are frequently used. When they build nests of their own they are usually placed between the diverging limbs in a crotch of the tree or saddled on some smaller limb close to the trunk. This last is most often the case when they nest in thick and bushy conifers. No preference seems to be given to any particular kind of tree, and their nests are generally found at no very great height from the ground, ranging from 20 to 50 feet up, rarely lower or higher.

Their nests vary considerably in bulk, according to the locations in which they are placed, and are composed of sticks lined with finer twigs, and scattered among these are generally found small scales and flakes of the outer bark of different species of trees, those of the yellow pine being preferred when obtainable. Nests built by the birds themselves compare favorably with those of other Raptores.

On the plains, where, from scarcity of suitable timber elsewhere, they are confined to the shrubbery of the creek bottoms, consisting mainly of cottonwoods and willows, they sometimes nest as low as 10 feet from the ground, and I have here found some of their nests fairly well lined with the dry inner bark of the cottonwood and with weed stalks; while in the vicinity of Grand Forks, North Dakota, according to information furnished me by Mr. G. G. Cantwell, they are said to nest occasionally directly on the ground.

Even in the more southern parts of their range nidification rarely begins before April 15. In the southern New England States, as well as in New York, Pennsylvania, Indiana, Michigan, Minnesota, Nebraska, Cooper's Hawk occasionally begins laying about May 1, but usually not before the second week of this month and frequently later. Even in Lower California and southern Arizona it rarely lays as early as April 20, and in the mountain regions of Colorado and Montana the time is protracted until the beginning of June.

Mr. Denis Gale, of Gold Hill, Boulder County, Colorado, writes me: "On June 25 I found a nest of a Cooper's Hawk containing four unmarked bluish white eggs, resting upon some thin flakes or scales of spruce bark, which alone constituted the lining of the nest, the available contrivance for which was a large bunch of matted scrub, an excrescence upon a horizontal limb, about 18 inches from the trunk and about 20 feet from the ground. This bunch consisted of a wonderful growth of very densely interlaced twigs, the surface of which offered a commodious nesting site, having not only an ample flat area, but a sufficient depression in its center to meet every requirement for a nest. On July 2 these four eggs were represented by four bright lively Hawklets, densely covered with white down. Neither food nor feathers of any kind were found in or about the nest. The character of this species is bold and brave, especially so in defense of its young. It has a rapid and graceful flight. The skill with which it pursues its devious path along a creek bottom, noiselessly following the intricate opening close to the ground through the trees and undergrowth is remarkable.

"From the date at which they have their young to provide for, until late in the season, this seems to be their favorite mode of hunting, but especially is it so after the hen Grouse, accompanied by her young brood, has sought such localities for food and water, and where they loiter throughout the summer."

On the Pacific coast, in Oregon, Washington, and Idaho, I found Cooper's Hawk rather rare, while in southwestern Montana it is more abundant than any of the other Raptores. It is also common in Arizona.

Throughout the greater part of the year Cooper's Hawk is usually a silent bird and rarely utters any notes, but during the mating season it frequently emits a cackling or chattering sort of noise, also a note something like "tick, tick" frequently repeated. When disturbed on the nest, it usually screams shrilly and violently, but generally flies off and watches proceedings from a safe distance. Others, however, show a good deal of courage at times, and boldly dash at the intruder.

As far as known only a single brood is raised in a season, incubation lasting about twenty-four days, the male assisting to a slight extent only in these duties, but keeping its mate supplied with food. The number of eggs laid varies from two to six. In the more southern parts of its range it usually lays three or four, and sometimes but two eggs, while along our northern border and the southern New England States, they range from four to six in number, sets of five being rather common.

Cooper's Hawks are persistent layers. The loss of their first set does not at all discourage them; in due time they will have a second, and if this is taken, even a third one, each one usually smaller in number than the first clutch. Some birds show great attachment to the original nesting site, and will continue laying in the same nest, even after being repeatedly robbed; others abandon the rifled nest each time and select a new site, generally in the same woods, however. Mr. C. J. Pennock records taking a set of four eggs from the nest of a Cooper's Hawk on April 24, 1874; May 5 two more eggs were taken from the same nest, and on May 11 two others. Later in the season (about August 1), on visiting the same locality, two young Hawks of this species were seen, but it is not known if they were reared in the old nest.[1]

The eggs of Cooper's Hawk are deposited at intervals of one or two days, and incubation does not begin until the set is nearly completed.

Their ground color varies from a pale bluish white to a greenish white tint, which fades out considerably in time. Occasionally a much higher tinted set is found. Mr. C. J. Pennock has a set of five eggs in his collection, in which the ground color is a rich bright green, and four of these eggs are distinctly and handsomely marked. They were collected by himself near Kennett Square, Pennsylvania, May 2, 1887.

While many of these eggs are perfectly immaculate, fully one-half of those in the U. S. National Museum collection are spotted with irregular markings or scrawls of different shades of brown, drab, or fawn color. These markings, in most cases, are rather faint and irregularly scattered over the egg, usually heaviest about the larger end.

The average size of sixty-two eggs in the U. S. National Museum collection, from various parts of the United States, is 49 by 38.5 millimetres. The largest egg of this series measures 51.5 by 42, the smallest 43 by 34 millimetres.

Of the type specimens No. 23003, selected from a set of four eggs (Pl. 5, Fig. 18), taken by First Lieut. H. C. Benson, Fourth Cavalry, U. S. Army, near Huachuca, Arizona, May 12, 1887, shows the scrawls referred to above; and No. 23306 (Pl. 5, Figs. 19 and 20), both from a set of five, taken by Mr. Jerome Trombly, near Petersburg, Monroe County, Michigan, May 14, 1885, show one of the better marked types, and an unspotted egg. Four eggs of this set are distinctly spotted. These eggs were taken from an old Crow's nest placed in the forks of a pin oak, a trifle over 40 feet from the ground. The Hawks had simply repaired the inside of the nest, and lined it slightly with a few dry leaves, a little moss, and shreds of bark.

[1] Bulletin Nuttall Ornithological Club, Vol. III, 1878, p. 41.

67. Accipiter atricapillus (WILSON).

AMERICAN GOSHAWK.

Falco atricapillus WILSON, American Ornithology, VI, 1812, 80, Pl. 52, Fig. 3.
Accipiter atricapillus SEEBOHM, History of British Birds, I, 1883, iv.
(B 14, C 340, R 433, C 496, U 334.)

GEOGRAPHICAL RANGE. Northern, central, and eastern North America; south in winter to the Middle States and southern Rocky Mountains, straggling west into Oregon (Fort Klamath).

The Goshawk, one of our handsomest birds of prey, breeds principally north of the United States, occurring more or less commonly, though nowhere abundantly, in suitable localities throughout the British Dominion, from the shores of the North Atlantic Ocean to the Arctic and Bering Seas.

Within the United States its breeding range is confined to the extreme northern border, breeding sparingly but regularly from central Maine northward, and probably also in the northern portions of Vermont, New Hampshire, and New York. In the Rocky Mountain region it is said to breed as far south as Colorado. West of the Rocky Mountains it occurs as a straggler only, but possibly breeds in limited numbers in the Bitter Root Mountains and the spurs thereof in Idaho. In eastern Washington and Oregon it is replaced by the western race, a much darker-colored bird. A typical specimen of *Astur atricapillus* has, however, been taken on the eastern slopes of the Cascade Range, near Fort Klamath, Oregon, by Dr. James C. Merrill, U. S. Army, on March 11, 1887, the most westerly record I know of. Throughout the United States it is by far more abundant during the winter months than in the breeding season. At this time it keeps more to the mountains, where game is abundant, while in the fall and winter it may often be seen in the fertile and cultivated valleys adjacent to these, in search of prey, and is naturally much more readily noticed. Five out of six of the birds seen at this time are the young of the year, which have been left by the parents to their own resources.

The Goshawk is the boldest and by far the most destructive of the North American Raptores; infinitely more injurious to our game birds, and the poultry yard as well, than any other species, the large Gyrfalcons not excepted. Notwithstanding its comparatively short wings, its flight is powerful and swift; it is strong and active in body, shy and keensighted, savage and bloodthirsty in disposition, a veritable terror to all smaller birds, and more than a match for others considerably larger than itself. It loves to destroy life for the sake of killing. In northern Alaska, along the shores of the Yukon River, it is perhaps more abundant than anywhere else within its range, feeding principally on the numerous Ptarmigan, lemmings, and Arctic hares to be found there at all times in abundance.

Mr. L. M. Turner says: "The American Goshawk is a common species throughout the Yukon Valley, and apparently confines itself entirely to the mainland, although plentiful along the seashore. Specimens were obtained from Fort Yukon, the Yukon Delta, and the vicinity of St. Michael. The tracts preferred by this Goshawk are the narrow valleys, borders of streams, and open tundras, which it constantly scans for Ptarmigan and small animals, the lemming forming a considerable portion of its food. It will sit for hours in some secluded spot waiting for a Ptarmigan to rise on its wings. No sooner does its prey rise a few feet from the ground, than with a few rapid strokes of the wing and a short sail, the Goshawk is brought within seizing distance; it pounces upon the bird, grasping it with both feet under the wings, and after giving it a few blows on the head they both fall to the ground, often tumbling several feet before they stop, the Hawk not relinquishing its hold during the time. * * * I have seen this Hawk sail, without a quiver of its pinions, until within seizing distance of its quarry, and suddenly throw its wings back, when with a clash they came together, and the vicinity was filled with white feathers floating peacefully through the air. I secured both birds, and found the entire side of the Ptarmigan ripped open. * * *

"It is a resident of the interior, and comes to the coast quite early in spring, as is attested by the fact that I killed one specimen April 28, and a fine example was brought to me from the mouth of the Uphún (part of the northern Yukon Delta) where it was killed April 25. It was a female, and contained an egg ready for extrusion, which had already received a pale bluish green color on the shell. The bird was shot on the nest which was placed in a small poplar tree. The nest was composed of sticks and a few blades of grass. The size was quite bulky, measuring nearly 2 feet in extreme diameter, and having but a slight depression. The bird was extremely vicious, choosing to remain on the nest rather than desert it. The male attacked the native, tore his cotton shirt into shreds, and snatched the cap from the head of the astonished man, who was so surprised at the impetuosity of the attack that he struck wildly at the bird with his arm, and before he could reload his gun the bird took flight.

"This Goshawk breeds wherever found in summer, placing its nest in a tree or shrub, or even on the edge of a cliff, inaccessible to foxes or other enemies."[1]

Mr. Manly Hardy, of Brewer, Maine, writes me: "According to my observations the Goshawk does more damage to game and poultry than all other hawks combined. They live almost entirely on Ruffed Grouse, domestic fowls, and rabbits. I have known one to destroy five Ruffed Grouse in one morning, tearing them to pieces and leaving them; and have also known one to tear out a Grouse's leg and hip and to swallow it whole, but never knew one to be taken with an empty crop.

"There is both an old and new nest of this species on my own land. The first about 140 yards from a blacksmith's shop. These are the only two I ever heard of having been found in this vicinity. I think they occupied the first

---

[1] Contributions to Natural History of Alaska, Turner, 1886, p. 157.

nest at least six or eight years, as ten or twelve adults and young were shot in the near vicinity; but so shy were they that no one ever suspected their nesting until the nest was accidentally found. A female was shot from it in 1877, and another in 1878. They then deserted the locality, and I have recently found where they rebuilt. The nesting sites were in both instances in white birches, about 20 feet from the ground. A very small young one was taken from the last nest June 5, 1887. They probably commence laying here about May 1."

Mr. J. W. Banks, of St. Johns, New Brunswick, writes me: "I know of two nests of the Goshawk taken in this vicinity. One nest about 10 miles away, and from which one of the parents was shot, was built in the forks of a large birch tree. It contained two fresh eggs, and was taken about the last of April. Another nest was found about 25 miles from this city; both parents were shot, and the two young birds it contained were brought here to be stuffed. The male only was in the mature plumage. The young were about as large as Brahma chicks six weeks old, and covered with a thick coat of pure white down without a sign of pin feathers. This nest was also placed in the forks of a birch tree. It was found May 5."

According to Mr. W. L. Bishop's observations in Nova Scotia, this Hawk breeds on high ground, mostly in hard wood timber, and nearly all the nests found by him were placed in tall slim beech or birch trees, about 40 feet from the ground, built of sticks, and lined with a few green hemlock twigs, yellow birch bark, and a few feathers.

Mr. L. M. Turner says that this Hawk is rather rare in Labrador, but as he observed it near Fort Chimo in May, it certainly breeds there. On May 1, 1876, he took a handsome set of five eggs in the Yukon River Delta, Alaska, and another incomplete set of two eggs on the same date a year later. These eggs were fresh when taken and are now in the U. S. National Museum collection. The nests were placed in poplar trees. Another set of four was taken by Capt. W. H. Dall, of the U. S. Coast Survey, on April 28, 1868, near Nulato, Alaska.

I cannot add any personal observations concerning the life history of this species. It nests early, even in the northern regions; beginning to lay about April 1, or even earlier, in the more southern portions of its range, and in Alaska from about April 20 to May 10. Incubation, as with most of the larger Raptores, lasts about four weeks. But a single brood is raised in a season.

The number of eggs laid by the Goshawk varies from two to five. In the North, where an abundance of food can be readily procured for the young, the larger sets of four or five eggs seem to predominate. They are pale bluish white in color and unspotted. An occasional specimen shows slight traces of brownish-buff markings, which are probably old blood stains. The shells of these eggs are somewhat rough to the touch, deeply pitted, and granulated. They vary in shape from ovate to elliptical ovate. The average measurement of seventeen specimens in the U. S. National Museum collection is 59 by 45.5 millimetres, the largest egg measuring 64 by 47.5, the smallest 54 by 45 millimetres.

The type specimen (No. 20684, Pl. 6, Fig. 1), selected from a set of five, was taken by Mr. L. M. Turner, U. S. Signal Service, May 1, 1876, in the Yukon River Delta, Alaska.

68. Accipiter atricapillus striatulus RIDGWAY.

WESTERN GOSHAWK.

Astur atricapillus var. *striatulus* RIDGWAY, in History of North American Birds, III, 1874, 240.
Accipiter atricapillus striatulus RIDGWAY, Proceedings U. S. National Museum, VIII, 1885, 355.

(B —, C —, R 433*a*, C 497. U 334*a*.)

GEOGRAPHICAL RANGE: Western North America: north to Sitka, Alaska; south to California; east to Idaho.

The breeding range of the Western Goshawk extends from Sitka, Alaska, south through the mountains of the coast region in British Columbia and the States of Washington and Oregon, to about latitude 38° in California. Mr. Ridgway's statement, in his Manual of North American Birds, "breeding at least south to latitude 30°," is evidently a typographical error, and should read "39°."

According to my observations, the general habits of the Western Goshawk are very similar to those of its eastern relative; it is equally destructive to small game of all kinds, particularly the Sooty, Ruffed, and Sharp-tailed Grouse, as well as to the fowls of the poultry yard. While nowhere abundant, it seems to be pretty generally distributed throughout the Blue Mountain region of Oregon and Washington, and breeds in suitable localities where food is plenty. During spring and summer it is seldom seen in the more open districts, though it is abundant enough later on, when the heavy snows drive the game into the foothills and lower valleys. I have shot quite a number of these birds at various times, and all, as far as I am aware, are referable to this subspecies, with one exception, which is intermediate between it and the preceding.

My first acquaintance with the Goshawk dates back to 1870, and on April 21, 1871, while hunting in Lawyer's Cañon, 30 miles south of Fort Lapwai, Idaho, I found a nest of this subspecies containing a single egg. It was placed in the forks of a large cottonwood tree about 50 feet from the ground, and was a bulky affair, fully 2 feet in diameter and quite as deep. The nest was composed of sticks, some of them quite large and loosely put together. It was rather shallow on top and lined with weed stalks, a species of wild nettle, and a few pine needles. The parent on the nest was quite fierce, and although she did not actually strike the man who climbed to it, she came several times very close to him, uttering shrill screams of anger and protest. While stationed at Camp Harney, Oregon, with the assistance of woodchoppers and Indians, several of their nests were found, none nearer than 2 miles from each other. One, found on May 26, 1875, contained two young just hatched and an egg already chipped. It was in a cluster of pines a few miles northeast of the post. Both parents were exceedingly aggressive, and several shots fired close to them did not seem to intimidate them in the least. I could have easily killed both, but refrained on account of the young. Only one of the birds was in the adult plumage.

A nest found on April 18, 1876, was placed in the top of a tall and bushy juniper tree, only about 20 feet from the ground. It was not as large as the two former, and looked as if it had been newly built. It was situated in a fork of the main trunk and was well hidden. The female was on the nest and commenced screaming before we came within 20 feet of the tree, which caused the discovery. She defended her eggs valiantly, and did not cease her attacks on the climber till he finally succeeded in hitting her with a club, which caused her to leave. The male was not seen. The nest contained three slightly incubated eggs, and was sparingly lined with the dry inner bark of the juniper trees growing in the vicinity. On April 9, 1877, I found another nest not far from where the first was taken in 1875. This was built in a tall pine, at least 50 feet from the ground, and in addition to the usual juniper bark lining it contained a few green fir tops. This also contained three eggs, and incubation had already commenced. I shot the female, a handsome bird in the adult plumage, while it was circling about the climber and trying to strike him. The largest set obtained was one of five eggs. The nest was placed in a bushy pine in a cañon of the Blue Mountains, close to the road from the Umatilla Indian Agency to Grande Ronde Valley, Oregon. This nest, evidently used for years, was well out on one of the larger limbs and placed in a fork of it. It was quite large, and slightly lined with grass, tree moss, "usnea," and a few scales of pine bark; distance from the ground about 50 feet. Both parents were present, and the female was shot, as she was too aggressive for the comfort of the climber. The male was also rather demonstrative, but not to the extent of his mate. The eggs were nearly hatched when found, April 17, 1881. All the cavities of the nests were very shallow, none being over 1½ inches deep. While none of the nesting sites were in the denser portions of the forests, they were all found in the heavy timber, and generally on the slopes of cañons not far from water.

A pair of these birds bred within a mile or so of Fort Klamath, Oregon, in the spring of 1883, but I never succeeded in locating the nest on account of the dense timber found all around the post and the large size of most of the trees. Both birds, but especially the male, a handsome specimen in the adult plumage, paid regular daily visits to the different poultry yards of the garrison, and almost invariably managed to get away with a chicken. I, as well as others, tried time and again to put an end to these altogether too frequent depredations, but no one succeeded in killing either of the birds. They were too smart to be caught, knew just when to make their raids, and were successful nearly every time.

The following incident will illustrate the perfect fearlessness and audacity of these Hawks. I was returning from a short hunt one afternoon in September, 1882, my breech-loader charged with dust shot. At the outskirts of the garrison, near the cavalry stable, was an old brush corral, much frequented by the fowls kept in the neighborhood. While walking past this fence I suddenly heard a great outcry and saw quite a commotion among a number of chickens in the place, which were squeaking and scattering in all directions at a lively rate. At the same instant a large Goshawk, an adult female, dashed through the inclosure, failing to get a chicken this time, however. I fired at her at short range, and, as it subsequently proved, peppered her well with dust shot as she went by, which possibly disconcerted her aim a little. Never dreaming for an instant that the bird would return after such a reception, I nevertheless inserted a heavier cartridge in my gun, and had scarcely done so when she came back to make a second and last attempt at a too venturesome chicken. This time I brought her down with a broken wing, and her flight was so suddenly arrested that she rolled over several times after striking on the ground. I never saw more vindictive fury expressed in a bird's eyes than was shown by hers. She tried to attack me, and would have done so had she not been so badly wounded. The will and courage to do so were there, but her strength failed her. On skinning her I found a number of dust shot imbedded under the skin, showing that she had been hit the first time I fired. This, though, was not sufficient to cause her to leave without her intended victim, notwithstanding the fact that she saw me plainly enough the second, if not the first time. When its appetite for blood is once excited, the Goshawk is certainly devoid of all fear and discretion as well, while under ordinary circumstances there is no shyer bird to circumvent and bring to bag.

Mr. L. Belding met with this bird in Calaveras County, California, where it seems to be a summer resident.

The Western Goshawk becomes strongly attached to the locality once chosen for a breeding ground. I noticed, especially at Camp Harney, Oregon, that considerable discrimination was shown by these birds in the selection of such places only in which the game they fed on was most abundant.

Besides a shrill scream of anger they have a call note resembling the word "keeah, keeah," or "kree-ah," frequently repeated, this note being often uttered in the early spring. Their food consists of the different game birds found in the country they inhabit, especially the Sooty Grouse, as well as hares and smaller mammals. One of the nests found by me contained the partly-eaten remains of a Columbian Sharp-tailed Grouse.

Nidification begins early, usually about the latter part of March or the beginning of April, long before the snow has disappeared from the mountains and while the hillsides are still saturated with moisture, making it anything but easy work to look for their nests. These are usually built in tall trees, and no particular preference seems to be shown in their selection. The nests are mostly placed close to the trunk and generally well hidden from view. Occasionally one is placed some distance out, or between the forks of one of the larger limbs, and on that account can be more readily seen. I believe each pair of birds has its regular hunting range, from which all other species of Raptores are driven off. At any rate I never found the Western Red-tailed or Swainson's Hawks, the most common kinds found, breeding in the vicinity of a pair of Goshawks.

The eggs are from three to five in number and indistinguishable from those of the eastern bird. I believe but one brood is raised in a season, and do not know whether the male assists in incubation.

The eggs in the U. S. National Museum collection, nine in number, average a trifle larger than those of the preceding subspecies, the mean being 60 by 45.5 millimetres. The largest specimen measures 65.5 by 49.5, the smallest 55 by 44 millimetres. One of these eggs shows some peculiar discolorations of the shell, which are not stains, neither are they spots; and another has very minute and scarcely perceptible markings of pale, rusty brown near the smaller end.

The type specimen (No. 20696, Pl. 6, Fig. 2), selected from a set of five, from the Bendire collection, was taken by the writer near the Umatilla Indian Agency, Oregon, April 17, 1881.

69.   Parabuteo unicinctus harrisi (AUDUBON).

HARRIS'S HAWK.

Falco harrisi, AUDUBON, Birds of America, v, 1839, 30, Pl. 392.
Parabuteo unicinctus var. harrisi RIDGWAY, in History of North American Birds,
    III, January, 1874, 254.
           (B 46, C 348, R 434, C 512, U 335.)

GEOGRAPHICAL RANGE: Middle America; north to southern border of United States (Louisiana to southern Arizona), Lower California.

The breeding range of Harris's Hawk within the borders of the United States is rather restricted, it being found only along our southern frontier from southern and southwestern Texas to southern Arizona and Lower California. It is said to occur also occasionally in Louisiana, but it is doubtful if it breeds there. It extends through Mexico to Central America.

Within the United States it is most abundant in southern and southwestern Texas, where it is a resident and where it breeds commonly.

According to Mr. G. B. Sennett, "it is a resident on the Lower Rio Grande, and more abundant than any other of the family. I found in the crops of those I obtained mice, lizards, birds, and often the Mexican striped gopher (*Spermophilus mexicanus*), proving them to be active hunters, instead of the sluggish birds they appeared the year before at Brownsville. They are silent and not very shy. Young from the egg are covered with down, more plentifully on crown and back than elsewhere; are colored white on under parts, shading to buff and ochraceous on back and head, and are very pretty little chicks. Dissection of a female on April 11 disclosed an egg almost ready to be laid. On April 22, two sets of eggs were taken, 4 or 5 miles from Lomita, one 25 feet high in an ebony tree, the other 20 feet high in a mesquite. Each contained three eggs hard sat upon."[1]

Capt. B. F. Goss found this species very abundant in the vicinity of Corpus Christi, Texas, in the spring of 1883, and took quite a number of their nests and eggs. He found one containing eggs as early as February 18. He says: "They build a substantial nest of sticks and weeds, lined with grasses and small roots, after the fashion of the Buteos, but do not finish it as nicely as the latter. I examined eighteen of their nests, and there was nothing remarkable about them, except that they were often placed quite low, when higher sites were equally available. The average number of eggs found by me to a set was three, but sets of four are not uncommon."

Personally I met with the nest of this bird on but three occasions during the spring of 1872, while stationed near Tucson, Arizona. One of these nests, containing two fresh eggs, was found on May 17. It was a bulky structure, placed in a low bushy cottonwood tree, in a fork about 20 feet up, about 10 miles below Tucson, near the Laguna, the sink of the Santa Cruz River. It was composed of sticks, and sparingly lined with pieces of the dry inner bark of the cottonwood, and grasses. The bird made no hostile demonstrations, but sailed slowly around above the nest out of gunshot range. The inner cavity of the nest was slight.

The two other nests, each containing but two eggs, were found in low mesquite trees, about 15 feet above the ground, on June 4 and June 6, respectively. The first nest was a very slight affair, composed of mesquite sticks, as well as the dry seed pods of this trees, and a little grass. While standing directly under the nest I could see the eggs through the bottom of it. The third one was similarly situated, and both were found on the barren plains west of the camp.

I know but little about the habits of this species, and while in other regions they seem to associate with the Turkey Vultures and the Caracaras, I never saw them doing so in southern Arizona, where both were also found by me. They appeared to be a lazy sluggish bird, their flight slow, and not graceful.

Dr. James C. Merrill, U. S. Army, found a nest of this species near Brownsville, Texas, placed on the top of a Spanish bayonet some 8 or 9 feet above the ground.

[1] Further Notes on the Ornithology of the Lower Rio Grande of Texas, U. S. Geological and Geographical Survey, Hayden, Vol. v, No. 3, 1879, pp. 419, 420.

Mr. L. Belding reports Harris's Hawk as common at the Cape region of Lower California, where he frequently met with it in May, along the route from San José del Cabo to Miraflores. He also found it within 40 miles of San Diego, California. Mr. Walter E. Bryant saw one at San Jorge, and again near San Juan, where a pair had built in a giant cactus, *Cereus*. On April 6, 1889, he found a nest at San Gregorio, built on the top of a bush, *Atamisquea emarginata*. The nest was rather flat, composed of sticks and lined with grass and orchilla. It measured about 2 feet in diameter. It contained two eggs, which were secured, one quite fresh, the other with a small embryo. One of the eggs is white, the other pale greenish white.[1]

The number of eggs varies from two to four, usually three, and these are mostly oval in shape, a few are ovate, and an exceptional one is short ovate. The shell is lusterless and fairly smooth. The ground color is a dead dirty white; perfectly fresh specimens show a slight greenish tint occasionally. The eggs are usually more or less nest-stained, and some of these stains might readily be mistaken for markings.

A careful examination of twenty-eight specimens in the U. S. National Museum collection shows that about one-half of these eggs are unmarked, the remainder are spotted with small irregular blotches of pale cinnamon in some cases and fawn color in others, while some, again, are lavender colored. Only one shade of markings is found on each egg, and none are heavily marked. One of the specimens figured shows the most pronounced markings in the series; in the others they are less distinct, and in some so faint as to be barely noticeable. The eggs are deposited at intervals of several days, but incubation commences as soon as the first egg is laid, and lasts about four weeks. In southern Texas sets of four eggs are by no means rare, while in Arizona and Lower California two seem to be the rule.

The average measurement of the specimens in the U. S. National Museum collection is 54 by 42 millimetres. The largest egg of the series measures 57.5 by 44.5, the smallest 49 by 38.5 millimetres.

The type specimen, No. 20757 (Pl. 6, Fig. 3), from the Merrill collection, is the most distinctly marked egg in the series, and was collected by Asst. Surg. James C. Merrill, U. S. Army, near Fort Brown, Texas. No. 22572, selected from a set of four, an unmarked specimen (Pl. 6, Fig. 4), was taken near Corpus Christi, Texas, on March 30, 1883, and obtained in exchange from Capt. B. F. Goss, of Pewaukee, Wisconsin.

[1] Proceedings Academy Sciences of California, 2d series, Vol. II, 1889, p. 279.

70. Buteo buteo (LINNÆUS).

EUROPEAN BUZZARD.

Falco buteo LINNÆUS, Systema Naturæ, ed. 10, 1, 1758, 90.
Buteo buteo LICHTENSTEIN, Nomenclator Museo Berolinensis, 1854, 3.
(B —, C —, R 435, C —, U 336.)

GEOGRAPHICAL RANGE: Northern portions of eastern hemisphere; accidental in the United States, Michigan (?)

The common European Buzzard is admitted to a place in the North American avifauna on a somewhat questionable record. It certainly does not breed with us. It is a common species throughout the greater part of the continent of Europe, a resident in the southern portion of its range, and only a summer visitor in the northern parts.

Mr. Henry Seebohm, in his excellent work on "British Birds," describes the breeding habits of the Buzzard as follows: "The Buzzard breeds on the outskirts of the forests, whence it issues in search of food, and may often be seen perched on the bare hillsides, waiting for mice and other small mammals, or may be observed crossing the open fields with somewhat heavy and indolent flight. It is equally common in dry as in swampy woods, and breeds in pine forests as freely as in those of beech and oak. In the forests near Brunswick, I found the nests mostly in beech and oak, but in Pomerania many were in Scotch firs, one in a birch, and one in an elm. * * * The Buzzard builds a nest from 1½ to 2 feet in diameter, and if it is in a fork of a tree, sometimes nearly as high. The foundation is of large twigs, finished at the top with slender twigs. It is very flat, the hollow in the middle, containing the eggs, about the size and depth of a soup plate. The final lining is fresh green leaves, generally beech; but in one nest, although it was in a beech tree, the lining was green larch twigs. This lining must be renewed from time to time. Out of eleven nests near Brunswick, five of which contained eggs, five young birds, and one three eggs and a young bird, all but one were lined with fresh leaves. * * *

"The nests varied from 50 to 90 feet from the ground, but some, to which we did not attempt to climb, were higher. In Pomerania I saw several nests in Scotch firs, not more than 25 feet from the ground. My friend Dr. Holland, who has paid great attention to the birds of prey in Pomerania for many years, informs me that the Buzzard begins to lay about the middle of April, that the period of incubation lasts three weeks, and that the male relieves the female at her duties. He tells me that, besides small mammals, the Buzzard will eat grasshoppers and other insects, reptiles, and occasionally small birds, if it gets a chance of catching them sitting. The spines of the hedgehog have been found in the stomach of the Buzzard, and Dr. Holland also mentioned an instance of a female bird having been found dead on the nest, with a live viper under her.

"The Buzzard returns year after year to the same nest, but is said not to breed a second time the same year if the eggs are taken. When the eggs are very much incubated she sits very close."[1]

The eggs are usually from two to four in number, generally three, and according to Seebohm the ground color varies from milky blue to a pale reddish white. They are blotched, streaked, spotted, or clouded with rich brown surface spots and pale lilac shell markings. There is considerable variation in both size and color, some specimens being spotless. They also vary considerably in shape; some are said to be almost round, others oval, some elongated, and, more rarely, elliptical. Their size is given as varying from 57.2 to 50.8 millimetres in length by 48.3 to 41.9 millimetres in breadth.

The U. S. National Museum collection contains a series of eighteen eggs whose average measurement is 55.5 by 44.5 millimetres. The largest of these measures 58.5 by 47, the smallest 52 by 44 millimetres. None of these eggs are figured, as this species does not breed on the North American continent.

71. Buteo borealis (GMELIN).

RED-TAILED HAWK.

Falco borealis GMELIN, Systema Naturæ, I, ii, 1788, 266.
Buteo borealis VIEILLOT, Nouveau Dictionaire, IV, 1816, 478.
(B 23, C 351, R 436, C 516, U 337.)

GEOGRAPHICAL RANGE: Eastern North America, west to border of Great Plains; occasional in eastern Mexico; Panama (casual?).

The Red-tailed Hawk is generally distributed, and breeds more or less abundantly in suitable localities in all that portion of the United States east of the Mississippi River. West of this stream it occurs sparingly in northern Louisiana, and more frequently in Arkansas, the eastern portion of Texas, and in the Indian Territory, Kansas, Nebraska, North and South Dakota. It is fairly common throughout the area indicated, excepting Florida and the Gulf coast generally, where, although very common in winter, it can, from our present knowledge at least, be considered only as a rare summer resident.

North of our border it is found throughout the southern parts of the Dominion of Canada, ranging from Newfoundland, Nova Scotia, and New Brunswick, where along the coast it is generally rare, but common enough, probably, in the interior, through the provinces of Quebec and Ontario, west to Manitoba and the Northwest Territory. Mr. W. L. Bishop found a single nest of this species placed in a large birch tree near Kentville, Nova Scotia, on June 8, 1890, containing two young in the down. He considers it rare in this province.

[1] History of British Birds, 1883, Vol. I, pp. 117-122.

On the Atlantic seaboard it has not, according to Chamberlain, been taken north of latitude 49°, but during the cruise of the U. S. Fish Commission schooner *Grampus*, in the months of July and August, 1887, Mr. William Palmer, the taxidermist of the U. S. National Museum, saw two of these birds on one of the Mingan Islands in the Strait of Belle Isle, off the southern coast of Labrador, which point marks, as far as known, the northern limit of its range in this direction. In the interior it reaches a much higher latitude, having been observed at Fort Churchhill, Hudson Bay, by Dr. Bell, near latitude 59° north.[1]

The Red-tailed Hawk is one of the commonest and best known of the larger Raptores, and occurs more or less abundantly throughout the eastern portion of the United States. It is partial to moderately timbered districts, swampy woods and the bottom lands along water courses being its favorite abiding places; but it is also found, though less frequently, in the more extensive tracts of woods on uplands and in the mountain regions. In the extreme northern parts of its range it is a summer resident only, but nevertheless it is a hardy bird, and can readily endure great degrees of cold. It usually passes south, at times in considerable flocks, during the latter part of September and the month of October, while quite a number linger along our northern borders until November, and not a few winter regularly north of latitude 42°.

It is one of the less active of our Raptores, generally slow and deliberate in its movements, and though larger and far stronger in proportion to its size, it lacks the dash and courage of Cooper's Hawk, and willingly contents itself with more humble prey. It may frequently be seen sitting, for hours at a time, on a dead limb of a tall tree growing at or near the outskirts of a piece of woods, preferably near water, watching for its quarry. One might think them asleep at such times, but they are wide enough awake, as the would-be collector would soon find out should he try to get within gunshot of one. Each bird seems to have its favorite perch, and this is resorted to pretty constantly from day to day. They are very shy and wary, from being more or less shot at and molested, and will not often allow themselves to be approached closely.

Unfortunately the Red-tailed Hawk has a far worse reputation with the average farmer than it really deserves; granting that it does capture a chicken or one of the smaller game birds now and then, and this seems to be the case only in winter when such food as they usually subsist on is scarce, it can be readily proved that it is far more beneficial than otherwise, and really deserves protection instead of having a bounty placed on its head, as has been the case in several States.

This statement is fully borne out by the careful examination of more than three hundred stomachs of this species, made by and under the direction of Dr. C. Hart Merriam, chief of the Division of Economic Ornithology and Mammalogy, U. S. Department of Agriculture. His report shows that the

[1] Canadian Birds, Chamberlain, p. 56.

Red-tailed Hawk, instead of living largely on poultry and game birds, as is generally believed, feeds mostly on mice and shrews, as well as on frogs, toads, crawfish, snakes, lizards, and insects of various kinds. Comparatively few of their stomachs, less than one in ten, contained the remains of poultry or game birds, while fully two-thirds contained mice. In the Mississippi and Ohio Valleys, where squirrels abound, they feed very largely on these, and in the more open prairie regions in Iowa, Illinois, and Wisconsin they live on ground squirrels, gophers, and meadow mice; rabbits also are often caught by them, and small birds but rarely. Pairs of these Hawks frequently hunt together, and in such case it is difficult for even so nimble an animal as a squirrel to escape them.

In the Middle and Northern States such birds as have migrated, generally return from their winter haunts by the middle of February or the beginning of March, and somewhat later farther northward. During the mating season they are rather noisy, like most Raptores. Their principal call note at these times— generally uttered during their aërial gyrations, while circling about, high in air, chasing and pursuing each other in the vicinity of their future home—is a shrill and far-reaching "kee-aah," repeated at short intervals. Another note, somewhat like "chirr" or "pii-chiir," is also uttered during this season, when perched on some dead limb near their nest. At other seasons they are much more silent.

There is considerable difference in their nesting habits in some sections, so that while in certain localities most of them nest on high ground, in other places the majority prefer the heavily timbered bottom lands.

Dr. William L. Ralph has kindly furnished me the following notes on this species, based principally on observations made by him in Oneida and Herkimer Counties, New York. He says: "The Red-tailed Hawks are migrants in this locality, although a few may remain during mild winters. They begin to arrive from the south from the first to the middle of March, and at once commence repairing their old nests or building new ones, and by the 1st of April the first eggs are laid. I think that, with the exception of the Bald Eagle, *Haliæetus leucocephalus*, this bird nests earlier in this locality than any other member of the family *Falconidæ*. The places chosen by them for nesting are rather small woods, and when they nest in large forests, which is seldom, they will be found on the extreme borders. Once in a great while they nest in isolated trees growing in the vicinity of woods. One pair that I have watched with great interest has nested for several years in a large elm tree that is standing in a meadow about 20 rods from a wood where they formerly nested, and from which they were driven by a pair of Great Horned Owls, *Bubo virginianus*, taking possession of their nest. I have known other instances of the Great Horned Owl appropriating the nest of the Red-tailed Hawk. In one case the Hawks had built a nest which they were occupying at the time I found it, and they used it for two or three years after. It was then taken possession of by the Owls, which held it for two years and then for some cause left it, and after remaining unused for a year it was again taken by the Hawks which had, in the mean time, nested in the same wood only about 20 rods away.

"Both this and the Red-shouldered Hawk will sometimes, when driven from a nest, build a new one, which they may occupy for a year or two and then return to the old one. In this vicinity the Red-tailed Hawk prefers birch trees above all others to build in, and about 80 per cent. of their nests will be found in such situations. The remaining 20 per cent. is about equally divided among beech, maple, hemlock, elm, and basswood trees. Why these birds should prefer birch trees I do not know, for they are usually not very hard to climb, while the most difficult of their nests to reach were built in elm, hemlock, and basswood trees. They generally select the largest and tallest trees they can find to build in, and their nests are situated near the tops, in crotches formed by two or more large limbs, or at the junction of large limbs with the trunks. They are usually placed from 60 to 70 feet from the ground.

"The nests are composed of sticks lined with strips of bark and twigs from coniferous trees, usually the hemlock, and feathers from the birds themselves, which become more and more numerous as incubation advances. They are large structures, but not out of proportion to the size of the birds. If not molested they will occupy the same nest for a number of years.

"About one-half of the nests of the Red-tailed Hawk found in this region contained four eggs or young, and nearly all the remainder two. At least I have never seen a nest with more than four, or less than two young birds, or partly incubated eggs, and but one or two that contained three.

"Their eggs vary considerably in size, more so than those of any other bird of prey with which I am acquainted, and the size of these seem to have no connection with the size of the sets, as the largest sets will often contain the largest eggs, and again both large and small ones are often found in the same nest. But one brood is raised in a season; but like most other birds of prey the female will lay several sets of eggs when the first have been destroyed. Both sexes assist in incubation, and usually are not very solicitous about their eggs and young, and after they once leave their nests they seldom come within gunshot, or make much ado.

"The places where Red-tailed Hawks like best to live are small woods with open swamps, or with meadows and pastures near by, and indeed all grounds that are frequented by mice, for these little mammals are their favorite food, as is the case with nearly all other Hawks and Owls.

"Their call note is weak for the size of the bird, and has a rather disagreeable sound, resembling the squealing of a pig more than anything else. I found the Red-tailed Hawk rather rare in Putnam and St. Johns Counties, Florida, even in winter, and very few remain there to breed. On March 18, 1891, one of my assistants found a nest of this species in a wild and desolate place, about 20 miles southeast of San Mateo and 4 or 5 miles from the nearest house. I did not examine this nest till March 26, when it contained a young bird just hatched and an egg from which the young was on the point of emerging. The parents made much ado and frequently flew at my climber, coming sometimes within 10 feet of him. The nest was placed in a good-sized pine tree in low pine woods and about 30 yards from a cypress swamp. It was situated 56 feet from the ground, and about 2 feet from the body of the tree, on a large limb. It was composed of sticks with pine needles scattered among them, and lined with cypress bark, pine needles, Spanish moss, rabbit hair, and feathers from the sitting birds. It also contained a dead mouse and the remains of a large rabbit. The nest, though of large circumference, was very shallow and had hardly any depression for the eggs to lie in; from the fact that another large nest laid on the ground below this one, which appeared to have been blown from the same tree, I think the female was obliged to lay in the new one before it was completed. The egg was badly nest-stained, fairly well marked, and measured 62 by 45 millimetres."

According to Mr. P. Smith, jr., the Red-tailed Hawk is very common during the breeding season in portions of southwestern Illinois, and especially so in Bond County, where they nest in considerable numbers. Here they prefer high dry woods to the swampy bottom lands for the purpose of nidification. No special preference is shown for particular species of trees; oaks, hickory, or elms are most often used to nest in. They place their nests at various distances from the ground, anywhere from 20 to 90 feet up, generally from 50 to 60 feet. Usually they nested in or near the edges of timbered tracts, and an occasional nest might be placed in an isolated tree in a pasture. A pair of these Hawks were always sure to be found in any grove inhabited by squirrels, whose worst enemies they are. They also catch many rabbits.

In southwestern Illinois nidification ordinarily commences about March 20; full sets of eggs, however, have been found as early as March 3, and again as late as June 10, the latter no doubt being second or even third layings. The number of eggs to a set is two or three, some seasons the smaller sized sets predominating, in others the larger. In more than a hundred nests examined by Mr. Smith, but one contained four eggs.

The nest of the Red-tailed Hawk is a large bulky structure, measuring on an average about 24 inches in diameter and from 10 to 15 inches in depth. The inner cavity is rather shallow, usually not over 2 inches deep. There is considerable variation both as regards size and bulk in their nests, those which have been used for several years in succession, and to which slight additions are made yearly, exceeding such as are newly built. Generally they are very indifferently lined and not models of neatness. Sometimes the eggs lie directly on the coarse twigs of which the nest is composed; other nests are fairly well lined with strips of bark, corn husks, dry leaves and grasses, weed stalks, moss, or fine hemlock twigs.

In the southern part of its range the Red-tailed Hawk commences nesting in the latter half of February or the first week in March, and even at more northern points nidification sometimes begins equally early. Mr. J. W. Preston informs me that he took a set of eggs of this species on March 5, 1890, near Baxter in central Iowa, which were, even at that early date, slightly incubated. Heavy snow fell while the eggs were being deposited, and this pair of birds were seen repairing the old nest some time in the latter part of January. He found a pair of these birds nesting in a most unusual situation, their domicile being placed in a cottonwood tree, not over 12 feet from the ground and close to a public road. In the western parts of its range it nests frequently in large cottonwood and sycamore trees.

In the New England and Northern States, the Red-tailed Hawk nests usually in April, apparently as often during the first as the last half of this month, and correspondingly later northward.

Usually but a single brood is raised in a season, but in the hill country of South Carolina, the Piedmont region, where this species is not uncommon, Dr. Leverett M. Loomis, well known as an exceedingly accurate observer, informs me that two broods are not infrequent.

If the first set of eggs is lost, a second and sometimes a third is laid, and on very rare occasions even a fourth. A pair of birds will sometimes build two nests, and if one of these is robbed the second is made use of. The eggs vary from two to four in number. Sets of two are most often found throughout the greater part of its range, while three are not rare, and in northern New York sets of four appear to be as common as smaller ones. Incubation lasts about four weeks, the male assisting to some extent in this duty, as well as providing his mate with food while on the nest. The eggs are deposited at intervals of about two days.

Generally the Red-tailed Hawk shows but little courage or devotion in defense of its eggs or young, contenting itself with uttering its protest by shrill screams from a safe distance, but as in everything else, there are frequent exceptions to this rule, and many pairs show much devotion in the defense of their homes, darting down close to the intruder, and screaming fiercely at him.

The ground color of these eggs is usually a dull or pale creamy white. Occasionally a faint bluish white tint is perceptible, which fades more or less with age.

In the series before me about one-fifth of the eggs are unspotted, and a heavily marked one may be found in the same set with an unspotted egg. About four-fifths of the eggs are more or less marked with irregularly shaped spots and blotches of different shades of reddish and yellowish browns, which vary considerably in size, shape, and intensity. A few specimens have longitudinal markings, like No. 23150, *Buteo swainsoni* (Pl. 8, Fig. 6), and some show shell markings of lavender and écru-drab. These markings are often of rather uniform size, sometimes large and bold, in others small and pretty evenly distributed over the entire surface of the egg, and again they are confined principally to either end. The majority are moderately well marked and only occasionally is a heavily blotched set found. In shape, some are nearly oval, one end however, is generally a trifle smaller. Many are ovate or short ovate.

The shell in most specimens is coarse and strongly granulated, but now and then an egg is found that is much closer grained and feels rather smooth to the touch. The average measurement of a series of eighty-three eggs is 60 by 47.5 millimetres. The largest egg in this series measures 65.5 by 50, the smallest 55.5 by 45 millimetres.

Inasmuch as there is no perceptible difference in the eggs of the Red-tailed Hawk and its geographical subspecies, the type specimens have been selected from the whole series to show as nearly as possible the variations in markings irrespective of race. Those figured (of *Buteo borealis* proper) show the unmarked type and a fairly well-spotted egg.

Of these, No. 12740 (Pl. 6, Fig. 5), a plain colored egg, was collected in Richland County, Illinois, by Mr. Robert Ridgway, of the Smithsonian Institution, on March 6, 1887, and No. 23610 (Pl. 6, Fig. 6), a distinctly and fairly well-marked egg, was taken by Mr. J. W. Preston, near Baxter, Iowa, April 7, 1888. They are from sets of two eggs.

72. Buteo borealis kriderii HOOPES.

KRIDER'S HAWK.

Buteo borealis var. *kriderii* HOOPES, Proceedings Academy Natural Sciences, Phila., 1873, 238, Pl. 5.

(B —, C 351e, R 436a, C 519, U 337a.)

GEOGRAPHICAL RANGE: Great Plains of the United States, from Minnesota to Texas; east, irregularly or casually to Iowa and northern Illinois.

Very little is yet known regarding the exact limits of the breeding range of Krider's Hawk, a light-colored race of the Eastern Red-tail. It seems to be restricted to the plains proper, and up to the present time has been found more frequently in western Minnesota during the breeding season than anywhere else, though nowhere abundant. Col. N. S. Goss states that it breeds occasionally in western Kansas. Mr. George G. Cantwell gives it as a summer resident, in his recently published list of the "Birds of Minnesota," and he writes me that he found a nest of this Hawk on May 18, 1889, containing three eggs. It was placed in an elm tree about 50 feet from the ground, situated in a grove which fringed the Lac-qui-parle River, near Dawson, Minnesota. The structure of the nest, as well as the eggs, he says, were similar to those of the common Red-tail.

Messrs. Thomas S. Roberts and Franklin Benner, in speaking of Krider's Hawk, say: "On the 17th of June we took from one of the large cottonwood trees on the border of the Minnesota River, in Brown's Valley, a young Hawk, not more than a week old, which we brought back to Minneapolis with us.

The parent bird soared above the nest while the young bird was being taken, and her noticeably white appearance attracted our attention at once, and we judged her to be of this species. The growth of the young bird has gradually confirmed this idea, as it now, at the age of nearly three months, shows unmistakable evidences of being this light variety of Red-tailed Hawk. * * * This bird has become very tame and is a great pet, allowing itself to be handled, and distinguishes persons."[1]

Mr. Robert Ridgway, of the Smithsonian Institution, has also examined two females of this race which were shot from their nests in Minnesota, and as far as I have been able to learn this State is the only one in which Krider's Hawk is as yet positively known to breed. It is likely, however, that it also nests in the prairie regions of Wisconsin, Iowa, and northeastern Illinois. Mr. H. C. Cole, of Chicago, Illinois, writes me that he collected a bird of this race at Half Day, Lake County, Illinois, 34 miles northwest of Chicago, on July 25, 1876, which evidently was a summer resident. Its mate was also seen, but was too wary to be shot. This specimen is now in the U. S. National Museum collection.

It has been repeatedly reported as breeding in western Texas and likewise among the cliffs of Colorado, but these records are somewhat questionable and require confirmation. No specimens actually taken during the breeding season in these States have as yet found their way into any of our larger ornithological collections.

There are no positively identified specimens of the eggs of Krider's Hawk in the U. S. National Museum collection, but they are not likely to differ materially from those of the preceding species. Nidification seems to take place a little later than with the common Red-tailed Hawk. The winters are spent in more southern latitudes, in Texas and southward.

73. Buteo borealis calurus (CASSIN).

WESTERN RED-TAIL.

Buteo calurus CASSIN, Proceedings Academy Natural Sciences, Phila., VII, 1855, 281.
Buteo borealis var. *calurus* RIDGWAY, Bulletin Essex Institute, V, November, 1873, 186.
(B 20, 24, C 351a, R 436b, C 517, U 337b.)

GEOGRAPHICAL RANGE: Western North America, from the Rocky Mountains to the Pacific; south into Mexico; casual east to Illinois.

The breeding range of the Western Red-tail, a darker colored race than the two preceding, extends from northern and northwestern Texas, through the Rocky Mountain regions of Colorado, New Mexico, and Arizona, northward through Wyoming, central and western Montana, Utah, Nevada, and Idaho, and on the Pacific coast from Lower California, through California, Oregon, and Washington well into British Columbia. A single specimen, a young bird, was shot by Lieutenant Rockwell, U. S. Navy, in the vicinity of Hot Springs Bay, on Baronof Island, near Sitka, Alaska, latitude 56° 30′, on June 5, 1880; this was probably raised in the neighborhood, and marks the most northern point at which it has yet been found.

The Western Red-tail, though nowhere very abundant, is pretty generally distributed over western North America at large, and is, next to Swainson's Hawk, the commonest of the larger Raptores found in these regions. Its habits are in many respects similar to those of the common Red-tailed Hawk of the Eastern States; it is fond of the tall timber bordering the banks of streams, and is as often found far in the mountain passes and deep cañons as in the more open country in the foothills and the adjacent plains, but seems to shun the dense and extensive forests, and is rarely seen excepting on the borders of these. In some of the desert regions of western Texas, southern New Mexico, and Arizona, it is not infrequently met with at long distances from water, and has even been found breeding in such localities. It is only a summer resident in the more northern parts of its range, wintering along the southern border of the United States, or passing south into Mexico. Col. N. S. Goss has found it not uncommon in winter, in Kansas, and I have met with it during every winter month at Fort Klamath, Oregon, but not as frequently as in the summer.

It is one of the earliest migrants to return to its breeding grounds, arriving about the latter part of February or the first week in March, and is readily noticed then, both on account of its size and its shrill squeals, uttered during the greater part of the day while circling high in the air, in proximity to its future summer home. They appear to be very much attached to certain localities and return to them from year to year. Their call notes are very similar to those of the Eastern Red-tail.

In Washington, Idaho, Oregon, and California it lives principally on the different species of ground squirrels so common and destructive in these States, as well as on chipmunks, mice, snakes, lizards, frogs, rabbits, and now and then a chicken. I have never seen one molesting chickens during a number of years' residence where these birds were not uncommon. Mr. Charles A. Allen, of Nicasio, California, corroborates this. He says: "I have never known them to disturb domestic fowls. They subsist on small animals and catch large numbers of snakes." Dr. E. A. Mearns, assistant surgeon, U. S. Army, found remains of rattlesnakes in the crops of three specimens, indicating that they are a match for even these poisonous reptiles. In the late summer and fall they live to a great extent on grasshoppers, wherever they are abundant, and seem to be very fond of them. Birds shot at such times rarely contain any other food in their stomachs.

Mr. William Lloyd writes me that in western Texas they feed on prairie dogs, rock squirrels, cottontail and jack rabbits, an occasional Scaled Partridge, and in winter sometimes on the carcasses of goats, sheep, and cattle, sitting around like true Vultures waiting for sick animals to die. Mr. W. Otto

Emerson informs me that in one of their nests near Haywards, California, containing a young Hawk still in the down, he found two gophers and a steel trap containing a squirrel; and he also saw a pair of these birds fighting or playing over a snake, which would be dropped by one of them and caught by the other before it reached the ground, until finally the prize was carried off.

In his paper on "Birds from the Farallon Islands," Mr. Walter E. Bryant makes the following statement regarding this subspecies: "Every spring the island is visited by numbers of these Hawks. In 1882 they came in April, about the time of the arrival of the Murres, leaving again in May. During their short stay they fed almost exclusively upon the Murres, killing, in the estimation of Mr. Emerson, several dozen a day."[1]

On the whole the Western Red-tail, viewed from an economic standpoint, is far more beneficial than otherwise.

Nidification begins rather early, and where the birds are not persistently disturbed the old nests are resorted to and repaired from year to year. In southern California full sets of eggs of this subspecies have been found as early as February 20; usually, however, they do not lay much before March 10, and the majority not before April. Most of them nest in this month, not only in California but in Arizona, New Mexico, and northwestern Texas also. Even at points considerably farther north, as at Fort Lapwai, Idaho, I have found full sets of eggs by April 10. In the mountains of Oregon, Washington, Idaho, Montana, and Wyoming they usually nest somewhat later, generally in the last week of April or the beginning of May. Mr. F. Stephens has found this subspecies nesting as late as June 12, but in this case the birds had most likely lost their first clutch.

The nests are placed at various distances from the ground, sometimes in quite low situations, and again at a great height, especially when placed near the tops of tall redwood and pine trees, often fully 100 feet from the ground, and practically inaccessible. Mr. W. E. D. Scott mentions finding nests of this species in southern Arizona as low as 10 feet, and one only 7 feet from the ground. Their average height is from 30 to 50 feet. In their choice of nesting sites, large cottonwoods, sycamores, and live oaks are generally selected, while pines, redwood trees, junipers, mesquites, willows, and aspens come in the order named.

In southern Arizona they also nest occasionally in a sahuara, the giant cactus peculiar to that region. On March 24, 1872, I found a nest of this subspecies thus situated, which contained two partly incubated eggs. It was a very large and bulky one, and had evidently been used for a number of years, the sticks in the bottom of the nest being quite rotten. It was placed between the main trunk and one of the arms of the cactus, about 12 feet from the ground, and was found near the source of Rillitto Creek, Arizona. The nest was rather flat on top, and fairly well lined with the inner bark of

[1] Proceedings California Academy Sciences, 2d ser., Vol. I, 1888, p. 45.

the cottonwood and a few dry leaves. It was fully 30 inches in diameter on top and 20 inches deep outside. In southern Colorado and northwestern Texas they nest occasionally on cliffs and the sides of perpendicular bluffs. Mr. William Lloyd presented a fine set of three eggs of this species to the U. S. National Museum collection, taken from the side of a chalk bluff, about 75 feet from the bottom and 20 feet from the top of the bluff, on April 2, 1890, in Presidio County, Texas. Such sites are probably selected on account of the absence of suitable trees.

I believe they rarely build a new nest in the West, the old ones being generally repaired by the addition of a little fresh material. The fibrous bark of the redwood is often used for lining when obtainable, and the fine inner bark of the cottonwood likewise.

In western Texas the same nest has occasionally done duty for two different species of Raptores, the Western Horned Owls using it in January and February, and the Red-tails subsequently.

But a single brood is raised in a season; a second and sometimes a third set of eggs is laid, however, should the first be taken or destroyed. The usual number of eggs is two or three, sets of one or four being rarely found. I have, however, found the Western Red-tail incubating a single egg in which the embryo was well advanced and where the birds had not been previously disturbed, and also found a set of four, near Fort Lapwai, Idaho, in the spring of 1871.

Mr. Charles H. Townsend took another set of four near Red Bluff, California, in April, 1884. The nest from which these eggs were taken was placed in a scrubby oak, about 20 feet from the ground. It was built of heavy twigs and had a uniform lining of soap root fiber.[1]

Incubation lasts about four weeks, and the eggs are deposited at intervals of a couple of days. There is no marked difference in the size, shape, and color of these eggs as compared with those of the common Red-tailed Hawk, and they are practically indistinguishable from each other; but among those before me the proportion of spotted ones is larger than in the former, very few being entirely unmarked. The average measurement of thirty-six eggs of the Western Red-tail is 59 by 46.5 millimetres. The largest egg of the series measures 66 by 46.5, the smallest 53.5 by 44 millimetres.

The type specimens selected show the more heavily marked styles. No. 20745 (Pl. 6, Fig. 7), from a set of three in the Bendire collection, was taken by the writer near Camp Harney, Oregon, May 29, 1877, and incubation was far advanced when found. It shows a considerable amount of the lavender and écru-drab shell marking, not commonly found among the eggs of the Red-tailed Hawks. No. 20763 (Pl. 6, Fig. 8), from a set of two, Bendire collection, is the darkest colored egg of the entire series. The ground color in this specimen is a rather light fawn, while in the other it is of the normal tint. It was taken by myself from a nest in an oak tree near Camp Critten-den, Arizona, April 29, 1872.

---

[1] Proceedings U. S. National Museum, 1887, Vol. x, p. 202.

74. Buteo borealis lucasanus RIDGWAY.

SAINT LUCAS RED-TAIL.

Buteo borealis var. *lucasanus* RIDGWAY, in Cones's Key to North American Birds, 1872, 216 (under *B. borealis*).
(B —, C 351b, R 436c, C 518, U 337c.)

GEOGRAPHICAL RANGE: Cape St. Lucas, Lower California.

The St. Lucas Red-tail, a scarcely tenable race, very similar to *B. borealis calurus* excepting that the black bars on the tail of the adult are generally wanting, is confined to the Cape St. Lucas region of Lower California. Its habits, nests, and eggs, are like those of the Western Red-tail which occupies the same localities.

Since this was written, Mr. William Brewster has made a careful study of this supposed subspecies, and in a letter to me, dated March 15, 1891, he makes the following remarks: "In my opinion there is no such bird as *Buteo borealis lucasanus*. The type and the mounted specimen in the U. S. National Museum collection are simply lightly banded specimens of the Western Red-tail. Both have the band, but it is indistinct. I can match the mounted bird by specimens from both Lower California and Arizona. I brought together nearly one hundred skins, including over thirty from Lower California; the latter as a series show no peculiarities so far as I can see. The type of *Buteo borealis lucasanus* is certainly peculiar in respect to the tail, but in no other way." I fully agree with Mr. Brewster's conclusions.

75. Buteo borealis harlani (AUDUBON).

HARLAN'S HAWK.

Falco harlani AUDUBON, Birds of America. I, 1830, 441, Pl. 86.
Buteo borealis harlani RIDGWAY, Auk, VII, 1890, 205.
(B 22, C 350, R 438, C 515, U 337d.)

GEOGRAPHICAL RANGE: Gulf States and Lower Mississippi Valley; north (casually) to Kansas, Iowa, Illinois, and Pennsylvania; east to Georgia and Florida.

Harlan's Hawk, which till recently figured as a distinct species, is now considered only a variety of *Buteo borealis.* It is the darkest of the different geographical races of this species, and ranges from northern Florida, Georgia, the Gulf States, and the Lower Mississippi Valley, north to Kansas, Iowa, Illinois, and Pennsylvania.

All we know about the breeding habits of this subspecies is Audubon's statement, who first described this Hawk from a pair obtained by him near St. Francisville, Louisiana, which had bred in that neighborhood for two seasons; were shy and difficult of approach, and for a long while eluded his pursuit.[1]

---
[1] History of North American Birds, 1874, Vol. III, p. 294.

My friend, Dr. William L. Ralph writes me: "Harlan's Hawk is not very uncommon, during the winter, in St. Johns and Putnam Counties, Florida, but until this season I had taken it for the Caracara, which is quite rare here, and I now believe that most of the Black Hawks seen by me during past years were referable to this subspecies. They are exactly like the Red-tailed Hawks except in color, and their call note is also the same, only being longer drawn out. The call of the latter bird as already stated, sounds like the squealing of a pig, or 'kee-ee-e,' and that of Harlan's Hawk like 'kee-ee-ee-e-ee.'

"They are not uncommon here in wild and unfrequented places, but are seldom seen in the more settled parts. They are well known to the native hunters and cattle men, who call them Goshawks. I have seen six or eight different birds this season, 1891, that I am sure belonged to this sub-species. The last one noticed by me was a few days before these notes were written, on March 26. With the exception of the first pair, observed on February 3, which acted as having taken possession of a large nest in a pine tree, but which, on a subsequent visit, was found occupied by a pair of Florida Barred Owls, I have seen no evidence of their breeding, nor have I found an occupied nest, although a friend and myself, as well as two assistants, have looked for them to the exclusion of other nests for the past month.

"If the grass in the low pine woods or prairies be fired, and these birds are near enough to see the smoke, they will come to it at once, like the Red-tailed and other Hawks already mentioned, and fly just ahead of the blaze to catch the small mammals driven out by the heat. They are very wild indeed, and unless concealed from their view it is almost impossible for a person to get nearer to one than 300 yards. One of my assistants tells me that on March 29 he saw five of these Hawks that came to a big prairie fire that had been started. He said that a pair of these birds were chasing one another as if they were mating, and that one flew into the smoke and caught some small mammal which it gave to its mate. I have examined several nests that I was told were occupied by these birds last year, but have not found anything in them, and believe they have not begun nesting yet."

The nest and eggs of Harlan's Hawk are probably very similar to those of the common Red-tailed Hawk.

76. Buteo lineatus (GMELIN).

RED-SHOULDERED HAWK.

Falco lineatus GMELIN, Systema Naturæ, I, ii, 1788, 268.
Buteo lineatus JARDINE, ed. Wilson's American Ornithology, II, 1832, 290.
(B 25, C 352, R 439, C 520, U 339.)

GEOGRAPHICAL RANGE: Eastern North America; north to Nova Scotia and southern Canada; west to Texas and the Great Plains; south to the Gulf coast and Mexico.

The breeding range of the Red-shouldered Hawk extends through eastern North America from Maine to Florida, and westward to the borders of the Great Plains. It reaches central Texas, where it appears to be common, and, according to Mr. William Lloyd, it is a rare resident of western Texas; it is also found in the eastern portions of the Indian Territory and in eastern Kansas and Nebraska. It appears to be rare in Minnesota, and probably occurs in small numbers in the more heavily timbered parts of South and North Dakota. Mr. W. E. D. Scott, in his "Birds of the Gulf Coast of Florida," reports it as common and breeding there.[1]

While perhaps the majority of the Red-shouldered Hawks found in the central parts of Texas bear a closer resemblance both in size and coloration to *B. lineatus* than to the two geographical races, *B. lineatus alleni* and *B. lineatus elegans*, they are not typical representatives of either. Specimens from Hale County, central Alabama, shot during the breeding season, are typical *B. lineatus*. North of the United States it reaches the southern borders of the Dominion of Canada from northern Nova Scotia, where it is, however, very rare, and westward to eastern Manitoba. Mr. T. McIlwraith reports it as a common summer resident in southern Ontario,[2] and Mr. M. Chamberlain, in his "Catalogue of Canadian Birds," makes the following remarks about this species: "A rather common summer resident of the eastern provinces, probably more abundant in Ontario than elsewhere. Dr. Bell reports its occurrence at York Factory, on Hudson Bay; and Mr. Thompson, on the authority of Mr. Hunter, gives it as rather common in eastern Manitoba."

The habits as well as the ranges of the Red-tailed and Red-shouldered Hawks are very similar; the former being slightly the hardier bird of the two, and reaching somewhat farther north. Like the Red-tailed Hawk, it is only a summer resident in the northern parts of its breeding range, migrating south in winter, some passing into Mexico. It winters commonly in latitude 39°, and some remain throughout the year in favorable localities in the southern New England States.

Dr. William L. Ralph writes me: "The Red-shouldered Hawk is the commonest bird of prey in Oneida and Herkimer Counties, New York. There has

[1] Auk, Vol. VI, 1889, p. 247.
[2] Journal and Proceedings Hamilton Association, Vol. II, 1886, p. 160.

hardly been a year in the past twelve that, with the help of one or two assistants, I have not found from twenty to thirty of their nests while looking for
those of other birds. Although they occur sometimes during the winter months,
these birds are migratory, and usually arrive in this locality from the South
about the middle of March, occasionally a little earlier or later, being influenced
by the condition of the weather. Like the Red-tails, they commence working
on their nests almost immediately after their arrival, but do not begin laying
quite so early. The majority of their eggs are deposited here between the
20th of April and the 1st of May; I have found full sets, however, on the 10th
of the former month, and as late as the last day of the latter. Those laid after
May 12 were probably second layings. The sets here vary in the following
proportion: One-half of the nests contain three eggs, one-third four, and one-
sixth only two. The second and third layings never consist of more than three
eggs, and generally of but two.

"The Red-shouldered Hawk raises but one brood in a season, but they
generally lay two or three sets of eggs should the first be destroyed. Both
birds assist in incubation and make a great fuss when their nests are disturbed,
and although they seldom approach very near to the persons examining them,
they keep up their loud calls as long as any one remains in the vicinity. Like
all birds they show more solicitude just before and just after their eggs are
hatched, and no matter how wild and suspicious they may have been, or how
many times they may have been shot at, they usually appear to lose all sense
of danger at such times.

"The young birds when just out of the shell look like balls of yellowish
down, and their eyes are as bright and clear as those of the adults. In a
remarkably short time they learn what food is for, and will fight among themselves for it, and peck at one another when they are but a few hours old.

"The nest of the Red-shouldered Hawk is composed of sticks, lined
with birch bark, strips of the inner bark of various trees, and hemlock
twigs. Occasionally evergreen twigs, dry grass, or dead roots are added to
to, or substituted for, some of the materials of which the lining is composed;
also more and more of the feathers from the lower parts of the sitting birds
drop out and are added to the lining as the eggs advance in incubation.
There are no birds of prey that I am acquainted with, of whose eggs I cannot
tell almost the exact stage of incubation by the quantity of feathers in the
lining of its nests, they increase so regularly.

"The nest is either located in a crotch of a tree or against the trunk
on limbs growing out from it, at an average height of about 50 feet; 37
and 60 feet from the ground are the lowest and the highest locations of the
nests according to measurements made by me. The trees chosen by the
Red-shouldered Hawk to nest in are birch, ash, maple, and beech. If there
is any preference shown in their selection it is in favor of the birch and ash.
The places they like best to breed in are small woods, located in the neigh-

borhood of open fields and swamps, and they frequent such localities most of the time they remain with us.

"I have noticed one habit of this Hawk which is common with most birds of prey that I am familiar with—while they will breed in close proximity to birds of the same order, they will never breed very near one another—each pair occupying a piece of woods by themselves, unless it should be quite a large one, and excluding all others of the same species.

"Among other cases of the kind, I have known the Red-tailed and Red-shouldered Hawks and the Great Horned Owl to nest near one another in a small wood, and on one occasion I found a pair each of the Sharp-shinned, Cooper's, and Red-shouldered Hawks, and of the Long-eared Owl breeding so near together that I could stand beside the nest of a Ruffed Grouse, which was close by also, and throw a stone to any of the others.

"According to my observations the food of this Hawk consists principally of mice, with the addition of moles, squirrels, young rabbits, and once in awhile a small bird. They catch very few of the latter, but destroy perhaps more poultry than the Red-tails, probably because they are more numerous here and not so wild, but I am certain that the few fowls they take are more than compensated for by the good they do in destroying noxious rodents."

According to a careful investigation of one hundred and two stomachs of this species, made under the direction of Dr. C. Hart Merriam, in charge of the Division of Economic Ornithology and Mammalogy, U. S. Department of Agriculture, during the year 1887, the following summary was obtained: Of 102 stomachs examined, 1 contained poultry; 5, other birds; 61, mice; 20, other mammals; 15, reptiles or batrachians; 40, insects; 7, spiders; 8, craw-fish; 1, earth-worms; 1, offal; 1, catfish, and 3 were empty. The majority of these birds were taken between the months of October and March. This shows conclusively that the Red-shouldered Hawk is one of the beneficial species, and does little injury.

In the southern part of its range it begins laying early in March; in North Carolina, Tennessee, Missouri, and Kansas about the latter half of this month; in the New England and other Northern States rarely before the second week in April, usually about the middle of this month, and correspondingly later northward. They build more frequently in small woods than the Red-tails, and the majority of their nests are found in lower situations, varying in height from 20 to 65 feet, rarely more, averaging about 40 feet from the ground.

Besides the species of trees already mentioned, they nest in elms, oaks, hickories, pines, and chestnuts. On Long Island, New York, Mr. William Dutcher informs me that the latter are almost invariably selected. Their nests are usually smaller than those of the Red-tail, measuring generally from 18 to 24 inches in outer diameter by from 5 to 8 inches in depth. The bulk of the nest varies considerably, according to its location in the tree and as regards its component parts. Some nests are almost entirely constructed of pine or hemlock twigs, both green and dry, and lined with leaves of the pine, moss, lichens, bits of bark or corn husks, besides the materials already mentioned. They nest later than the Red-tails and build new nests more frequently than does that species.

They are certainly the most prolific of the Buteos; sets of four eggs are not at all rare, others of five and even six eggs are occasionally found. The late Dr. William Wood, of East Windsor Hill, Connecticut, records having taken a set of six, and Mr. R. B. McLaughlin, of Statesville, North Carolina, writes me that he took a set of six eggs of this Hawk on April 5, 1889. He says: "I saw these birds at work on the nest early in March; it was an old one remodeled, and was placed in the crotch of a good sized pine. I expected to find eggs in it much sooner than I did. Incubation had just commenced. The eggs were of the usual size, without any perceptible variation either in shape or size. There was considerable difference in their markings, however, but no greater than sometimes occurs in sets of three."

The Red-shouldered Hawk is far more common in the lowlands than in the mountain regions, in parts of which it seems to be entirely absent. It prefers the borders of streams, lakes or swampy woods for a permanent residence. Its flight is easy and graceful, and it is seen on the wing mostly in the early and late hours of the day. Like the Red-tail, each of these birds has its favorite perch on a dead limb of some tall tree on the outskirts of a piece of woods, from which it can obtain a good view of the surrounding country; this it occupies for hours at a time when not disturbed.

According to Dr. Ralph "its call notes consist of a loud whistle-like sound, which resembles the syllables 'whee-ee-e,' with once in awhile a 'ca-ac' added to or rather mixed with it." The full-grown young after having left the nest, and when still depending on the parents for food, utter frequently, according to my observations, a call sounding, as near as I can interpret it, like "yeeh-ack, yeeh-hack," rapidly repeated. A note uttered in the early spring, especially during the mating season, sounds like "kée-yooh, kée-yooh," the last syllable drawn out.

Incubation sometimes begins with the first egg laid, young birds of different sizes being frequently found in the same nest, and this difference is too great to attribute to sex alone. Mr. Austin F. Park, of Troy, New York, writes me on this subject as follows: "On June 4 and June 11, 1888, I received three young Red-shouldered Hawks in the down, all taken from the same nest. These birds varied greatly in size. I kept the three chicks, fed them well, and noted the rates of growth of the wing and tail quills. I estimated from the length of these that there was about four days' difference in the stages of growth. Further experiments with others confirmed this opinion."

Incubation lasts about four weeks, and the eggs are deposited at intervals of two and sometimes three days. As said before, the number laid by this species varies from two to six. Sets of three are most frequently found, and are a fair average number. About one set in three contains four eggs.

In shape these range from ovate to short and rounded ovate, and a few are perfect oval. The ground color varies from a dull white to a pale yellowish, and again a pale bluish white. Unspotted eggs are very rare, and even faintly marked ones are not common. The majority are all more or less heavily smeared, blotched and spotted, with different shades of reddish brown, fawn color, écru-drab, vinaceous buff, and pearl gray. Some of the lighter tints are shell-markings. In some eggs the markings are few, but large and bold; in others they are finer, of smaller size, and more profuse, hiding the ground color to a considerable extent. They are generally irregularly blotched, and in a limited number of specimens these run longitudinally, as in the egg of Swainson's Hawk (figured on Pl. 8, Fig. 6). They show an almost endless variety of patterns, and there is practically no difference in this respect between the eggs of this species and its two geographical races, excepting that those of typical *Buteo lineatus alleni* from Florida are considerably smaller, as will be seen hereafter. The shells of these eggs are close grained, finely granulated, and without luster.

The average measurement of a series of one hundred and seventeen eggs of this species, from various points of eastern North America, is 54.5 by 43 millimetres. The largest egg measures 59 by 47, the smallest 51 by 41 millimetres. The latter is from Locust Grove, New York, the former from East Hampton, Connecticut, both from Dr. C. Hart Merriam's collection.

Ten specimens, collected by Mr. J. A. Singley, near Giddings, Lee County, in central Texas, give an average measurement of 54 by 44.5 millimetres, being fully as large as eastern specimens. I am inclined to think that though perhaps not referable to typical *B. lineatus*, they resemble more closely the eggs of this subspecies than those of the Florida Red-shouldered Hawk.

The type specimens selected to show some of the various patterns of coloration, are as follows: No. 12873 (Pl. 7, Fig. 1), a single very dark colored egg, was collected by Mr. Robert Ridgway, of the Smithsonian Institution, near Mount Carmel, Illinois, April 1, 1867; No. 21688 (Pl. 7, Fig. 2), from a set of four collected by Dr. C. Hart Merriam, near Locust Grove, New York, April 24, 1878; No. 23942 (Pl. 7, Figs. 3 and 4), two eggs from a set of four, were collected near Giddings, Lee County, Texas, by J. A. Singley, on April 11, 1888; No. 23943 (Pl. 7, Fig. 5), from the same source and locality, from a set of three, and taken April 16, 1888.

77. Buteo lineatus alleni RIDGWAY.

FLORIDA RED-SHOULDERED HAWK.

Buteo lineatus alleni RIDGWAY, Proceedings U. S. National Museum, VII, January 19, 1884, 514.

(B —, C —, R —, C —, U 339a.)

GEOGRAPHICAL RANGE: South Atlantic and Gulf coasts; south from middle South Carolina (along the coast only?) to Florida; west to southeastern Texas.

The Florida Red-shouldered Hawk is a smaller bird than the preceding subspecies, and as far as yet known its breeding range is confined to Florida and the extreme southern borders of the Gulf coast States, west to southern and (occasionally) central Texas.

Among an interesting collection of birds, made recently by Mr. J. E. Benedict, along the South Atlantic coast in the vicinity of Charleston, South Carolina, is a typical specimen of this form, shot near Georgetown, on Winyah Bay, South Carolina, in about latitude 33° 25′, on January 4, 1891, extending its range considerably to the northward, and probably the breeding range also.

I am indebted to Dr. William L. Ralph for the following notes on this subspecies: "Before the commencement of the disgraceful business of plume hunting which bids fair to exterminate many of the Florida birds, the little Sparrow Hawk was the commonest bird of prey in that State, but since then its numbers have been growing rapidly less, and now this position is held by the subject of this article.

"The following notes were taken in St. Johns and Putnam Counties, where, excepting that part of the former which borders on the Atlantic Ocean, the Florida Red-shouldered Hawk is very common. In fact, I have seen very few of these birds—and none during the breeding season—in the immediate vicinity of either the east or west coast of this State. It is a resident species, and so far as I can learn, if not molested, it seldom goes very far from the place where it nests. There has been no apparent diminution in their numbers during the last fifteen years, as they are seldom disturbed except in the vicinity of the main routes of travel, where they are more or less shot at by tourists, who, by the way, seem to be their only enemies. However, they will probably before many years have to succumb to the fate that is fast overtaking nearly all Florida birds.

"The localities where they are most commonly found are the flat pine woods, and during the breeding season they frequent these most of the time, nesting in or near the small cypress swamps which are very numerous in these forests, building nearly always in pine and rarely in cypress trees. At other seasons they wander more, and are often seen in the vicinity of houses, but rarely far from their nesting places. During the time they are sitting, these birds move around but little, when not in search of food, and those not on their nests will usually be found perched sleepily on a branch of some shady tree. One can at this time often approach within a few feet of them before they will fly.

"This variety of the Red-shouldered Hawk seems to be much more sociable than the northern species. I have often during the mating season seen flocks, consisting of from two to four pairs of birds, playing together. They chased one another around in apparent sport, circling through the air and uttering their shrill calls after the manner of the Swallow-tailed Kite.

"Their food consists principally of mice, Florida rats, young rabbits, the small gray squirrel found in this State, and probably an occasional frog or small snake. A very few of the inhabitants of this locality—chiefly those that formerly lived in the North—informed me that these Hawks would catch domestic fowls. I have never seen an occurrence of the kind, although I have lived for twelve winters and springs where both these birds and poultry were common.

"The eggs of this bird average considerably smaller than those of the northern subspecies, being but little larger than those of the Broad-winged Hawk. Undoubtedly in the majority of cases two is the number laid, as all their nests, and I have examined a dozen or more, contained either two eggs or two young birds, and most of the eggs found were partly incubated. Where the same species of bird nests both in Florida and in the North, or where a northern bird has a Florida variety, the eggs of the southern bird usually average fewer in number and less in size, and there is less variation in the numbers found in a set.

"I could never distinguish any difference in the call notes of the Florida Red-shouldered Hawk and the northern bird. Both use the same whistle—like 'whee-ee-e,' and the mated birds also assist each other in incubation. They make much ado when disturbed at their nests, but generally take good care not to come too near while these are being examined. I believe but a single brood is raised in a season. I first noticed these birds mating during the present season, about February 1, 1891. The earliest date of nesting was February 20, and young birds just hatched were found by me on March 20.

"The following description of several nests will give a fair idea of their mode of building. I found one on March 8, 1888, four miles east of San Mateo, Florida, in a small cypress swamp. It was situated in a pine tree, in a crotch 57 feet from the ground, and was composed of sticks and lined with pine needles, bits of pine and cypress bark, Spanish and green moss, and a few feathers from the birds themselves. It contained two nearly fresh eggs. Another, found on March 20, 1888, not far from the former locality, was placed in the top of a cypress tree about 65 feet from the ground, and resembled the first one in every respect, except that no Spanish moss was used in the lining of the nest. It likewise contained two eggs, which were about one-fourth incubated. On April 2, 1888, I found a third, about 6 miles southeast of San Mateo, also near the edge of a small cypress swamp. This was placed in a pine tree about 40 feet above the ground and similarly constructed. Among the materials used for lining, this nest contained, in addition to those already mentioned, some dry grass and a piece of snake skin; it contained but two eggs, one of these half incubated, the other unfertilized. Several other nests were found, showing little if any difference from those already described, so far as location and the materials used in their construction are concerned. All these nests contained but two eggs, and this seems to be the usual number laid by this race, in that part of Florida, at least."

Mr. W. E. D. Scott says that the Florida Red-shouldered Hawk is common in the vicinity of Tarpon Springs, Hillsborough County, Florida, where he found them mating and nesting generally in March, frequenting flat woods and ponds. The eggs are indistinguishable from those of *Buteo lineatus*, but, as before mentioned, they average considerably smaller. Their ground color is usually grayish white, and the markings appear to vary fully as much as in that subspecies, but are similar.

The average size of sixteen eggs from Florida, all collected by Dr. Ralph, is 50.6 by 42.5 millimetres. The largest of these eggs measures 53.3 by 42.5, the smallest 49.8 by 40.1 millimetres. None are figured, as they are similar to those of *Buteo lineatus* in every respect except size.

78. Buteo lineatus elegans (CASSIN).

RED-BELLIED HAWK.

Buteo elegans CASSIN, Proceedings Academy Natural Sciences, Phila., 1855, 281.
Buteo lineatus var. *elegans* RIDGWAY, in History of North American Birds, III, January, 1874, 257, 277.

(B 26, C 352a, R 439a, C 521, U 339b.)

GEOGRAPHICAL RANGE: Pacific coast of United States; north to southern Oregon; south to Lower California; (and Mexico?)

The breeding-range of the handsome Red-bellied Hawk, as far as yet known, is exclusively confined to the Pacific coast proper, extending into northern Lower California, from about latitude 29° north, through California, to central Oregon, where it is a rare summer resident. While not common in any locality, it is perhaps more abundant in the middle counties of California than anywhere else.

Mr. L. Belding, in his "Paper on the Birds of the Pacific District," makes the following remarks about this bird: "Upper Sacramento Valley. Apparently rare; probably resident, though not seen by me later than October 20. It is very common about Stockton in summer; nearly as common in the breeding season as the Red-tailed Hawk. I knew of a pair nesting within less than 200 hundred yards of a residence where poultry was plentiful and easily obtained. They nested there three consecutive seasons unmolested by the occupants of the dwelling. I shot the female as she flew from the nest, April

4, 1880. Her stomach contained several small lizards, a tree frog (*Hyla*), grubs, and insects. May 11, 1879, there were three young in the nest that weighed about a pound each. Mr. Charles Moore, who climbed the large oak in which the nest was placed, reported a lining of green, but dried and broken leaves in the nest, about 3 inches deep in the center. On April 4, 1880, there were three nearly fresh eggs in the nest, which this year had a lining of the lace-like lichen (*Ramalina reticformis*) found on the oaks in the vicinity, a sample of which was brought down from the nest by Mr. George Ashley, who, with great difficulty, secured the eggs.[1]

"I saw one of these Hawks at Stockton, January 25, 1885, repairing an old nest. This, too, was near a farmhouse where fowls were abundant, but I doubt if they often attack poultry, though I have known them to catch small birds."[2]

Mr. A. M. Ingersoll found it nesting near Los Angeles, California, usually in large sycamore or cottonwood trees, and says the nest is generally placed farther out on the branches than is the case with other Buteos.

Mr. F. Stephens writes me: "I find the Red-bellied Hawk rather rare in the region I have lived in, in southern California, but I think it is probably more common in the lower valleys near the seacoast. I found a single nest on April 7, 1882; this was placed near the end of a large limb of a cottonwood tree, about 35 feet from the ground, in San Mateo Cañon, 15 miles east of Colton. It was composed of twigs and lined with the inner bark of cottonwood and a few feathers; inside diameter 6 inches, depth 2 inches. The nest contained three eggs, and incubation had begun."

Mr. B. T. Gault found it common among the oak groves skirting the banks of the San Joaquin River, and noticed several also among the sycamores bordering the little stream running through the Santa Margarita Valley, on the road from San Diego to San Bernardino, California.

Prof. B. W. Evermann reports it not uncommon in the vicinity of Santa Paula, Ventura County, California, where he found their nests in sycamores, live oaks, cottonwoods, and willows, near the borders of streams. He says that he found as many as five eggs in a nest, and gives four as the average number laid to a set.

Mr. A. W. Anthony sends me the following notes on this subspecies: "The Red-bellied Hawk is not uncommon along the coast ranges of Lower California, nesting as far south at least as latitude 29°, and reaching here an elevation of about 2,000 feet. Nests were frequently seen in giant cactus and candlewoods, near San Fernando, Lower California. The only one I took was built in a small sumac scrub, 10 feet from the ground, near San Quintin, Lower California. It was simply a platform of small sticks and twigs, 12 or 14 inches in diameter and about 3 inches in depth, lined with a few dried leaves

[1] These eggs (No. 18082) are now in the U. S. National Museum collection, having been kindly presented by Mr. L. Belding, with numerous other valuable specimens, and measure as follows: 54 by 43, 55.5 by 43, 53 by 43 millimetres.
[2] Occasional Papers of California Academy of Sciences, II, 1890, pp. 34, 35.

of the sumae. The eggs, which were taken on April 30, three in number, contained large embryos."

I met with the nest of this Hawk on two occasions only during the spring of 1878, in the vicinity of Camp Harney, Oregon, where it is a summer visitor. I found the first one on April 17, 1878, and as the bird was unknown to me, shot the female, whose lower parts were of a uniform chestnut red throughout, and proved to be of this race. The nest was placed in a young pine on some limbs close to the top and the trunk of the tree, near the sources of Archies Creek in the foothills of the Blue Mountains and on the outskirts of the heavy timber. It was principally composed of sagebrush twigs, lined with pine needles, the inner bark of juniper trees, and a few green willow twigs with the leaves on. It measured about 18 inches in outer diameter and about 6 inches in depth. Inside it was about 8 inches in diameter by 2 inches deep. It contained two handsome fresh eggs, and another would have been laid, as I found on skinning the parent.

On May 6, during another visit to this locality, I discovered a second nest of this race still farther up this creek. This was placed in a tall juniper tree, likewise near the trunk and about 20 feet from the ground, composed of similar materials, and contained a single egg.

The principal call note when disturbed about its nest is a shrill "yee-ak, yee-ak," uttered rapidly and in a high key.

The number of eggs to a set varies from two to five; sets of three seem to be most frequently found. Judging from the limited number of these eggs seen by me, they appear to be not as heavily blotched and spotted as are those of *B. lineatus*, and the neutral tinted shell markings, such as pearl gray and lavender, predominate in them over the bolder brown and russet markings. Otherwise they do not differ materially in size, shape, or general appearance from the eggs of the Red-shouldered Hawk.

The average measurement of six eggs in the U. S. National Museum collection is 54 by 43.5 millimetres. The largest of these eggs measures 55.5 by 44.5, the smallest 53 by 43 millimetres. The type specimen, No. 20746, from a set of two, Bendire collection (Pl 6, Fig. 9), was taken by the writer, on April 17, 1878, near Camp Harney, Oregon.

79. Buteo abbreviatus CABANIS.

ZONE-TAILED HAWK.

Buteo abbreviatus CABANIS, in Schomburgk, Reise in British Guiana, III, 1818, 739.
(B —, C 353, R 440, C 522, U 340.)

GEOGRAPHICAL RANGE: Middle America ; north to Lower and southern California, Arizona, Texas, etc. ; south to northern South America.

The breeding range of the Zone-tailed Hawk is rather a restricted one within our borders. As far as is yet known it is confined to the southwestern portions of the United States, from Comal and Presidio Counties in central and southwestern Texas, through southern New Mexico, and central and southern Arizona. It probably breeds occasionally in southern California, where Dr. J. G. Cooper obtained the first specimen taken within the limits of the United States. It was shot 30 miles north of San Diego and 5 miles from the coast, on February 23, 1862.

It has been noted by Mr. L. Belding at the Cape St. Lucas region of Lower California; and Mr. A. W. Anthony writes me that he found it not uncommon on the San Pedro Martir Mountains in latitude 31°, at an elevation of about 7,000 feet. Two pairs of these birds were seen by him on April 24, both nesting in tall pines. He states: "The birds were greatly worried at our presence, flying about overhead and constantly uttering a loud querulous cry, not unlike that of *Buteo borealis*. One of the nests, examined from the ground, was rather a bulky affair of sticks, and placed in the very top of a pine about 70 feet up. Several shots from our rifles failed to drive the birds away. Shortly afterward a second pair were seen, and one of these was secured."

They are a common Mexican and Central American bird, and only summer visitors with us.

The first description of the supposed nest and eggs of the Zone-tailed Hawk, published in any of our ornithological papers, is that of Mr. William Brewster in the Bulletin of the Nuttall Ornithological Club (Vol. IV, April, 1879, p. 80), from specimens collected by Mr. William H. Werner in Comal County, Texas, on May 17, 1878. Mr. Brewster, in speaking of this find, says: "The nest, a large bulky structure, composed of coarse sticks, with rather a smooth lining of Spanish moss, was built in a cypress tree on the bank of the Guadalupe River. It was placed on a large and nearly horizontal branch, about 15 feet out from the main stem, and at least 40 feet above the ground. It measured as follows: External diameter 20 inches; external depth 6 inches; internal diameter 7 inches; internal depth 4 inches. The two eggs which it contained were slightly incubated. One is still preserved with the nest, the other is in Mr. Ricksecker's collection. The latter measures 2.09 by 1.55 inches (about 53.1 by 39.4 millimetres). It is marked with blotches of reddish brown upon a dull white ground. These blotches occur most thickly about the larger end, where they tend to form a nearly confluent ring. In Mr. Werner's specimen, which is similar in color, the markings are most numerous around the smaller extremity. Its dimensions are 2.06 by 1.53 inches (about 52.3 by 38.9 millimetres). Although the parent birds belonging to this nest successfully eluded all attempts at capture, their identity can scarcely be doubted. As Mr. Werner was climbing to their eyrie, they swept down about his head, repeatedly passing within a few feet of him. As but a few days previously he had shot the specimen above referred to [which is quoted in the beginning of this article and not copied by me] it is not likely that he could have mistaken a species so distinctly marked."

LIFE HISTORIES OF NORTH AMERICAN BIRDS.

Mr. Stephens published subsequently the following account: "May 28, 1876, I found a nest of *Buteo zonocercus* in a very large cottonwood tree, in a grove of the same, in the mouth of a cañon of the Gila River, in New Mexico, about 20 miles above the Arizona line. I saw the parent fly from the nest and with its mate circle around overhead. One alighted on the cliff overhanging the grove, which I succeeded in killing. It proved to be the male. I had no climbers and could not then get to the nest, but the next day I returned with a rope and succeeded in getting near enough to work my hand up through the nest and reach one egg, which was all there was. The nest was quite bulky, composed of twigs lined with strips of the inner bark of the cottonwood.

"The egg was very near hatching, and in attempting to extract the embryo I broke it, and it has since been broken into small pieces. It was marked with large reddish brown blotches, irregularly distributed on a dirty white ground. I still have the male parent."[1] Mr. Stephens sent me the pieces of this egg, which were all small, but as described above.

The next published account of the nesting habits of the Zone-tailed Hawk, and by far the most complete, is one by Asst. Surg. Edgar A. Mearns, U. S. Army. It is too long to quote fully, and I extract only such portions as are of special interest.

Dr. Mearns found his first nest of this species while camped in a grove of cottonwood trees upon the banks of New River, Arizona, May 16, 1885. While he was resting in the shade, a shrill whistle drew his attention to a Hawk that came gliding toward him through the dark shadows of the dense foliage. A quick shot brought the bird to his feet. On searching the vicinity he was not long in discovering a bulky nest fixed in the forks of a large cottonwood branch across the stream, at an elevation of about 25 feet, and the female parent standing upon it. She gave a loud whistle and came skimming toward him, and was also shot. The nest was coarsely built of rather large sticks, with considerable concavity, lined with a few cottonwood leaves only, and contained a single egg of a rounded oval shape, slightly smaller at one end, in color clear bluish white, immaculate, and measuring 55 by 43 millimetres. On dissecting the female he discovered that two would have been the full complement for this pair.

The doctor encamped the next day at the Aqua Frio. He says: "Here I again found the Zone-tailed Hawk. A female was shot as she flew screaming at me, and the nest was soon found in a cottonwood tree near by. The male parent sat upon the eggs and flew away when I got close up to the tree and shouted. It disappeared after circling over the cañon a few times, and did not return while I was there, although I spent several hours in the vicinity. I climbed with vast exertion to the nest, which was built in a fork about 50 feet from the ground and was exactly like the first one. It was composed of sticks,

[1] Bulletin of the Nuttall Ornithological Club, Vol. IV, July, 1879, p. 189.

lined only with green leaves of cottonwood attached to the twigs. It was rather concave, and contained two eggs which differ considerably in size, shape, and markings from those first found, but there can scarcely be any doubt about the identification, for the female was shot close to the nest, while the other bird was distinctly seen when flying from it, and was black, having its tail barred with white below. Perhaps, however, it is safest to say that these eggs are not absolutely free from the suspicion of being those of *Urubitinga anthracina*, as the parent seen to leave the nest was not shot. They are oval, considerably smaller at one end, ground color white, with yellowish weather stains in spots. One measures 63 by 45 millimetres. It is finely sprinkled with dark sepia-brown specks, and a few paler brown and lavender spots, having a smeary granular appearance. All the marks are most numerous at the large extremity. The other measures 61 by 43 millimetres. It is evenly blotched with very pale yellowish brown and lavender. Both contained large embryos and were emptied of their contents with difficulty."[1]

Several years previous to these accounts I met with the Zone-tailed Hawk on Rillitto Creek, Arizona, and found my first nest of this species on April 22, 1872, but as neither of the parents was procured with the eggs I did not at the time describe them. They were summer residents only, and I saw the first pair of "Black Hawks" on April 4, 1872. On the 6th I noticed another pair, which were just commencing to build a nest in a tall cottonwood tree, in a large grove about a mile above my camp. Neither pair of these birds showed any shyness, allowing me to approach closely to them; and, with the assistance of an excellent field glass, I took careful observations of them at different times while they sat at rest on some dead limb where they could be plainly seen.

On April 22, while riding along the banks of Rillitto Creek, which even at that early date had dried up, leaving only a stagnant water hole here and there, I noticed one of these Black Hawks flying up the creek bed, and being at leisure I followed it. Some 5 miles above my camp, near the entrance to Sahuaritto Pass, it perched on a dead limb of a large cottonwood tree on the west side of the creek. On nearing this, I saw an old and bulky nest placed in a fork close to the main trunk of the tree, about 40 feet up, and the mate of the bird I had been following sitting on the nest. As my principal object was to study the nesting habits of our birds, as well as to collect their eggs, I refrained from shooting either of them, which I might easily have done at the time. On climbing to the nest I found it contained but a single pale bluish white unspotted egg. The old birds during this time were circling around above the tree giving vent to shrill screams. Being some distance from camp I took this egg, and had not moved more than a hundred yards away from the tree before one of the birds, presumably the female, settled on the nest again as if nothing had happened. As the set was certainly not complete, I concluded to pay them a second visit and secure the other eggs and one of the parent birds also. On reaching camp and blowing the egg, I found it quite fresh.

On May 3, I paid a second visit to this locality and found one of the birds on the nest, where it remained until I rode up to the tree and rapped on it with the butt of my shotgun. This caused it to fly off about 50 yards farther up, on the opposite side of the dry creek bed, where it alighted in a smaller tree. As the bird appeared so very tame I concluded to examine the nest before attempting to secure the parent, and it was well I did so. Climbing to the nest I found another egg, and at the same instant saw from my elevated position something else which could not have been observed from the ground, namely, several Apaché Indians crouched down on the side of a little cañon which opened into the creek bed about 80 yards farther up. They were evidently watching me, their heads being raised just to a level with the top of the cañon.

In those days Apaché Indians were not the most desirable neighbors, especially when one was up a tree and unarmed; I therefore descended as leisurely as possible, knowing that if I showed any especial haste in getting down they would suspect me of having seen them; the egg I had placed in my mouth as the quickest and safest way that I could think of to dispose of it—and rather an uncomfortably large mouthful it was, too—nevertheless I reached the ground safely, and, with my horse and shotgun, lost no time in getting to high and open ground. I returned to the place again within an hour and a half looking for the Indians, but what followed has no bearing upon my subject. I only mention the episode to account for not having secured one of the parents of these eggs. I found it no easy matter to remove the egg from my mouth without injury, but I finally succeeded, though my jaws ached for some time afterward. On blowing it the next day I found it slightly incubated. It was unspotted like the first. The nest was evidently an old one which had been used for many years. It was quite flat on top and sparingly lined with cottonwood bark.

On May 17, a set of three eggs were brought to me by two of my men who were familiar with the appearance of these birds. The nest was in a similar situation in a large cottonwood tree near the sink of the Santa Cruz River, about 10 miles northeast of Tucson. These also were unspotted and nearly fresh.

At that time I was not aware that the Mexican Black Hawk was found in Arizona, but on a subsequent visit to Washington, District of Columbia, in the fall of 1874, while examining specimens of both *Buteo abbreviatus* and *Urubitinga anthracina*, I saw plainly that I had accurately described a peculiar plumage of the latter species which is not found in *Buteo abbreviatus*, and on showing my original notes to Mr. Robert Ridgway, of the Smithsonian Institution, he fully agreed with me that my description of a bird seen on April 4, 1872, plainly referred to the Mexican Black Hawk. On the strength of this, as well as of the description of the eggs of *Buteo abbreviatus* being spotted, based on Stephens's and Werner's identification, I naturally came to the conclusion that the eggs taken by me in Arizona were clearly referable to the Mexican Black Hawk and none other.

Even Dr. Mearns's subsequent article on these two species published in the Auk (Vol. III, January, 1886, pp. 60–73), did not shake my faith in my correct identification of these eggs, after having examined specimens of the two species. This was further strengthened by a letter from Mr. William Lloyd, who wrote me that he had found a nest of *Urubitinga anthracina* on April 13, 1890, in Presidio County, Texas, and having shot the female, he found a fully developed egg in her oviduct. This was badly broken; he sent me the pieces, however, and Mr. F. A. Lucas kindly restored the specimen very skillfully. This egg was likewise bluish white in color, unspotted, and measured 54 by 42 millimetres; it was in fact an exact counterpart of Dr. Mearns's egg of *Buteo abbreviatus*, and of my supposed eggs of *Urubitinga anthracina*. I was naturally anxious to see the parent, and this, on examination, proved to be a Zone-tailed Hawk.

I am satisfied now that Dr. Mearns was perfectly right in his surmise, that the birds found by me breeding in Arizona in 1872 were really *Buteo abbreviatus* as I originally supposed, and I am equally certain that some of the notes taken by me at the time, and which will be quoted in their proper place, are positively referable to *Urubitinga anthracina*, and show clearly that I met with both species.

The Zone-tailed Hawk is pretty generally distributed over the greater portion of Arizona during the breeding season, and is not especially rare. A pair or two may be found inhabiting all the larger cottonwood groves in the Territory, and it seems to be especially partial to such trees to nest in, they usually being found only along the banks of the few streams in this poorly watered country. According to Dr. Mearns, its food consists of lizards, frogs, and fishes, but small mammals are undoubtedly also included in its bill of fare.

It usually makes its appearance from the south in the beginning of April, and but a single brood is raised in a season. Both sexes assist in incubation, which, as with our larger Raptores, lasts about four weeks. The eggs vary from one to three in number, usually two, and seem to be for the most part unspotted. They are oblong oval in shape, pale bluish white in color, and the shell is rather smooth and finely granulated.

According to Messrs. Stephens and Werner the eggs are occasionally spotted, and I have no reason to doubt their identification. Nidification commences in the latter part of April and lasts through May.

The egg taken by Dr. Mearns near New River, Arizona, on May 16, 1885, about which there can be no possible doubt as to its identification, both parents being shot at the same time, and which is now before me, measures 54 by 43 millimetres, and is figured on Pl. 7, Fig. 6, as the type of this species, having been kindly loaned to me by the American Museum of Natural History, New York City, New York, to which collection it belongs. The eggs of this species from near Tucson, Arizona, measure as follows: First set, 53 by 43 and 54.5 by 42.5 millimetres; second set, 53.5 by 42, 53.5 by 42, and 55 by 41 millimetres. Mr. William Lloyd's specimen, from Presidio County, Texas, measures 54 by 42 millimetres. On none of these eggs are any markings visible.

80. Buteo albicaudatus Vieillot.

WHITE-TAILED HAWK.

Buteo albicaudatus Vieillot, Nouveau Dictionaire, IV, 1816, 477.
(B —, C —, R 441, C 513, U 341.)

GEOGRAPHICAL RANGE: Whole of Middle America; north to southern Texas; south to portions of eastern South America.

This common Mexican and South American species, easily recognized by its pure white underparts, is an abundant resident of the Gulf coast of Texas and the lower Rio Grande Valley, the only localities within the United States where it has been found to breed, as far as known at present.

I believe that my friends, Dr. James C. Merrill, U. S. Army, and Mr. George B. Sennett, who have both contributed so much to our knowledge of the ornithology of the Lower Rio Grande, of Texas, were the first naturalists who obtained the nest and eggs of this handsome Hawk within our borders. The former, in speaking of this species, says: "This fine Hawk is rather a common resident on the extensive prairies near the coast, especially about the sand ridges covered with yuccas and cactus. Its habits appear to be like those of the allied species of Prairie Hawks. On the 2d of May, 1878, I found two nests, each placed in the top of a yucca growing in Palo Alto prairie about 7 miles from the fort. The nests were not more than 8 feet from the ground, and were good sized platforms of twigs with scarcely any lining. While examining these nests, the parents sailed in circles overhead, constantly uttering a cry much like the bleating of a goat. Each nest contained one egg. The first was quite fresh and measures 2.35 by 1.91 inches (about 59.7 by 48.5 millimetres). It is of a dirty white color with a few reddish blotches at the smaller end. The second egg was partly incubated. It resembles the first one, but the reddish blotches are rather sparingly distributed over the entire egg. It measures 2.35 by 1.85 inches (about 59.7 by 47 millimetres)."[1]

In the spring of 1882, while on a collecting trip on the Gulf coast, near Corpus Christi, Texas, Capt. B. F. Goss found this species breeding abundantly in that vicinity. He writes me: "We found the favorite breeding places of the White-tailed Hawks to be a strip of open bushy land lying between the thick line of timber and chaparral along the coast and the open prairie. Any bush rising a little above the surrounding level seemed a suit-

[1] Proceedings U. S. National Museum, Vol. 1, 1878, pp. 156, 157.

able nesting site, and no attempt was made to conceal the nest. In most cases it was very prominent, and could be seen for a long distance. I examined fifteen; they were all placed in low bushes, generally not higher than 6 feet. In a few cases I had to stand upon the wagon to reach them. They were composed of sticks, dry weeds, and grasses. A coarse dry grass entered largely into the composition of most of them. They were poorly constructed, but moderately hollowed, and usually lined with a few green twigs and leaves. Taken as a whole, the nests looked ragged in outline and slovenly finished. About one nest in four contained three eggs, the rest but two. This Hawk is wary and difficult of approach at all times. They would leave their nests as soon as we came in sight, sometimes when still half a mile away, and generally they kept entirely out of sight. An occasional pair sailed high over our heads, uttering a faint cry while we were at their nest. Only a single one came within range of our guns. The earliest date on which eggs were found was March 5; the latest April 25." During the spring of 1884 Captain Goss's collector took eggs of this species on February 1, and in 1885 as late as July 4. Only one nest was 11 feet from the ground; the others were all lower, generally from 4 to 8 feet up.

Nothing is mentioned about the food of this species by any of our ornithologists who have met with it in southern Texas. Mr. W. H. Hudson says, in speaking of the habits of the White-tailed Hawk in the Argentine Republic: "I have dissected a great many and found nothing but Coleopterous insects in their stomachs; indeed they would not be able to keep in such large companies when traveling if they required a nobler prey. At the end of one summer, a flock numbering about one hundred birds appeared at an 'estancia' near my home, and though very frequently disturbed they remained for about three months roosting at night on the plantation trees and passing the day, scattered about the adjacent plain, feeding on grasshoppers and beetles. This flock left when the weather turned cold; but at another 'estancia' a flock appeared later in the season and remained all winter. The birds became so reduced in flesh that after every cold rain or severe frost numbers were found dead under the trees where they roosted; and in that way most of them perished before the return of spring."[1]

The usual number of eggs laid by this species is two; sets of three are not uncommon however, and occasionally but a single egg seems to be laid. The eggs are large for the size of the bird. Nidification, some seasons at least, begins very early, full sets of eggs having been taken on February 1, 1884, and again it is protracted well into July, as several fresh sets of eggs in the collection here attest. This may have been caused by repeatedly robbing them of their eggs, or perhaps they raise two broods in a season.

The eggs of the White-tailed Hawk are dull dirty white in color, faintly and sparingly marked with irregular small blotches of pale brown and drab. An occasional specimen shows a few small lavender shell markings also.

About one-third of the eggs are unspotted. In shape they vary from an elliptical oval to an oval. Now and then one occurs that may be designated as elliptical ovate. The shell is fairly smooth and close grained.

The average measurement of fifteen eggs of this species in the U. S. National Museum collection is 60 by 47.5 millimetres; the largest egg measuring 65 by 50, the smallest 55 by 44 millimetres.

The type specimen No. 22581, selected from a set of three, taken March 8, 1882 (Pl. 7, Fig. 8), is the heaviest marked egg of the series, but in another egg of this set the spots, while fewer, are darker colored, and No. 22583, from a set of two, taken April 4, 1885 (Pl. 7, Fig. 9), shows but a trace of a faint spot here and there, and resembles an unspotted egg. Both these sets were obtained in exchange from Capt. B. F. Goss, and were taken near Corpus Christi, Texas.

81. Buteo swainsoni BONAPARTE.

SWAINSON'S HAWK.

Buteo swainsoni BONAPARTE, Geographical and Comparative List, 1838, 3.
(B 18, 19, 21, 28, C 334, R 442, C 523, U 342.)

GEOGRAPHICAL RANGE: Western North America; north to Alaska and western side of Hudson Bay; east to Wisconsin, Illinois, and Arkansas (casually to Massachusetts); and south through Central America and greater part of South America to the Argentine Republic.

Swainson's Hawk has a wide distribution during the breeding season. Commencing with the southern portion of its range east of the Rocky Mountains—this includes portions of the less densely timbered and prairie regions of Texas and Arkansas—thence northeastward in similar localities to Illinois and Wisconsin; north and westward through the intervening States and Territories into the Dominion of Canada from Manitoba, westward and north to the Arctic regions in about latitude 65°. West of the Rocky Mountains it is found in New Mexico and Arizona, and over the entire Pacific coast region north into British Columbia and Alaska.

On the arid wastes and table lands of southern Arizona, as well as in the sage and bunch grass districts of Nevada, Oregon, Washington, and Idaho, Swainson's Hawk is especially abundant, outnumbering, perhaps, all the other Raptores of these regions combined. It is eminently a prairie bird, shunning the densely timbered mountain regions, and being more at home in the sparingly wooded localities usually found along the water courses of the lowlands.

Compared with the majority of our Hawks it is gentle and unsuspicious in disposition, living in perfect harmony with its smaller neighbors. It is no unusual sight to find other birds, such as the Arkansas Kingbird, *Tyrannus verticalis*, and Bullock's Oriole, *Icterus bullocki*, nesting in the same tree; and the first-mentioned species goes even further than this, sometimes constructing its home immediately under the nest of these Hawks or in the sides of it. Two such instances came under my personal observation.

The food of Swainson's Hawk consists almost entirely of the smaller rodents, principally striped gophers and mice, as well as grasshoppers and the large black cricket, which is very common as well as destructive in certain seasons, and the bane of the farmers in eastern Oregon, Washington, Idaho, Nevada, and other localities in the Great Basin, destroying and eating up every green thing as they move along. Even the bitter leaves of the sage bush, *Artemisia*, are not despised by these pests, and while these last they constitute their principal food. From almost daily opportunities enjoyed by me for years of observing this bird, I do not hesitate to say that aside from an occasional half-grown hare or rabbit, which from their abundance in some of the regions referred to are themselves considered as quite a nuisance, its daily fare consists almost exclusively of such food as I have mentioned.

I cannot recall a single instance where one of these birds visited a poultry yard; and if other food is procurable it will seldom molest a bird of any kind. From an economic point of view I consider it by far the most useful and beneficial of all our Hawks. It is found in a great variety of plumages and in some stages it is an exceedingly handsome bird.

In the more northern portions of its range it is only a summer resident, migrating regularly in large straggling flocks. In the fall of 1881, while encamped near the Umatilla Indian Agency, in Oregon, I noticed numbers of these birds passing; I should think that not less than two thousand of them flew by in straggling bands and settled down on the foothills a few miles south of the agency. They were evidently returning from their breeding grounds in the North, and as they flew rather low could be readily identified, although they varied greatly in plumage. As near as I could tell this body of Hawks seemed to be almost exclusively composed of this species. .

On the eastern slopes of the Rocky Mountains it winters from about latitude 39° southward, a few remaining in favorable localities still farther north. On the Pacific coast I have observed a few wintering in southeastern Oregon in about latitude 42°, the majority passing southward, and the birds remaining are probably such as breed much farther north, replacing the regular summer residents, which in turn move south on the approach of cold weather.

Swainson's Hawk, as a rule, nests late in the season, even in some of the southern portions of its range. This, however, does not seem to hold good everywhere, as Mr. William Lloyd informs me that on the prairies west of Chihuahua, Mexico, he took eggs of this species on March 6; Mr. William Cobb took some in central New Mexico, March 20, and Mr. F. Stephens in southern California on April 16.

In the large series of eggs of this species in the U. S. National Museum collection, consisting of one hundred and sixty-six specimens, which represent nearly every State and Territory west of the Mississippi River, from Arizona to Alaska and the Arctic regions, there is but a single set taken as early as

April 10, this being from southern California. Even the numerous records from southern Arizona fail to show a single date earlier than May 12, by far the greater portion falling in the last ten days of this month and the first week in June. These same dates correspond equally well with the commencement of the breeding season in Oregon, Washington, and Idaho, where I took a number of their nests.

In Wyoming, Montana, the Dakotas, Manitoba, and farther northward, most of the records are in June, a few in July, and a single specimen was taken by Dr. Elliott Coues on St. Marys River, Montana, on August 17, 1874, but this was probably an addled egg.

Swainson's Hawk arrives in its summer home fully a month before nidification commences, and during this period these birds spend considerable time on the wing, sailing and circling around high in the air, uttering their not unmusical notes; at other times they are much more sluggish and are fond of sitting on a dead limb of a tree; a telegraph pole on the plains is a particularly favorite perch of theirs, or a sage bush or some hillock on the prairie, from whence they watch the surroundings for their humble fare. This latter habit of sitting on the ground is rather characteristic of this species.

They rarely build a new nest, the one used the previous year, if not already occupied by some of the earlier breeders, like the American Long-eared Owl, *Asio wilsonianus*, is slightly repaired by the addition of a few sticks and a little lining, or an old Crow's nest is taken possession of and reconstructed to suit; these repairs seldom take more than a day or so, and the nest is then ready for the eggs. The nesting sites of this species vary greatly according to circumstances and locality. In southern Arizona, especially in the vicinity of Fort Huachuca, where this Hawk is a resident and exceedingly common, Lieut. H. C. Benson, Fourth Cavalry, U. S. Army, found forty-one of their nests between May 12 and June 18, 1887. All of these were placed in low mesquite trees, from 3 to 15 feet from the ground. A few found by me near Tucson, in the spring of 1872, were located in similar trees from 10 to 18 feet from the ground.

In southeastern Oregon, as well as in Washington, Idaho, and Montana, I frequently found nests of this species placed in some small bunch of willows growing here and there in marshy places or along the banks of streams, and again in isolated pine trees, straggling patches of junipers, and now and then in mere bushes on side hills of cañons, or near the rims or edges of the sagebrush-covered table lands making out from the mountains proper; never in what might be called a forest country. Solitary trees, no matter how small in size, giving a good outlook over the surrounding country, are, in those regions at least, their favorite nesting sites. With but few exceptions the nests were easily reached and usually placed in a fork close to the main stem, seldom out on one of the larger limbs. Their average height from the ground was not over 20 feet. But a single nest, placed on a large limb of a good sized cottonwood on the Umatilla River, Oregon, of all those found by me, more than thirty in number, was difficult to get at and much higher from the ground than usual. It was fully 50 feet up, and contained but a single fresh egg on May 28, 1882. The nests varied considerably in size and bulk, but rarely averaged more than a foot and a half wide by a foot in depth. The top of the nest was usually rather flat, and in most instances fairly well lined with either dry grass, weed stalks, dry cottonwood, or juniper bark. A few were lined with green willow tops, and now and then I found one with scarcely any lining. They are not neatly constructed and have a ragged looking appearance from the outside; frequently a streamer or two of juniper bark hangs down from the sides of the nest, giving it a very dilapidated appearance, and many times, had I not been able to see the bird on the nest, I should certainly have taken it for an abandoned one.

Capt. B. F. Goss says: "I found this species breeding in North Dakota in the high timber along the streams, from 40 to 60 feet up, and in low brush patches on the prairies in the lake region, where the nests were but 2 to 4 feet from the ground; I also found nests on the ground in the open prairie. I think from appearances there has been brush growing where I found these nests, and the birds became attached to the locality, and continued to use it after the bushes had been burned off by the prairie fires so common in those regions. I also found it breeding in the heavy timber along the Nueces River in southern Texas."

Mr. F. Stephens says: "Swainson's Hawk arrives in southern California in the early spring, in flocks, many remaining in the larger unsettled valleys to breed. I have seen hundreds in sight at once, scattered over the plain, on the ground, where they sit most of the day, after gorging themselves on beetles and especially grasshoppers. At night they roost in the nearest trees or on rocky hillsides. On one occasion I took a set of eggs of this species, and a set of *Icterus cucullatus nelsoni* from a nest pendant from some of the twigs composing the Hawk's nest; another time I found nests of *Tyrannus verticalis* and *Carpodacus mexicanus frontalis* built in the mass of the Hawk's nest, all occupied at the same time."

Mr. Lynds Jones states that in Iowa this Hawk usually constructs its nest in moderately timbered tracts, usually in oak or elm trees, from 30 to 80 feet up, the nesting season beginning in the first half of May, their nests being often lined with leaves. Their call note, aside from a peculiar gurgling sound made while diving through the air, resembles the word "pi-tick, pi-tick," frequently repeated.

In Montana, according to Mr. R. S. Williams, fresh eggs may be looked for in the last week in May. The nests here are usually placed in low cottonwoods from 12 to 15 feet from the ground and near the center of the tree.

Mr. J. W. Preston found a pair of these birds nesting in a Blue Heron's deserted nest, near Baxter, Iowa. He says: "Although a bird of graceful form and often of pleasing movement, it is the most careless in habits of any of our Hawks. 'Tame' is an appropriate epithet. A pair spent many days in rebuilding an old nest of the common Crow 15 feet up in a small tree, directly over a road in a wood and in plain sight, yet the birds evinced no more concern about the passers-by than a Bantam hen would do. It is popularly known as the Prairie Hawk here, and so it is, but it often nests in the bordering groves." According to Mr. W. H. Cobb, in New Mexico they occasionally nest in cañons or the sides of cliffs under an overhanging rock. Here they feed principally on hares and wood rats, and remains of snakes have been found in one of their nests.

A nest whose contents are now in the U. S. National Museum collection, taken at the forks of Milk River, Montana, on July 17, 1874, by Dr. Elliott Coues, was placed against the face of a perpendicular earth bank 100 feet high, on a slight projection of the ground about half way up. It contained two eggs in which incubation was far advanced. Nests have also been found in live oaks, sycamores, and aspens.

Lieutenant Benson found Swainson's Hawk living in harmony with other birds and writes me that after the Arkansas Kingbirds began to build he invariably found one of their nests in each tree that contained one of the former. In one case a pair of these Flycatchers had placed their nest directly under and but 8 or 9 inches from that of the Hawk. A pair of White-rumped Shrikes nested also directly under one of these Hawk's nests. On another occasion he found rather a strange occupant in a nest of this species, and it is questionable if, in this instance at least, the usually good-natured Swainson's Hawk would not have resented the intrusion. This visitor was a good-sized rattlesnake which had managed to climb up the low bushy tree and coiled itself comfortably in the nest.

But one brood is raised in a season. If the first set is taken they will frequently lay a second one consisting of one, seldom two eggs, and use the same nest again as a rule. Incubation lasts about twenty-eight days, both sexes assisting, and the eggs are deposited at intervals of about two days.

The number of eggs to a set varies from two to four. According to my own experience, and that of Lieutenant Benson as well, the number most often found is two. About one nest in four contains three eggs, and a set of four is rarely met with. I found but one such in over thirty nests, and the only other set in the collection was taken by Mr. Ernest E. Thompson, near Carberry, Manitoba, June 8, 1883. My set of four was taken near Fort Lapwai, Idaho, May 28, 1871.

In shape these eggs vary from a short ovate to an oval, and their shells are rather smooth and close grained. Their ground color when fresh is a very distinct greenish white which in course of time fades into a dull yellow white. When not closely looked at many of the eggs of Swainson's Hawk appear to be unspotted, but on careful examination there are in reality very few that are immaculate. The majority are more or less distinctly spotted and blotched with different shades of brown, such as burnt umber, liver brown, hazel, and tawny, and of clay color, French gray, and drab gray. With but very few exceptions none of the eggs are heavily marked; about one-half are, however, moderately well spotted.

The average size of one hundred and sixty-six specimens in the U. S. National Museum collection is 56.5 by 44 millimetres. The largest egg of the series, one taken by myself at Fort Lapwai, Idaho, measures 62 by 46.5 millimetres, the smallest, taken by Lieut. H. C. Benson, at Fort Huachuca, Arizona, measures but 50.5 by 40.5 millimetres. As is usually the case, the eggs from the northern breeding ranges average larger in size than those from the southern.

The type specimens selected to show the principal variations found in those eggs are: No. 20713 (Pl. 8, Fig. 1), from a set of two taken near Camp Harney, Oregon, May 12, 1875; No. 20733 (Pl. 8, Fig. 2), a very peculiarly marked egg from a set of four, taken near Fort Lapwai, Idaho, May 28, 1871; No. 20740 (Pl. 8, Fig. 3), a single egg, and very distinctly marked, was taken near the Umatilla Indian Agency, Oregon, May 28, 1882; these are from the Bendire collection. No. 23139, two eggs from same set, very handsomely marked (Pl. 8, Figs. 4 and 5), taken near Fort Huachuca, Arizona, May 27, 1887, and No. 23150 (Pl. 8, Fig. 6), from the same locality, taken June 18, 1887, were collected by Lieut. H. C. Benson, Fourth Cavalry, U. S. Army, who made an exceedingly interesting collection of both birds and eggs of that then slightly explored region; and he generously presented them all to the U. S. National Museum.

82.　Buteo latissimus (WILSON).

BROAD-WINGED HAWK.

Falco latissimus WILSON, American Ornithology, VI, 1812, 92, Pl. 54, Fig. 1.
Buteo latissimus SHARPE, Catalogue of Birds in British Museum, I, Accip., 1874, 193.
(B 27, C 355, R 443, C 524, U 343.)

GEOGRAPHICAL RANGE: Eastern North America; north to New Brunswick and Saskatchewan; west to edge of Great Plains; south (in winter only ?) through Central America and West Indies to northern South America.

The breeding range of the Broad-winged Hawk includes the whole of the eastern United States, from the Mississippi River to the Atlantic Ocean. West of the Mississippi Nehrling found it breeding near Houston, in eastern Texas. Col. N. S. Goss gives it as rare in Kansas, but probably breeding, and it is known to be common in portions of Iowa and Minnesota. It ranges beyond our border into southern Canada, and breeds from Nova Scotia and New Brunswick west to the province of Assiniboia and north to the Saskatchewan plains.

In Florida, South Carolina, and the Gulf coast generally, it seems to be a rare resident, but it has been reported as breeding near Lake Harney,

Florida, and an egg collected by Mr. H. B. Moore, near Manatee, Florida, in the spring of 1872, and now in the U. S. National Museum collection, entered as one of *Buteo lineatus*, is so small that it would seem much more likely referable to this species. Mr. W. E. D. Scott and other well known ornithologists who have collected extensively in Florida during the past ten years, do not appear to have met with it during the breeding season, but as it is reported to breed in Cuba it is reasonable to suppose that the same is true of our southern border. Dr. A. K. Fisher tells me that he saw a pair near Mobile, Alabama, in May, 1886, which acted as if they had a nest in the vicinity.

Throughout the Central and Northern States it is somewhat irregularly distributed, being common in some sections and rare in others. As it is eminently a bird of the larger forests, and seldom seen in the more open and cultivated country, its abundance in certain regions and scarcity in others can be readily accounted for.

Dr. William L. Ralph, of Utica, New York, writes me as follows: "The Broad-winged is the Hawk of the Adirondack wilderness, and it replaces in this locality the Red-tailed and Red-shouldered Buteos which are so common in the smaller woods of the more settled parts of this State. I think that with the exception of the Pigeon Hawk and Goshawk, which probably breed here also but are very rare, they are the only Hawks that nest in the interior of these woods, but along the borders they are sometimes found breeding in the same situations as the more common species.

"They are to be found in considerable numbers in this wilderness, or at least that part of it lying within the counties of Hamilton, Herkimer, and Oneida, but they are evidently rare in the more open country.

"I never met with them during the breeding season in any other locality, with the exception of a small district a few miles south of Utica, where three or four nests have been taken during the last fifteen years. They are very fond of living near water, and their nests are always to be found in close proximity to the lakes and streams which are so numerous in the Adirondack region. The smaller lakes especially are favorite places of resort, and when a pair takes possession of one they apparently hold it against all intruders of their kind.

"I have never seen more than a single pair in close proximity to one of these small lakes. They frequent them on account of the mice, chipmunks, shrews, and frogs that usually live in large numbers along the shores, and which seem to form their principal food.

"Most writers on this species say that it is a very quiet bird, and should one of their nests be disturbed, it will fly silently away and make no protest whatever. Now, while the statement that it is a quiet bird evidently holds good for the greater part of the year, my experience shows that it is just the opposite during the breeding season; then it is as noisy as any Hawk I know of. When one is driven from its nest it at once utters a shrill call, which soon brings its mate to the spot, and together they will keep up their noise as long as anyone remains in the vicinity. They are very tame in this locality, and frequently when one is started from its nest it will not even leave the tree, but alight on a limb near by. They are gentle in disposition and never attempt to strike at a person, although they are very solicitous about their eggs and young. For days after they have been robbed these birds will utter their complaints when anyone approaches their homes.

"The only note I have heard them utter is a whistle which sounds almost exactly like that given by the Killdeer Plover (*Ægialitis vocifera*). They do not seem to decrease in numbers, probably because they do no harm and consequently are not much hunted; and their usually quiet habits also render them rather inconspicuous. Both parents assist in incubation and seem to be equally solicitous for their eggs and young.

"I took two nests with eggs during the season of 1890. The first was found May 28, about half a mile south of Wilmurt, New York. It was situated in a crotch of a birch tree, 50 feet above the ground, which was growing on a wooded hillside sloping back from a small stream. The materials composing it consisted of sticks and hemlock twigs, and it was lined with hemlock and birch bark. It measured 6 by 9 inches in depth, 15 by 21 inches in diameter, and contained two nearly fresh eggs.

"The second nest was found on May 30, and contained three eggs about three-fourths incubated. It was situated 56 feet above the ground, also in a crotch of a birch tree growing near the bank of West Canada Creek, about 1 mile north of Wilmurt, New York. It was composed of sticks and lined with birch bark, hemlock twigs, and feathers from the sitting bird, and it was a trifle smaller than the other nest."

Mr. J. W. Preston, who has taken a number of the nests of the Broad-winged Hawk in Becker County, Minnesota, writes me as follows: "This Hawk is a common summer resident here, and nests from May 12 to May 20. I have found a nest and eggs, however, as late as June 18, probably a second laying after losing the first set. Their nests are usually placed in the crotch of a tree, such as basswood, elm, oak, or larch, from 15 to 45 feet from the ground. Occasionally it is placed on a leaning trunk or a large branch, generally in dark woods near lakes. It is a close and compactly built structure, composed of small sticks gathered from the ground near by, no long pieces being used. It is lined with small shreds of bark, but as incubation advances, downy feathers dropped from the breast of the bird become a factor in the lining; a green, leafy twig of basswood or poplar is also added occasionally.

"Their call note is a peevish 'chee-é-é-é' prolonged at pleasure and. uttered in a high key. They will sit for hours in high dead trees calling to each other during the breeding season, and they are certain to be heard from should anyone come near the nest. In this locality their food consists mostly of the red squirrel.

"Though usually a sluggish bird, they will at times show considerable courage and dash at an intruder. I have noticed two such instances. Once, while I was in a tree watching a Swallow-tailed Kite, a male Broad-winged Hawk which was guarding a nest fought another bird of this species, driving and pursuing it a great distance. Then suddenly it turned back and almost struck me in the face as it came on with arrow-like swiftness."

Mr. H. H. Brimley, of Raleigh, North Carolina, found a nest of this species, containing two fresh eggs, and sent me an accurate description of it, made before it was removed from the tree. It was taken on April 25, 1890, and is thus described: "It was placed in a small pine in original woods, about 38 feet from the ground. About 6 feet from the top of the tree the main trunk ended in five wide spreading limbs. At the junction of these limbs the nest was built in the crotch. The nest itself was the roughest kind of platform, made entirely of oak sticks, not a single stick of any other kind being used. It was lined with a good sized double handful of pieces of bark from large yellow pines, flat, thin, and smooth scales, not the rougher, thicker bark found on younger trees. This bark lining was 4 or 5 inches deep in the center of the nest; a very few twigs with green leaves of both pine and oak were scattered through the structure apparently accidentally. The bark center of the nest went right down to the limbs of the crotch supporting it. The depth of the depression was more than that in a nest of *Buteo lineatus*, and the whole nest was much rougher and looser than any of those of the latter I have ever seen. No attempt at using any soft lining had been made. While I was up in the tree taking the eggs and making these notes the old birds soaring high above gave vent to notes much like those of the Killdeer."

Mr. J. C. Cairns, of Weaverville, North Carolina, took a set of three eggs of this species on April 25, 1890, and kindly sent me the nest, which was placed in the forks of a large oak tree about 56 feet from the ground. Considering the size of the bird the nest is large and a somewhat loosely built structure. Its outer diameter is 19 by 13 inches; depth, 9 inches; the inner cavity appears to have been shallow and not over 3 inches deep, if so much. It is principally composed of small oak twigs, and among these a few slender pine tips are mixed. The lining consists of thin pieces of pine bark scales. Mr. Cairns says: "In this portion of North Carolina they usually nest on a wooded ridge or at the foot of a mountain, never on or near the top of these, and they return each year to the same localities to breed, nearly always building a new nest each season. Occasionally they make use of an old Crow's nest, or one abandoned by some other Hawk. As a rule they nest in somewhat lower situations than the majority of Raptores, and now and then so low that the nest can be reached from the ground."

Mr. George G. Cantwell, of Lake Mills, Wisconsin, writes me that he found one of their nests in the crotch of a large tree, only 3 feet from the ground. From 25 to 30 feet up is perhaps a fair average.

Besides the different species of trees already mentioned, they nest in maples, poplars, black walnut, beech, chestnut, hemlock, different kinds of pines, oaks, and birches. In the northeastern parts of their breeding range birches, especially the yellow birch, seem to be most often selected for nesting purposes.

Ordinarily the nest is lined only with thin scales of bark, that of the yellow pine perhaps predominating. Sometimes green fir and other leafy twigs and the fine inner bark of the white cedar is added to the lining. Their nests are most often found in the more extensive woods near water and in swamps, and much less frequently in the more open and cultivated sections. The Broad-winged Hawk is only a summer resident north of about latitude 36°, migrating in straggling bodies during September and October to their winter homes in central and northern South America, moving sometimes together in quite large numbers and returning north in March and April.

Their food consists to a great extent of small rodents, such as mice, gophers, and squirrels; shrews, small snakes, frogs, grasshoppers, beetles, larvæ of insects, and very rarely small birds. It is one of the most harmless of our Raptores and of great benefit to the farmer. It is a late breeder; in the more southern portions of its range nidification begins about the second week in April and correspondingly later northward. In the New England States, northern New York, Pennsylvania, Iowa, and Minnesota, generally in the latter half of May, and in New Brunswick and the southern portions of Canada about the beginning of June and sometimes later. Incubation lasts from twenty-one to twenty-five days, and the eggs are deposited at intervals of one or two days. Both parents assist in incubation and in the care of the young. A single brood is raised in a season.

The number of eggs laid by the Broad-winged Hawk is usually two or three, very rarely four. Mr. O. C. Poling informs me, however, that in the vicinity of Quincy, Illinois, sets of four are not especially uncommon, and that he found a nest of this species containing five eggs. These vary considerably both in shape and markings, sometimes even when taken from the same nest.

The average measurement of forty-five specimens I find to be 48.5 by 39 millimetres. The largest eggs, three in number, one from New Jersey, another from New Brunswick, and a third from Pennsylvania, measure each 52 by 40 millimetres. The smallest, from Maine, measures 45 by 38.5 millimetres. In shape they range from short and rounded to elliptical ovate, the greater number being short ovate.

The ground color of the majority of these eggs is dull grayish white; in a few a faint trace of pale gray green is perceptible, and in rare instances this greenish tint is rather pronounced and readily noticed. In some of the eggs faint lavender, pearl gray, or écru-drab shell markings predominate, scattered either in fine dots or irregularly shaped blotches and in variable amounts over the greater or smaller axis of the egg; or again they may be evenly distributed over its entire surface, relieved here and there by an occasional dark lilac or rich chestnut colored spot. In others, these shell markings are either entirely absent or but faintly perceptible, and the egg is spotted and blotched with different shades of brown, hazel, drab, and fawn color. These markings are usually heaviest on either the large or small end. One specimen in Dr. William L. Ralph's collection is very finely marked with small dots not larger than dust shot, rather evenly distributed over nearly the entire surface, giving it a flea-bitten appearance; occasionally an egg is almost entirely unspotted.

The type specimens selected to show some of the variations are as follows: No. 7271 (Pl. 7, Fig. 13), a single egg collected by Mr. T. B. Richie, near Brookline, Massachusetts, May, 1861; No. 23084 (Pl. 7, Fig. 11), from a set of two taken by Dr. A. K. Fisher, near Sandy Spring, Maryland, May 10, 1887; No. 23272 (Pl. 7, Fig. 10), from a set of two collected by Dr. A. K. Fisher, near Sing Sing, New York, May 24, 1888; and No. 23979 (Pl. 7, Fig. 12), from a set of two taken by Mr. Manly Hardy, near Holden, Maine, May 21, 1890.

83. Buteo brachyurus VIEILLOT.

SHORT-TAILED HAWK.

Buteo brachyurus VIEILLOT, Nouveau Dictionaire; IV, 1816, 477.
(B —, C —, R —, C —, U 344.)

GEOGRAPHICAL RANGE: Tropical America in general, except West Indies, north to eastern Mexico and Florida.

Within the United States the breeding range of the Short-tailed Hawk, as far as is known at present, is confined to the State of Florida. It can no longer be considered as only an accidental visitor to our southern borders, having been met with in various portions of Florida, and is known to breed there regularly, at least as far north as Tarpon Springs and St. Marks on the Gulf coast. It may be confidently looked for as a rare summer resident throughout that State, and possibly also in the southern portions of the other States bordering on the Gulf of Mexico.

Mr. W. E. D. Scott states: "The observations already recorded   *   *   * and other records here given, lead to the conclusion that this species is of regular occurrence on the Gulf coast of Florida, at least as far north as the vicinity of Tarpon Springs, and that it breeds regularly, though rarely, in this region there can be no doubt. The birds that have been met with in the immediate vicinity of Tarpon Springs have usually been seen in pairs; once three were observed together.

"During March and April, 1888, within a radius of 10 miles of the town in question, there were observed by me on March 17 a single bird, on April 6 a pair, on April 10 a pair, these last two pairs being probably the same individuals. On two other days late in March and April, and several times in

May, 1888, I saw pairs of Hawks that were certainly the same birds. They were always very shy and wary, and difficult to approach in any way. About 200 yards in the open was as near as one could generally approach. They frequented the vicinity of hammocks, and their habits, except the extreme shyness, appeared much like those of the common Red-shouldered Hawk of this region. * * *

"On the 16th of March, 1889, near Tarpon Springs, I found a pair of these Hawks just starting to build a nest. The locality was on the edge of a hammock, and the nest, the foundation of which was finished, was in a gum tree some 40 feet from the ground. Both birds were seen in the act of placing additional material on the structure. As the birds were rare and I could not risk their being killed or driven away, with the aid of a native hunter both were secured, though before killing them I was certain of their identity. * * * The female (No. 6392) of this pair had eggs with the yolks almost developed and would have laid within a week. From the appearance of the ovary and oviduct, I believe that three eggs would have been laid."[1]

Dr. William L. Ralph has also met with this species occasionally in the vicinity of San Mateo, Florida, and informs me that he saw two or three specimens during the spring of 1891. He says: "The native hunters and cattlemen seem to know this bird, and say that it breeds here. They call it the 'Little Black Hawk,' and state that it is more common in the spring and summer than in winter."

Mr. C. J. Pennock, in a short article on the nesting habits of this species at St. Marks, Florida, makes the following statement: "Early in April, 1889, while on a collecting trip at St. Marks, Florida, I spent several days in the swamps that line the Gulf coast. April 3 I noticed a small Black Hawk fly to a nest in a pine tree, about 3 miles from the coast. On climbing to the nest, I found that the tree had formerly been occupied by Herons, there being three old nests besides the one occupied by the Hawk, which I also took for an old Heron's nest. It had evidently been recently repaired, and contained two or three fresh twigs of green cypress on the bottom. At this time there were no eggs, but I again visited the nest April 8. The old bird was seen near, and this time showed some concern, flying around us above the tree tops as we approached, and several times uttering a cry somewhat resembling the scream of the Red-shouldered Hawk, but more shrill and not so prolonged. The nest had received further additions of cypress twigs, but was still empty. My boatman wrote me, May 2, stating that after three visits he had shot the bird on the nest and taken one egg. He skinned her but found no more eggs."[2]

Mr. Pennock has kindly loaned me this egg, and I am thus enabled to figure it. He describes it as dull white, showing blue when held against a strong light. It is spotted on the larger end with reddish brown in small spots and blotches over about one-fourth of the surface. A few finer spots extend to the middle of the smaller end, where, however, they can hardly

[1] Auk, Vol. VI, 1889, No. 3, pp. 243-245. [2] Auk, Vol. VII, 1890, No. 1, pp. 56, 57.

be seen unless closely examined. The egg measures 55 by 41 millimetres, and the ground color is a pale greenish white. The egg is ovate in shape and it is figured on Pl. 8, Fig. 7.

I have been unable to gather any information additional to that already given. From the foregoing it would appear that these Hawks commence laying from the last half of March to the beginning of May, and that from one to three eggs constitute a set.

84. Urubitinga anthracina (LICHTENSTEIN).

MEXICAN BLACK HAWK.

Falco anthracinus LICHTENSTEIN, Preis-Verzeichniss, 1830, 3.
Urubitinga anthracina LAFRESNAYE, Review Zoölogique, 1848, 241.
(B —, C —, R 444, C 528, U 345.)

GEOGRAPHICAL RANGE: Tropical America in general, north to central Arizona, and the Lower Rio Grande Valley in Texas.

As far as is known at present, the breeding range of the Mexican Black Hawk is confined to central and southern Arizona, and the Lower Rio Grande Valley in Texas, and it does not appear to be a common species anywhere within our borders.

While encamped on Rillitto Creek, 7 miles northeast of Tucson, in the spring of 1872, I first noticed this species on April 4, apparently just returning to their summer homes. Knowing but little about our birds I supposed it at first to be one of the dark forms of *Buteo swainsoni*, then known as *Buteo insignatus*, and I subsequently wrongly identified it as *Buteo abbreviatus*. My notes taken on that day read as follows: "Saw a pair of Hawks to-day which I take to be of this species (No. 21, *Buteo insignatus*, Baird's Cat., 1859). They were quite tame, and let me come within 30 feet. I should not exactly call them black, but rather a dark and uniform slate color, with the wings a little darker, possibly. The cere and bill appeared to be yellow, also the feet, and the tail was banded by a white stripe at the end. One of them as it flew off appeared to have every feather of the upper parts of the breast edged with ferruginous. The pair kept circling around and over me, uttering at the same time repeated cries, exact counterfeits of the piping in the spring of *Numenius longirostris*. I could easily have shot both, but they evidently meant to build in the neighborhood."

The ferruginous edgings of the feathers of the breast, noticed by me, clearly indicate that the specimen in question is referable to this species, being a bird still in the immature plumage, and, as far as known, this plumage is not found in *Buteo abbreviatus*.

Col. A. J. Grayson, in his "Notes on the Birds of Northwestern Mexico," says of this species: "Common at all seasons, usually found about the esteros and marshes near the seacoast, subsisting chiefly upon land crabs."

Asst. Surg. Edgar A. Mearns, U. S. Army, published the following account, comprising about all we know as yet about the nesting habits of this species within our borders. He says: "When hunting along a sluice of the Verde River, beneath a dense growth of willows and cottonwoods, I first discovered the Anthracite or Mexican Black Hawk, perched among the thickest foliage of a low willow overhanging the shallow water. The imperfect view obtained as it flew off through the trees led me at first to suppose that it was an immature Golden Eagle, a species that I had several times encountered thereabouts in similar situations. A snap shot proved unsuccessful, as was the case on several subsequent occasions, and, although I frequently saw them along the river, it was long ere I succeeded in procuring a specimen. Always extremely shy, they were usually found hidden in the foliage near the water in some low situation, whence, when surprised, they generally managed to escape through the foliage of the cottonwoods without affording a good opportunity for a shot. Their flight is swift and powerful. Occasionally one was seen eating a fish upon the sandy margin of the river. They were present throughout the summer, but departed in the autumn, my absence in the field during the months of October and November having prevented me from determining the date of departure.

"On the 26th of March, 1885, I found one of these Hawks upon the Agua Fria, about 30 miles southwest of Fort Verde, at a considerably greater altitude; and on Oak Creek, a mountain stream 30 miles north of Fort Verde, in the foothills of the San Francisco Mountains. I wounded an immature example on the 12th of August of the same year, it having probably been reared on that stream, which abounds with trout and other fishes.

"On the 19th of June, 1885, Capt. T. A. Baldwin and I set out to visit Fossil Creek, 30 miles east of Fort Verde, with an escort of two soldiers. We carried some rations and mining implements packed upon a mule and two burros. We found the trail to the cañon without difficulty, but when nearly at the bottom took the wrong fork of the trail, which finally led us to the spring and forks of Fossil Creek, both branches of which we explored for several miles, finding tracks of wolves, bears, deer, raccoons, and beavers. A pair of Mexican Black Hawks were found at the forks of the stream close to the place where we had pitched our camp. Their loudly whistled cry is different from that of any bird of prey with which I am acquainted, and is difficult to describe, although rendered with great power.

"They circled about us a few times, then retreated to some tall piñons upon the hillside, where they continued to cry vehemently until I essayed to force my way through the thick scrub oak towards them, when both birds flew with loud screams to a tall pine tree down the stream, where I succeeded in obtaining a long shot at the male bird, which, although mortally wounded, flew beyond my reach before dropping to the ground. His mate flew to the piñons far up the steep banks of the cañon, out of reach, and continued screaming, following me up the cañon. Towards nightfall I came up with Captain Baldwin, and he told me that he had discovered the nest of my *rara avis* in a tall cottonwood down the cañon, and said if we hastened we might procure the eggs before dark and secure the other parent.

"The nest was built in a cottonwood tree in the same grove in which we first found the birds. The nest had evidently been the birthplace of many generations of these Hawks, for it measured 4 feet in depth by 2 feet in width. It was lined with a layer of cottonwood leaves several inches deep, was very slightly concave, and composed of large sticks, much decayed below, showing that they had been in position for a number of years. The nest was about 30 feet from the ground. The female parent remained too shy to return to the nest until I began to climb the tree. At first I attempted to ascend by means of some grapevines, which gave way; then I managed to reach the upper part of the huge bole by swinging from a tall, slender box-elder tree, and scrambled with much exertion to the lowest branch. Meanwhile the Hawk had shown much uneasiness, fluttering in the air and screaming lustily. As I approached her treasure her parental solicitude overcame her terror, and she sailed over the tree top. I saw the gun at the captain's shoulder and feared he would miss; but he wisely held his fire until the bird wheeled and rushed directly toward me, when a well-directed shot dropped her just at his feet. A minute later I reached the nest and discovered a single half-grown nestling, having the quill feathers webbed terminally, and leaden gray down covering the greater part of the body. It fought fiercely and evinced great pluck and ability to defend itself. The wounded parent was also savage, and tried to reach its assailant. After it was dispatched the captain proposed that we should attempt to find my wounded Hawk; but the locality was too dangerous, so we abandoned it with regret."[1]

On May 20, 1887, the doctor found a nest and two eggs of this species on Beaver Creek, 6 miles northeast of Fort Verde, in central Arizona. The eggs had been incubated when found, and are now in the American Museum of Natural History, New York City. One of these is figured.

Mr. D. B. Burrows writes me that he found a nest of this species in Starr County, Texas, on April 25, 1891, containing a single egg. The female was shot from the nest, and dissection showed that no more eggs would have been laid. The nest, a newly constructed one, was placed in a dense willow grove in the main forks of a tree of this species, about 30 feet above the ground, and growing about 80 yards from the banks of the Rio Grande. It was about 15 inches wide by 8 inches deep and rather shallow. It was composed of dry twigs and was well lined with green willow leaves.

Mr. Burrows describes the egg as ovate in shape, the ground color as dull white, with a faint greenish tinge, and as marked over the entire surface with small and irregular blotches, varying from reddish brown to burnt umber, with a few spots of purplish drab. The markings are heaviest near the larger end,

---

where some of them are drawn out into irregular lines. The egg measures 61.5 by 47 millimetres.

An egg sent by Lieut. H. C. Benson, Fourth Cavalry, U. S. Army, from near Fort Huachuca, Arizona, identified as that of *Buteo abbreviatus*, and so described by me in the Proceedings of the U. S. National Museum (Vol. x, 1887, pp. 551–552), is, according to our present knowledge of the eggs of these two birds, much more likely to belong to the Mexican Black than to the Zone-tailed Hawk, and although not absolutely certain of this, I figure it under this species, especially as it is somewhat different from any Hawk's egg in the U. S. National Museum collection. This nest was found May 6, 1886, in a sycamore tree, in a deep arroyo near the base of the Huachuca Mountains. The nest was a large and bulky one, and lined with a few leaves only. It contained but a single egg, which was slightly incubated. This is ovate in shape, has a ground color of pale greenish white, and is sparingly spotted and blotched, with small irregular markings, lines, and tracings, varying in color from burnt umber to tawny olive. It measures 59.5 by 46.5 millimetres. It is No. 22930, U. S. National Museum collection, and is figured on Pl. 8, Fig. 9.

The Mexican Black Hawk is only a summer resident along the southwestern border of the United States, and nowhere common. Nidification begins in the southern part of Arizona in the latter part of April, and a little later northward. But a single brood is raised in a season, and one or two eggs constitute a set.

A fully identified egg of this species, taken by Dr. Mearns, on May 20, 1887, and already referred to, now in the collection of the American Museum of Natural History, is oval in shape, dull white in color, and irregularly blotched, principally about the larger end, with small markings of different shades of brown. This egg measures 56.5 by 46 millimetres, and is figured on Pl. 8, Fig. 8.

85. Asturina plagiata SCHLEGEL.

MEXICAN GOSHAWK.

Asturina plagiata (LICHTENSTEIN) SCHLEGEL, Musé de Pays Bas, Asturinæ, 1862, 1.
(B 33, C 358, R 445, C 527, U 346.)

GEOGRAPHICAL RANGE: Middle America, south to Panama; north to southern border of United States, in southern New Mexico and Arizona.

As far as known at present, the breeding range of the Mexican Goshawk includes that portion of Arizona south of the Gila River and southern New Mexico (Fort Bayard), and it is only a summer visitor within the limits of the United States, arriving in the vicinity of its breeding grounds early in March, and in late seasons not until the beginning of April. As it is known to breed in the province of Tamaulipas, Mexico, close to our southern border, it will probably yet be found nesting along the Lower Rio Grande Valley in Texas as well.

The first pair of these birds seen by me were circling high in the air above the timber in the Rillitto Creek bottom near Tucson, Arizona, on April 9, 1872. After sailing around for sometime, they finally perched on a dead limb of a cottonwood. During the next week I noticed several other pairs and watched them carefully in order to locate their nesting sites; they were not at all shy, and, had I been so inclined, could have secured a number of specimens with but little trouble. From that time on not a day passed without my seeing two or three pairs of these handsome little Goshawks (which were readily recognized by their light color) engaged in sailing gracefully over the tree tops, now sportively chasing each other, or again circling around, the female closely followed by the male, uttering at the same time a very peculiar piping note, which reminded me of that given by the Long-billed Curlew in the early spring (while hovering in the air in the manner of a Sparrow Hawk), rather than the shrill cries or screams usually uttered by birds of prey. To my ear, there was something decidedly flute-like about these notes. After they were paired they became more silent.

When in search of food their flight is powerful, active, and easily controlled. I have seen one of them dart to the ground with arrow-like swiftness to pick up some bird, lizard, or rodent, continuing its flight without any stop whatever. A good proportion of their food consists of beetles, large grasshoppers (a species of which about 3 inches long was especially abundant), and other insects; these are mostly caught on the wing, and I believe small birds also form no inconsiderable portion of their food, as I have seen them chasing such.

Mr. F. Stephens compares their cry to a loud "creer," repeated four or five times, and says that at a distance it sounds much like the scream of a peacock. The stomachs of the specimens examined by him contained lizards, small squirrels, fish scales, the wing covers of beetles, and fur and bones of small unrecognizable rodents.

About the last week in April several pairs had selected their nesting sites within a radius of 10 miles from my camp, and commenced building. All the nests found by me, four in number, were placed in cottonwood trees, usually the largest to be found in the vicinity, and as near their tops as they could be placed with security.

The first nest was obtained on May 17, and the male, who was sitting on a limb close by, was shot. This nest was located in the topmost branches of a large cottonwood tree near the laguna, the sink of the Santa Cruz River, not less than 70 feet from the ground, and contained three fresh eggs, the only set I found which contained this number. The nest, not a very substantial affair, consisted of a shallow platform, composed principally of small cottonwood twigs, a number of which were green and had been broken by the birds themselves. I have seen them do this, selecting a suitable twig, then flying at it very swiftly, grasping it with their talons, and usually succeeding in breaking it off at the first trial.

Many of these twigs had the leaves attached to them and only partly dried. The inner lining of the nest consisted of dry cottonwood leaves and the tops of willows, the latter also taken while green. No bark nor material other than that mentioned was used in the construction of this nest.

A second one, found June 6, 1872, contained but two eggs, on which the bird had been sitting for about a week. This was in a similar situation to the first, and resembled it in structure, but was lined with a few strips of the soft, dry inner bark of the cottonwood and with dry leaves of the same tree. A third egg was taken from this nest June 18, but whether laid by the same bird or not, I am unable to tell.

On June 19 I took another set of two eggs, which contained small embryos. The nest, like the first one, was composed principally of small green twigs, many with the leaves still on them, and lined with green willow tops. The last nest was found on June 20, within a mile of my camp, and though positive that a pair had a nest somewhere in this grove—a rather dense one— I failed to locate it, although I had, as I supposed, carefully examined this very tree several times previously. I also saw the birds about, but the nest was so well hidden among the dense foliage of the top that only a small portion was visible from below, and this only from a certain point of view. It contained two eggs with good-sized embryos, and the lining consisted of partly dried cottonwood leaves.

The nests are rather frail structures, and were all apparently newly built. They were shallow and but slightly hollowed, not more than $1\frac{1}{2}$ inches deep. The last two found were very difficult to get at, resting as they did on very slender limbs, and from the fact that they were composed principally of green twigs it was no easy matter to detect them. The birds made but little demonstration when the eggs were taken, beyond circling above the tree tops and uttering a few shrill screams. I believe that but one brood is reared in a season. The male assists in getting the nesting material and perhaps in incubation as well. By the latter part of October most of them had departed for their winter homes.

Mr. F. Stephens has also found their nests in Arizona and southern New Mexico, and considers them as common about Tucson, especially in some of the large mesquite groves on the Santa Cruz River. He found them nesting between May 2 and June 2, usually finding two eggs to a set. The nests were placed in cottonwoods and large mesquite trees.

I consider this one of the handsomest Hawks we have; graceful and quick in all its movements, a swift flyer, and resembling the Goshawk in many respects, but it prefers more open country than the latter. It seems to be found only in the vicinity of water courses, and not, like many of the other Raptores, on the dry and comparatively barren desert-like plains. It nests later than most Hawks found in Arizona, usually during May, and even as late as the middle of June.

Mr. Otho C. Poling writes me from Fort Huachuca, Arizona, regarding this Hawk, as follows: "I first met with this species on March 3, 1890, when a male was shot in a deep wooded cañon in this vicinity, at an elevation of about 7,000 feet. It had a squirrel in its talons about two-thirds eaten. During the month of June, 1890, I was camped in a cañon of the Huachuca Mountains, among some thick spruce and sycamore woods, and had not been long in camp when I heard a faint squeaking noise overhead, and on investigating, found a nest of young Raptores in the top of a high sycamore, directly in front of my tent. To my great pleasure I found, on waiting for the parent to arrive, that it was the Mexican Goshawk. She made half a dozen or more trips daily to the nest, and whenever she arrived her presence was at once hailed by the hungry nestlings. I watched her closely; she would make daily trips to the mesquite plains for cotton tails (*Lepus arizonæ*), some 6 or 8 miles out in the valley. After the first week a neighbor came to my camp and during my absence shot the female, and presented it to me on my return.

"Up to this time I had not seen the male, or at least had seen only one individual at a time, but noticed on the following day that another bird, evidently the male, appeared and carried on the feeding of the family as regularly as if nothing had happened. The young were now growing rapidly, and their cries were much louder while being fed. One day, on glancing up at the nest, I saw one of them perched upon a limb beside it. The parent bird was near by with some game, and seemed to be urging the young one to fly to it, if it would have its meal. Although it demanded its regular allowance loudly, I observed it was left out of reach by the old bird until its first lesson of flying was learned. The young were three in number, and all were out of the nest the following day, but returned to it at night. They remained about for several days and finally disappeared."

Incubation lasts from three to four weeks. The eggs are usually two in number, seldom more; about one set in four contains three eggs. Their ground color is a pale bluish white; and all the eggs I have taken, with a single exception, are unspotted, but always more or less stained with yellowish, which is difficult to remove. These stains are probably caused by the green leaves on which the eggs are usually laid. The shell is fairly smooth, close grained, slightly pitted, and without luster. In two sets of eggs of this species' taken by Mr. F. Stephens, near Fort Bayard, New Mexico, both found on April 23, 1876, and which are now in the collection of the American Museum of Natural History, in New York City, one egg in each is marked with a few buffy brown spots about its larger end. These, although few, are readily noticeable, while the markings on the specimen taken by myself are scarcely perceptible to the naked eye. Only a single brood is raised in a season. The eggs vary considerably in shape, the majority are a perfect oval, a few are elongate ovate, and one may be called ovate pyriform.

The average measurement of the eggs of the Mexican Goshawk in the U. S. National Museum collection is 51 by 41 millimetres, the largest egg measuring 54 by 40, the smallest 48 by 41 millimetres.

The type specimen, No. 16525 (Pl. 7, Fig. 7), U. S. National Museum collection, from a set of two, Bendire collection, was taken by the writer on June 6, 1872, on Rillitto Creek, near Tucson, Arizona.

86. Archibuteo lagopus (BRÜNNICH).

ROUGH-LEGGED HAWK.

Falco lagopus BRÜNNICH, Ornithologia Borealis, 1764, 4.
Archibuteo lagopus GRAY, List Genera of Birds, ed. 2, 1841, 3.
(B —, C —, R —, C —, U 347.)

GEOGRAPHICAL RANGE: Northern parts of the Old World; (Alaska?)

The Rough-legged Buzzard has been included in the "A. O. U. Code and Check-list of N. A. Birds," based on specimens from Alaska, but Mr. Robert Ridgway, in his "Manual of North American Birds, 1887," p. 240, in a foot-note on this species, writes as follows: "So far as evidence to date tends to show, the typical form of this species, if a distinctly American race be recognized, must be expunged from the 'List of North American Birds.'"

According to Mr. H. Seebohm, "the true home of the Rough-legged Buzzard Eagle is in the northern portions of the European and Asiatic continents. It breeds throughout Arctic Europe and Asia, being a very common species in Norway and Sweden, up to the North Cape, becoming rarer in Russia, yet more plentiful in Siberia, where it ranges as far to the east as the watershed of the Yenesay and Lena. In the winter it retires southward to various parts of central and southern Europe and the steppes of Russian Turkestan."

Mr. Harvie Brown found it breeding in south Norway in 1871, the nests usually being placed in clefts of more or less inaccessible rocks. In Lapland, according to Wolley, they often breed in firs. The number of eggs vary from three to five. The nests are large, composed of sticks and lined with grasses; when placed on cliffs, sticks are frequently dispensed with, and it consists of a slight hollow lined with grasses. In its general habits it resembles our American Rough-legged Hawk in every respect, and the differences in plumage are but very slight in the majority of specimens.

According to Mr. Seebohm, the eggs vary greatly in size and markings, some being poorly marked while others are very richly blotched with dark red, or clouded and mottled with pale brown. In some eggs the coloring is confined to a few large rich blotches of red, others are evenly spotted with color just as intense over the entire surface. A more uncommon variety is delicately streaked and penciled with a few irregular dashes of pale brown, something like the egg of a Kite. Other varieties are seen in which all the coloring is distributed in pale purplish shell markings, with perhaps a few streaks of rich brown. They vary from 2.25 by 2.1 inches in length, and from 1.8 to 1.65 inches in breadth (equal to 57.15 to 53.34 millimetres in length and 45.72 to 41.91 millimetres in breadth).

[1] History of British Birds, Seebohm, 1883, Vol. I, pp. 111-115.

There seems to be no difference whatever in the eggs of this species and our own, and none are figured for that reason. Ten eggs in the U. S. National Museum collection from Lapland average 57 by 44.5 millimetres, the largest measuring 60 by 46, the smallest 53 by 43.5 millimetres.

87. Archibuteo lagopus sancti-johannis (GMELIN).

AMERICAN ROUGH-LEGGED HAWK.

Falco sancti-johannis GMELIN, Systema Naturæ, i, ii, 1788, 273.
Archibuteo lagopus var. sancti-johannis RIDGWAY, in Cones's Key to North American Birds, 1872, 218.

(B 30, 31, C 356, R 447, C 525, U 347a.)

GEOGRAPHICAL RANGE: Whole of North America.

Excepting the Territory of Alaska, the American Rough-legged Hawk does not breed within the United States, the various records to the contrary notwithstanding. Not a single one of these is absolutely unquestionable. Its most southern breeding range, so far as I am able to learn, is perhaps that given by Dr. C. Hart Merriam, based upon the observations of Mr. Napoleon A. Comeau, made in the vicinity of Point de Monts, province of Quebec, Canada, in about latitude 49°, who states that it is rather common, and breeds there.[1] It is questionable if it breeds anywhere south of the river St. Lawrence, and if it does such instances must be of rare occurrence.

Mr. L. M. Turner, of the U. S. Signal Service, found it breeding abundantly in southern Labrador and at Fort Chimo, Ungava Bay, while stationed at the latter point, and he considers it one of the most common of all the birds of prey in that region.

In the interior it is said to be not uncommon, and to have been found breeding on the Saskatchewan plains, in the similarly named district, about latitude 53°. It is most common, however, from the Anderson River and Rendezvous Lake country up to the Arctic coast, where that indefatigable naturalist and explorer, Mr. R. MacFarlane, of the Hudson Bay Company, took not less than fifty-eight nests with eggs while traveling through this scarcely known and ice-bound wilderness nearly thirty years ago. Quite a fine series of the eggs of this species collected by this gentleman are in the U. S. National Museum collection.

On the west coast it is not uncommon during the breeding season in the vicinity of Fort Yukon, Alaska, as several sets of eggs taken there attest. Mr. E. W. Nelson, of the U. S. Signal Service, took their eggs at Saint Michael; and Mr. C. L. MacKay on the Nushagak River, Alaska; but it appears to be rarer near the coast there than in the interior. It probably breeds in the northern portions of British Columbia as well, but I find no reliable records from there.

---

[1] Bulletin Nuttall, Ornithological Club, Vol. VII, 1882, p. 237.

In the United States it is only met with as a migrant, arriving regularly from its breeding grounds in the far north in the fall and returning early in the spring. It winters chiefly in the middle and more open prairie States, not being partial to heavily timbered regions. In the East it is generally found along the flat, open country adjacent to the coast and the larger rivers, avoiding the mountainous and heavily wooded districts of the interior.

Although a large and powerful bird, it is of a peaceful disposition, and its food is humble, consisting principally of meadow mice and small rodents. It is doubtful if it ever catches a bird, its flight usually being slow, deliberate, and apparently laborious.

I found this species very abundant during the fall on the Umatilla Indian Reservation, and in the Harney Valley, Oregon, where they principally feed on small rodents and grasshoppers and occasionally on rabbits. I have often seen a dozen or two in a few hours' ride, usually standing singly on a little hillock on the open prairie, or perched upon a sage bush watching for prey. They are a perfectly harmless bird and deserve to be fully protected. When mounted they allowed me to approach closely, but when on foot they kept well out of range.

Mr. L. M. Turner, in his "Notes on the Birds of Labrador and Ungava," says: "The American Rough-legged Hawks arrive here, at Fort Chimo, about the last week in May and remain until the first week in October. Immediately on their arrival a site is secured for a nest, as mating has evidently occurred before they appear in this vicinity. Oftentimes the same place is resorted to, where the same pair have reared their young for many seasons. All the nests discovered by me were placed on a ledge or projection of a high bluff. Should there be several ledges, apparently suitable in all respects on the same bluff, the one nearest the top is selected. I surmise this is done in order to allow the birds a greater view of the surrounding country for the purposes of searching for food and to avoid danger.

"The nest is composed of sticks of various sizes, together with a few grass or weed stalks placed irregularly crosswise. The particular location of the nest modifies the amount of material used. A flat rock usually has but sufficient of these materials to prevent the eggs from rolling about. Where the place slopes the nest is usually higher in front, and often with nothing at the rear portion of it except the side of the cliff. In locations where the nest has been used for several years the amount of material accumulated is astonishing. I have seen several nests which would form a good load for a wheelbarrow. It often occurs that the nest material is increased considerably each year, and other nests appear to have been only rearranged. The depression containing the eggs is quite shallow and in some instances nearly flat. The accumulations around the nest, such as refuse food, is also surprising in quantity, and when this decomposes forms a soil in which grow grasses and other plants, which, from the character of the soil and favorable position (rarely to the northward), often attain a most luxurious growth and thus indicate the site of a nest which otherwise might have been over-looked. * * * The parent birds usually denote the proximity of a nest by sailing high in the air over the locality. Occasionally a nest may be discovered in an unexpected location. I found one, some 3 or 4 miles from the mouth of the Kotsoak River, on a grassy ledge, near the top of the side of a short but deep ravine. The nest was easily approached from the top without other labor than that of walking down to it. The positions of nests of various birds of prey were often a matter of wonder to me how they escaped the ravages of foxes and other prowlers. * * * At other times the nest is placed in a most inaccessible spot. I thought these were probably the nests of birds which may have been more persecuted than others and had not profited by experience. At no time did I observe any-thing like fierceness exhibited by these birds, either when wounded or when their nest was approached. The male can seldom be secured near the nest, while the female is sometimes heedless of distance, although rarely approach-ing very near."

In regard to the nesting sites most frequently made use of in the Fur Country, Mr. R. MacFarlane's experience was quite different, showing that the nesting habits of this, as well as of many other species, differ greatly in places, the birds adapting themselves to the immediate surroundings. Out of the fifty-eight nests found by him, forty-six were placed in trees, usually pines, and at an average height of about 20 feet from the ground. The remaining twelve were built on the edges of steep cliffs of shaly mud, on the banks of creeks, rivers, or lakes. Such nests as were found in trees were usually placed in a crotch not far from the top, composed of sticks and warmly lined with dry grasses, down, and feathers. Those on cliffs were similarly constructed, but usually with a smaller base of sticks and better lined.

The eggs vary from two to five in number, usually three or four, and are deposited at intervals of two or three days, and were often found in different stages of incubation. With but few exceptions, most of Mr. R. MacFarlane's specimens were found during the month of June. The earliest date on which he took eggs of this species, according to the records here, was on May 23. Mr. L. M. Turner took a set on May 24, evidently not complete, and the remainder in June. The middle of June seems to be the proper time to look for full sets.

Incubation lasts about four weeks, and by the middle of July most of the young are hatched and they leave the nest about the beginning of Sep-tember. According to Mr. Turner, the young appear to be able to take care of themselves as soon as they leave the nest. He says the Eskimo apply the name of "Kin-wi-yuk" (in imitation of its notes) to this species, and that the people of Labrador term it the "Squalling Hawk," from the noise it makes when alarmed.

But one brood is raised in a season. The eggs of the American Rough-legged Hawk vary greatly in size as well as in shape. Some are ovate, many short ovate, and others rounded ovate. The ground color in the more recently collected specimens is a pale greenish white, which appears to fade out in time, leaving the egg a dull dingy white. The shell is close grained and strong. There is an endless variety in the markings, both in regard to size and amount, in different specimens. In some they are fairly regular in shape as well as size, in others exactly the reverse. In some they are well defined, evenly colored throughout; in others quite clouded and of different tints. A few specimens are streaked and the markings run longitudinally from end to end. The spots and blotches consist of various shades of brown, the predominating tints being burnt umber and claret brown, and among these are mixed lighter shades of ochraceous, clay, fawn color, and ceru-drab. Quite a number of specimens show also handsome shell markings of a rich helio-trope purple and pale lavender, mixed in and partly overlaid with darker tints. In many eggs the blotches are large and irregular in outline, and usually heaviest on the large end, but in no case do they hide the ground color. Others are regularly and sparingly marked over the entire egg, with fine dots of different shades of brown and lavender, giving the egg a flea-bitten appearance. While some eggs are but slightly marked, none are entirely unspotted. To sum it up in a few words, they show a great variety of styles. Compared with the eggs of other Raptores, they perhaps resemble those of *Buteo lineatus* in coloration more than any other species.

The average measurement of sixty-three specimens in the U. S. National Museum collection is 56.5 by 45 millimetres. The largest egg of the series measures 62 by 47, the smallest 51 by 41.5, and a runt egg but 42.5 by 38 millimetres.

The type specimens selected to show some of the different styles of mark-ings are No. 8818 (Pl. 8, Fig. 10), from a set of two, taken May 23, 1863, on the Anderson River, Arctic North America, by Mr. R. MacFarlane, of the Hudson Bay Company; No. 8831 (Pl. 8, Fig. 11), from a set of three, taken June 16, 1863, on the same river and by the same collector, and No. 22393 (Pl. 8, Fig. 12), from a set of three, taken June 10, 1883, by Mr. L. M. Turner, U. S. Signal Service, near Fort Chimo, Labrador.

88. Archibuteo ferrugineus (LICHTENSTEIN).

FERRUGINOUS ROUGH-LEG.

Falco ferrugineus LICHTENSTEIN. Abhandlungen der Koeniglichen Akademie Berlin, 1838, 428.
Archibuteo ferrugineus GRAY, Genera of Birds, fol. ed., 1849, 12.
(B 32, C 357, R 448, C 526, U 348.)

GEOGRAPHICAL RANGE: Western United States; east to and across the Great Plains (occasionally to Illinois); north to Saskatchewan; south into Mexico.

The Ferruginous Rough-leg, a large and handsome species, is an inhab-itant of the open prairie country of the West, and breeds in suitable localities from eastern Colorado and Wyoming, northern Utah, central and western

Kansas and Nebraska, northward through the Dakotas and eastern Montana, passing beyond our border through Manitoba to the Saskatchewan plains in about latitude 55°. It possibly breeds very rarely in northwestern Texas, and has been reported as nesting near Grinnell, Iowa, but this record is not fully verified.

Dr. Elliott Coues records it as common and resident at Fort Whipple, Arizona, and on the Pacific coast it has frequently been taken in California, but does not seem to breed there, and I have personally obtained it in the vicinity of Walla Walla, Washington, in 1880–1882, apparently migrating southward, and I believe it nests occasionally in that vicinity, as well as farther north. It is seldom met with east of the Mississippi River, and does not seem to breed in Minnesota, as far as known at present. In the northern portions of its range it is a summer resident, wintering abundantly in western Texas, and many passing south into Mexico.

This handsome Hawk, easily recognized by its large size, pale ashy-colored tail, and generally light colored underparts, strongly contrasting with its rufous legs, seems to be essentially a prairie bird, and while not particularly common anywhere, is perhaps more so on the extensive prairies of the Dakotas than anywhere else within our limits. In the early days, when California was not as thickly settled as it is now, the Ferruginous Rough-leg, or the California Squirrel Hawk, as it was then called, was not uncommon on the extensive Tularé Plains, as well as in the vicinity of Los Angeles, but in more recent years it seems to have been but rarely noticed in these regions. I have observed it in northern Nevada, southern Oregon, and Washington, where it is by no means common. In the northern part of the latter State, I believe, it is a summer resident, shunning the settled regions, and probably breeding on the extensive dry plain in the great bend of the Columbia River, and the Okinakane Valley, north of this stream. I have also met with it in southern Arizona in winter, and it may possibly breed there in small numbers.

Its flight is rather slow, but graceful nevertheless; it seems to take life easy and to be but seldom in a hurry. Its food, like that of the Rough-legged Hawk, consists principally of rodents of different species abounding in the prairie regions which it inhabits.

My friend Capt. B. F. Goss writes me as follows: "The Ferruginous Rough-leg is not uncommon in the high broken prairie and lake regions of northwestern North Dakota. In May, 1880, I took four sets of their eggs, the nests being all placed on the ground, on rocky hillsides, generally near large bowlders. They were constructed of bones, turf, and dry grasses, usually quite bulky and rather poorly finished. Bleached buffalo skeletons were scattered over the country in considerable numbers, and the ribs of these had principally been used by these Hawks for the groundwork of their nests. While traveling over the country, I had several times seen circles of these ribs, lying on the ground, all pointing toward the center like the spokes of a wheel, and I wondered at their regular arrangement. The finding of these nests solved the mystery; the annual prairie fires had burned the rest of the nest, and left the bones as placed by the bird. Three of the sets found contained three eggs each, the other one four. From reliable information received, I think sets of four eggs are common with this species. Some of these nests, besides being lined with weed stalks and dry grasses, contained also small pieces of dry turf, which the birds must have pulled or dug up with a great deal of labor, as the sod was very tough. The buffalo ribs were used in lieu of sticks, there being no timber of any consequence in the Coteau Hills, where I found them breeding, within a distance of 15 miles. The only trees of any size in that country are found along the borders of the large streams. The settlers call this bird the Eagle Hawk; they were so very wild, I could not get within rifle shot of them in the open, and it was only by hiding in a washout that we were able to get specimens for identification."

In eastern Colorado, and in other localities as well, the Ferruginous Rough-leg builds in trees, where such are available. Mr. F. M. Dille, of Greeley, Colorado, found one of their nests on Lone Creek, and describes it as follows: "When we located our ranch on this creek, we noticed in a cottonwood tree about 100 yards from the tent an immense nest, and in a few days a large pair of these Hawks took possession. They did not mind our presence at all, and, with the aid of a good field glass, I watched the pair pretty closely. The nest was lined with immense tufts of dry grass, roots and all, and an egg was laid on Monday, April 13. The male bird never visited the nest after this, but would sit out upon the prairie catching gophers, etc., which were turned over to the female. I visited it again while they were away, on Tuesday, and found three or four gopher heads and tails in it, but no more eggs. This egg resembled a Fish Hawk's egg very much, having a light blue background with a rich rufous brown and chocolate markings, blotched all over, and measured 2.50 by 1.95 inches (equal to 63.5 by 49.5 millimetres). Another egg was laid on Wednesday afternoon, and this was considerably lighter and less marked than the first. What blotches it had were gathered around the smaller end, and ran back in long lines.

"These birds did not attempt to drive me off while I was up the tree, and I waited till Monday, the 20th, for another egg, but thinking that the set was complete, and wishing to obtain fresh eggs, I took the set on the afternoon of the 20th.

"I thought at first that the bird was a Fish Hawk, so I shot the female in order to make sure. In the bird's fall a large egg about to be laid was broken, the shell of which was clear sky blue, and there were ten or twelve distinct eggs of all sizes in the Hawk, three of them quite large. The nest was composed of dry limbs and sticks, some as large as broomsticks. It was 3 feet in diameter and about 2 feet deep, situated about 12 feet from the ground."[1]

[1] Young Oölogist, 1885, pp. 44, 45.

This species has also been found breeding in northern Utah by Mr. C. S. McCarthy, while attached as naturalist to Capt. J. H. Simpson's exploring expedition, and the parents as well as the eggs are now in the U. S. National Museum collection. The nest was found in Rush Valley, on May 3, 1859, and was placed in a cedar tree 15 feet from the ground ; it contained four eggs.

The first eggs of the Ferruginous Rough-leg brought to scientific notice are a set of four, taken by Capt. T. Blakiston, on April 30, 1858, and one of these is figured. The nest was placed in an aspen tree 20 feet from the ground, and was composed of sticks, lined with buffalo wool. It was a large, bulky affair, measuring 2½ feet across. It was found between the north and south branches of the Saskatchewan River, Manitoba. Another set of eggs, taken in the same locality, five in number, came from a nest only 10 feet from the ground. None of the eggs of this set are in the U. S. National Museum collection.

Probably but one brood is raised in a season; the eggs, from two to five in number, usually three or four, are deposited about April 15 in the more southern portions of its range, and from two to four weeks later farther northward. Nesting sites seem to be as often selected on the ground as in trees, and if in the latter they are frequently quite near the ground. Incubation, as with most of the large Hawks, probably lasts about four weeks. The eggs are laid at intervals of two or three days. They are large and among the handsomest of those of the Raptores, and in shape usually ovate, seldom elliptical ovate. The shell is close grained and compact, dull creamy or pale greenish white in color, irregularly blotched and spotted with various shades of brown, and a few pale lilac and lavender shell markings. In some specimens the darker markings predominate, and in others the reverse is the case. An occasional egg is almost unmarked, and the larger sets usually contain one or two eggs much less marked than the others, and sometimes almost entirely unspotted.

The average measurement of fifteen specimens in the U. S. National Museum collection is 63.5 by 49 millimetres. The largest egg of the series measures 66 by 50.5, the smallest 61 by 48 millimetres.

Of the type specimens, No. 2662 (Pl. 9, Fig. 2), selected from a set of four, was collected by Capt. T. Blakiston, April 30, 1858, on the Saskatchewan plains, British North America; the other two eggs, both from the same set, No. 22717 (Pl. 9, Figs. 1 and 4), were collected by Capt. B. F. Goss in the Coteau Hills, North Dakota, May 12, 1880, and obtained in exchange. They show the different styles of coloration often found in eggs of the same set.

89. Aquila chrysaëtos (Linnæus).

GOLDEN EAGLE.

Falco chrysaëtos Linnæus, Systema Naturæ, ed. 10, 1, 1758, 88.
Aquila chrysaëtus Dumont, Dictionaire Sciences Naturelle, 1, 1816, 339.
(B 39, C 361, R 449, C 532, U 349.)

Geographical range: Northern portions of northern hemisphere, chiefly in mountainous regions.

It is questionable if at the present day the Golden Eagle breeds to any extent within the more thickly settled portions of the United States east of the Mississippi River. An isolated pair, here and there, may perhaps still be found in the wildest mountain regions of the New England States, the Adirondacks of northern New York, the mountains of the two Virginias, Kentucky, Tennessee, northern Georgia, and North Carolina. In the last mentioned State they are to-day far more likely to be found than in the other localities. It probably occurs in Minnesota. In the West this bird, while nowhere especially common, seems nevertheless to be pretty generally distributed from northwestern Texas, through New Mexico, Arizona, and California, northward to the Arctic Ocean. In the interior Rocky Mountain region it is fairly common; while in portions of California it may be called common, and it is likewise so in Alaska and the adjacent islands. In the eastern part of its range, as well as in the Rocky Mountains and the neighboring ranges on either side, the Golden Eagle resorts almost exclusively to the most inaccessible cliffs for the purpose of nidification. In the extensive prairie regions of the West, where there are no such localities to be found, steep perpendicular bluffs on the banks of streams, and occasionally trees, are utilized. This appears also to be the case in the fur countries in British North America, where Mr. R. MacFarlane took a number of their nests in such situations. On the Pacific coast, especially in California and Oregon, trees seem to be the favorite sites; usually large pines or oaks are preferred to high cliffs, which in many instances are available, in close proximity to the trees. This applies more particularly to the Blue Mountain region of Oregon, Washington, and Idaho, where I have personally observed such to be the fact.

Notwithstanding the many sensational stories of the fierceness and prowess of the Golden Eagle, especially in the defense of its eyrie, from my own observations I must confess that if not an arrant coward, it certainly is the most indifferent bird, in respect to the care of its eggs and young, I have ever seen. This may possibly be due more to utter parental indifference than to actual cowardice, as three of these birds, an adult male caught in a trap, and a pair of young, male and female, taken from the nest when about three weeks old and raised by me, did not seem to be deficient in spirit, by any means, and were always ready to attack anything and everything, on the slightest provocation.

The Golden or "Mountain" Eagle, as it is frequently called in the West, is a clean, trim-looking, handsome bird, keen-sighted, rather shy and wary at all times, even in thinly settled parts of the country, swift of flight, strong and powerful in body, and more than a match for any animal of similar size. In the West, where food is still plenty, their bill of fare is quite varied. This, I am informed, includes occasionally young fawns of antelope and deer, but more frequently small mammals of different kinds, as the yellow-bellied marmot, prairie dogs, hares, wood rats, squirrels, and smaller rodents, waterfowl, from Wild Geese to the smaller Ducks and Waders, Grouse, and Sage Fowl. On the extensive sheep ranges in the West, they are said to be occasionally quite destructive to young lambs.

Capt. Platt M. Thorne, Twenty-second Infantry, U. S. Army, writes me: "On December 10, I saw a Golden Eagle brought in by an Indian who had shot it about 20 miles south of Fort Keogh, Montana. He told me, 'I had badly crippled a black-tailed deer just before dark, and as soon as it was light the next morning I started to look for it. I saw this bird hovering pretty high up, and all at once drop down like a stone, and I heard the deer bleat. I was not far off and saw that it had struck its talons in the deer's flanks, and seemed to force it to the ground. As soon as the deer was down it changed its hold to the throat. Judging by its motions it was choking and tearing it. When I got close up, the deer was dead, and the bird stood on the ground with head stretched out, wings extended, the tips touching the ground. It seemed ready to spring if the deer moved. It appeared to have little fear of me and acted as if it meant to fight.' Of course I cannot vouch for this story, but I know the Indian well and have no reason to doubt it."

Birds are usually well plucked, but the smaller mammals are eaten hair and all. They are exceedingly cleanly at all times, and bathe frequently; mine took their bath daily, as soon as fresh water was given them, but while apparently fond of this for bathing purposes, I have seldom seen them drink.

One March morning in the spring of 1878, I was hunting on a small plateau in the Rattlesnake Cañon, near Camp Harney, Oregon, over which a number of large bowlders were scattered (that had tumbled down from a high, basaltic cliff above), and in passing around one of the larger ones, I suddenly came face to face with one of these birds feasting on a large yellow-bellied marmot (*Arctomys flaviventer*), which it had just caught. We were within 3 feet of each other before either of us was aware of it, and it would be hard to tell which was the most surprised. As the bird was one of a pair that furnished me regularly with a handsome set of eggs each year I did not further disturb it. It made no hostile demonstrations, notwithstanding I had spoiled its breakfast, which it left behind, and was soon out of sight, much to the delight of several Magpies which had been watching from a safe distance for a chance to pick up such stray morsels as the Eagle might leave behind. As these birds are usually only seen in pairs at all times of the year, I am inclined to believe that they remain mated for life, notwithstanding the fact that the eggs differ very greatly in markings from year to year, although coming from the same nest and evidently from the same pair of birds.

In southern Oregon each pair of Eagles seem to confine themselves to a certain district, over which they hunt, and no others are allowed to encroach on their ground. All the nests in that region which I know of were about 20 miles apart.

Nidification begins early. In southern Arizona Mr. W. E. D. Scott saw them carrying nesting material on December 10, 1884, and they probably lay in January or February. In California full sets of eggs are usually found between the 10th and 20th of March and occasionally in the last week of February. In southern Oregon they lay between April 1 and April 10; in Colorado about the middle of March; in the far North, in the Arctic regions, from about the 1st of May to the beginning of June, Mr. R. MacFarlane taking eggs but slightly incubated on June 23, 1862, near Franklin Bay, British North America, within the Arctic circle. There they are summer visitors only, and in the United States constant residents wherever found.

The nest of the Golden Eagle is a large structure; one near Camp Harney, Oregon, situated in a large pine tree close to the trunk and about 50 feet from the ground, was 3½ feet high by 3 feet wide. It consisted of large sticks, some of these over 2 inches in diameter, and it was sparingly lined with bits of juniper bark, pine needles, and green fir tops, evidently broken off by the birds. This nest when first found, on May 18, 1875, contained two young, probably two weeks old. These I took three weeks afterwards and kept them over two years. The top of the nest was nearly flat, and contained, besides the birds, a medium-sized marmot (*Arctomys flaviventer*), partly eaten. The parents made no resistance when the young were taken, although the latter protested considerably while being put in a gunny bag. The male flew at once out of sight and the female circled around at a great height, uttering shrill cries resembling the syllables "kiah-kiah-kiah." The young Eagles were very cleanly, fed readily on fresh meat and such birds and small mammals as I could obtain for them, and they at first used water freely. They grew finely, and the difference in size between these two birds, in their second year, and an old male caught about that time was astonishing; the latter looked dwarfed alongside of them, due possibly to their always having an abundance of food and plenty of clean water, as well as a roomy stable to exercise in.

Mr. Denis Gale writes me: "Here in Colorado, in the numerous glades running from the valleys into the foothills, high inaccessible ledges are quite frequently met with which afford the Eagles secure sites for their enormous nests. I know of one nest that must contain two wagon loads of material. It is over 7 feet high and quite 6 feet wide on its upper surface. In most cases the cliff above overhangs the site. At the end of February or the beginning of March, the needed repairs to the nest are attended to, and the universal branch of evergreen is laid upon the nest, seemingly for any purpose save that of utility. This feature has been present in all the nests I have examined myself or have had examined by others; it would seem to be employed as a badge of occupancy.

"I am familiar with a case in which the female was shot in the early part of the winter. When the nesting time came around the male was constantly about the nest for days, screaming in the most frantic manner; when the time arrived for the eggs to be there, as was the case the previous season, with the assistance of a rope let down from above, I climbed to the site to find nothing new about the nest but a small fresh branch of evergreen. I afterwards learned that the female had been shot.

"My experience has been that the birds are in no way aggressive or even demonstrative while being robbed of their eggs; circling round at a great height, from which they watch the proceedings with seeming indifference."

In an interesting article on the nesting habits of the Golden Eagle in Zoe (Vol. I, April, 1890, No. 2, pp. 42–44), Mr. H. R. Taylor makes the following statements: "The sitting bird is said to leave the eggs uncovered for several hours after 12 m. on sunny days, while it takes recreation in flying with its mate. * * * After several years' study of these Eagles the writer feels that he has formed the acquaintance of a number of individuals of this species, and this purely from the regularity of their habits. The first Eagle I ever saw in Santa Clara County was moving about the grassy top of a big hill, and on nearly every day thereafter I observed him at his favorite playground. On my visit the year following he was still doing "lookout" duty at his old post. The nest of this Eagle had a curious ornament to the interior in the shape of a large 'soap root.' The new nest, built the year after, also contained a soap root (probably *Chlorogalum pomeridi?*), which fact is of interest as showing the individuality of my feathered friend. Another Eagle I know has a singular predilection when nest-building for grain sacks, which it uses chiefly in the lining. When I first discovered this Eagle's nest there was one of these large sacks inside. The heavy storms of the next winter dislodged its nest and in the new one built I was surprised and interested to observe a grain sack. * * *

"A curious circumstance about the Eagles that make their home near Sargents is that several pairs always seen there apparently do not nest. The nature of the country in some of the hills is such that one accustomed to riding about might actually know every tree where the birds could build, so that an undiscovered nest would be an impossibility. My friend showed me a pair of Eagles that had lived in the hills just back of his house for many years. He also pointed out to me their nest, which the Eagles repaired last year but did not use. He says they have not laid since 1884, when they had three eggs. This pair stay about the place all the year, living largely (like the other Eagles thereabouts) on ground squirrels. They are accustomed to roost in one particular tree. I heard them uttering their peculiarly plaintive whistle in the mornings several times during my stay.

Their nest was not much over 300 yards from the house and was a large structure built on a horizontal limb about 40 feet from the ground."

Mr. W. Steinbeck, of Hollister, California, wrote me in April, 1885, as follows: "All the Golden Eagles' nests I found about here have always been placed in trees, and there were plenty of cliffs within 10 miles of these nests. I have taken eight sets of eggs, found two nests with young, and also examined nine old nests, which I am certain belonged to this species. With one exception I found all these nests on hillsides, and one of them within 200 yards of a large cliff; there were, however, a number of nests some distance back in the mountains among the rocks which I have not been able to take. The nests, with the single exception of one which was placed in the bottom of a small ravine, usually commanded a good view of the entire valley before them, and they were generally placed in oak trees from 20 to 50 feet from the ground. Some of the nests were quite large, measuring $5\frac{1}{2}$ feet in diameter, and all were nearly flat on top, with just enough depression to bring the top of the egg on a level with the sides. Some were lined with grass and straw, others with the hair-like tops of the Spanish soap root, one was finished off with Spanish moss, and in another I found quite a number of feathers, evidently from the breast of the parent.

"These Eagles nest where they can most easily procure food. Each pair has its range, and will drive any outsider away from it. These ranges are usually from 2 to 6 miles wide, and the birds become so attached to them that it seems impossible to drive them away. In one case, where I took three sets of eggs in successive years and killed the female, the male procured another mate and occupied the same nest again next season. In no case was I molested by one of these birds when taking their eggs. Sometimes I sat for hours in the nest waiting for one of the birds to come within gunshot, but with only one or two exceptions the birds left the vicinity, and I did not even obtain a second sight of them. Generally on approaching a nest the male will fly over, as if he had been watching, and come within 300 or 400 yards of it, he will then sail slowly out of sight. When about half a mile from the nest the female will also leave, come a little closer to the intruder than her mate, and then disappear.

"On two occasions I have flushed the bird from the nest and in both cases found the eggs well incubated. The last nest I took was from a bird which did not leave the nest until I almost put my hand on it. I think she was in the act of laying, at least the egg had that appearance. A nest found on March 1, 1885, contained a single egg, and on a second visit, March 4, I found the set complete. Of course the first egg may have been laid a day or two before I found it. Both were perfectly fresh. I am satisfied that incubation does not begin till the set is complete, and I know that both young leave the nest at the same time, and all the eggs taken by me which had been sat on were both in the same stage of incubation.

"Here they feed principally on ground squirrels and rabbits, and occasionally on carrion. In the winter when other game is scarce they hunt Wild Geese and Ducks, making a kind of 'swoop' when in chase, and should their first attempt be unsuccessful, they will rise in the air and make a second trial. My nests have all been found in the southern extremity of the Santa Clara Valley, and were within 2, 4, and 6 miles of each other. There are a number of other nests of this species in the mountains near by, which I have not been able to find, and in one locality there I have frequently seen as many as six of these Eagles sailing about; possibly this may be a bit of neutral ground, but their nests cannot be far away.

"I have a set of Golden Eagles' eggs taken March 17, 1885, which are, with the exception of a few dirt stains, pure white. They are unquestionably identified. The previous year I obtained a very dark colored and well marked set of eggs from the same nest and apparently the same bird. The eggs vary greatly in markings. I have one egg that is as thickly covered with light brown spots as that of a Marsh Wren (*Cistothorus palustris*). Scarcely two sets are marked alike."

On July 12, 1890, Mr. Steinbeck wrote me as follows: "A very interesting fact which I noticed this season for the first time is that a few days before nidification commences these birds will sit perched closely together on a limb of some tree near the nest.[1] Of sixteen nests taken by me, twelve were found on hillsides commanding an extensive view. Two were placed on the sides of gulches at the bottom of deep ravines, and two out on the plains; fifteen of these nests were in live oaks and one in a white oak. These nests were of all sizes, some being not much larger than those of the Western Red-tailed Hawk (*Buteo borealis calurus*), others measured $5\frac{1}{2}$ feet in diameter, and were large enough to fill a wagon.

"I believe that in excessively wet seasons like the past one (1890), the majority of these birds do not nest. When disturbed in their nesting they will likewise occasionally refrain from laying that season, although they remain in the same neighborhood throughout the year. They are a very suspicious bird, and it takes very little to disturb them."

My friend Dr. James C. Merrill, U. S. Army, writes me: "A nest found May 22, 1883, was placed on a ledge of rocks on Indian Creek, a small tributary of Pryor's Creek, near its junction with the Yellowstone River, about 33 miles from Fort Custer, Montana. It was located about 15 feet from the top and 25 feet from the bottom of the ledge, and had apparently been occupied for several years. It was a large platform of branches and twigs, a felted mass of cattle's hair forming the lining. The day before my visit it contained two eggs, but these had just hatched when I was there. A dead Sharp-tailed Grouse was lying in the nest. In the Big Horn Mountains, Montana, a pair of these Eagles were seen repairing their nest April 10, 1883."

---

[1] This is a peculiarity with many Raptores, and has been **noticed** by myself.

Quite a number of eggs have been taken in California during the last ten years, and the nests were nearly all placed in trees. Dr. Arthur Lemoyne, in his "Notes on Some Birds of the Great Smoky Mountains of North Carolina," published in the Ornithologist and Oölogist (Vol. xi, 1886, p. 148), mentions taking a set of these eggs in the same year on Bald Mountain, North Carolina, the only recent eastern record I have seen.

But one brood is raised in a season, and if the first set of eggs is taken the birds will not lay a second one that year. Mr. Steinbeck, however, believes that the same pair of birds from which he took a set of eggs on February 26, 1885, repaired another old nest in the vicinity, and this contained a single egg on March 28, when it also was taken. The pair of birds from which I obtained several sets of eggs while stationed at Camp Harney, Oregon, used two nests about a mile and a half apart in the same cañon, but on opposite sides. The year after I took the young they used the second nest, and the following spring returned to the first one again. At no time, however, was a second set of eggs laid the same season, although the birds remained in the vicinity the entire year.

The Golden Eagle generally lays two eggs, rarely three, and I believe these produce birds of different sexes. The U. S. National Museum collection contains a set of three eggs of this species taken by Mr. James McDougall, of the Hudson Bay Company, near Fort Yukon, Alaska, in the spring of 1868, and Mr. William Steinbeck informs me that he also took a set of three on March 21, 1891. These last eggs had been incubated for probably three or four days and were all fertile. One of the eggs is almost invariably a trifle larger than the other. Several days, sometimes a week passes between the laying of the eggs. Incubation lasts about four weeks, and from personal observation I believe the male does not take part in this to any great extent, but supplies his mate with food while she is so engaged. The young when first hatched are covered with white down and grow very rapidly, but it takes fully two months or more before they are able to fly and leave the nest. They remain in company with the parents but a short time, and are cast off as soon as they are able to take care of themselves. The usual call note is a shrill "keē,-keē,-keē," uttered in a high tone; it is often heard in the early spring before nidification commences. Another note, not so frequently used—one of alarm—is "kiah-kiah," repeated a number of times.

The eggs vary from an ovate to a short ovate. Their shell is coarse, thick, and roughly granulated. The ground color is a dirty white, in some specimens approaching a pale cream color. The markings of these eggs vary greatly, one set in the U. S. National Museum collection being almost pure white, without a spot of any kind on them, excepting a few stains. Others are thickly blotched and spotted with various shades of brown, claret, walnut and ferruginous brown predominating. Some are principally marked with fine spots and blotches of drab color and vinaceous rufous. In a few specimens pearl gray and lavender shell markings, sparingly overlaid with darker tints, predominate, but in the majority of specimens these lighter tints are entirely absent. Except the unspotted eggs, no two are exactly alike. An egg kindly sent by Mr. William Steinbeck for examination is heavily and uniformly blotched with fawn color, hiding the ground color almost completely. At the larger end this color is confluent and obscures it entirely. It is the heaviest marked egg of this species I have ever seen, and is figured. In the majority of specimens the markings are heaviest about the larger end; in a few the reverse is the case, and in others the markings, generally small in size, are regularly distributed over the entire surface. There is considerable difference in size. The twenty specimens in the U. S. National Museum collection—mostly from Arctic regions, excepting a few collected by myself in Oregon—average 74.5 by 59 millimetres. The largest of these eggs from Fort Yukon, Alaska, measures 81 by 64, the smallest 71.5 by 54 millimetres. This was taken at Anderson River Fort, Arctic America.

The measurements of twenty-eight specimens, all taken by Mr. Steinbeck, near Hollister, California, give an average of 75 by 58 millimetres. The largest of these eggs measures 76 by 63.5, the smallest 71.5 by 56 millimetres. These figures are based on Mr. Steinbeck's measurements.

Of the type specimens, No. 20699 (Pl. 9, Fig. 5), selected from a set of two eggs from the Bendire collection, was taken by the writer in Rattlesnake Creek Cañon, near Camp Harney, Oregon, April 9, 1877. It is one of the finer and more uniformly marked specimens in the collection. The heavily marked specimen (Pl. 9, Fig. 3) was taken by Mr. William Steinbeck on February 28, 1886, near Hollister, California, and kindly loaned for figuring.

90. Thrasaëtus harpyia (LINNÆUS).

HARPY EAGLE.

Vultur harpyia LINNÆUS. Systema Naturae, ed. 10, 1, 1758, 86.
Thrasaëtus harpyia GRAY. Proceedings Zoölogical Society, 1837, 108.
(B —, C —, R 450, C 631, U 350.)

GEOGRAPHICAL RANGE : Tropical America in general, south to Bolivia and Paraguay, north to Mexico, and rarely to the mouth of the Rio Grande (and in Louisiana?).

The Harpy Eagle if not the largest is certainly the most powerful of all the birds of prey found on the North American continent, and can only be considered as a straggler, having been noticed in the Lower Rio Grande Valley in Texas. No specimens have as yet been taken within our borders. It breeds in southern Mexico and thence southward as far as Bolivia and southern Brazil.

In an interesting article in the American Naturalist (Vol. XII, 1878, pp. 146–157), Dr. Felix L. Oswald gives the following account of the nesting habits of this species. While evidently misinformed as to the size of its eggs, and its occurrence in southern California, his statements otherwise are seemingly correct. He says: "The Harpy Eagle (*Harpia destructor*) has been shot in the mountains of southwestern Bolivia, in the Mornes du Diable of San Domingo, and in the valleys of southern California; but a hunter may range those regions for years without getting a chance to add to his trophies the feather coronet of the *Aquila real*, the King Eagle, as the Spaniards call him, while every farmer's boy of an Oaxaca Mountain village knows an eyrie or two in the neighboring crags, which he is ready to rob of its eaglets or large white eggs for a couple of reals. From the projecting rocks of the Lower Sierra on any bright morning of the year one may see the hovering form of the Destructor suspended in the clear sky or wheeling in ascending circles over the misty ocean of foliage; and from March to the end of June the tree tops of the *Tierra caliente* resound with the screams of the ever-hungry eaglets. * * *

"The *Lobo volante*, or Winged Wolf, as Quesada translates the old Aztec name of the Harpy, attacks and kills heavy old Turkeycocks, young fawns, sloths, full-grown foxes and badgers, middle-sized pigs, and even the black Sapayou monkey (*Ateles paniscus*), whose size and weight exceed its own more than three times. * * *

"As soon as the lengthening days of the year approach the vernal equinox the hen Harpy begins to collect dry sticks and moss, or perhaps only lichens, with a few clawsful of the feathery bast of the *Arauca* palm, if her last year's eyrie has been left undisturbed. Her favorite roosting places are the highest forest trees, especially the *Adansonia* and the *Pinus balsamifera*; the more inaccessible rocks of the foothills are also commonly chosen for a breeding place, and it is not easy to distinguish her compactly built eyrie on the highest branches of a wild fig tree from the dark colored clusters of the Mexican mistletoe (*Viscum rubrum*) which are seen in the same tree tops. The eggs are white, with yellowish brown dots and washes, and about as long though not quite as heavy as a hen's egg. Of these eggs the Harpy lays four or five, but never hatches more than two, and, if the Indians can be believed, feeds the first two eaglets that make their appearance with the contents of the remaining eggs.

"The process of incubation is generally finished by the middle of March, if not sooner, and from that time to the end of June the rapacity of the old birds is the terror of the tropical fauna, for their hunting expeditions, which later in the year are restricted to the early morning hours, now occupy them the larger part of the day."

Judging from the size of several specimens of the Harpy Eagle in the U. S. National Museum collection, the egg of this species should at least be as large as that of our Golden Eagle (*Aquila chrysaëtos*), and in fact considerably larger. I have been unable to find a correct description of the egg of this species.

91.   Haliæetus albicilla (Linnæus).

GRAY SEA EAGLE.

Falco albicilla Linnæus, Systema Naturæ, ed. 10, 1, 1758, 89.
Haliæetus albicilla Leach, Systematic Catalogue of Mammals and Birds in the
       British Museum, 1816, 9.
             (B 42, C —, R 452, C 533, U 351.)

GEOGRAPHICAL RANGE: Northern portions of eastern hemisphere and south-eastern Greenland.

The Gray Sea Eagle is included in our fauna from its being found in southern and southeastern Greenland, where it breeds on the rocky cliffs of the seashore of Davis Straits.

Mr. Henry Seebohm, in his "History of British Birds," speaks of this Eagle as follows: "The haunt of the White-tailed Eagle is not necessarily a maritime one, although the bird is more attached to the coasts and the sea cliffs than the Golden Eagle. It may, however, be often seen far away from the ocean, choosing for its haunt some large inland lake, especially if there be lofty cliffs and rocky islets on which it can perch to scan the surrounding country.

"The haunts of this noble looking bird are the brown hills of the Hebrides and the adjacent isles, and the wild mountain country of the mainland in the West. On the bold and rocky headlands of this wild, rugged coast, whose hoary peaks are washed by the treacherous waters of the Minch, the Sea Eagle finds a congenial home. The scenery of Skye is typical of this Eagle's favorite haunt. On that bleak and desolate isle it occurs in probably larger numbers than in any other place in Great Britain.

"In Pomerania, especially between Stettin and the Baltic, the Sea Eagle is a common resident, breeding in forests. It builds an enormous nest, sometimes 6 to 8 feet in diameter, near the top of a pine, or on the horizontal branch of an oak or beach, preferring forests near inland seas and large lakes. Instances have been known of its breeding in the same 'horst' for twenty years in succession. Every year some addition is made to the nest, until it becomes 5 or 6 feet high. Occasionally a pair of Sea Eagles have two 'horsts,' which are used alternately. They are shy birds and leave the nest at the least alarm, but do not easily forsake their old home. If the eggs are taken early in the season they will frequently lay again in the same nest. They make a very flat nest, and generally line it at the top with moss. The male and female are said to sit alternately, and the female is said to be shyer than the male at the nest. Two is the usual number of eggs, but frequently only one is found; in rare cases as many as three are laid. Eggs may be taken from first week in March to the middle of April.   *   *   *

"The White-tailed Eagle is undoubtedly mated to its partner for life, and even should one of the birds be destroyed the survivor will obtain a fresh companion in an incredibly short space of time—a habit peculiar to most if not all rapacious birds. For many seasons in succession this bird returns to its old eyrie, merely making a few necessary alterations each season, adding to the structure, or making good what damage it may have sustained during the storms of the previous winter.

"The site is varied according to locality, and may be on rocks, in trees, or on the ground. In the inland districts the birds usually select a rocky islet in the middle of a loch, where they either build their bulky nest on some ledge of the sloping ground, in a tree, or on the rocks, as occasion offers. Sometimes a site is chosen at some distance from the water in small open woods, but such instances are rare; inland rocks, too, are often selected in similar places to those which the Golden Eagle frequents—broken cliffs, often quite easy of access from above or below. But the most characteristic eyries of this bird in our islands are on the coast, built high up in the almost inaccessible rocks, hundreds of feet above an ever turbid sea, and in situations to which none but the most intrepid climbers dare venture. Some nests in these situations are indeed quite inaccessible, and the birds have remained in undisturbed possession from time immemorial. * * *

"Several instances are recorded of the Sea Eagle breeding upon the ground. Herr Tancré describes a nest which he found upon the island of Hiddensoe, on the southern shores of the Baltic, near Stralsund, on the naked meadow among the reeds. The nest was carefully made of sticks and was about 2 feet high.

"The food of the Sea Eagle consists of fish, sea fowl, and occasionally carrion. Dixon states: 'Within my own observation the favorite food of this Eagle is the stranded fish and shore garbage on the beach of its maritime haunts; while farther inland a dead carcass or a weakly bird or animal are shared with the Ravens and the Crows.'"

According to Seebohm, the eggs of the Gray Sea Eagle vary from 69.8 to 83.8 millimetres in length, and from 53.3 to 60.9 millimetres in breadth They are pure white in color and ovate to rounded ovate in shape. The shell is coarse, granulated, and usually considerably nest stained, giving it a yellowish appearance.

Three eggs from southern Greenland, in the U. S. National Museum collection, measure, respectively, 80 by 59, 76 by 57.5, and 75.5 by 61 millimetres.

They are scarcely distinguishable from the eggs of the Bald Eagle (*Haliæetus leucocephalus*), excepting that they are somewhat larger. None are figured.

26957—Bull. 1——18

92. Haliæetus leucocephalus (LINNÆUS).

BALD EAGLE.

Falco leucocephalus LINNÆUS, Systema Naturæ, ed. 12, I, 1766, 124.
Haliætus leucocephalus BOIE, Isis, 1822, 548.
            (B 41, 43, C 362, R 451, C 534, U 352.)

GEOGRAPHICAL RANGE : Whole of North America and across the Aleutian chain
to the Commander Islands, Kamchatka.

The Bald or American Eagle, our national emblem, is pretty generally
distributed over the entire United States, and breeds more or less abundantly according to food supply along the Atlantic seacoast, from northern
Maine to Florida and from the Gulf of Mexico throughout the total length
of the Mississippi Valley and the larger streams and lakes of the interior,
as well as British North America, to the Arctic coast. It is quite abundant
on the Pacific coast, and especially common at the mouth of the Columbia
River, the shores of British Columbia and the Alaska mainland, as well as
on all the Aleutian Islands. It appears to be equally indifferent to extreme
heat or cold, but in the northernmost portions of its range it is only a
summer resident, leaving these inhospitable regions and retiring to a warmer
climate as soon as the rivers and lakes freeze up, which furnish it with
most of its food supply.

Within the United States, it is perhaps more abundant in Florida than
anywhere else. Dr. William L. Ralph furnishes me the following observations
on this species, made principally in the immediate vicinity of Merritt Island,
Indian River, Florida, during February, 1886, and the two succeeding winters. He says: "Before I discovered this paradise for these noble birds, I
would not have believed there were so many east of the Mississippi River
as I found there within a radius of a few miles; for I not only saw them in
great numbers, but found, with the help of an assistant, nearly one hundred
occupied nests and took thirty-five sets of eggs.

"Frequently when returning to the hotel at Rockledge, just before dark,
I would while crossing the island opposite the village, a distance of about a
mile, see fifteen or twenty Eagles, most of them birds in young plumage,
roosting in the trees, and it was no uncommon thing to see six or eight in
one flock.

"Notwithstanding these birds were so very common, I concluded from
the great number of deserted nests found that they must have been more
abundant formerly, and, on inquiry among the inhabitants, found this to have
been the case.

"These Eagles seem to breed earlier than those in other parts of Florida, due no doubt to the immense number of waterfowl, especially Coots
(*Fulica americana*), that frequent this vicinity during the winter, and which
seem to form the principal article of their food, though they will sometimes

condescend to eat fish, like their more northern brothers and sisters. I have
often seen them catch wounded birds, and I visited one nest that contained in
addition to two well grown young birds the remains of thirteen Coots and
one catfish. Most of their eggs are hatched by the middle of December,
and some must be laid as early as the 1st of November, as my assistant
found a nest containing two eggs on the point of hatching on December 5,
and I found young birds two or three weeks old on December 15. The
latest sets I collected were one of two fresh eggs, taken January 26, and
another of two, one egg of which was nearly hatched and the other addled,
taken February 3. The eggs of this species from this vicinity are more
elongated and on the average smaller than the descriptions and measurements usually given of the eggs taken farther north.

"From what I could learn these birds invariably lay two eggs, as I
never found a larger set, and where I found but a single egg, it was always
fresh. On several occasions I found but one young bird in a nest, and
as I took quite often a set of two eggs, of which one was addled, I concluded that this was the reason. One peculiarity of the Bald Eagle that
I have never noticed in other birds of prey is, that when a pair are
robbed of their eggs or young, they will not lay again until the next season.
I watched a great many nests after they were robbed, visiting some close
by where I lived every few days for a period of two months or more, yet,
notwithstanding I almost always found one or both birds at home, I never
succeeded in getting two sets of eggs from the same nest during the same
season, though the next year these nests were again occupied.

"The nests are immense structures, from 5 to 6 feet in diameter and
about the same in depth, and so strong that a man can walk around in one
without danger of breaking through; in fact my assistant would always get
in the nest before letting the eggs down to me. They are composed of
sticks, some of which are 2 or 3 inches thick, and are lined with marsh
grass or some similar material. There is usually a slight depression in the
center, where the eggs are placed, but the edge of the nest extends so far
beyond this that it is almost impossible to see the bird from below, unless
it has its head well up. I have frequently found foreign substances in their
nests, usually placed on the edges of it, the object of which I cannot
account for. Often it would be a ball of grass, wet or dry, sometimes a
green branch from a pine tree, and again a piece of wood, bark, or other
material. It seemed as if they were placed in the nests to mark them.
From its frequent occurrence, at least, it appeared to me as if designedly
done.

"I believe these birds have a certain time for laying, and that their
eggs are deposited within a few days of that time every year. In the vicinity of San Mateo, Florida, the Bald Eagle is now quite rare, and I know
of but one nest that is occupied by these birds. This season they began
sitting on January 31, the earliest date on which I have ever known them

to begin here, and, during the several years I have known this nest to be occupied, I do not remember them ever commencing to lay later than February 6 or 7. I also think if one of a pair be killed after they have eggs or young, if nothing further happens, that the remaining bird will perform all the duty of incubation and care for the young.

"Two of my assistants watched the nest previously referred to during the present season, 1891, and on the day the birds were first found sitting both my men had visited the place, and on their return I saw them separately when they both told the following story. They said 'that they found one of the Eagles on the nest and that on pounding on the tree it flew to another near by and gathered a bunch of Spanish moss in its claws. It then flew slowly back over the nest, and, when just above it, poised for a moment on its wings, and dropped the moss, which nearly fell into the nest.' I believe this to be true, for I questioned both of the men very closely and they told exactly the same story, and they know nothing about the habit these birds have of marking their nests.

"Nearly all the nests I found were in pine trees, and generally the highest and thickest the birds could find, but as pines in this locality are not very tall, the majority of the nests were only 50 or 60 feet above the ground. The highest and lowest locations of nests were 75 and 30 feet, but these were extremes, at least so far as occupied nests were concerned. The places the Bald Eagle likes to frequent are forests of tall heavy timber in the immediate vicinity of some good-sized body of water, and they almost always build their nests in such situations. Both sexes assist in incubation, and are equally solicitous in the care of their eggs and young. They show great distress when their nests are disturbed, but are very careful to keep just out of gunshot, and I can recall but one instance of the Bald Eagle's attacking anyone, and that was a pair which had a nest in a large pine tree near the south shore of Crescent Lake, Florida. These birds would swoop down and almost strike the head of my climber, and were so very savage that one of my party became frightened, and thinking they might injure him, shot the male, which was the fiercer of the two. The nest contained two young but a few hours old, and as I was afraid they would either starve, or that Vultures or Crows would get them while the mother bird was absent after food, I thought I would try to raise them myself.

"They throve splendidly on a diet of meat and fish, and the amount they would consume in the course of twenty-four hours was something wonderful. I would stuff them, until I was almost afraid they would burst, in hopes of keeping them quiet for a few minutes; but it was no use, for the first noise they heard would set them yelling as loud as ever. I kept them until I left Florida, a period of almost two months, and they were then nearly half grown, and the family with whom I lived promised me to care for them. Soon after my departure one of them was injured and died shortly afterward, but the other got along nicely, and on my return the next winter,

I found it still alive. One peculiar thing about this bird was that it never learned to fly, though it was not confined during the first year. Its wings did not grow to the proper length and the feathers on them were twisted in all directions. It had a very rough and dilapidated appearance generally, which I cannot account for, as it was never handled much. While not afraid of man, this bird was a great coward in other respects, and would run from any other animal of whatever size; even a chicken would greatly frighten it.

"It would seem from what little chance I had of noticing the growth of these young Eagles, that unless they grow much faster in a state of nature than in captivity, birds of this species must remain in their nests from three to four months. They are very much attached to their chosen homes, and although their eggs and young may be taken from them for several successive seasons, and even one of the old birds killed, the survivor will find another mate and return to the old eyrie another season.

"The cry of the male is a loud and clear 'cac-cac-cac,' quite different from that of the female, so much so that I could always recognize the sex of the bird by it; the call of the latter is more harsh and often broken. Bald Eagles are constant residents of this vicinity, as they are of most parts of Florida, unless, as the inhabitants say, they go away for awhile during the summer. This I believe to be true, for there must be a scarcity of food then, when the waterfowl go north, as most of them do early in the spring. I believe the Bald Eagle does not breed before getting the adult plumage, and that this is the case generally with birds of prey, with few exceptions."

In the vicinity of Corpus Christi, Texas, the Bald Eagle breeds sometimes on the ground. Capt. B. F. Goss writes me as follows: "Of six nests of this species examined by me, near the above mentioned locality, all within 25 miles of it, four were found in trees, but the two others deserve special mention, as both were placed on the ground on small islands in Nueces Bay. An assistant, whom I hired to help me in collecting, showed me a nest from which he had taken the two young about five days previously, and had them in his possession at the time. The nest was placed on a small island, not more than 2 feet in its highest part above high water mark, and, with the exception of a little grass growing in the central part, it was a bare sand reef. The nest site, for it could hardly be called a nest, was located in the center of the island. It consisted simply of a few sticks laid on the bare ground, not enough to make a single tier even, and these were covered with bones, feathers, and fish scales, and the ground in the immediate vicinity was littered with the remnants of their food and the excrement of the young. We also found a small armadillo on the island, which was evidently brought there by the Eagles to feed their young, who probably found the shell too hard for them to crack, as it appeared uninjured. The owners of this nest were said to have been in the immature plumage, which accounts for the poorly constructed nest, probably being a first attempt.

"About the middle of the month I found another nest of this species on an island in the upper part of the bay, about 5 miles from the former. This was a massive structure, also built on the ground, at least 6 feet high and 5 feet in diameter. I saw it fully 2 miles away, and from that distance it looked like a monument. Although out of our course, and in a secluded part of the bay, it so aroused my curiosity that I ordered my boatman to pull to it. It contained a single young Eagle, nearly half grown. This island was larger and also a little higher than the first, and a solitary small tree was growing on it some 2 rods from the nest. Otherwise it was quite bare. This nest was quite a prominent object and was visible for miles. It was built with surprising regularity; appeared to be a perfect circle, and the sides smooth and almost perpendicular. The top sloped slightly toward the center where the eaglet sat, which viciously snapped at me as I looked over the edge of the nest. It was built of sticks and had evidently been raised to the present height by successive yearly additions, as the lower half had begun to decay, and a few inches of the top had evidently been recently added. Both parent birds attacked us with great fury, screaming and striking at us with their talons; while examining the nest they came within a few feet of me and I was glad to retire. This pair were both adult birds. Sometime later, as my assistant was taking the eggs from a nest in a tree, he was set upon by both the Eagles, and if he had not had a good stick to defend himself, I feel sure they would have struck him; as it was, they approached within 3 feet of him. These are the only two instances coming under my observation where any attempt was made to defend the nest."

That the Bald Eagle shows considerable courage at times in the defense of its nest, even without any provocation, I can corroborate from personal observation. In the months of March and April, 1883, I repeatedly visited a low marshy swamp at the head of Wood River, about 2½ miles northeast of Fort Klamath, Oregon; the object of these visits being to locate the nest of a pair of Goshawks, which committed daily depredations among the poultry at the post, and which always disappeared in this direction. A number of large aspens and several dense groves of conifers were scattered through this marsh, as well as a few immense pine trees.

A pair of Bald Eagles nested in one of the largest of these pines, and at no time could I approach this tree nearer than 100 yards without one of these birds, probably the male, swooping down at me, sometimes as close as 20 feet, lustily screaming, and giving me plainly to understand that I had no business in that particular vicinity.

These Eagles seem to nest in trees by preference, and only where such are wanting will they resort to cliffs or to the shelves found occasionally on the river bluffs. They are far more abundant along the seashore than in the interior, but they are by no means uncommon in suitable localities on the larger inland lakes.

At the Klamath Lakes they are especially numerous, and I have repeatedly seen a dozen within a distance of 3 miles. Some of our earlier writers speak in rather uncomplimentary terms of our national bird, stigmatizing it as a robber and tyrant, and as feeding principally on fish, stolen from the Osprey, and on carrion. This is not strictly true. According to my observations the Bald Eagle lives to a great extent at least on prey captured by its own exertions, principally on wounded waterfowl. When engaged in the chase of a flock of Geese, Brant, Ducks, or other water birds, on which it subsists almost entirely when such are procurable, it is by no means the sluggish, lazy bird some writers would have us believe, but the peer in swiftness, dash, and grace of any of our Raptores.

While it undoubtedly has occasionally to resort to an exclusive fish diet, some of which is captured from the Osprey, this habit is by no means universal, and carrion, in my opinion, is only used when other kinds of food are not available.

On May 1, 1886, Mr. S. B. Ladd found a nest of the Bald Eagle containing young in Lancaster County, Pennsylvania. The ground directly under it was covered with numbers of land terrapins in various stages of decay. These reptiles had evidently been carried to the young eaglets to feed on, were found unmanageable, and pushed out of the nest: probably but little other food was to be procured at the time.

Sometimes also when apparently fishing the Bald Eagle is after a different sort of game, as the following incident will show. Mr. W. W. Worthington writes me from Darien, Georgia: "The other day I noticed a Bald Eagle hovering over the sound, much the same as the Fish Hawk does when about to strike a fish. Suddenly he plunged down and grappled with what I supposed to be a large fish, but was unable to raise it from the water, and after struggling awhile he lay with wings extended and apparently exhausted. After resting a minute or two he again raised himself out of the water and I saw he had some large black object in the grasp of one of his talons, which he succeeded in towing along the top of the water toward the shore, a short distance, and then letting go his hold. He was then joined by two other Eagles, and by taking turn they soon succeeded in getting it to the shore. Investigation proved it to be a large Florida Cormorant, on which they were about to regale themselves."

Nidification begins early. In Florida and other parts of the Gulf coast eggs are sometimes deposited in the early part of November, but generally from the 1st to the 15th of December. In the Middle States they nest occasionally in the beginning of February, Mr. Thomas H. Jackson taking a full set of eggs in Lancaster County, Pennsylvania, on February 11. Usually they do not commence to lay till March, and correspondingly later as they advance northward.

On the Pacific coast in California, they nest about the middle of February; in Oregon and Washington, about April 1; in Alaska about the middle of the month, and in the interior, in the Arctic regions, as late as the latter part of May, and occasionally even in June.

As already stated, trees seem to be preferred for nesting sites, and large pines are oftener made use of for this purpose than other kinds. In certain portions of Florida they resort to a considerable extent to the mangroves, occasionally to live oaks, and in the West I have seen them nest in huge cottonwood trees.

The height from the ground varies considerably also, the extremes being probably from 20 to 100 feet. Cliff and bluff sites are generally used when no suitable trees are to be found in the vicinity, and nests on the ground must be considered as unusual locations and of very rare occurrence.

Incubation lasts about a month and both sexes take part in this duty. The usual number of eggs laid by this species is two, rarely less, and very seldom three; they are laid at intervals of three or four days. One of them is always somewhat larger than the other, and occasionally this difference in size is quite marked. But a single brood is raised in a season. The young at the end of the first year are considerably larger than the parents, and were for sometime considered as a distinct species, and named by Audubon *Haliætus washingtonii*—the Bird of Washington. In this plumage they are dark brown throughout, mixed with dull fulvous; they do not attain the adult plumage until the third year.

The nests vary greatly in size and are usually almost flat on top. Some are fairly well lined with dry grasses or seaweed. The eggs are pure white in color, but frequently nest stained; in very rare instances slight traces of markings are observable of a pale buffy brown, and in the handsome series of eggs of the Bald Eagle in Dr. Ralph's collection are two such specimens. The shell is strong, granulated, and without any luster. The shape varies from a rounded ovate to an ovate, the former predominating. Eggs from the more northern breeding grounds are considerably larger than those from Florida and the Gulf coast.

Sixteen specimens in the U. S. National Museum collection, principally from the Arctic regions, give an average measurement of 73.5 by 57.5 millimetres, the largest egg measuring 76.5 by 58, the smallest 69.5 by 56.5 millimetres.

Forty-five Florida specimens, all collected by Dr. Ralph, and now in his collection, give the following average: 69 by 53.5 millimetres, the largest egg of this series measuring 74.5 by 55, the smallest 61.5 by 50 millimetres. About a fair average would be 71 by 54 millimetres.

The type specimen, No. 20697 (Pl. 9, Fig. 7), from a set of two, Bendire collection, was obtained near Alden, Iowa, April 18, 1873, and was slightly incubated when found.

93. Falco islandus Brünnich.

WHITE GYRFALCON.

Falco islandus Brünnich, Ornithologia Borealis, 1764, 2.
(B 11, C 341a, R 412, C 501, U 353.)

GEOGRAPHICAL RANGE: Circumpolar regions, including Greenland, Arctic America, Commander Islands, etc.

With our present limited knowledge of the White Gyrfalcon but little can positively be stated about its breeding range, excepting that during the season of reproduction it inhabits the circumpolar regions. It breeds along the coast of northern Greenland, the ice-bound and inhospitable shores of eastern North America bordering on Baffin Bay and Smith Sound, and probably also along the numerous islands of the Arctic Sea, and the coast of northern Labrador and Hudson Strait. According to Dr. Stejneger, it also breeds on Bering Island, one of the Commander group of northeastern Asia. He says: "A pair had their nest in a steep and inaccessible rock in the so-called 'Nakovahaga,' a couple of miles from the main village." A male was shot by him on August 27, 1882, and a female on May 2, 1883.[1]

This species is the lightest colored of the Gyrfalcons. Gen. A. W. Greely, Chief Signal Officer, U. S. Army, says: "The Greenland Falcon was probably seen by the Polaris Expedition about May 22, 1872. Mr. Hart noticed these birds nesting in cliffs near Cape Hayes, latitude 79° 42′ N., and a Falcon was seen near Cape Frazer, latitude 79° 44′ N., August 24, 1876. The latest specimen in Smith Sound was recorded by Hayes during his boat journey on Northumberland Island, near Cape Alexander, September 10, 1854. We saw an example August 4, 1881, just north of Carl Ritter Bay, and another near Fort Conger on July 30, 1882, the attention of Sergeants Jewell and Ralston being called to the latter through the great alarm shown by the Buffon's Skuas in their vicinity. Three Falcons were seen August 13, 1882, in the valley at the head of St. Patrick Bay, about latitude 81° 55′ N., attracted by offal. They were observed in the same locality August 15, and a pair of them flew around Fort Conger two days later."[2]

Nelson found the White Gyrfalcon rare in Alaska and only secured a single specimen of this bird, an Eskimo skin, taken at Cape Darby, on the Alaskan shore, near Bering Strait. The natives also told him that these White Hawks sometimes bred on the mountains of the Kaviak Peninsula, but he had no means of personally verifying this statement, although his solitary specimen came from there.

According to Kumlien, it is very rare on the Cumberland Gulf side of Davis Strait, and much more numerous on the Greenland shore. He says that it is apparently much slower on the wing than the common Duck Hawk,

[1] Bulletin 29, U. S. National Museum, 1885, p. 205.
[2] International Polar Expedition to Lady Franklin Bay, Grinnell Land, 1888, Vol. II, p. 26.

that it was seen pursuing the Gulls until they were exhausted before the Hawk could secure its prey, showing that their endurance exceeds their speed.[1]

Turner obtained a single specimen at St. Michael, Alaska, May 15, 1877, and says that it is not a common bird in that vicinity, and oftener seen in the spring than at other seasons. In his notes "On the Birds of Labrador and Ungava," he states: "This beautiful Falcon is common throughout the entire region, although more numerous in the sparsely wooded tracts and in the neighborhood of high rugged hills. This species is known as the Partridge Hawk by the English-speaking people, who apply the name of Partridge to both species of *Lagopus* and to *Dendragapus* indiscriminately. In the vicinity of Fort Chimo it is not at all rare in winter, but so wary that but one specimen was secured.

"A pair was building their nest on the side of the bluff known as Hawk's Head, some 2 miles north of Fort Chimo. I undertook to ascend the bluff by means of a rope drawing me up a distance of 168 feet, where I had nothing to cling to but the rope, and after whirling around and around dozens of times (as the nest was on a part which could not be reached from above), reached the spot where the nest was located. I was then so dizzy that I was glad to find a resting place, and when I attained the site I put my foot directly on the half completed nest, composed of a few sticks and a great quantity of dry grasses, forming a bulk about 15 inches across the top and 3 inches high. The birds were circling and screaming a short distance off while a man was firing at them. They deserted the locality and were not again seen.

"This Falcon is extremely shy, and when sitting with its pure white breast toward the hunter will often escape detection on the snow, as it is far more numerous from September to April than at other times. The natives assured me that they repair to the rugged mountains in the vicinity of Cape Chidley (the northernmost point of Labrador) to breed, and that they fed their young on the Rock Ptarmigan, which also seek that region for the same purpose. The manner of flight is by rapid beats of the wings, followed by a short sail."

The nesting habits of the White Gyrfalcon would appear, from what little we know on the subject, to be in no way different from those of the other members of this genus, excepting that they probably nest almost exclusively on rocky cliffs near the seacoast. Few suitable trees, even in the more southern portions of their summer range, would be found in these northern regions sufficiently large to hold their nests. As a rule, these rocky cliffs are the summer homes of innumerable waterfowl, on whose young, as well as on Ptarmigan, they prey to a great extent during the season of reproduction.

The usual number of eggs to a set is probably three or four, and they are not likely to be deposited much before May 1.

[1] Extracts from Nelson's Report on Natural History Collections made in Alaska, 1877–1881, pp. 145, 146.

Of the two specimens, both from different nests, in the U. S. National Museum collection, and about whose correct identification I am by no means satisfied, No. 2606, marked *Falco candicans*, collected in Greenland in 1858, and purchased from Madam H. Drouet, measures only 57 by 45 millimetres, and seems to be too small an egg to belong to this bird. No. 13231, obtained from Dr. T. M. Brewer, and also from Greenland, is more of the probable size, measuring 59.5 by 46.5 millimetres.

They resemble the eggs of the Gyrfalcon both in shape and color, and are not distinguishable from these, making a detailed description unnecessary. None of these eggs are figured.

94. Falco rusticolus LINNÆUS.

GRAY GYRFALCON.

Falco rusticolus LINNÆUS, Systema Naturæ, ed. 10, I, 1758, 88.
(B 12, C —, R 412*a*, C 500, U 354.)

GEOGRAPHICAL RANGE: Extreme northern portions of Europe (except Scandinavia), Asia, and North America, including Iceland and southern Greenland; south in winter to northern border of United States.

The Gray Gyrfalcon is another species whose breeding range is still rather imperfectly known. According to Ridgway, it inhabits the extreme northern portions of Europe (excepting Scandinavia), Asia, and North America, including Iceland and southern Greenland, and it probably breeds throughout the range indicated. In south Greenland and Iceland it is said to nest mostly on the inaccessible cliffs along the seashore, usually in close proximity to bird rocks, which supply them with an abundance of easily procured food.

The eggs of the Gray Gyrfalcon are said to be three or four in number, and indistinguishable, both in color and size, from those of the other large Gyrfalcons found in northern North America. In south Greenland they are usually deposited during the latter part of April and throughout May. There are no specimens in the U. S. National Museum collection.

95. Falco rusticolus gyrfalco (LINNÆUS).

GYRFALCON.

Falco gyrfalco LINNÆUS, Systema Naturæ, ed. 10, I., 1758, 91.
Falco rusticolus gyrfalco STEJNEGER, Auk, II. 1885, 187.
(B —, C 341, R 412*b*, C 498, U 354*a*.)

GEOGRAPHICAL RANGE: Northern Europe and Arctic America, from northern Labrador and coasts of Hudson Bay to Alaska.

The breeding range of the Gyrfalcon, a slightly darker colored bird than the preceding, is somewhat better known than that of the two allied subspecies already mentioned. It includes Arctic North America, from northern Labrador and the coasts of Hudson Bay, throughout the so-called Fur

Country in the interior, to Alaska. It is also found in northern Europe (Scandinavia). It apparently does not reach quite such high northern latitudes as the other subspecies already mentioned.

Mr. R. MacFarlane, chief factor of the Hudson Bay Company, states that this Falcon is common in the wooded country on both sides of the Anderson River, and from the fact that over twenty nests were secured by him this must certainly be the case. All of these nests, with but two exceptions, were placed close to or near the tops of the tallest trees in the neighborhood, generally in pines. One nest was built on a ledge of rocks and the other against the side of a deep ravine. The nests were composed of sticks and small branches and lined with mosses, hay, deer hair, feathers, etc. They were similar in composition but smaller in size than those of the Bald Eagle, and while the number of eggs was either three or four, their contents were frequently found in different stages of development.

Both parents manifested much anger and excitement when interfered with, or even distantly approached. They made a great noise, and indeed more than once their folly in coming so near and screaming so loudly over our heads attracted attention to some that would otherwise have escaped notice. The earliest date of finding a nest was May 10, 1863, at Anderson River Fort. The eggs, three in number, were quite fresh. In another, taken five days later, the eggs contained partially formed embryos. In a few cases young birds were found in the same nest with eggs, the contents of which were but little changed, and in another nest a perfectly fresh egg was found with several ready to hatch. In nearly every case the eggs seemed to be in different stages of development, and incubation seems to begin as soon as the first egg is laid.[1] The latest date on which eggs were found by Mr. MacFarlane, according to the records in the U. S. National Museum, is June 12, 1864, when two sets of four eggs each were taken.

Nelson, in speaking of this species, says: "Throughout Alaska, from the Aleutian Islands north, both along the coast and in the interior, extending from Bering Strait across the northern portion of British America, the present Falcon is the commonest resident bird of prey. It was observed by Murdoch at Point Barrow, though it was not common. It frequents the vicinity of cliffs and rocky points about the seacoast, or the rocky ravines of the interior during the breeding season; and the remainder of the year, especially in the fall, it is found wandering over the country everywhere that food can be obtained; it is especially numerous during the migration of the Ptarmigan along the seacoast. * * *

"Along the seacoast in the vicinity of St. Michael it breeds rarely, choosing rocky cliffs, facing the sea. Along the Lower Yukon, and the Kuskokwim River, it is numerous in winter, and finds an abundance of Ptarmigan, upon which it preys. At this season it is frequently seen perching on a stout branch of a tree overhanging the river bank, and I have seen it on several occasions

[1] From R. MacFarlane's Manuscript Notes on the Land and Water Birds Nesting in British America.

allow a train of dog sledges to pass within 40 or 50 yards, only noticing their presence by slowly turning its head.

"It was seen in the vicinity of Bering Strait and around the shore of Norton Sound during the cruise of the *Corwin* in the summer of 1881, as also upon the northeast shore of Siberia, in the vicinity of East Cape and Plover Bay."[1]

Mr. L. M. Turner obtained several specimens of this Gyrfalcon in the vicinity of St. Michael, where he says it is a constant resident, except during protracted periods of severe weather in winter only. According to the natives it breeds on the high hills, either on a rocky ledge or on the moss-covered ground. He failed to obtain the nest and eggs.

Mr. James Lockhart found the Gyrfalcon breeding on the Yukon River 100 miles above the mouth of the Porcupine River, and took a nest and three eggs there in June, 1862. Messrs. MacDougall and Jones also took their eggs near Fort Yukon, Alaska, in 1865, and all of these specimens are now in the U. S. National Museum collection.

From our present knowledge it appears that the Gyrfalcon breeds north of latitude 65°. It has not been found nesting farther south than this, so far as I can learn, although in winter these birds straggle sometimes into Canada and the northern borders of the United States. In Scandinavia the Gyrfalcon usually commences laying about the end of April, and the nests are generally placed on cliffs, rarely in trees.

The eggs are three or four in number. The ground color, when distinctly visible, which is not often the case, is creamy white. This is usually hidden by a pale cinnamon rufous suffusion. In an occasional specimen it seems to be pinkish vinaceous. The eggs are closely spotted and blotched with small, irregular markings of dark reddish brown, brick-red, ochraceous rufous, and tawny. These markings—usually pretty evenly distributed over the entire egg—are generally small in size, and more or less confluent. Some specimens show scarcely any trace of markings, the egg being of a nearly uniform color throughout. One of the eggs figured, from an incomplete set of two (No. 10172), both alike, is a most peculiarly marked specimen. In the general pattern of markings the eggs of the Gyrfalcon approach those of the Prairie Falcon (*Falco mexicanus*) much closer than those of the Duck Hawk (*Falco peregrinus anatum*), which as a rule are much darker. In shape they vary from ovate to rounded ovate. The shells of these eggs feel rough to the touch, are irregularly granulated, and without luster.

The average measurement of thirty-two specimens in the U. S. National Museum collection is 59.5 by 45 millimetres, the largest egg in the series measuring 63 by 46.5, the smallest 57.5 by 43 millimetres.

The type specimens figured were obtained as follows: No. 10172 (Pl. 9, Fig. 9), from an incomplete set of two, was taken June 12, 1864; No. 13790 (Pl. 9, Fig. 8), a single egg, taken in June, 1865; No. 13792 (Pl. 9, Fig. 6),

[1] Report on Natural History Collection made in Alaska, Nelson, 1877-1881, pp. 146, 147.

from a set of three, also taken in June, 1865. They were all collected by the indefatigable Mr. R. MacFarlane, near Anderson River Fort, Arctic America, situated in latitude 68° 35′ N.

96.   Falco rusticolus obsoletus (GMELIN).

BLACK GYRFALCON.

Falco obsoletus GMELIN, Systema Naturæ, 1, i, 1788, 268.
Falco rusticolus obsoletus STEJNEGER, Auk, ii, 1885, 187.
(B —, C —, R 412c, C 499, U 354b.)

GEOGRAPHICAL RANGE : Coast of Labrador ; south in winter to Canada, Maine, and New York.

As far as known at present the breeding range of the Black Gyrfalcon, the darkest colored bird of this genus, is confined to the coast of Labrador.

Mr. L. M. Turner, of the U. S. Signal Service, gives the following account of this subspecies in his "Notes on the Birds of Labrador and Ungava," and he has kindly permitted me to make the following extracts from the same. He says: "A number of specimens of this large Hawk were procured at Fort Chimo. * * *

"A pair of these Hawks had been frequenting the Chapel for the last six weeks and were occasionally seen throughout the winter. The nest site first selected was deserted and a new one chosen, some 200 yards to the south, but on the same side of the bluff. The Chapel is an immense rock, some 300 feet above the surrounding rocks, and gradually slopes upward to the north end, which is almost precipitous and absolutely inaccessible. The west side of the north end is nearly perpendicular for half the length, but varying some 40 to 90 feet in height along the perpendicular portion. The eastern side is more abrupt and higher, being in places over 200 feet, almost perpendicular. Here are several ledges on which these Hawks have built their nests for many years. The southern end of the hill is low and forms an easy slope to the top. The nests of these Hawks were attainable only by a person being lowered over the side of the cliff. On April 7 I observed beneath the nest site first selected a number of sticks and other refuse lying on the snow below, as though the locality had been subjected to a rearrangement or cleaning process, and such material as appeared unnecessary was rejected and cast over the side of the ledge. The site of this nest was a narrow ledge of rock which projected from the main wall and embraced an area of not over 3 superficial feet. Here were a number of spruce and larch twigs and branches of various sizes imbedded in what appeared to be the accumulated débris of many generations. Among this a few grass seeds had found enough soil to enable them to send forth a rank growth, which was now appearing. This mass or accumulation was about 10 inches deep and covered nearly the entire surface of the ledge. The new nest, forming an irregular truncate cone, was placed on this.

"On the 20th of May I again visited the locality and succeeded in putting a charge of shot into the female, which had since the 11th of the month taken up with the mate of the bird shot on that date. The one shot on the 20th fell nearly a mile distant after tumbling headlong several times through the air, and I could not find her. I supposed that the nest would now be abandoned by the male, as he was particularly wary and gave warning even when I was over a thousand yards distant.

"On May 22 I went with a party of four to lower me over the cliff to secure the eggs which might remain in the nest. To my great astonishment I found a pair of these Falcons launching into the air from the same site. I descended to the nest. In front of it huge icicles stood joined with the slightly projecting roof above the ledge; some of these ice columns were 2 or 3 inches thick and 4 inches wide, forming an icy palisade around the edge of the nest and permitting approach to the interior only by a narrow space or doorway next the main wall of rock, and I was compelled to detach the ice before I could reach the four eggs I saw within the nest, which was composed of a few twigs and branches of larch and spruce, irregularly disposed on the outer side of the rim of the nest to prevent the eggs from rolling out, forming only a semicircular protection, while the rear portion was a part of the bare rock of the ledge. Below these twigs were the remains of former nests. Some of the sticks were so rotten that they would not support their weight when held by one end. The eggs were placed nearly touching each other. They contained small embryos and had been incubated about twelve days. The parent of these is No. 94308 in the U. S. National Museum collection. * * *

"This subspecies is the most abundant of any Hawk observed in this region. During the excessively cold periods of winter but few are to be seen. About the middle of March they are more abundant and become plentiful by the last of April. * * * Their food consists almost exclusively of Ptarmigan, little else being found in their stomachs. They seize their prey while on the wing, depending doubtless on their sudden appearance among a flock of Ptarmigan to put their prey to flight, when it may be secured. Their food is devoured on the ground; I have never seen them carry it in their talons.

"The young birds are able to leave the nest by the middle of August, and in September of some years they are quite abundant, flying over the houses at Fort Chimo with but little fear. * * * Their manner of flight is by extremely rapid wing beats followed by sailing for a few rods. They pass through the air with great rapidity, no bird of prey in those regions flying more rapidly. * * * The only note ever heard from this Hawk was a chattering scream of the syllables 'ke-a, ke-a, ke-a,' repeated a number of times, more rapidly toward the fifth or sixth utterance, and finally so blended that the sound is a rattling scream. I have never seen one of these Hawks alight in or fly from a tree. In the Ungava district they invariably select a ledge of rock on the cliffs for nesting places."

"On August 6, 1833, Mr. John W. Audubon found a nest of this Falcon among some rocky cliffs near Bras d'Or, Labrador, containing four young birds ready to fly, two of which were secured. The nest was placed among the rocks about 50 feet from their summit and more than 100 feet from their base. It was composed of sticks, seaweed, and mosses, was about 2 feet in diameter, and almost flat. Its edges were strewed with the remains of their food, and beneath the nest was an accumulation of the wings of the Ptarmigan, Mormons, Uriæ, etc., mingled with large pellets of fur, bones, and various substances."[1]

Four eggs seem to be the number usually laid by this Falcon, and they are deposited about the middle of May. The shells of these eggs are roughly granulated and without luster. In shape they are ovate, and apparently indistinguishable from those of the preceding subspecies; the set under consideration, and the only one in the U. S. National Museum collection, is a trifle lighter colored.

The ground color is a creamy white nearly hidden by fine irregular markings of light reddish or rusty brown. Three of the specimens are thus marked; in the fourth and lightest colored one these markings are more of a clay color, with scarcely a trace of red. They all differ slightly in color from each other.

The measurement of these eggs is as follows: 58 by 45.5, 58 by 44.5, 57.5 by 45, and 57 by 45.5 millimetres.

The type specimen, No. 22389 (Pl. 10, Fig. 1), was collected near Fort Chimo, Labrador, May 22, 1883, by Mr. L. M. Turner, U. S. Signal Service.

97. Falco mexicanus SCHLEGEL.

PRAIRIE FALCON.

Falco mexicanus SCHLEGEL, Abhandlungen aus dem Gebiete der Zoölogie, 1841, 15.
(B 10, C 342, R 413, C 502, U 355.)

GEOGRAPHICAL RANGE: Western United States, from eastern border of the Great Plains to the Pacific, south into Mexico; casual eastward to Illinois.

The breeding range of the Prairie or Lanner Falcon includes the more open country from eastern Texas in the vicinity of Houston, where Nehrling reports it as resident but not common, north through the Indian Territory, Kansas, northwestern Missouri, and Nebraska to South and North Dakota, which seems to form the northern limit of its breeding range, as far as known at present. Prof. J. Macoun has taken it at Rush Lake, Northwest Territory, in latitude 51°, and it probably occurs and breeds in small numbers along the southern border of the Dominion of Canada from western Manitoba to southern British Columbia. From the Dakotas it reaches westward through the intervening States, excepting the densely timbered and higher mountain regions, to the

[1] History North American Birds, 1874, Vol. III, p. 121.

Pacific coast, where it seems to be fairly common in suitable localities throughout Washington, Oregon, and California. I have no doubt that it crosses our border into British Columbia, as I have found them breeding not uncommonly on the Upper Columbia River, close to the boundary line. South it extends through Arizona, New Mexico, Lower California, and northwestern Texas into Mexico. It seems to be equally and possibly more abundant in the central portions of its range, including portions of Colorado, Wyoming, Montana, Utah, Nevada, and Idaho.

It breeds in all the localities mentioned, but is only a summer resident in the northern portions of its range, wintering east of the Rocky Mountains from about latitude 37° southward. On the Pacific coast some of these Falcons at least remain throughout the winter as far north as latitude 45° 30', where I have met with them at Fort Walla Walla, Washington, and also shot an adult male on February 19, 1885, at Fort Custer, Montana, while the weather was still intensely cold.

Though on the whole perhaps more of a prairie-loving species, this Falcon does not confine itself exclusively to open country, but inhabits and breeds in the lower mountains as well; sometimes even where heavily timbered. I have repeatedly seen them during the breeding season in the Blue Mountains, near Camp Harney, as well as in the Cascade Range, near Fort Klamath, Oregon, where they nested in some of the numerous cañons, and on inaccessible cliffs abounding in these regions.

In the latter part of May, 1879, while ascending the Upper Columbia River in a steamer from Wallula to Priest's Rapids, Washington, which was then the head of navigation, I noticed a nest of this Falcon in a small cavity of a chalk bluff, rising perpendicularly for about 120 feet directly out of the water. The nesting site was nearly midway up the bluff (called White Bluff), and as the main channel ran along the foot of it, I had rather a good opportunity to examine it closely.

In those days steamers seldom ascended the river so far, and the surrounding country was almost uninhabited, excepting here and there by small roving bands of Indians. As we approached the bluff both the old Falcons circled around above the nest, screaming loudly in a high key. Their alarm note was a rapidly repeated "kéé, kéé, kéé," and a sort of cackle. Standing on the upper deck of the boat I could see the young plainly; there were certainly four and possibly five of them, and they appeared to be sitting on the bare ground. I saw but little nesting material lying about the outer edge of the cavity, and this seemed to me to look more like scraps of refuse than anything else. The cavity was rather shallow and not much more than large enough to hold the young. The site was practically inaccessible, excepting by rope from above. The surrounding country was utterly devoid of trees and consisted of open grass and sagebrush-covered table-lands.

Landing at the foot of Priest's Rapids, the journey was continued to Wenatchee on horseback along the north side, and from there on the south side of the Columbia to a point opposite the mouth of the Okinakane, following the shores of the former wherever practicable. On this march I saw several more of their nests, all in similar locations on small projecting ledges of perpendicular cliffs, most of them at a considerable height from the ground and nearly all practically inaccessible, unless one was provided with strong ropes and lowered from above. As the season was too far advanced I did not attempt to examine any of these nests for eggs. These Falcons seemed to be common enough throughout the entire region traveled over, and were about the only Raptores seen in this, for the most part, scantily timbered country.

At Fort Walla Walla, Washington, I noticed the Prairie Falcon more often in the late summer and fall, a few during the winter, and again in the spring on the return from their migrations. Without especially looking for them, I usually obtained half a dozen specimens each season, young birds greatly predominating.

While here they fed principally on Brewer's Blackbirds (*Scolecophagus cyanocephalus*), which always congregated in large flocks about the cavalry stables, except during the breeding season, and were constantly harassed by this Falcon, and also, but to a less extent, by the Pigeon Hawk and the Black and Richardson's Merlins.

The dead top of one of the largest cottonwood trees growing on the banks of the little creek flowing past the garrison was selected as a favorite perch by nearly all these birds, and fully three-fourths of the specimens obtained were killed from it. As affording them the best outlook over the surrounding country it was no doubt selected on that account. Most of their hunting was done early in the morning and again about an hour before sunset. Mourning Doves, Western Meadow Larks, and a domestic Pigeon furnished them now and then with a meal, and I have also found the remains of a Columbian Sharp-tailed Grouse in the crop of one specimen. Poultry was rarely molested; and although one of these Falcons would sometimes make a dash at some of the fowls, it seemed to me that it was done more to scare and to see them run than to capture them. Not a single instance came under my observation where a chicken was actually struck by one of them. I have no doubt whatever that they are fully capable of killing a full-grown hen and of carrying her off, but they do not seem to care for poultry, and I have more than once seen chickens feeding under a tree in which one of these birds was sitting.

On the plains where hares are abundant they are said to live partly on them, and, should other food be scarce, no doubt some of the various species of rodents found in such localities are also captured. In the vicinity of Walla Walla they subsisted almost entirely on such birds as mentioned.

They are strong and powerful for their size, very tenacious of life, and ugly customers to handle when wounded, invariably throwing themselves on their backs and fiercely striking out with their sharp talons. Their flight is easy and gracefully accomplished by rapid wing beats and alternate sailing, and when once launched after the selected victim they are as swift as an arrow.

I believe that the Prairie Falcon nests almost invariably on rocky cliffs or perpendicular bluffs, on ledges or in small cavities in the same, and but rarely in other situations. Col. N. S. Goss, however, in his "Birds of Kansas," mentions that two eggs of this species now in his brother's collection, were taken April 28, 1880, at Marysville, Missouri, from a tree, the nest being 35 feet from the ground; it was not stated whether the nest was an open one in the forks of branches, or in a hole of the tree, but doubtless the latter. Mr. Walter E. Bryant, of San Francisco, California, who has taken a number of sets of eggs of this species, found them invariably on cliffs from 30 to 100 feet from the bottom. A handsome set of five eggs, which he kindly presented to the writer, was taken on March 25, 1882, at Pine Cañon, near Mount Diablo, California, and were fresh when found; they were laid in a cavity of a cliff about 4 feet in depth on some sand and bits of fur, feathers, etc., ejected by the birds, there being no nest. The cliff was about 100 feet high and perpendicular, and the site about 30 feet from its base.

Sets of eggs of the Prairie Falcon now in the U. S. National Museum collection, from near Gilmer, and others from the Wind River Mountains, Wyoming, as well as from Battle Mountain, Nevada, were all taken in similar situations. The first eggs of this species brought to the attention of naturalists were a set of three taken on Gros Ventre River, in the Wind River Mountains, by Dr. F. V. Hayden, on June 8, 1860.

Mr. Walter E. Bryant also found a pair of these Falcons nesting in a high cliff near San Esteban, Lower California, on April 18, 1889. The earliest date on which he took full sets of eggs of this species in California was March 24, 1881, and fresh eggs were taken by him as late as April 7. These dates are probably as early as they nest anywhere within their range. Farther north nidification is protracted through April and the beginning of May, and occasionally even into the first week in June. As a rule there is but little of a nest, if any, the eggs being usually laid on the bare rock, among some of the refuse carried there as food, such as bones, bits of fur, and feathers.

The eggs number from three to five, sets of five seeming to be the usual number found on the Pacific coast, while in the interior sets of three or four are perhaps more common. Occasionally but two eggs have been found, but it is doubtful if such so-called sets were complete when taken. Should they lose their first clutch of eggs a second and smaller one is sometimes laid. The eggs are deposited at intervals of a day or two, and nidification lasts probably from three to four weeks. The young when first hatched are covered with a thick white fluffy down, and they grow rapidly. As soon as they are fully grown and able to care for themselves they are turned adrift to make their own living, and fall easy victims to the guns of collectors. The old birds are shy and wary, and much harder to bring to bag. If the eggs are near the point of hatching, or if there are young in the nest, the parents will occasionally defend them, showing at times considerable courage.

The majority of the eggs of the Prairie Falcon in the U. S. National Museum collection are lighter colored than those of any other Falcons found on the North American continent, excepting the eggs of the Sparrow Hawk. The ground color is creamy white, and easily visible in a majority of the specimens before me. In a few, however, this is entirely obscured by the overlying pigment, giving the egg a vinaceous cinnamon color, and in others a dirty clay color. They are blotched and spotted with different shades of reddish brown, tawny, and chocolate. As a rule these markings are pretty evenly distributed over the entire egg; only in an occasional specimen are they heaviest on one end, and in such cases usually confluent. The average measurement of fourteen specimens in the U. S. National Museum collection is 53 by 41.5 millimetres, the largest egg measuring 56 by 41, the smallest 52 by 40 millimetres.

The type specimens No. 15596 (Pl. 10, Fig. 2), selected from a set of five, representing one of the heavier marked eggs, was taken by Mr. H. R. Durkee, near Gilmer, Wyoming, May 13. 1870, and No. 20664, from the Bendire collection (Pl. 10, Fig. 3), also from a set of five, was taken in Pine Cañon, Mount Diablo, California, by Mr. H. E. Bryant, March 25, 1882. This is one of the more sparingly marked specimens in the collection.

98. Falco peregrinus anatum (BONAPARTE).

DUCK HAWK.

Falco anatum BONAPARTE, Geographical and Comparative List. 1838, 4.
Falco peregrinus β anatum BLASIUS, List of the Birds of Europe, 1862, 3.
(B 5, 6, C 343, R 414, C 503, U 356.)

GEOGRAPHICAL RANGE : Whole of America, south as far at least as Chile ; eastern Asia.?

The breeding range of the Duck Hawk, or Peregrine Falcon, includes the greater portion of the United States and the Dominion of Canada, excepting perhaps the extreme southern portions of our domain, such as Florida and the cotton States bordering on the Gulf coast, where it appears to be a winter visitor only. It nests sparingly in suitable localities throughout the thinly settled mountainous portions of South Carolina from about latitude 35° N., northward, and generally among the more or less inaccessible cliffs found near the shores of the larger mountain streams and lakes of the Atlantic watershed. In the middle and western prairie regions, where there are no suitable cliffs, it nests in hollows of trees in the heavily timbered bottom lands. Throughout the Rocky Mountains, north to the shores of

the Arctic Sea, as well as near the numerous lakes of the Great Basin, and
westward over the entire Pacific coast from Lower and southern California
northward to Oregon and Washington, it seems to be more abundant than
east of the Mississippi River, and breeds throughout the entire range indicated.
While it cannot be considered a common species anywhere, still it is the
most numerous of our larger Falcons.

In the northern portions of its range it is only a summer resident, following the immense flocks of waterfowl during their migrations, on which
it principally subsists. Hares, Ptarmigan, Grouse, Bob Whites, Pigeons, as
well as smaller birds, enter largely into its bill of fare, and the poultry
yard as well occasionally suffers. Next to the Goshawk the Peregrine Falcon
is our most audacious bird of prey. Its flight, when once fairly started
in pursuit of its quarry, is amazingly swift; it is seemingly an easy matter
for it to overtake even the fleetest of birds, and when once in its grasp resistance is useless. I have seen this Falcon strike a Cinnamon Teal (*Anas
cyanoptera*) almost within gunshot of me, kill it apparently instantly from the
force of the shock, and fly away with it as easily, and without visible struggle,
as if it had been a Sparrow instead of a bird of its own weight.

Within recent years quite a number of the nests of this species have
been taken in the New England States and other localities in the North,
records of these captures being noted in various ornithological publications.
In a letter now before me, written December 18, 1887, Dr. John W. Det-
willer, of Bethlehem, Pennsylvania, states: "It may interest you to know
that the Duck Hawk breeds abundantly in this State. I found a nest containing four eggs on a cliff on the west shore of the Delaware River, 12
miles below Easton, Pennsylvania; and about four weeks later a second
containing two eggs slightly incubated, which doubtless belonged to the
same birds." In the same letter several other localities are mentioned along
the Susquehanna River from which the doctor has taken full sets of eggs of
this species, and the bird is evidently not uncommon there. From the observations of Mr. L. M. Loomis "On the Summer Birds of Pickens County,
South Carolina," in the Auk (Vol. vii, 1890, p. 37), there is little reason
to doubt that these birds had a nest with young there, and if other suitable
mountain regions throughout North Carolina, the Virginias, northern Alabama, Georgia, Tennessee, and Kentucky were equally well examined the
Duck Hawk would be found breeding there, too.

Mr. Robert Ridgway, of the Smithsonian Institution, found the Duck
Hawk not at all rare in the heavy timber of the river bottoms in the vicinity
of Mount Carmel, Illinois. Not less than three nests were found in the
immediate vicinity of town, all placed in cavities in the tops of very large
sycamore trees, and inaccessible. One of these trees was felled and four
fully feathered young were taken from the nest.[1] Col. N. S. Goss also found
the Duck Hawk nesting in trees near Neosho Falls, in southeastern Kansas,

[1] Bulletin Nuttall Ornithological Club, Vol. iii, 1878, pp. 165, 166.

a pair occupying a trough-like cavity in a large sycamore about 50 feet from the ground. The nest contained three fresh eggs on March 27, 1875, which were laid on the fine, soft rotten wood in a hollow, worked out of the same to fit the body of the bird. There was no other material or lining except a few feathers and down mixed with the decayed wood. On March 17, 1876, a pair were found nesting on the opposite side of the river from the above described nest, in a cottonwood at least 60 feet from the ground, the birds entering a knot hole in the tree apparently not over 5 or 6 inches in diameter. The tree was very straight and without limbs up to the nest, and this was out of reach. The birds were very noisy, but shy. On April 30, 1877, he found another pair about 4 miles farther up the river, breeding in a broken hollow limb of a giant sycamore, and from the actions of the birds he thought they had young.

Colonel Goss further states: "The males, so far as noticed, sit upon the eggs in the fore part of the day, and the females during the latter part; each, while off duty, occasionally feeding the other, but putting in a good share of the time as sentinels, perched upon a favorite dead limb near the nest, ready to give the alarm in case of approaching danger. At such times they scold rapidly and manifest great anxiety and fear, circling overhead, occasionally alighting, but taking good care to keep out of reach. Their fear of man is not without cause, for our hunters never lose an opportunity to shoot at them, knowing how destructive they are to the waterfowl found in the sloughs along the river bottoms."[1]

In Cleburne County, Arkansas, it nests in the sandstone bluffs along the Little Red River. Mr. B. T. Gault noticed them in such locations there in the spring of 1888, and Mr. O. Widmann informed him that they also nested quite abundantly in the rocky bluffs and ledges along the Mississippi River, a few miles south of St. Louis, Missouri.

In Montana, Dr. James C. Merrill, U. S. Army, says they are quite common along the Missouri and Yellowstone Rivers, near cliffs or cut banks, where their nests were placed, sometimes in a cavity in some rock high above the river, and then, again, on a shelf or projection of a clay bank, so low that he could see the contents of the nest from the upper deck of the steamer. Their abundance on the Upper Missouri is further confirmed by Mr. R. S. Williams, who writes me: "Quite a number of these birds breed in the high sandstone cliffs above the Falls of the Missouri. I found a nest containing two birds just hatched and two eggs, on June 5, 1885. The nest was situated in a small hollow in a perpendicular wall of rock, some 15 or 20 feet above the base of the wall, and consisted of a few coarse twigs and bits of grass, forming a ridge on the outer side barely sufficient to prevent the eggs from rolling out. The parents were both about and quite bold, dashing back and forth overhead and keeping up a constant succession of noisy screams as long as I was near."

---

[1] Bulletin Nuttall Ornithological Club, Vol. III, 1878, pp. 32–34.

A nest taken by me on April 24, 1877, from a small cavity in a basaltic cliff, 30 miles south of Camp Harney, Oregon, and about 3 miles from the eastern end of Malheur Lake, contained three slightly incubated eggs. They were lying on the bare rock, with a little rubbish and a few small bones placed around them to keep them from rolling out. The site was about 25 feet from the ground and about 10 feet from the top of the cliff. The female was very bold and was easily shot. Another set collected for me a few miles south of San Diego, California, on March 29, 1873, was placed in a similar situation, a basaltic cliff on the seashore. It contained three fresh eggs.

Occasionally the Duck Hawk will use the nest of another species to lay in, but such instances seem rare. Mr. Denis Gale took a very handsome set of eggs from an old Eagle's nest on April 30, 1889, which he generously donated to the U. S. National Museum collection. He writes me: "I discovered this nesting site first on April 4, 1889, in a ledge of rocks on the Cache la Poudre Creek. Some difficulty was experienced in getting to the nest, and no eggs were found on that occasion, it being too early. Nevertheless the birds were a good deal excited and flew around overhead, yet it might only have been caused by seeing their favorite hunting ground invaded. I think if they had eggs to guard at that time, the watching I bestowed upon them with my glass would have located the site exactly, but it did not. Probably it was when nearest to their nesting site, that the female flew overhead and cackled like a two-thirds grown duckling. This was unmistakable, and a peculiarity I never before met with.

"On the 24th the nest contained two eggs, which were left till the 30th. The site was in a rocky ledge about 80 feet high and about 50 feet from the foot of the cliff. The nest was reached by climbing down a rope from above. On April 30, six days later, four eggs were found in the nest and taken. They were all perfectly fresh. I do not think that the set was complete, and believe that another would have been added a day or two later."

The Duck Hawk must be fairly common in the Arctic regions and the interior of Alaska, as attested by a number of sets of eggs, now in the U. S. National Museum collection, taken by Mr. R. MacFarlane, on the Lockhart and Anderson Rivers; by Mr. S. Jones, at Fort Rae, Great Slave Lake, and Fort Yukon, Alaska; and by Mr. James Lockhart, at Fort Resolution and other localities. They usually nested on sandstone and clay banks along the rivers and numerous lakes of that region.

Mr. L. M. Turner found the Duck Hawk quite common near Fort Chimo, Labrador, in 1883 and 1884, and he took several sets of their eggs. He states: "Scarcely an island of large size but what has one or more pairs of these Hawks breeding on them. Those islands which are in the vicinity of the localities selected by Eider Ducks, Gulls, and other water birds, are the ones also chosen by these Hawks for breeding purposes, if a suitable ledge on a cliff can be found. * * *

"At the southern end of a cliff on Hunting Bay a pair of this species were building their nest; some 70 yards from it was the nest of the Black Rough-legged Hawk (*Archibuteo lagopus sancti-johannis*), while some 300 yards to the north was the nest of a Raven (*Corvus corax sinuatus*), and near it the beginning of the nest of the Labrador Falcon (*Falco rusticolus obsoletus*). Just beneath the latter were the deserted nests of the preceding year of the White-bellied Swallow (*Tachycineta bicolor*)."

Several eggs of this species from Greenland, where it is not uncommon, are also in the U. S. National Museum collection. I think the Duck Hawk remains paired during life, and that when a nesting site is once selected it is usually resorted to from year to year, even should the eggs be taken regularly each season. They appear very much attached to certain localities, and if persistent persecution during the nesting season should cause them to abandon the original site they nearly always select another on the same cliff, or if one cannot be found there, as near to it as possible. A site once selected is valiantly defended against interlopers, and as a rule no other pair of Hawks is allowed to nest in close proximity to it. Each pair seems also to have its proper hunting range during the season of reproduction, and any infringement on this is stoutly resisted. At other seasons they are usually seen alone, and they are at all times rather a silent bird, excepting when their homes are invaded; then they are noisy enough, and give vent to loud screams and cacklings.

Nidification in the southern and middle portions of its range begins in the latter half of March, in New York and the New England States usually about the beginning of April, in Colorado, Montana, and Oregon about the latter half of April, and correspondingly later in Labrador and the Arctic regions, where eggs have been taken throughout the month of June, and a set of four at Fort Chimo, Ungava, as late as July 6, 1884. But a single brood is raised in a season. Incubation lasts about four weeks and both parents assist in this duty. The eggs are usually four in number, occasionally but three, and very rarely five. These are deposited at intervals of two or more days, and generally laid on the bare rock or soil, if the eyrie is on a cliff or clay bank. The Duck Hawk can scarcely be said to build a nest, and if there is occasionally a semblance of one, it is but very slight. Sites in hollow trees are only used when cliffs or suitable bluffs are wanting, and old nests of other Raptores are still more rarely resorted to.

The eggs of the Duck Hawk vary considerably in shape as well as in coloration. The most common forms are short and rounded ovate, and occasionally a specimen is distinctly elliptical ovate.

The ground color when visible, which is not often the case, is pale creamy or yellowish white; in some specimens this is evenly overlaid with coloring matter, giving it a light chocolate colored appearance, in others it approaches a pale yellowish brown, and in one it is a decided fawn color. The eggs are irregularly blotched, streaked, smeared, and spotted with various shades of claret brown, vinaceous rufous, and brick red. In some the markings are well defined, in others confluent, giving them a clouded appearance; the majority are pretty evenly marked throughout, in a few they are more heavily spotted on one of the ends. Compared with the eggs of other Falcons they are decidedly darker colored, resembling the heavier marked and darker specimens of the eggs of Audubon's Caracara (*Polyborus cheriway*).

The average measurement of sixty-one specimens in the U. S. National Museum collection is 52.5 by 41 millimetres. The largest of these eggs measures 57 by 43, the smallest 48.5 by 38.5, and a runt egg but 38.5 by 30 millimetres.

Of the type specimens selected to show some of the styles of coloration, No. 6191 (Pl. 10, Fig. 5), a single egg, was collected by Mr. James Lockhart, near Fort Yukon, Alaska, in June, 1863; No. 10181 (Pl. 10, Fig. 6), by Mr. R. MacFarlane, on the Lockhart River, Arctic America, June 5, 1866, from a set of four; and No. 23890 (Pl. 10, Fig. 7), by Mr. Denis Gale, on Cache la Poudre Creek, Colorado, April 30, 1889, likewise from a set of four.

99. Falco peregrinus pealei Ridgway.

PEALE'S FALCON.

Falco communis var. *pealei* Ridgway, Bulletin Essex Institute, v, December, 1873, 201.
Falco peregrinus pealei Ridgway, Proceedings U. S. National Museum, iii, August 24, 1880, 192.

(B —, C 343*a*, R 414*a*, C 504, U 356*a*.)

GEOGRAPHICAL RANGE: Pacific coast region of North America, from Oregon north to the Aleutian and west to the Commander Islands.

This dark colored race of the Duck Hawk, Peale's Falcon, seems to be a fairly common resident of the Aleutian as well as the Commander Islands.

Mr. W. H. Dall, U. S. Coast Survey, in his "Notes on the Avifauna of the Aleutian Islands," especially those west of Unalaska Island, published in the Proceedings of the California Academy of Sciences, 1874, p. 3, speaks of this subspecies, under the name of *Falco gyrfalco* Linn., as follows: "A male was obtained in the harbor of Kyska, June 30, 1873, being one of several which had their nests on the brow of a precipitous and inaccessible cliff at the west end of the harbor, perhaps 150 feet above the water. The same species was observed flying over the low island of Amchitka a little later in the season. It does not appear to be common, but was the only Hawk observed in the islands west of Unalaska. It appeared to pass most of its time near the nest and raised a loud outcry when anyone approached the base of the cliff on the beach below. The fragments of several Ptarmigan, probably remnants from some of its meals, were noticed at the foot of the bluff below the nest."

Mr. L. M. Turner, in speaking of this Falcon, says: "Peale's Falcon was frequently observed on Amchitka Island in the month of June, 1881, and on several occasions on Attu Island during 1880 and 1881. It breeds on nearly all the islands of the chain and is a winter resident on the nearer group, at least. On Agattu it is reported to be very common, and on Amchitka I knew of three nests on the ledges of the high bluffs hanging over the sea. Any approach to the cliffs was heralded by the bird darting from the nest and circling high in the air, screaming fiercely all the while, and any attempt to shoot the birds while flying over the water would have resulted in the loss of the specimen, for they always flew in front of the cliffs out of range.

"At Attu Island I frequently saw one of these birds join the Ravens when the latter were performing their aërial gymnastics on the approach of a gale. The Hawk endeavored to imitate the Ravens, which paid but little attention to the antics of the intruder. At Attu this Hawk is not common, though the natives assert that it is common enough at Agattu and the Semichi Islands, and they told me that where this Hawk breeds there will also be found the nests of the Eiders. I could not believe this until a short stay at Amchia Island forced me to recognize it as a fact; for in each instance the Eiders were very abundant in every locality where the nest of this Hawk was known to be. It is quite probable that the Hawk selects the place with special reference to prospective young Eiders."[1]

Nothing is as yet known about the eggs of Peale's Falcon, but they are doubtless indistinguishable, both in size and color, from those of the preceding subspecies. It appears to nest in localities similar to those of the latter when breeding on the seacoast.

100. Falco columbarius Linnæus.

PIGEON HAWK.

Falco columbarius Linnæus, Systema Naturæ, ed. 10, i, 1758, 90.
(B 7, C 344, R 417, C 505, U 357.)

GEOGRAPHICAL RANGE: The whole of North America; south in winter to the West Indies, middle America, and northern South America.

Although the well known Pigeon Hawk is pretty generally distributed over the entire United States during the fall and winter seasons, by far the greater number breed north of our border, and comparatively few remain as summer residents, at least east of the Mississippi River, and those that do, generally confine themselves to the mountain districts and to the thinly settled and heavily wooded regions along our Northern States. In the Rocky Mountains, as well as in the Sierra Nevada and Cascade Ranges, and their spurs, the Pigeon Hawk is probably quite a common summer resident, but as yet its nest and eggs have been rarely taken, and even where they have been found, there remains more or less doubt as to their proper identification, as the two closely allied forms, Falco columbarius suckleyi and

---

[1] Contributions to the Natural History of Alaska, 1886, p. 160.

Falco richardsonii, occur in some of these mountains as well, and are very liable to be mistaken for the true Pigeon Hawk, even by fairly well posted ornithologists, and almost certainly by the average collector. Unless the parent is secured with the eggs the latter cannot be identified with any certainty. Its range northward is quite extended, reaching north of latitude 68° in the interior of British North America to Norton Sound and to the shores of Bering Sea, in Alaska.

In the eastern United States the Pigeon Hawk has been occasionally observed during the breeding season in Maine, Vermont, New Hampshire, and the northern parts of New York and Pennsylvania. It has likewise been noticed at this time in Michigan, Iowa, Wisconsin, and Minnesota, and doubtless breeds here also, as well as in the mountainous portions of some of the Southern States.

Mr. William Brewster records the finding of a nest and four eggs on Amherst Island, in the Gulf of St. Lawrence, on June 9, 1882, by Mr. Frazar, who while passing a spruce-clad knoll, surrounded by a boggy swamp, noticed a pair of Pigeon Hawks circling above the trees, and approaching quickly, discovered the nest built in a dense spruce at the intersection of a horizontal branch with the main stem, and at a height of about 10 feet. As he climbed the tree the Hawks, now thoroughly alarmed for the safety of their home, dashed wildly about his head, frequently passing within a few feet and uttering shrill screams of anger and dismay. After taking the eggs he made a close examination of the nest, which was found to be very bulky, in fact "as large as a Crow's," and composed chiefly of bark, with some coarse sticks surrounding the exterior, and a neat soft lining of finer bark and horse-hair. From its general appearance he felt convinced that it was constructed by the Hawks themselves. This was June 9; returning five days later he found both birds flying about the knoll, and their actions indicated that they had built another nest somewhere near, but it could not be found. As he was then on the point of leaving the island he shot the male, a fine adult specimen, which accompanies the eggs.[1]

Mr. Lynds Jones writes me that he found a nest of this species near Grinnell, Iowa, on April 28, containing four eggs. They were placed in a hole in an American linden tree about 8 feet from the ground. The nest was made of dry grasses, fibrous bark, and a few feathers. The birds hovered near when the nest was disturbed, but did not offer any resistance. Mr. J. W. Preston, of Baxter, Iowa, informs me that a pair of these birds remained one season near Iowa City under circumstances which led him to believe they were nesting. Mr. Charles D. Gibson states that the Pigeon Hawk is a resident of the Maryland and Delaware Peninsula, and that it breeds in both localities. I have tried to verify this record, but was unsuccessful.[2]

Mr. Audubon found three nests of this bird in Labrador, each containing five eggs. These nests were placed on the top branches of the low firs so

[1] Bulletin Nuttall Ornithological Club, Vol. VII, 1882, No. 4, p. 255.
[2] Ornithologist and Oölogist, Vol. VIII, September, 1883, No. 9, page 72.

common in that country, composed of sticks, and slightly lined with moss and a few feathers.[1]

Mr. L. M. Turner observed a few of these birds in the vicinity of Fort Chimo, Labrador, in 1882 and 1883, but failed to find their nests or eggs.

Mr. R. H. Taylor describes a nest of the Pigeon Hawk taken by himself on April 6, 1888, in Santa Clara County, California. He says: "I first saw the nest late in May one year ago, when it contained three young birds ready to fly. It was located on a steep mountain side on the stock ranch of Mr. J. P. Sargent, in the above named county, on a ledge of a precipitous bluff about 35 feet high. While near the nest last year the old bird was astonishingly fearless and would swoop down close to my head, uttering ear-piercing cries. These angry demonstrations, while they made me somewhat apprehensive of damaged eyes, afforded an excellent opportunity for jotting down a description of this Falcon.

"When I visited the nest this season the bird flew off as I came up and winged its peculiar flight across the cañon, when it settled quietly on a hillside. My friend Mr. R. C. Sargent, adjusted the end of a rawhide lariat to my body, and as soon as the rope had been placed around a small shrub which grew above, I swung over the ledge and was slowly lowered toward the nest, and as it was rather in from the face of the cliff I found it difficult to get a foothold, but, when I did at last, saw that it was composed simply of pieces of friable rock, and in it, to my delighted surprise, were five eggs, which contained large embryos."[2]

While I was stationed at Camp Harney, Oregon, a woodchopper working in the lower foothills of the Blue Mountains, about 5 miles from the post, found a nest, probably belonging to this species, on April 20, 1876. It contained five well incubated eggs, three of which he broke before I received them from him, nearly a week after they were first taken. I made him show me the nest at once in the hope that the parents might still be found in the vicinity, but a lengthy and careful search failed to discover them. The two eggs left are indistinguishable, both in size and markings, from fully identified specimens of this species; still they are just as likely to belong to Richardson's Merlin (*Falco richardsonii*), which is also found there and apparently equally abundant. The nest, evidently built by the birds themselves, was placed in a young spruce tree about 25 feet up, and close to the trunk. It was well concealed and the woodchopper's attention was drawn to it by the uneasiness of the birds while he was working in the vicinity. The nest appeared to me considerably smaller than a Crow's nest and was loosely constructed of small fir and juniper twigs, and slightly lined with dry juniper bark and a little moss.

With but few exceptions, nearly all the eggs of this species in the U. S. National Museum collection were obtained north of the United States. Mr. R. MacFarlane, who took several of their nests, says: "This Falcon ranges

[1] History of North American Birds, 1874, Vol. III, p. 152.
[2] Ornithologist and Oölogist, Vol. XIII, December, 1888, No. 12, p. 185.

along the Anderson River to near the Arctic coast of Liverpool Bay. Several of their nests had apparently been built on pine trees and others on ledges of shaly cliffs. The former were composed externally of a few dry willow twigs and internally of withered hay or grasses, etc.; and the latter had only a very few decayed leaves under the eggs. In one instance the oviduct of a female contained an egg almost ready for extrusion; it was colored like the others, but the pigment was still so soft that it adhered to the fingers on being touched. * * * Other specimens taken from the oviduct were perfectly white.

"I would also mention the following interesting circumstance: On May 25, 1864, a trusty Indian in my employ found a nest placed in the midst of a thick branch of a pine tree at a height of about 6 feet from the ground. It was rather loosely constructed of a few dry sticks and a small quantity of coarse hay; it then contained two eggs; both parents were seen, fired at, and missed. On the 31st he revisited the nest, which still held but two eggs, and again missed the birds. Several days later he made another visit thereto, and, to his surprise, the eggs and parents had disappeared. His first impression was that some other person had taken them; after looking carefully around he perceived both birds at a short distance, and this led him to institute a search which soon resulted in finding that the eggs must have been removed by the parent birds to the face of a muddy bank at least 40 yards distant from the original nest. A few decayed leaves had been placed under them, but nothing else in the way of lining. A third egg had been added in the interim. There can hardly be any doubt of the truth of the foregoing facts."[1]

Mr. J. Sibbiston took a nest and eggs at Fort Yukon, Alaska, in June, 1864; Messrs. Robert Kennicott and Alexander McKenzie, each one, at Fort Resolution, Great Slave Lake, on June 6, 1860, and Mr. A. Flett, one at Fort Peele, Arctic North America, in June, 1863. In all these cases the parent was taken and sent on with the eggs.

Dr. R. B. Hitz took two fresh eggs at Sun River, Montana, on July 5, 1867, but failed to secure the parents. The nest was in the hollow of a tree about 20 feet from the ground.

From the foregoing accounts it will be seen that the Pigeon Hawk breeds in open nests and in natural hollows of trees as well as on cliffs and in cavities and projections of river bluffs. Its flight is swift and powerful, resembling that of the larger Falcons. During the breeding season these birds are rather retiring and are seldom noticed unless their nesting site is very closely approached; but in the early fall and spring, as well as during the winter, excepting along our northern border, they are not at all uncommon, especially during their migrations. I found them moderately abundant in Oregon, Washington, and Idaho, and have no doubt that they regularly breed in the mountains there, but the majority go farther north.

R. MacFarlane's Manuscript Notes on the Land and Water Birds Nesting in British America.

Their food to a great extent consists of small birds, and no hesitation is shown in attacking one fully as large or even larger than themselves. In the summer grasshoppers and crickets, as well as other insects, furnish a considerable portion of their daily fare.

In the southern portions of their range nidification begins probably about the end of March or the first week in April, and correspondingly later farther north. In central Alaska and the Anderson River country it is protracted till the latter part of May or the beginning of June, and occasionally even later. Incubation probably lasts about three weeks, and but a single brood is raised in a season. The eggs are generally four or five in number, and these vary considerably in shape; some are short ovate, others rounded ovate, some nearer oval, and a few specimens elliptical ovate. The shell of the egg is close grained, and without luster. The ground color when visible is pale creamy white as a rule, and is hidden by a reddish brown suffusion of various degrees of intensity, and this, again, is finely marked or boldly blotched, with different shades of burnt umber, claret brown and vinaceous rufous. These markings are generally equally and profusely distributed over the entire egg, and are superficial; occasionally they are most distinct about one of the ends, being disposed in the shape of a wreath. Compared with the eggs of other Falcons, they resemble those of the Duck Hawk (*Falco peregrinus anatum*) closer than any others as far as coloration is concerned.

The average measurement of twenty-seven specimens in the U. S. National Museum collection, is 40.5 by 31.5 millimetres. The largest egg of the series, from Fort Yukon, Alaska, measures 44.5 by 33, the smallest 37.5 by 33 millimetres.

Of the type specimens, No. 6192 (Pl. 10, Fig. 8), from an incomplete set of two eggs, was taken June 6, 1860, near Fort Resolution, Great Slave Lake, by Mr. Alexander McKenzie, and No. 8808 (Pl. 10, Fig. 4), from a set of four eggs, was collected by Mr. R. McFarlane at Anderson River Fort, Arctic North America, in June, 1863.

101. Falco columbarius suckleyi RIDGWAY.

BLACK MERLIN.

Falco columbarius var. *suckleyi* RIDGWAY, Bulletin Essex Institute, v, December, 1873, 201.

(B —, C 344*a*, R 417*a*, C 506, U 357*a*.)

GEOGRAPHICAL RANGE: Northwest coast region from California to Sitka, Alaska; east to eastern Washington and Oregon.

Scarcely anything is known about the breeding habits of this the darkest colored and handsomest of our Merlins. Its range probably includes the mountain regions from northern California to Alaska, along the coast; and in the interior it is found in southern Oregon (Fort Klamath) and eastern Washington (Fort Walla Walla), in both of which localities I obtained specimens, but am not aware that its eggs have ever been taken. I am quite positive, however, that an occasional pair breeds in the vicinity of Fort Klamath. On May 9, 1883, while en route from this post to Linkville, Oregon, I observed a pair of these birds in the large open pine forest about midway between the two points. I had halted my party to let the horses graze, and, while resting, my attention was attracted to the male, by its incessant screaming in the trees overhead; this also brought the female around, and she was equally as noisy. It was clear that they had a nest in the vicinity, either in the tall pines or in a cliff about 400 yards distant, but a careful search instituted by the members of my party and myself failed to reveal it. Both birds were rather shy, but I finally succeeded in killing the male, a very handsome adult specimen. This is now in Mr. Manly Hardy's collection at Brewer, Maine, to whom I presented it. Where I observed Suckley's Merlin, it was much rarer than *Falco columbarius* at any time. At Fort Walla Walla, Washington, I obtained both subspecies as well as Richardson's Merlin; all of the specimens taken here, however, were shot during the migration, or in winter. It is probably common enough along the coast of Oregon, Washington, and British Columbia, and the mode of nidification as well as the eggs are not likely to differ from those of the preceding subspecies.

102. Falco richardsonii RIDGWAY.

RICHARDSON'S MERLIN.

Falco (Hypotriorchis) richardsonii RIDGWAY, Proceedings Academy Natural Sciences, Philadelphia, December, 1870, 145.
(B —, C 345, R 418, C 597, U 358.)

GEOGRAPHICAL RANGE : Interior and western plains of North America, from the Mississippi River to the Pacific coast; north to Saskatchewan, British America, south in winter to Texas and Arizona (and probably into Mexico).

Richardson's Merlin, a slightly larger and paler colored bird than the Pigeon Hawk and resembling very much the common European Merlin (*Falco regulus*), with which it was confounded by some of our earlier ornithologists, is another species about which little is known respecting its eggs, nests, and breeding range.

From an examination of the series of skins of the subgenus *Æsalon* in the U. S. National Museum collection, it plainly appears that the northern range of Richardson's Merlin is not nearly so extended as that of the common Pigeon Hawk. Among the numerous specimens received through the exertions of some of the gentlemen of the Hudson Bay Company from the Great Slave Lake, the Mackenzie and Anderson river region, there is not a single one referable to this species, and it is doubtful if it reaches a higher latitude than 55° in the interior, though on the Pacific coast it may possibly be found somewhat farther northward during the breeding season.

Dr. Elliott Coues, in speaking of this species, says: "Dr. Richardson's bird was an old female killed at Carlton House, on the Saskatchewan, May 14, 1827, while flying with her mate; in the oviduct there were several full-sized white eggs, clouded at one end with a few bronze colored spots."[1]

Carlton House is situated on the line of the Canadian Pacific Railroad, near latitude 53° N. Within our own limits, it seems to be most common along the timbered river bottoms of the Missouri and its tributaries and along the borders of the various streams having their sources in the Rocky Mountains from Colorado northward. It is known to breed near Fort Rice, Dakota, where young birds just out of the nest were obtained on July 10, 1867. Among the specimens in the U. S. National Museum collection are some from Colorado, Montana, Wyoming, the two Dakotas, the Indian Territory, and New Mexico; also from Washington, Oregon, and California, on the Pacific coast. I have taken several at Walla Walla, Washington, but do not believe that it breeds in the immediate vicinity of this place.

There are no absolutely identified eggs of Richardson's Merlin in the U. S. National Museum collection. Its mode of nidification and the eggs as well, are doubtless very similar to those of the common Pigeon Hawk, and indistinguishable from them, but may possibly average a trifle larger in size.

103. Falco regulus PALLAS.

MERLIN.

Falco regulus PALLAS, Reise, Russischen Reichs, II, Anhang, 1773, 707.
(B —, C —, R —, C —, U 358.1.)

GEOGRAPHICAL RANGE: Europe, Asia, and Africa; accidental on the coast of Greenland.

This little Falcon, one of the smallest found in Europe, breeds throughout the more northern portions of the Old World, and is entitled to a place in our avifauna, a specimen having been shot at Cape Farewell, Greenland, May 3, 1875, which is now in the collection of the Public Museum at Milwaukee, Wisconsin.

According to Mr. Henry Seebohm, "The Merlin breeds throughout north Europe, Iceland, and the Faroes, wintering in south Europe and north Africa, where, according to Loche, a few remain through the summer, retiring to the highest districts to breed. Eastward it breeds throughout northern Siberia, passing through Mongolia and Turkestan on migration, and wintering in south China, northwest India and Scinde."

Speaking of this species in Great Britain, he says: "The Merlin, in those districts frequented by it from North Derbyshire to the Shetlands, is a resident species, living on the moorlands and the mountains in summer, and retiring to

[1] Birds of the Northwest, 1874, p. 349.

more cultivated districts for the winter, in a manner similar to that of the Meadow Pipit. Even in the wild country of the Shetlands, the Western Isles, and the Highlands the Merlin is found throughout the year; in summer on the mountains and in winter lower down, in more sheltered districts on the seashore. * * *

"Although the Merlin arrives on the moorlands from its winter haunts late in March or early in April, it is a somewhat late breeder. The date of nidification is evidently chosen with relation to an abundant supply of food for the young. * * * The site selected for the nest varies in different localities; for in Lapland, both Wolley and Wheelwright mention instances of nests being found in trees, and Collett says that in south Norway it frequently takes possession of an old nest in a tree, like the Kestrel. On the Faroes it is said to breed on the cliffs. On our own moorlands a site is chosen on the ground, in the tall heather, or in some flat spot among the rocks on the steep slopes at the foot of the precipitous ridges so often met with in these localities. The site usually slopes down to a stream, and is one that commands a good view of the surrounding country. In most cases a small hole is made; whatever roots and dry grass may chance to be upon the spot are scratched into the rudiments of a nest, and the only materials actually selected by the bird appear to be a few slender twigs of 'ling' to form the outside of the structure, and which are generally broken from the heather overhanging the nest. When on the rocky slopes, it is usually made under a heather tuft, or beneath a mass of coarse herbage, and is then but a mere hollow in the scanty soil, as often without the ling twigs as with them.

"The eggs of the Merlin are usually five in number, sometimes only four; and are somewhat rounded in form. * * * Some specimens are deep reddish brown, so richly colored as to hide all trace of the ground color; others are pale red with most of the deep brown confined sometimes to the large and sometimes to the small end. Some specimens are pale cream in ground color, evenly and beautifully marbled with deep purplish red, or finely dusted over the entire surface with minute specks of blackish brown, and vary from 1.65 to 1.50 inches in length, and from 1.20 to 1.15 inches in breadth [equal to 41.9 to 38.1 in length, and 30.5 to 29.2 millimetres in breadth]."[1]

The eggs of the Merlin are scarcely distinguishable from those of the Pigeon Hawk. The average measurement of eight eggs in the U. S. National Museum collection is 39 by 31 millimetres. The largest egg measures 40 by 31.5, the smallest 38 by 31 millimetres.

[1] History of British Birds, 1883, Vol. i, pp. 34-39.

104. Falco fusco-cœrulescens Vieillot.

APLOMADO FALCON.

Falco fusco-cœrulescens Vieillot, Nouveau Dictionaire, xi, 1817, 90.
(B 9, C 317, R 419, C 541, U 359.)

Geographical range: Tropical America in general (except West Indies), north to southern Texas and the Territories of New Mexico and Arizona.

The breeding range of the handsome Aplomado Falcon, a common and widely distributed Central and South American species, as far as known at present, is along the southwestern border of the United States, through the valley of the Rio Grande from southern and southwestern Texas to southwestern New Mexico and southern Arizona. It has not as yet been met with in Lower California, and it is doubtful if it occurs there.

In its habits, compared with other Falcons, it is said to be rather spiritless and does not appear to possess the dash of the Pigeon Hawk or the Duck Hawk, between which it is intermediate in size.

Dr. James C. Merrill, U. S. Army, is the first naturalist who found it breeding within our borders. In his "Notes on Birds observed in the vicinity Fort Brown, Texas," he says: "During 1876 and 1877, I had occasionally seen a Hawk that I felt confident was of this species, but did not succeed in obtaining any specimens. On the 16th of June of the latter year, I found a nest placed in the top of a low Spanish bayonet, growing in Palo Alto prairie, about 7 miles from Fort Brown. After waiting a long time I wounded the female, but she sailed off over the prairie and went down among some tall grass, where she could not be found; the male did not come within gunshot, though he rose twice from the nest on my approach. The nest was a slightly depressed platform of twigs, with a little grass for lining. The eggs, three in number, were rotten, though containing well developed embryos. They measured 1.81 by 1.29, 1.77 by 1.33 and 1.88 by 1.33 inches, respectively [equal to about 46 by 32.8, 45 by 33.8, and 47.8 by 33.8 millimetres].

"On May 7, 1878, a second nest was found within 100 yards of the one just mentioned and the parent secured. The nest in situation and construction was precisely like the other, except that the yucca was higher, the top being about 12 feet from the ground. The eggs were three in number, all well advanced, but one with a dead embryo. They measured 1.78 by 1.34, 1.82 by 1.29, and 1.73 by 1.32 inches, respectively [equal to about 45.2 by 34, 46.3 by 32.7, and 44 by 33.5 millimetres]. The ground color is white, but so thickly dotted with reddish brown as to appear of that color; over these are somewhat heavier markings of deeper shades of brown."[1]

Among an exceedingly interesting collection of birds, nests, and eggs, made by Lieut. Harry C. Benson, Fourth Cavalry, U. S. Army, while stationed at Fort Huachuca, in southern Arizona, and which he generously presented to the U. S. National Museum collection at Washington, D. C., are three sets of eggs of this Falcon. He writes me that he found it exceedingly shy and difficult to approach, but fairly common in that vicinity. According to his observations it often alights on the ground when pursued, and prefers the open plains, covered here and there with low mesquite trees, yuccas, and cactuses, to the more mountainous regions. He does not consider it a resident throughout the year, but he observed it as late as January. Five nests were found by the lieutenant during the spring of 1887, all of them placed in low mesquite trees from 7 to 15 feet from the ground. These nests were apparently old ones of the White-necked Raven (*Corvus cryptoleucus*), and used without any repairs being made to them. A nest found on April 25, 1887, contained three young birds, which were taken by him and raised, becoming quite tame; one found on April 28 contained three fresh eggs; another found on May 5 likewise contained three eggs, two with large embryos, the third addled. A fourth and fifth nest, both found on May 14, contained each two fresh eggs, possibly a second laying of some of the birds previously despoiled.

Their food consists of small reptiles, mice and other rodents, grasshoppers and insects of various kinds, and occasionally a bird.

The usual number of eggs laid by this species seems to be three; but nothing is known about the length of incubation. Nidification commences, occasionally at least, by the latter part of March, continuing through April and the first half of May, and it is not likely that more than one brood is raised in a season. It seems to be only a summer resident in the United States.

The ground color of the eggs of the Aplomado Falcon is a dirty yellowish white, and this is thickly sprinkled with reddish and chestnut brown blotches and spots of various sizes, almost completely obscuring the ground color. In one of the sets of eggs these markings are very fine, nearly of the same size throughout, and of a delicate reddish buff color, giving them quite a different appearance from the others. Judging from the limited number of specimens in the collection, their variation in color is fully as great, if not greater, than that found in the eggs of any of our Falcons. In general appearance they approach the eggs of *Falco rusticolus gyrfalco* nearer than any others.

The average measurement of nine specimens in the U. S. National Museum collection is 45 by 35 millimetres. The largest specimen measures 46.5 by 35, the smallest 43 by 35 millimetres.

Of the type specimens, No. 23001 (Pl. 10, Fig. 9), from a set of three eggs, was taken May 5, 1887, and No. 23026 (Pl. 10, Fig. 10), also from a set of three taken on April 28, 1887. Both were collected by Lieut. Harry C. Benson, U. S. Army, near Fort Huachuca, Arizona.

105. Falco tinnunculus Linnæus.

KESTREL.

Falco tinnunculus Linnæus, Systema Naturæ, ed. 10, I, 1758, p. 90.
(B —, C —, R 422, C —, U 359,1.)

Geographical range: Europe, Asia, and Africa; accidental in eastern North America (Massachusetts).

The Kestrel, one of the most common birds of prey of the Old World, is admitted as a bird of our fauna, a specimen having been shot at Strawberry Hill, Nantasket Beach, Massachusetts, September 29, 1887, and which is now in the collection of Mr. Charles B. Cory, of Boston, Massachusetts.

Mr. Henry Seebohm says: "The Kestrel breeds in almost every part of the Palæarctic region and is common up to latitude 60° N. Farther north it rapidly becomes rarer, and north of the Arctic circle its appearance is only accidental, though there seems to be good reason to believe that Wolley once obtained a nest in Lapland as far north as 68°. North of the Alps it is principally a summer migrant, but in the countries south of the Baltic a few remain during the winter. South of the Alps it appears to be a resident.

"Mice form the chief part of the Kestrel's food; but occasionally small birds are taken, although very rarely and only when its usual fare is wanting; frogs, moles, caterpillars, lizards, and earthworms are also eaten. * * * The Kestrel appears to delay its nesting season until field mice and insects are plentiful. It generally breeds in the thickest woods and rarely in nests built in isolated trees. It also rears its young on the cliffs by the seaside, and some of the best places to seek for its eggs are the rocks on the moors and the cliffs of limestone districts. The Kestrel will also not infrequently lay her eggs in holes of buildings, notably among the ivied ruins and the Gothic architecture of cathedrals in company with Doves and Jackdaws. When the eggs are laid in the crevices of rocks a little cavity is, if possible, scratched in the soft earth or vegetable refuse, or, failing in this, some natural cavity in the rock itself is chosen in which to deposit the eggs. In the wooded districts a Crow's or Magpie's nest is the usual situation chosen by the Kestrel in which to rear its young, and sometimes the nest of a Ring Dove is used, and, more rarely still, that of a Sparrow Hawk. It is also worthy of remark that when a Magpie's nest is chosen the rooty lining is usually removed, probably from motives of cleanliness, and the eggs are laid on the hard lining of mud. As incubation advances the pellets containing the refuse of the bird's food accumulate and serve as a lining, beautifully soft, on which the eggs rest secure.

"Six eggs is the number usually found, although in some cases the number has been seven, and in others only four or five. They are rich reddish brown of various shades upon a dirty or creamy white ground. Most eggs of this bird when newly laid possess a purplish bloom which, however, soon fades after exposure to light. The eggs of the Kestral vary from 1.70 to 1.45 inches in length, and from 1.35 to 1.12 inches in breadth [equal to 43.1 to 36.8 in length and from 34.3 to 28.4 millimetres in breadth]. The female Kestrel when laying does not always deposit an egg each successive day, and sometimes sits upon the first egg as soon as laid. The female bird usually incubates the eggs, although the male is sometimes found upon them."[1]

The eggs of the Kestrel resemble those of our Pigeon Hawk very closely, although a trifle smaller. The average measurement of thirty-nine of these eggs in the U. S. National Museum collection is 38.5 by 32 millimetres. The largest egg of the series measures 41 by 32.5, the smallest 36.5 by 30 millimeters. None are figured.

106. Falco sparverius LINNÆUS.

AMERICAN SPARROW HAWK.

Falco sparverius LINNÆUS, Systema Naturæ, ed. 10, i, 1758, 90.
(B 13, C 346, 346a, R 420, 420a, C 508, 509, U 360.)

GEOGRAPHICAL RANGE: Whole of temperate North America, and south (in winter only?) through Central America to northern South America.

This handsome little Falcon, next to the Cuban Sparrow Hawk the smallest of our diurnal Raptores, is pretty generally distributed over nearly the entire North American continent, excepting the extreme Arctic portions thereof, breeding from Florida and the Gulf coast to the shores of Hudson Bay, and in the interior at least as far north as Fort Rae, Great Slave Lake, in latitude 62° N. Beyond this it does not appear to occur, otherwise that energetic naturalist and collector, Mr. R. MacFarlane, of the Hudson Bay Company, would certainly have met with and reported it. On the Pacific coast it is found from Cape St. Lucas, in Lower California, northward to Alaska, where it appears to be rare, however, at least in the interior of that Territory.

In the eastern United States it is not nearly so abundant as throughout the West, where I have found it a common summer resident almost everywhere, if suitable timber for nesting sites was available. Mr. L. M. Turner did not notice it in northern Labrador and the Ungava district, but it is known to be a summer resident of Newfoundland, Nova Scotia, and New Brunswick, and the more southern portions of the Dominion of Canada.

In winters from about latitude 38° S. and southward in the eastern United Stated, as well as in the Rocky Mountain regioun; on the Pacific coast from about latitude 41° S., though stragglers remain in sheltered and favorable localities at still higher latitudes throughout the country.

[1] History of British Birds, 1883, Vol. i, p. 45-50.

Like most of our Raptores the Sparrow Hawk, I believe, remains paired throughout life; at any rate they certainly appear to be already so on their return from their winter homes. They usually arrive on the old breeding grounds in the central portions of their range about the middle of March, some seasons not before the beginning of April, and at later dates farther to the northward. In Florida nidification begins about the middle of March, sometimes in the last half of February; in southern Arizona, southern Texas and southern California about the first week in April; in the Middle States from April 15 to May 10, and in the more northern States from May 1 to June 1; in the Rocky Mountain region and thence westward to Oregon, Idaho, and Washington rarely before May 15, and usually during the last part of this month and the first ten days in June, and in the more northern portions of its range during the first two weeks in June.

The most common nesting place of the Sparrow Hawk is in holes of trees, either natural cavities or the abandoned excavations of our larger Woodpeckers. In regions where such sites are not readily obtainable, it resorts to holes in sandstone cliffs and clay banks. Occasionally a pair will nest in some dark corner in a barn, and even dovecots have been known to be appropriated. Such an instance is mentioned in the "History of North American Birds, 1874" (Vol. III, p. 174). Several observers report their nesting in Magpies' nests in the West, and Mr. H. R. Taylor states: "Of twelve sets of eggs of the Sparrow Hawk taken this year, in San Benito County, California, by a friend and myself, all but two were found in Magpies' nests, and these were placed one in a hole in an oak and the other in a cavity in a bank on the San Benito River."[1]

While such nests may be resorted to in certain localities, it is by no means a constant habit with this species. I believe there are few places in the United States where the Black-billed Magpie (*Pica pica hudsonica*) is more abundant than in the vicinity of the Nez Percé Indian Reservation in Idaho and the Sparrow Hawk is also common there, yet I never found a pair occupying a Magpie's nest, although the ordinary nesting sites used by it are rare on account of the scarcity of large timber. According to some observers they are also said to occasionally occupy open nests, but such instances must be very rare. Mr. Lynds Jones, of Grinnell, Iowa, informs me that he has found them breeding in open nests, usually old Crows' nests, and that very little or no new material is used, the old lining being simply rearranged. I have had excellent opportunities to study the nesting habits of this species, and only in one single instance had I reason to suspect that an open nest was used, and this was placed in the extreme top of a tall cottonwood tree on Lapwai Creek, Idaho, and was inaccessible. I repeatedly saw one of these birds sitting on the edge of this nest, which appeared to be rather a frail structure, if a nest at all, but it is very probable that it was made use of more as a perch to rest on, and that there was a cavity somewhere in the upper part of the tree which I failed to notice.

---

[1] Ornithologist and Oölogist, Vol. XIII, 1888, p. 95.

There seems to be a great difference in the manner of lining their nests. Strictly speaking, the Sparrow Hawk ordinarily makes no nest, depositing its eggs on whatever rubbish may be found in the bottom of the cavity used. Occasionally the eggs are laid on a few leaves or grasses, scarcely deserving the name of a nest. In some localities, however, they are credited with greater energy in this respect than is usually the case. Mr. J. W. Preston, of Baxter, Iowa, writes me as follows: "The amount of dry grass and leaves that this species sometimes carries into a hollow for its nest is prodigious. In one case a pair selected a hole in the end of a decayed branch which they filled with the dry leaves of the post oak a foot in depth, and then enough grass on this to fill a patent bucket."

Near Camp Harney, Oregon, Sparrow Hawks are very abundant, and I examined a great many of their nests, which usually were very accessible, the majority being placed in natural hollows or the excavations made by *Colaptes cafer* or *Melanerpes torquatus* in junipers, from 5 to 15 feet from the ground. In nearly every instance the four or five eggs were laid on the few chips usually found in the bottom of these burrows or on the decayed wood and rubbish in the natural cavities which had accumulated therein.

Dr. James C. Merrill, U. S. Army, found it nesting on ledges and in holes of cliffs and cut banks in Montana, the birds breeding along the lower streams usually laying five eggs, while those found in the mountains generally laid but four. Even when persistently disturbed the Sparrow Hawk will return to the same nesting site from year to year. They are diligent layers, usually depositing a second set and occasionally a third should they lose the first.

Mr. C. J. Pennock states: "In the spring of 1872 three sets of five eggs each, evidently from the same pair of birds, were taken at intervals of ten days each, from a partly decaying chestnut tree in southeastern Pennsylvania. In the spring of 1873 the same pair of birds probably occupied the old nesting site again, and on April 24 a set of five eggs was taken; on May 6, another set of four, and on May 23 the nest contained four more eggs, two of which were taken. On May 29 another egg had been deposited, making fourteen eggs laid by the same bird. The last varied greatly from the first eggs laid, being much smaller; the greatest difference, however, is in their color, the last eggs (the smallest) being but slightly marked, and one was almost white."[1]

Incubation lasts about three weeks; the young when first hatched are covered with fine white down and their heads, as is the case with most young birds of prey, are nearly as large as the remaining part of the body. Both parents assist in incubation and are very solicitous in the care of their family. No other birds are allowed to come in the vicinity of their nest at such times without subjecting themselves to a vicious attack, and it makes no difference if the intruder has greatly the advantage in size, as they will attack a Swainson's or a Red-tailed Hawk as readily as any other bird.

---

[1] Bulletin Nuttall Ornithological Club, Vol. III, 1878, p. 41.

Usually but one brood is raised in a season. Sometimes, however, fresh eggs of this species are found so late in the season that it seems as if two broods might possibly now and then be hatched. Mr. Frank Robinette, of Washington, District of Columbia, found a set of five fresh eggs of this species as late as the first week in August, 1889.

The nesting sites vary greatly, as has already been mentioned. I have seen their nests less than 4 feet from the ground, and again in the dead tops of pine trees fully 80 feet and more up. Oaks, sycamores, cottonwood and buttonwood trees, pines and other conifers, large willows, chestnuts, and, in the interior, junipers furnish them favorite sites, and where I have principally observed them, they are not at all shy and usually allow themselves to be closely approached. In the West they are oftener found in the narrow strips of timber bordering the streams or in the scattered juniper groves found in the foothills than in the heavier forests.

Mr. H. W. Henshaw tells me of a peculiar incident regarding this species which came under his observation in the spring of 1884, while collecting in the vicinity of Colorado Springs, Colorado. He found a nest of this little Falcon in a low pine stump, not more than 4 feet from the ground. The female was on the nest, her tail partly sticking out of the hole. As the bird could not be dislodged, and he did not want to pull her out by the tail, he left her. Coming by the place again sometime later in the day, he found her absent and saw on examination that the nest contained several eggs which were just ready to hatch, some of the eggs being chipped, and the young about to emerge from the shells. Wishing to procure a couple of young birds just hatched, he did not disturb the nest any further that day, but to his surprise on visiting it the next morning, the burrow was empty and no indications were visible to prove that it had been despoiled by any predatory animal. No sign of empty or broken shells was to be seen in the vicinity, and he came to the conclusion that the parents themselves had made the change, and carried the eggs or young to some other suitable burrow, a number of which were available in the immediate vicinity.

While in search of food, these handsome little Falcons frequently arrest their swift flight instantly, hovering suspended over the spot where their prey is supposed to be found. Their food consists principally of small rodents, grasshoppers, and other insects, and larvæ of various kinds; lizards and small snakes are also eaten by them, and occasionally, when other provender is scarce, especially in winter, small birds have to suffer. Grasshoppers when attainable form the bulk of their fare, and it is amusing to watch them catch and dispose of the latter, handling them as expertly as a squirrel does a nut, and no sooner has one been caught and swallowed than they are after another. They seize them with their talons both while on the wing and on the ground. After gorging themselves, they return to some favorite perch on a dead limb of a tree standing on the edge of a prairie or meadow, or to the top or the crossbars of a telegraph pole and sometimes to the wire itself. In the West, where these little Hawks are abundant, every such pole in sight stretching across a prairie may sometimes be seen occupied by this or some larger species; they appear to be very attractive to all the Raptores, affording them an unobstructed view of the surroundings. Now and then this species is charged with molesting a young chicken, which may possibly be true in rare instances, but I am inclined to believe this to be in most cases the work of the little Sharp-shinned Hawk, with which it is often confounded by the average farmer. Their common call note is a shrill "kee hee, kee hee" repeated several times.

The number of eggs laid by this species seems to vary from three to seven; the latter number is rare, however, five and four being the number most commonly found. Personally I have examined some forty nests of this species, and in no case have I found over five eggs to a set. They are deposited at intervals of a day. Their shape varies greatly, the majority ranging from a rounded ovate to an oval, and a few may be called elliptical ovate. A very peculiar shaped set in the U. S. National Museum collection I would call blunt cuneiform.

The ground color of these eggs ranges from a pure clear white in a few instances to pale buff or cream color in the majority, and to a bright cinnamon rufous in a few others. They are spotted, blotched, marbled, and sprinkled with different shades of walnut brown, chestnut, cinnamon rufous, and ochraceous in various patterns; frequently these markings are confluent, predominating in some specimens on either end; in others they are heaviest in the center, forming a wreath. Mixed among the various tints a few eggs show handsome lavender colored shell markings. Scarcely any two sets are exactly alike. In some the markings are regular and minute, in others they are coarse and bold, and occasionally a specimen is entirely unmarked, being pure white throughout. I have found two such eggs among first sets.

The average measurements of a series of one hundred and sixty-nine eggs in the U. S. National Museum collection is 35 by 29 millimetres. The largest egg measures 39 by 32, the smallest 31 by 28 millimetres. Both extremes were taken by myself near Camp Harney, Oregon.

Of the type specimens, No. 17926 (Pl. 10, Fig. 11), from a set of five, was taken by Mr. W. A. Cooper, near Santa Cruz, California, May 9, 1875; No. 20638 (Pl. 10, Fig. 16), from a set of five; No. 20640 (Pl. 10, Fig. 14), from an incomplete set of three; No. 20643 (Pl. 10, Figs. 12 and 15), two, from a set of four, were all taken near Camp Harney, Oregon, on May 23 and 24, 1875, and June 1, 1875, respectively. No. 20660 (Pl. 10, Fig. 13), is from a set of five taken near Fort Walla Walla, Washington, May 22, 1881. All are from the Bendire collection.

107. Falco dominicensis GMELIN.

CUBAN SPARROW HAWK.

Falco dominicensis GMELIN, Systema Naturæ, I, 1788, 285.

(B —, C —, R 421, C 510, U 361.)

GEOGRAPHICAL RANGE: Islands of Cuba and Haiti ; accidental or casual in southern Florida.

The Cuban Sparrow Hawk claims a place in our avifauna on the strength of stragglers having been taken in Florida. It is a much darker colored bird than ours, the breast and sides being a deep rufous color throughout, and it is a common resident of the Islands of Cuba and Haiti.

Dr. Jean Gundlach says: "The Cuban Sparrow Hawk, locally known by the name of 'Cernícalo,' is a common bird in the Island of Cuba, especially abundant about the borders of plantations, and is occasionally met with in the forests as well. It would not be an easy matter to find a locality covered with a few palms or other trees without seeing the little Cernícalo. They are generally met with in pairs; are peaceably disposed, but will not tolerate any others of their kind on their chosen range, which is somewhat limited. One may observe them perched either on some limb of a tree, the gable or roof of a building, or on a palm leaf, on the lookout for prey, which consists principally of lizards, grasshoppers, and other insects, which are caught by rushing at them with arrow-like swiftness, scarcely a movement of the wings being visible. It is astonishing how keen their sight is, as they seem to observe quite small objects at considerable distances. Small birds are also caught and eaten, and I have seen them about sundown successfully chase bats. They may also be seen almost daily during the evening twilight engaged in catching the large sphinx moths.

"During the mating season they often call and feed each other, and play together in aërial evolutions, circling about high in the air, then rapidly descending and rising again. While hunting, one of these birds may often be noticed to suddenly arrest its flight and hover for a moment or so over a certain point, then suddenly dart down on its prey; or, if it has disappeared, resume its hunt. After eggs have been laid or the young hatched, no large bird is tolerated about the neighborhood; Turkey Vultures and Herons are always chased, being especially obnoxious to them.

"The eggs, from three to five in number, are laid during March or April, and usually deposited in the hollow of a tree, a hollow palm, or a cavity in a wall or cliff. No regular nest is made."[1]

The eggs of the Cuban Sparrow Hawk are exact counterparts of those of our own species, with the exception that they are somewhat smaller. An egg in the U. S. National Museum collection, taken by Professor Poey in

[1] Journal für Ornithologie, Vol. XIX, 1871, pp. 373, 374.

Cuba in the spring of 1859, measures 32.5 by 27 millimetres. Two others, collected by Dr. Jean Gundlach in the spring of 1865, measures 33 by 29 and 32 by 29 millimetres. None are figured.

108. Polyborus cheriway (JACQUIN).

AUDUBON'S CARACARA.

Falco cheriway JACQUIN, Beiträge zur Geschichte der Vögel, 1784, 17, Tab. 4.
Polyborus cheriway CABANIS, in Schomburgk, Guiana, III, 1848, 741.
(B 45, C 363, R 423, C 535, U 362.)

GEOGRAPHICAL RANGE: Southern border of the United States (Florida, Texas, Arizona), and Lower California; south to northern South America, Ecuador, and Guiana.

This handsome bird, better known throughout its range as the Cara-cara Eagle, is generally a constant resident wherever found; this at least is the case in southern Texas, and also in southern Arizona, where I saw them in midwinter as well as during the summer months. It breeds in these localities, and also in Florida and Lower California.

Capt. B. F. Goss, who had excellent opportunities to observe these birds at various points on the Gulf coast of southern Texas, writes me as follows: "I found this bird quite abundant in the timber along the Gulf coast. Their nests were generally found in open spots in the woods, where the trees were low and scattering, and only a very few resorted to the heavier forests to breed. The earliest date on which I found them nesting was March 4, the latest on April 21. The nests were generally placed in low trees from 5 to 27 feet up, usually from 8 to 12 feet from the ground. These were largely, some of them wholly, composed of broom weed, an annual shrub growing about 2 feet high. This plant remains standing through the winter and dies, and the twigs are easily broken off in the spring. This material was piled up in a slovenly way in a crotch or on a horizontal limb, until a rough nest was formed; some of these were quite deeply hollowed, others slightly so; they looked unshapely, and many were insecurely placed, as I found several tilted over, so that the eggs had rolled out.

"Brown Pelicans bred in great numbers on an island in the Laguna Madre, off the coast of Texas. When these birds were returning to their breeding ground, with pouches filled with fish, the Caracaras would attack them until they disgorged, and then alight and devour their stolen prey. These attacks were made from above, by suddenly darting down on the Peli-cans with shrill screams and striking at them with their talons. I am not certain as to whether they caught any of their prey before it reached the ground. I saw this maneuver repeated a number of times by a pair of these birds that nested on this island and by others that came from the shore. They did not attack outgoing birds, but invariably waited for the incoming ones, and as soon as these were over land (so that the contents of their pouches should not fall in the water) they pounced on them.

"They were not especially shy, but rarely came within shooting distance, and were generally silent on the breeding grounds, but sometimes as they left the nest they uttered a prolonged cackling note."

"Mr. Herbert Brown informs me: "The Caracaras are common about Tucson, Arizona, in the vicinity of slaughter houses. On a hot day during the summer they can be seen frequently standing on the ground in the shade of bushes near where they feed. Some years ago I bought three young ones from a Papago Indian, who took them from a nest in a sahuara cactus, about 16 miles southwest of town. I kept them until they were full grown; they were extremely vicious, and would make a hissing noise and strike out with their feet whenever approached, notwithstanding the kindness that was shown them. This was their favorite method of settling disputes among themselves. They could inflict a very ugly wound, and I was much more afraid of their feet than their beaks."

According to Mr. William Lloyd they are very rare in Concho and Menard Counties, Texas, nesting from the last week in April until May 20. The nests here are usually placed in oaks or pines at a height of 18 to 50 feet. He says: "Although carrion feeding birds, they are very fond of live fish and frogs. I have seen them fishing repeatedly in Sonora, Mexico. In Concho County I have seen them hunting prairie dogs, in couples, and once showing a high degree of intelligence. One was hidden behind a tussock of grass while the other danced before a young lamb, trying to lead it from the place where its mother was grazing to where its companion was hidden. The ruse was nearly successful, as the lamb began to follow, but the dam, anxiously watching, finally called it back.

The nest is large and compact, with a depth of 4 or 5 inches. In one case at least, in Concho, they used the same nest for two successive seasons."

Mr. Nehrling says: "It is a showy bird and its flight is extremely elegant and quick. Although very shy, and not easily approached, it often builds its nest in trees not far from farmhouses. The farmers say they are as harmless as Turkey Buzzards. The nest is usually from 25 to 30 feet above the ground, and is built of sticks, sometimes lined with bits of cotton and Spanish moss; the cavity is very shallow. Often the birds, commonly single individuals, are to be observed with Vultures feeding together on carrion."[1]

Dr. James C. Merrill, U. S. Army, referring to this species, says: "I have seen a Caracara chase a jack rabbit for some distance through open mesquite chaparral, and while they were in sight the bird kept within a few feet of the animal and constantly gained on it in spite of its sharp turns and bounds. If one bird has caught a snake or field mouse its com-

---

[1] Bulletin Nuttall Ornithological Club, Vol. vii, 1882, p. 173.

panions tnat may happen to see it at once pursue, and a chase follows,
very different from what is seen among true Vultures. The nests are bulky
platforms of small branches, with a slight depression lined with fine twigs,
roots, and grasses, or sometimes altogether without lining; they are placed
in trees or on the tops of bushes at no great height from the ground. Both
sexes incubate. I have not found more than two eggs in one nest, and
these are laid at intervals of three or four days."[1]

In southern Florida the nests are usually placed on the tops of the
cabbage palmettos, nidification beginning the first week in April. In South
America, according to Mr. Darwin's statement, it nests occasionally in cliffs.
With us it does not seem to do this, their nests being usually found in trees
of various kinds, palmettos and sahuaras (*Cereus giganteus*). A nest brought
to Mr. Brown's notice in Arizona was placed in a large Palo Verde tree,
which contained three nearly fresh eggs on May 1, 1889.

I noticed these birds frequently about the outskirts of my camp on Rillitto Creek, near Tucson, Arizona, during the eighteen months I was stationed
there, from October, 1871, to March, 1873. They were generally seen in
pairs, foraging for such kitchen refuse as they could find in the vicinity of
the camp. Although I never allowed them to be molested in any way, they
were at all times exceedingly shy and difficult to approach closely, scarcely
ever coming within range of a shotgun. A great part of their time seemed
to be spent on the ground, walking around in search of food, and I believe
that a good deal of their hunting is done in this way. Their food, besides
rabbits and small rodents, consists largely of lizards, beetles, grasshoppers, and
snakes. I saw one of these birds engaged in quite an encounter with a good
sized snake which had partly coiled itself about its neck, both bird and snake
struggling for a few minutes at quite a lively rate. The Caracara had the
best of the fight, however, and before I could get to the place, the bird was
off with its quarry, the snake still squirming and twisting about in its talons.
I was disappointed in not being able to learn the species to which it belonged. On but a single occasion did I see more than a pair together; this
was on June 20, 1872, when a party of four were seen feeding a short distance below my camp.

Nidification begins in southern Texas sometimes as early as the middle
of February, but usually about the first week in March; in other localities
generally not until the beginning of April, and in Arizona about the latter
part of the month. But one brood is raised in a season; incubation, as with
the majority of the Raptores, lasting probably about four weeks. Both sexes
incubate, and the eggs are deposited at intervals of several days. They are
usually two or three in number, the smaller sets being somewhat more frequent. The shell is comparatively smooth and not as thick as is usual among
the larger Raptores. The eggs are rounded ovate in shape; the ground
color when visible, which is not often the case, is creamy white, and in the

[1] Proceedings U. S. National Museum, Vol. 1, 1878, p. 153.

majority of specimens is entirely hidden, the egg appearing to be of a uniform rufous cinnamon of different shades, some of the darker approaching vinaceous rufous. This is again overlaid with irregular blotches and spots of dark chocolate, claret, brown, and burnt umber. Most of these eggs are heavily marked, a few, however, only slightly, and in these the markings are usually small and more regular in outline; a few are unspotted, and although the ground color is not visible it is entirely overlaid with an even colored cinnamon tint. Others look clouded, as if smeared with the coloring matter, and a single specimen from Cape St. Lucas, Lower California, is a uniform creamy white, and spotted throughout with fine dots of reddish chocolate not much larger than pin points.

The average measurement of thirty-three specimens in the U. S. National Museum collection is 60 by 47 millimetres. The largest egg (abnormally large), from Comal County, Texas, measures 75.5 by 54.5, the next largest 63 by 48 millimetres. The smallest measures 55 by 44.5 millimetres, and comes from Matamoras, Mexico.

Of the type specimens selected to show the more common styles of markings, No. 21459, two eggs from the same nest (Pl. 11, Figs. 1 and 2), were collected in Comal County, Texas, March 7, 1876, and are from the Bendire collection; No. 22588 (Pl. 11, Fig. 3), a single egg taken near Corpus Christi, Texas, March 4, 1882; and No. 22592 (Pl. 11, Fig. 4), from a set of three taken at the same place February 15, 1884, were obtained in exchange from Capt. B. F. Goss, Pewaukee, Wisconsin.

109. Polyborus lutosus RIDGWAY.

GUADALUPE CARACARA.

Polyborus lutosus RIDGWAY, Bulletin U. S. Geographical and Geological Survey of the Territories, No. 6, 2d ser., February 8, 1876, 459.
(B —, C —, R 424, C —, U 363.)

GEOGRAPHICAL RANGE: Guadalupe Island, Lower California.

The Guadalupe Caracara, a much paler and browner colored species than the preceding, was and possibly still is a resident of the above mentioned island, which is situated some 220 miles south by southwest of San Diego, California, and is described as being about 15 miles in length and 5 miles in width, and has until recently been occupied as a goat raising station.

Dr. Edward Palmer, one of our Western pioneer naturalists, was the first ornithologist to visit this island in 1875. This visit resulted in important discoveries, not less than eight new species of land birds being added to our avifauna through his explorations there, and among them the one now under consideration. According to his observations the "Quelelis," as these birds were called by the inhabitants, were abundant on every part of the island, and no bird could be a more persistent or more cruel enemy of the poultry and domestic animals. He says: "It is continually on the watch, and, in spite of every precaution, often snatches its prey from the very doors of the houses. The destruction of the wild goats is not so great, as these animals are better able to protect themselves than the tame ones. No sooner is one kid born, and while the mother is yet in labor with the second, than the birds pounce upon it, and should the old one be able to interfere, she is assaulted also. No kid is safe from their attacks, and should a number be together, the birds unite their forces, and with great noise and flapping of wings they generally manage to separate the weakest one and dispatch it. * * * These birds are cruel in the extreme and the torture which is sometimes inflicted upon these defenseless animals is painful to witness. * * *

"Hundreds of these birds have been destroyed by the inhabitants, both with poison and firearms, without any noticeable diminution of their numbers. They are said to lay three eggs, speckled like those of a Gull. When surprised or wounded they utter a loud harsh scream, something like that of the Bald Eagle. In fighting among themselves they make a curious gabbling noise, and under any special excitement the same sounds are given forth, with an odd motion of the head, the neck being first stretched out to its full length, and then bent backwards until the head almost rests upon the back. The same odd motions are made and similar noises uttered when the birds are about to make an attack upon a kid. Besides the principal sources of food supply already indicated, the birds have other means of subsistence; they eat small birds, mice, shellfish, worms, and insects. To procure the latter they resort to plowed fields, where they scratch the ground almost like domestic fowls."[1]

Ten years later, in January, 1885, when Mr. Walter E. Bryant visited the island the number of these birds had very materially decreased. At a still later date, in his "Catalogue of the Birds of Lower California," published in the Proceedings of the California Academy of Sciences, second series, Vol. II, 1889, p. 282, Mr. Bryant states: "So effective has been the work of extermination carried on against this bird that Dr. Edward Palmer, who first discovered them in 1875, says that he visited the island this year (1889) and did not see a single individual. He tells me that when he landed fourteen years ago the 'Quelelis,' as they are known there, were so numerous and bold that men were obliged to stand over the Angora goats with sticks to protect them from attack, particularly the kids which were not defended by their mothers. The short-haired kind will drive off the birds, so Dr. Palmer says from his observations; and now that man has abandoned the island, I cherish the hope that a pair at least may still be living, and that some future explorer may succeed in finding the unknown eggs and give us an account of the nesting habits of this peculiar species."

Their nesting habits and eggs probably differed but little from those of the allied species, Audubon's Caracara.

[1] Hayden's Survey, 1876, Bulletin No. 2, pp. 192-195.

110. Pandion haliaëtus carolinensis (Gmelin).

AMERICAN OSPREY.

Falco carolinensis Gmelin, Systema Naturæ, 1, i, 1788, 263.
Pandion haliaëtus var. *carolinensis* Ridgway, Proceedings Academy Natural Sciences Philadelphia, December, 1870, 143.
    (B 44, C 360, R 425, C 530, U 364.)

GEOGRAPHICAL RANGE: Temperate and tropical America in general; north to Hudson Bay and Alaska.

The American Osprey, commonly called the Fish Hawk, breeds in suitable localities throughout the entire United States, and beyond our borders as far north as Labrador, the shores of Hudson Bay, and in the interior of British North America, where it has been found on the Mackenzie River, near the Great Slave Lake, by Mr. B. R. Ross, in about latitude 62°; but inasmuch as that careful observer, Mr. R. MacFarlane, failed to notice it in the Anderson River and Barren Ground regions, it is questionable if it occurs farther north there. In Alaska it is well known to attain a considerably higher latitude, its eggs having been secured at Fort Yukon by Mr. J. Lockhart, as well as by Messrs. S. Jones and J. Sibbiston on other points of the Yukon River, in about latitude 67°.

In these northern regions, as well as throughout the greater part of the United States, it is only a summer resident, arriving along the shores of the Chesapeake Bay about the middle of March and correspondingly later northward. In Florida and the Gulf States it is a constant resident. In many localities the Osprey may be said to breed in colonies, numbers of them nesting in close proximity to each other.

Mr. W. W. Worthington, well known as a close and accurate observer, writes me as follows: "Almost invariably on the 20th day of March the Osprey arrives at Long Island and is reckoned as the first harbinger of the breaking up of winter and of settled spring weather. At first a solitary individual will be seen circling slowly over some creek, in eager search of his first finny meal in his summer home. In a few days they become abundant and at once set to work to repair any damage their nests may have received from the previous winter blasts, for they occupy the same nests year after year.

"At Plum Island, where there are not enough suitable trees to go around, many pairs nest on the ground, on the tops of sand dunes, in such cases depositing the eggs on the sand, the nest consisting of a few sticks, bunches of seaweed, and pieces of various kinds of rubbish arranged in a circle. In other cases the nests are built up several feet, the height in all probability being regulated by the number of years the nest has been occupied and the amount added to it from year to year. In all other localities where I have observed these birds breeding, they nest in trees, both in deep woods and exposed situations, excepting in a few instances, where an unusual nesting place was chosen, either on the cross bar of a telegraph pole, on a large rock in Gardiner's Bay, or on an unused chimney of an occupied dwelling-house.

"The materials used in nest building consist principally of large sticks, small dead branches, and dry seaweed. I have found a sheep's skull, old shoes, the dried-up remains of a duck, a large stone, and other odd things in their nests. Three eggs are usually laid during the first week in May, and are hatched in about three weeks. The young are at first covered with whitish down and are fed on the changeless diet of their parents—fish, which are torn up and given to them in suitable sized pieces. They grow rapidly and soon feather out similar to the old birds. The Ospreys are very solicitous for the safety of their eggs and young, and with loud screams they will dart within a few feet of an intruder; one instance has come to my notice of their actually attacking the collector, a young friend of mine who was ascending to the nest of a pair of these birds, when one of them struck him on the back and nearly knocked him from the tree.

"It is a common habit of the Purple Grackles to nest in crevices of the Osprey's nests, and I have examined as many as half a dozen of their nests in one belonging to an Osprey. I have also observed the European House Sparrows taking advantage of these nesting sites. The Ospreys remain with us just about half the year and depart for the south about the 20th of September."

Mr. Moses B. Griffing, who has kindly collected a fine series of these eggs for the U. S. National Museum, several of which are figured in this work, informs me: "The earliest date on which I took a set of eggs of the Osprey on Shelter Island, was April 24, 1879, the latest June 7, 1882. The first set contained three fresh eggs, the last set, two, slightly incubated. They raise but one brood in a season, but will lay a second set, usually of two eggs, if the first one is taken. They nest generally in trees at a height of from 8 to 60 feet. I have seen their nests in the tops of cedars, the various species of oaks, hickory, poplar, buttonwoods, tupelos, wild cherry, black walnut, and pear trees. From one nest on Gardiner's Island, New York, I took sets of four eggs in two consecutive seasons; and in other nests I have seen four eggs one season and three the next. In the early spring they frequent the salt creeks to fish; later, mostly the bays and deeper water, where they catch menhaden or moss bunkers. On Gardiner's and Plum Islands this species may safely be said to breed in colonies, while on Shelter Island, and other localities near it, the nests are scattered, sometimes on isolated trees and again in woods and swamps. In the latter case a tree larger than the surrounding ones is chosen. I have seen nests in such small, low trees that they could be reached while standing on the ground."

Mr. Charles S. Shick says: "The Fish Hawk is one of the most familiar of the Raptores of southern New Jersey. Cape May County is noted for its many wooded islands lying between the mainland and ocean, and they afford these birds a congenial home. They are very abundant on Seven-Mile Beach, and several hundred pairs have nested on this island every season. It is interesting to watch the Fish Hawk obtaining its food. Sailing along from 50 to 100 feet above the water, with its keen eye it can easily see any fish swimming close to the surface of the water, and as soon as it sees its quarry, stops its flight, remains suspended motionless in the air for a moment, closes its wings, and then darts downward like an arrow. It disappears under the water for a few seconds, and when it rises and again takes wing a shining, wriggling fish can be plainly seen in the grasp of its powerful talons. It is a curious fact that this bird will never carry the fish with the tail to the front. Many times have I seen them turn the fish around in mid air. The nest from which I sent you the handsomely marked set of eggs, taken on May 12, 1890, was placed in the top of a dead cedar about 50 feet from the ground. It was composed of large sticks, dead branches of trees, pieces of driftwood, and oyster grass, neatly lined with fine sedge grass, cow dung, mud, and cedar bark; it had been occupied for a number of years. About 75 feet away from the nest was a platform on another old stump of a tree, which at one time had also served as a nest. This the male evidently used as a feeding perch, as beneath it a quantity of scales, bones, and skeletons of fish were scattered about, mainly those of the common menhaden or moss bunker. Quite a number of the nests here are lined with fish bones and cow dung."

Judge J. N. Clark writes me: "One curious fact in reference to the Osprey I noticed here on two occasions, was the building of nests late in the summer, either for next year's occupancy or for resting in during the season, long after breeding time was over. One such nest was used the next spring; in the other case the bird was shot. Another strange thing to me is that experience seems to teach nothing to these birds. I live about 2 miles from the seashore, and one of the matters of daily observation is to see an Osprey wearily bringing a heavy fish from the sea and passing on toward the woods where invariably one or more Bald Eagles are waiting to seize the prey it brings. A few futile efforts to escape, a few notes of remonstrance, and it surrenders to superior prowess, and again returns to the fishing grounds, only to repeat the same weary round over and over again. It often has the appearance of being purposely done for the accommodation of 'His Majesty the King of Birds.'

Mr. A. W. Anthony writes me that the Fish Hawk nests on the ground as well as on cliffs along the coast of Lower California. He says: "Near San Geronimo, Lower California, I found about a dozen of their nests on April 20, 1887; all of which were built on the ground or on small ledges of rock—none were over 4 feet in height—consisting of large piles of kelp and sea grass, etc. Some of these nests contained eggs from fresh to heavily incubated, and others young birds several days old.

"On St. Martin Island I have found young flying about by April 12, 1888, while other nests had young just hatched or fresh eggs. Nests on this island were placed on the ground, excepting a few cases where the birds had taken advantage of low bushes, raising them 2 feet or so. On Cerros Island, Fish

Hawks' nests were very common; but here the cliffs and rocky ledges had been taken advantage of, many of which were inaccessible, and all more or less difficult to reach."

The Osprey generally nests in the tops of trees of various kinds. In the West I have found their nests mostly in very tall pines, the tops of which had been broken off during a storm. Such trees were always selected in preference to any others, even if they were some distance from a body of water. The nest was invariably placed on the very top of the broken stump. Occasionally I have seen one in a cottonwood. On the Little Red River in central Arkansas, Mr. Gault informs me that the Fish Hawk nests in holes in the sandstone bluffs along that stream, and I have seen them using similar locations on the Upper Columbia River in Washington, where no timber is found in close proximity to the river.

The most picturesque nesting site of the Osprey I ever saw was located in the midst of the American Falls of Snake River, Idaho. Right on the very brink of these, and about one-third of the way across, the seething volume of water, confined here between frowning walls of basalt, was cleft in twain by a rocky obstruction which had so far withstood the ever eroding currents, and this was capped with a slender and fairly tapering column of rock rising directly out of the swirling and foaming whirlpool below. On the top of this natural monument, whose apex appeared to me to be scarcely 2 feet wide, a pair of Ospreys had placed their nest and were rearing their young amidst the never ceasing roar of the falls directly below them. The nests are often rather small considering the size of the bird, and usually not over 18 inches in outer diameter. Nests which have been in use for a number of years, however, are often quite bulky and very firmly built, in order to withstand the strong gales to which they are frequently exposed.

Mr. Manly Hardy found a nest of this species on a high bluff placed between three stumps; the nest was a large one and constructed principally out of shreds of cedar bark; it contained two fully grown young. Another nest found by him was placed on a bare ledge of rock just above high tide. This was constructed entirely of kelp stalks and contained one young bird just able to fly.

In central Florida nidification is said to begin, some seasons at least, in January, and probably earlier still in the southern portions of the State, and it continues into March and the beginning of April, while in the more northern parts, like St. Johns and Putnam Counties, according to Dr. William L. Ralph's observations, the Osprey rarely nests before March 1, and usually in the latter part of this month. Mr. C. J. Pennock found several nests of the Osprey near St. Marks, Florida, containing fresh eggs in the first week of April, 1889. These nests were all placed in high trees, pines or cypress, and all but one in living trees. In Lower California the nesting season begins in February and lasts until the beginning of May. In northern Idaho I usually found full sets of fresh eggs the first week in May, the same holding good in Oregon and Washington. On the middle Atlantic coast they also commence laying about the same date; in Maine about the latter part of May, and in Nova Scotia, as well as in the interior of Alaska, not until June, nearly fresh eggs having been taken by Mr. J. Lockhart, at Fort Yukon, on June 27.

The food of the Osprey consists entirely of fish, which are caught as already described, and these are usually the inferior species, such as are seldom used for the table. In Florida they live almost entirely on catfish, and on the interior lakes and streams of the Pacific coast subsist to a great extent on suckers, and frequently on some of the smaller species of salmon and whitefish (*Coregonus*).

But a single brood is raised in a season. Incubation is said to last about twenty-one days, but I am inclined to believe that it is nearer twenty-eight. The usual number of eggs is three, occasionally only two, and seldom four; they are among the handsomest of those laid by the Raptores, and subject to an endless variation in color, markings, and size. They are deposited at intervals of one or two days and the shell is strong and minutely granulated.

The eggs of the Osprey vary greatly in shape, ranging from an ovate to either a short, rounded, elliptical, or elongate ovate. The ground color is usually a creamy white, and this is sometimes so evenly and regularly overlaid with pigment as to give it a buffy or vinaceous appearance. Now and then a specimen is found showing a uniform cinnamon rufous or ferruginous color throughout, without any indications of blotches, thus strongly resembling certain types of eggs of the Falcons. The markings show an equally wide range of variation, both in amount and size. The majority of eggs are heavily blotched and spotted, but generally more thickly about the larger end, and these markings include nearly all the different shades of brown and vinaceous red. In some eggs lavender and pearl gray shell markings predominate, but in the majority of specimens before me these are either few or entirely absent; the beautiful vinaceous red tints found in some of the eggs of this species when fresh become darker with age.

The average measurement of sixty-nine specimens in the U. S. National Museum collection is 62 by 46 millimetres, the largest egg, from Cape St. Lucas, Lower California, measuring 68.5 by 49.5, the smallest, taken on Seven-Mile Beach, Cape May County, New Jersey, 59.5 by 42 millimetres.

Of the type specimens selected to show some of the many styles of coloration and markings, No. 20709 (Pl. 10, Fig. 17), from a set of three eggs, Bendire collection, was taken by Mr. A. R. Justice, near Cape May, New Jersey, June 1, 1875. No. 23965 (Pl. 11, Fig. 5), from a set of four eggs; No. 23967 (Pl. 11, Fig. 6), from a set of two, and No. 23969 (Pl. 11, Figs. 7 and 8), two specimens from a set of two, were collected on Shelter Island, New York, May 7, 1890, by Mr. Moses B. Griffing (and especially selected from a large number for the purpose of illustration); No. 23974 (Pl. 11, Fig. 9), from a set of two, was taken by Mr. Charles S. Shick, on Seven Mile Beach, Cape May, New Jersey, May 21, 1890. The other egg of this set is of nearly a uniform reddish brown color throughout, and resembles one of the types of *Polyborus cheriway* (Fig. 3, Pl. 11), very much, but cannot be figured from want of space. A normal colored set of three eggs had been previously taken from this nest.

Family STRIGIDÆ.   BARN OWLS.

111. Strix pratincola BONAPARTE.

AMERICAN BARN OWL.

Strix pratincola BONAPARTE, Geographical and Comparative List, 1838, 7.
(B 47, C 316, R 394, C 461, U 365.)

GEOGRAPHICAL RANGE: United States generally (rarer northward) and Mexico.

The northern limit of the breeding range of the Barn Owl extends from about latitude 40° 30' (Flushing, Long Island, New York) westward through the Middle States, but going southward these birds become more and more abundant, and north of latitude 41° it can only be considered as a rare straggler, though it is probable that a pair may breed now and then in favorable localities at a somewhat higher latitude. It has been met with near Hamilton, southern Ontario, Canada, at Sault St. Marie, Michigan, in Wisconsin, and Minnesota. In the New England States it has been taken in Connecticut and Massachusetts. It is not uncommon in Kansas and portions of southern Nebraska. On the Pacific coast it breeds from California southward, and according to Dr. Cooper its range extends through Oregon to the mouth of the Columbia River in latitude 46°. I have never met with it in southeastern Oregon, southern Idaho, and northern Nevada, and if it occurs at all in these regions it must be rare.

The Barn Owl is one of the most useful and harmless birds of prey, subsisting almost entirely on noxious vermin, such as ground squirrels, rats, pocket gophers, mice, and on shrews, bats, frogs, small reptiles, grasshoppers, and beetles. Very rarely small birds are caught by them, and occasionally a young rabbit varies the usual bill of fare. Looked at from an economic standpoint it would be difficult to point out a more useful bird than this Owl, and it deserves the fullest protection, but, as is too often the case, man, who should be its best friend, is generally the worst enemy it has to contend with, and is ruthlessly destroyed by him partly on account of its odd appearance and finely colored plumage, but oftener from the erroneous belief that it destroys the farmer's poultry.

It hunts during the evening and throughout the night, when its rather peculiar screaming may be frequently heard. During the day it remains hidden either in natural hollows in trees, cavities in the perpendicular bank of some ravine or cliff, burrows in the ground, abandoned buildings, old mining shafts, church steeples, barns, or similar retreats. In fact it does not object to abide near human habitations and frequently nests in the very center of cities of considerable size. Its flight, although accompanied by considerable flapping of the wings, is entirely noiseless, and the capture of its humble prey is thus greatly facilitated. The number of rats, mice, and other noxious vermin required by a pair of these Owls to feed their family, usually consisting of from five to seven young, is almost incredible, and I am certain exceeds the captures of a dozen cats for the same period. The young owlets are always hungry and will eat their own weight in food daily and even more if they can get it.

In the southern portions of the United States the Barn Owl is resident throughout the year, and at times somewhat gregarious during the winter. Mr. B. W. Evermann states that he saw a flock of more than fifty among the oaks in the Cañada de Largo, a few miles from San Buenaventura, California, and I believe it is more abundant in southern California than in any other portion of the United States. I met with it several times in the neighborhood of Tucson, Arizona, where they were rather rare, but they seem to be pretty generally distributed over this Territory, where they usually live in abandoned mining shafts and prospect holes. Mr. Herbert Brown writes me that he met with five of these birds in an abandoned mine at a depth of 50 feet I saw one actively engaged in hunting along the banks of Rillitto Creek during a cloudy day in December, 1872, and in April of the same year saw another on quite a bright sunny day being chased by either a pair of common Crows or White-necked Ravens. In this vicinity I believe they nest mostly in deserted burrows of badgers, at any rate more than once I saw them sitting in the mouth of such burrows.

Their nesting sites are quite variable and include all sorts of places, such as natural hollows in trees, holes and cavities in clay banks or cliffs, burrows under ground enlarged to suit their needs, in the sides of old wells, abandoned mining shafts, dovecots, barns, church steeples, etc., and sometimes, though rarely, in perfectly exposed and unprotected situations, such as the flat roof of an occupied dwelling-house in the midst of a village. Mr. W. O. Emerson, of Haywards, California, writes me: "A pair of Barn Owls nested the past season (1889) on the bare tin roof running around a cupola of a neighbor's house, which was surrounded by a low railing. Not less than twenty-four eggs were laid and none of them were taken away at any time. There was no nesting material on which the eggs were placed, not even a single twig, and they naturally rolled around on the roof, as it was impossible for the bird to cover them all. When taken down finally and examined, it was found they were all rotten, caused, no doubt, by the intense heat from the sun's reflection on the tin roof."

Where holes in clay banks along rivers and the sides of ravines are used, or the deserted burrows of ground squirrels or larger rodents, they are enlarged to suit their needs, and the birds live in them the year round, carrying most of their food to these places to be devoured at leisure.

In southern California nidification begins occasionally as early as January, and while usually but a single brood is raised by these birds in a season, now and then they will rear two. Mr. F. Stephens informed me that a pair hatched a brood of six young in January, 1885, at St. Isabel, California, and on March 25 the bird was sitting on a second set of ten eggs, using the same site, a dovecot in the barn. In southern Texas they begin laying about the latter part of February or the beginning of March, and correspondingly later northward.

At Washington City, District of Columbia, where the Barn Owl is by no means rare, they begin nesting from the last week in April to about the 10th of May, and I know of at least three broods having been raised within the city limits during the season of 1890. A pair of these Owls have been nesting off and on for years in one of the towers of the Smithsonian Institution building, and occupied this site again during the spring of 1890, rearing a family of seven young. As the supply of these birds in the zoölogical collection now forming at the national capital consisted of but a single specimen, which had been kept in confinement there for some months, the young above mentioned were taken from the nest before they were quite ready to fly and placed in the cage with the one already there; she at once adopted the orphans, and cared for and fed them as diligently as if they had been her own.

The Barn Owl, strictly speaking, makes no nest. If occupying a natural cavity of a tree the eggs are placed on the rubbish that may have accumulated at the bottom; if in a bank, they are laid on the bare ground and among the pellets of fur and small bones ejected by the parents. Frequently quite a lot of such material is found in their burrows, the eggs lying on and among this refuse. Incubation usually commences with the first egg laid, and lasts about three weeks. The eggs are almost invariably found in different stages of development and young may be found in the same nest with fresh eggs. Both sexes assist in incubation, and the pair may be sometimes found sitting side by side, each with a portion of the eggs under them.

Besides the peevish scream already mentioned, they utter at times a feeble querulous note like "quaick-quaick," or "äick-äick," sounding somewhat like the call of the Night Hawk (*Chordeiles virginianus*), frequently repeated, only not so loud. Like most Owls, they snap their mandibles when disturbed, producing a sort of clicking sound; at other times they make an unpleasant hissing noise like that of escaping steam. During the daytime they are sleepy, sad looking birds, but alert and active enough at night.

In disposition they are amiable, seldom fighting each other, even when feeding. Their quarry, if small, is firmly grasped with one foot; when larger, like a good sized Norway rat, the bird stands upon and holds it firmly with both feet, tearing it gradually to pieces, nearly always beginning with the head, which appears to be the part most liked.

The average number of eggs laid by this species is from five to seven, seldom less. Larger sets containing from nine to eleven eggs are by no means uncommon; it is questionable, however, if every egg in such large sets is usually hatched.

In shape the eggs are mostly ovate, a few are elliptical ovate, and a single specimen before me is elongate ovate. They are pure dead white in color, the shell is finely granulated, and they are decidedly more pointed than Owls' eggs in general.

The average measurement of twenty-seven specimens in the U. S. National Museum collection is 42.5 by 32.5 millimetres, the largest egg measuring 47.5 by 33.5, the smallest 40.5 by 27.5 millimetres.

The type specimen, No. 20627 (Pl. 12, Fig. 1), selected from a set of five from the Bendire collection, was taken April 4, 1876, near Santa Cruz, California.

Family BUBONIDÆ. Horned Owls, etc.

112. Asio wilsonianus (Lesson).

AMERICAN LONG-EARED OWL.

Otus wilsonianus Lesson, Traité d'Ornithologie, 1831, 110.
Asio wilsonianus Coues. Check List, ed. 2, 1882, 81, No. 472.
(B 51, C 320, R 395, C 472, U 366.)

Geographical range: Whole of temperate North America; south to the table-lands of Mexico.

The breeding range of the American Long-eared Owl covers the United States in general, but it is perhaps less abundant in the South Atlantic and Gulf States than in the central, northern, and western portions. It likewise breeds north of our border from Nova Scotia, New Brunswick, and southern Canada, west to the provinces of Manitoba and Saskatchewan, where it is reported as common in the woods skirting the Saskatchewan plains. Here it reaches latitude 54° N., and probably still farther north. On the Pacific coast it is met with from Lower California and Arizona, north through California, Oregon, and Washington, and extending well into British Columbia. It is equally common throughout the Rocky Mountain region.

From the nocturnal habits of the Long-eared Owl it might be entirely over-looked for years by the average observer, even in localities where it is fairly common; whereas another, thoroughly familiar with its haunts, would have no difficulty in detecting its presence at any time. Except during the mating season it is rather a silent bird, and the few notes which I have heard them utter, when at ease and not molested, are low toned and rather pleasing than otherwise. One of these is a soft toned "wu-hunk, wu-hunk," slowly and several times repeated, which really sounds much better than it looks in print; another is a low twittering, whistling note like "dicky, dicky, dicky," quite different from anything usually expected from or attributed to the Owl family. In the early spring they hoot somewhat like a Screech Owl, and may be often heard on a still evening, but their notes are more subdued than those of the latter.

Mr. J. W. Preston, of Baxter, Iowa, writes me that one of their notes resembles the "me-ow-ow-ow-ow" of a cat. Another is a subdued "hoo-hoo" or "oo-oo," often uttered for hours during the mating time. At a distance this sounds something like the lowing of a cow. He further states: "At the nest, when disturbed, the female ruffles her feathers, flies to the ground, curves her wings over her head, spreads her tail and feigns lameness, dragging herself along on the leaves, all the time snapping her mandibles, making a rapping noise as if two sticks were struck together. She will, at such times, also mew like a cat; if followed, she makes her way in a direction opposite to the nest. While this performance progresses the male is, perhaps, giving vent to his feelings by fluttering about and squealing like a half-grown rat in a trap, or muttering a mournful "hoo-maa-maa-voo" in a subdued tone. I have often mistaken the notes of this bird for those of human beings. On April 3, 1886, I took a fresh laid egg of the Long-eared Owl from a nest of Crow's eggs, and the parent of the latter did not seem to mind the intrusion."

Although I have examined quite a number of the nests of this Owl (some forty), in various parts of the West, I have never found the parents as demonstrative as Mr. Preston says they are; and in not a single instance did either of the birds fly to the ground when driven from the nest and feign lameness, or make much noise except that produced by the snapping of the mandibles. The female would simply ruffle her feathers, fly into a neighboring tree or some dense bushes, and watch my proceedings. On rare ocasions, she would utter a sound resembling the spitting and mewing of an angry cat. Like the Barn Owl, they are inoffensive and harmless birds, and on the whole far more beneficial than otherwise. Fully three-fourths of their food consists of the smaller rodents, such as squirrels, chipmunks, gophers, and mice; frogs also form a considerable part of their food where these batrachians are plentiful. Occasionally they make a meal of a small bird or a young rabbit, but this is the exception and not the rule. While by no means devoid of courage, I doubt if they ever molest poultry or any of our game birds, and if any of the remains of the latter are found in their stomachs it is more likely that they have picked up a badly wounded bird, or one that had been shot and not recovered by the hunter. The smaller rodents are swallowed entire, and the indigestible parts, consisting of bones and fur, are subsequently ejected in the form of pellets. This applies to the Owl family in general, excepting possibly the little Elf Owl.

In the more settled portions of the country their nests are found in both deciduous and evergreen woods, in swampy as well as high and dry locations, but usually at no great distance from water; and the gloomiest and densest parts of the forests are generally selected for nesting sites. In the thinly settled portions of the West, they frequently nest in quite open and exposed situations, as a clump of willows or a small pine sapling or in the lower shrubbery bordering small streams or springs. The height of the nests from the ground varies considerably in different sections. In the majority of cases it is not over 20 feet, rarely over 30, and in the West not infrequently as low as 10 and 12 feet.

The Long-eared Owl rarely constructs a nest of its own; usually the last year's nest of a Crow is slightly repaired by being built up on the sides and lined with a little dry grass, a few dead leaves, and feathers; some of the latter may nearly always be seen hanging on the outside of the nest. Fully

three-fourths of the nests found by me occupied by these Owls were those of
the Crow. Only a very few were evidently built by the birds themselves.
One such found near Camp Harney, Oregon, on April 4, 1877, was placed in
a thick bush of dry willows about 10 feet from the ground. This was tolerably well built, composed externally of small sticks and sprigs of willows
and aspens. Some of the latter had been peeled by beavers, which were common in the vicinity, and they were still green and pliable; these fresh looking
sticks drew my attention to the nest, which I mistook for that of a Raven or
Crow. The inner cup was about 5 inches deep and lined with dry grasses and
feathers; it contained four fresh eggs. An occasional pair where Magpies are
plentiful will now and then use one of their nests, and natural cavities in dead
trees are also sometimes used. In mountainous regions they are said to nest
occasionally in cliffs. Most of the nests found by me were placed in rather
open situations, in small willow thickets along some stream, or in an isolated
clump in a swampy meadow, or on some hillside near a spring, but usually
not far distant from other thickets. The most exposed nest I ever saw was
found on April 24, 1877, near Camp Harney, and it contained six eggs on the
point of hatching. It was evidently an old Hawk's nest—most likely Swainson's Hawk—placed in a small and very open scraggy juniper bush not over
6 feet from the ground; this bush stood entirely by itself and was quite a
prominent mark on the point of an extensive sagebrush-covered table-land.
It grew near the edge of the rim rock forming a perpendicular cliff, and
there were no other trees or bushes of any size within a mile of it. This
nest was in plain view from all sides and could be seen several hundred
yards away.

These owls seem certainly more sociable and peaceably disposed toward
each other than Raptores in general, as I have more than once found as
many as three pairs nesting within a narrow strip of bushes not more than
100 yards in length and bordering a small creek. On another occasion I found
a pair of Long-eared owls occupying a cavity in an old cottonwood stump
not over 12 feet high; a Red-shafted Flicker had that season excavated a
burrow directly over that of the Owl's and the two entrance holes, although
on different sides of the stump, were not more than 2 feet apart. The birds
seemed to live in perfect harmony with each other.

I believe these Owls are constant residents wherever found, although
they are not as often seen in winter as during the remainder of the year,
and they may migrate to a certain extent in severe seasons. Mr. Julius
Hurter, of St. Louis, Missouri, informs me that on January 30, 1873, a flock
of about thirty of these birds were seen by him resting in one tree in a
swampy place in the Mississippi bottom, and that they remained in the
neighborhood for several days.

On February 23, 1872, I saw about fifteen of these birds sitting close
together on a small mesquite tree in a dense thicket in the Rillitto Creek
bottom, near Tucson, Arizona. The fact that occasionally such numbers are
seen together looks as if they did at times migrate.

Their hunting is done almost entirely at night, while the days are spent in shady and dark places, among the heavier and denser undergrowth, or in bushy trees in the neighborhood of water.

In the daytime, particularly on a bright sunny day, the Long-eared Owl will allow itself to be closely approached, and on discovering the intruder will try to make itself look slender and long by pressing the feathers, which are usually somewhat puffed out, close to the body and sitting very erect and still. It might in such a position be readily mistaken for a part of the limb upon which it may be sitting.

Occasionally, while on the ground, for instance, and being suddenly disturbed at a meal, they throw themselves into quite a different attitude—one of defiance, making themselves look much larger than they really are, and presenting a fierce and formidable front. I nearly stepped on one of them once while it was busily engaged in killing a ground squirrel which it evidently had just caught. The Owl was sitting by the side of a fallen pine tree and as I stepped over it my foot was planted within 12 inches of the bird; she evidently had not heard me approaching, nor had I any idea of her presence until almost on her, and was consequently about as much startled as the bird itself, owing in part to the instantaneous transformation that took place before my eyes. All at once she seemed to expand to several times her normal size; every feather raised and standing at a right angle from the body; the wings were fully spread, thrown up, and obliquely backward, their outer edges touching each other over and behind the head, which likewise looked abnormally large, and this sudden change in appearance, combined with the hissing noise which she uttered, made it appear a very formidable object at first sight. I presume she intended at first to stand her ground, but changing her mind quickly and collapsing to her normal size, flew off, leaving her quarry behind.

The Long-eared Owl nests rather early. In the southern portions of its range nidification commences sometimes in the latter part of February. Mr. F. Stephens took a set of six eggs from an old Crow's nest in southern California on March 2, 1879, the earliest date known to me. In the Middle States it may begin laying in the latter part of March, but more often about the first week in April, and in late seasons sometimes not until the first week in May. These dates hold good also for Oregon, Washington, and Idaho. In Montana they rarely begin to lay before May 1.

I believe that but a single brood is raised in a season. Incubation usually commences with the first egg laid, the eggs being deposited at intervals of a day or two. If the first set is taken, a second and somewhat smaller one, and even a third is laid, frequently in the same nest. Incubation lasts about three weeks; the young are covered with thick grayish white down when first hatched and are usually of different sizes, showing that they are not all hatched the same day.

The eggs vary from three to six in number. Sets of five are most often found, and occasionally one may contain seven. Six is the largest number I have personally found. They are pure white in color and oval in shape, and the shell is smooth, finely granulated, and rather glossy.

The average measurement of one hundred and three specimens in the U. S. National Museum collection is 40 by 32.5 millimetres, the largest egg measuring 43.5 by 33.5, the smallest 37.5 by 31 millimetres.

The type specimen, No. 20615 (Pl. 12, Fig. 2), selected from a set of five eggs from the Bendire collection, was taken by me on March 16, 1882, near Fort Walla Walla, Washington. This is the earliest date on which I have found eggs of this species; and incubation had already begun.

113. Asio accipitrinus (PALLAS).

SHORT-EARED OWL.

Strix accipitrina PALLAS, Reise, Russischen Reichs., I, 1771, 455.
Asio accipitrinus NEWTON and YARROW, British Birds, ed. 4, I, 1872, 163.
(B 52, C 321, R 396, C 473, U 367.)

GEOGRAPHICAL RANGE: Entire western hemisphere, except Galapagos and part of the West Indies; also nearly throughout the eastern hemisphere, excepting Australia, etc.

The breeding range of the Short-eared Owl within our borders extends, as far as known at present, throughout the middle portions of the United States from about latitude 39° northward through the Dominion of Canada to the Arctic regions, where Mr. R. MacFarlane met with it quite commonly in the Anderson River country up to latitude 69°.

On the Pacific coast it is known to breed from southern Oregon, in the vicinity of Camp Harney, about latitude 42°, through Washington, Idaho, and British Columbia to northern Alaska; and it is not improbable that it may sometimes breed in California and Nevada.

In the northern portions of its range it is only a summer visitor, migrating south in winter. It is more than likely that it breeds, occasionally at least, in suitable localities along the borders of the extensive marshes on the seacoast of the South Atlantic States. By far the greater number of these birds, however, breed north of our borders.

In its general habits, the Short-eared Owl differs considerably from most of the other members of this family found with us, in being not nearly so nocturnal and in frequenting the more open country; for while most of our Owls inhabit timbered regions, this species shuns such sections and rarely even alights on a tree. Its home is amidst the rank grasses or weeds usually found along the borders of lakes and sloughs in the open prairie country, where it hides during bright sunshiny days. If the sky is clouded, this Owl may be frequently seen hunting in the early morning or evening and sometimes in the middle of the day, and at such times it flies very low, not more than a few feet from the ground, which it carefully scans for its humble prey. Its flight is remarkably easy, graceful, and perfectly noiseless, very similar to that of the Marsh Hawk, but accompanied with more flapping of the wings. On account of the great length of the latter it looks while flying much larger than it really is.

From the fact that these Owls are generally seen in pairs at all seasons of the year it is very probable that they remain mated through life. During the winter of 1881-'82 a pair of these birds took up their permanent quarters near an inclosure in which several hundred tons of hay were stacked within the limits of the military post of Fort Walla Walla, Washington. These hayricks were placed under open sheds, and the lanes between each stack formed the favorite hunting grounds of these birds and furnished them an abundance of mice. As shooting was absolutely prohibited in this vicinity and the birds were never molested, they became quite tame, flying within 3 or 4 feet of the sentinels on duty there at night. One of my men informed me of this, and I visited the place several times to watch their actions. In cloudy weather they might be seen flying in and out between the haystacks during the greater part of the day, but when it was clear they only made their appearance about sundown. They generally flew close to the ground, not more than 3 feet above it; their prey was caught without apparently arresting their flight an instant, and then carried on top of one of the stacks where I presume it was devoured at leisure. During the day they remained hidden among the tall weeds, wire grass, and rushes growing in a marsh close to the corral. Aside from a faint squeak, which might have been that of a mouse when suddenly pounced on, I heard no note that might be attributed to them.

The food of the Short-eared Owl consists almost exclusively of small rodents, such as meadow mice and gophers, as well as grasshoppers, insects of various kinds, and occasionally a small bird. Like the Barn and Long-eared Owls, it deserves the fullest protection, being far more beneficial than otherwise.

In the southern parts of its range in the East, nidification is said to begin sometimes in March, usually in April and again not until the beginning of May. In the Arctic regions it does not commence until the latter part of May, usually about the first week in June, and occasionally a month later, Mr. W. J. Fisher taking a set of five eggs on Kodiak Island, Alaska, on July 12, 1882. Incubation had advanced about one-half in these eggs.

While in the Eastern States these birds may nest, as before stated, in March and early April, in the West, where I found their nests, in Idaho and Oregon, as well as in Washington, they do not as a rule begin laying before the last week in April, and generally about the first week in May. The first nests I found of this species, two in number, near Fort Lapwai, Idaho, was on May 6, 1871. Both were placed on the ground, one in the center of a tall bunch of rye grass, the other by the side of one of these, and both were well hidden. These tall bunches of grass grew with others amidst small bushes and weeds on a little knoll in the center of a boggy place on a grass covered hillside, near the head of Tom Bell's Creek, about 4 miles east of the post. Both nests were in similar situations on opposite sides of the cañon, and not over half a mile from each other; they were simply slight depressions not more than 2 inches deep, lined with pieces of dry grass and a few feathers from the birds. One of the nests contained four eggs, two of which were slightly incubated; in the other one there were three eggs, which were fresh.

Both males were hidden in the tall grass close by the nest, and aside from an angry snapping of their mandibles they, as well as the females, made but feeble demonstrations in defense of their nests, merely circling around and alighting near by while I was in the vicinity. One of the birds uttered a weak whistling sort of note two or three times, the others were silent but kept bobbing their heads up and down and snapping their mandibles at the same time.

At Camp Harney, Oregon, two of their nests were found in similar situations on side hills in the last week in May, 1876, both containing three nearly fresh eggs, and two others in the following year, on May 1 and May 4. The last one was placed in the center of a thick but short bunch of grass on the level open prairie directly south of the post in daily use as a drill ground. I know that some of the men while drilling must have repeatedly passed within a couple of feet of this nest, if not directly over it, without flushing the bird. It was finally discovered by one of the lieutenants, whose horse almost stepped on the bird. This nest contained three eggs, two of which were broken either by the hoof of the horse or by the bird in its sudden start. It seems strange that this pair should have chosen such a site, as the drilling certainly commenced before any of their eggs had been laid and was continued daily, except on Sundays. None of the nesting sites found by me were resorted to again either during that or the succeeding year.

In Kansas and Nebraska they lay about the same time. Mr. H. A. Kline found a nest of this species on May 17, 1883, in Nebraska, containing eight eggs. It was placed on the ground and consisted of a lot of dry prairie grass and hollowed out 2 inches in depth. He says: "My dog was ranging a short distance in front of me when he was suddenly attacked by one of these Owls, which was soon joined by the other (the male), and together they succeeded in driving him from the field. They would swoop from right to left and strike him on the back with their wings. Not being used to such treatment by any member of the feathered tribe, he beat a hasty retreat, followed by both birds, and after chasing him some distance they returned to me and manifested great displeasure by swooping very close to me and snapping their mandibles, as many Owls do when angry."[1]

Mr. R. S. Williams, of Great Falls, Montana, writes me that he found a nest of this species containing six eggs nearly ready to hatch on June 13,

---

[1] Ornithologist and Oölogist, Vol. VIII, 1883, No. 8, p. 61.

1889. The nest was placed in a slight hollow surrounded by bunch grass, and it consisted of a few scattered straws and feathers from the bird.

According to Mr. W. H. Dall, U. S. Coast Survey, this species is resident and not uncommon in Unalaska, Alaska, and it is said to breed in burrows in the ground, usually on the side of a steep bank. The hole is horizontal and the inner end usually a little higher than the entrance and lined with dry grass and feathers. Those examined by him did not exceed 2 feet in depth.

On the Atlantic coast Mr. W. E. D. Scott found it breeding at Long Beach, New Jersey, taking a set of seven partly incubated eggs on June 28, 1873, and the National Museum collection contains a set of six taken by Mr. Thomas Beesley, near Cape May, New Jersey, in 1860.

It is gregarious at times, Dr. Elliott Coues observing a gathering of from twenty to thirty on the Colorado River, near Mojave, Arizona, and I have seen small parties of five or six, probably the old and young, along the marshes of Malheur Lake in Oregon.

Incubation probably lasts about three weeks, and ordinarily but a single brood is raised. The eggs are usually from four to seven in number, rarely more, though if Indian and Eskimo testimony is to be relied on, they lay as many as ten and even twelve, but this is scarcely probable. They are white in color, with a very faint creamy tint perceptible in most of the specimens; the shell is smooth, finely granulated, and not as lustrous as are the eggs of the preceding species. In shape they vary from oval to elliptical ovate, and a few are nearly equally pointed at each end.

The average measurement of fifty-six eggs in the U. S. National Museum collection is 39 by 31 millimetres, the largest egg measuring 42 by 32, the smallest 37.5 by 29.5 millimetres.

The type specimen, No. 13824 (Pl. 12, Fig. 3), selected from a set of five eggs, was taken June 30, 1865, by Mr. R. MacFarlane, near Anderson River Fort, Arctic North America.

114. Syrnium nebulosum (FORSTER).

BARRED OWL.

Strix nebulosa FORSTER, Philosophical Transactions, XXII, 1772, 386.
Syrnium nebulosum BOIE, Isis, 1828, 315.

(B 54, C 323, R 397, C 476, U 368.)

GEOGRAPHICAL RANGE: Eastern North America; north to the more southern British provinces; south to Georgia and northern Texas; west to eastern Nebraska and Kansas.

The range of the Barred Owl, next to the Great Horned Owl the largest of this family breeding within our borders, extends through that portion of the United States east of the Mississippi Valley from Georgia northward to the southern border of the Dominion of Canada south of latitude 50° N., where it has been met with in Nova Scotia, the southern parts of the provinces of Quebec and Ontario and southeastern Manitoba. Westward it reaches Minnesota, Iowa, Missouri, the eastern parts of Nebraska, Kansas, the Indian Territory, and northeastern Texas. Excepting in the more northern parts of its range it is a constant resident and breeds wherever found.

The Barred Owl is readily distinguished from the Great Horned Owl by its somewhat smaller size, conspicuous rounded head, due to the absence of ear tufts, its greenish yellow beak, and handsome dark colored eyes. In the central and southern parts of its range it is quite common, frequenting mostly the heavy timbered and, preferably, swampy tracts near water courses, and spending the days generally in natural hollows of trees or in dense shrubbery. Like most of the birds of this family, it is nocturnal in its habits, but nevertheless sees well enough and even occasionally hunts in the daytime, especially during cloudy weather. I believe that Owls in general prefer to remain hidden during the daytime on account of attracting the attention of nearly every feathered inhabitant of the vicinity, who instantly attack and annoy them in every possible manner the moment they leave their retreats.

The flight of the Barred Owl, like that of other members of this family, is easy, and though quite swift at times it is perfectly noiseless. A rapidly passing shadow distinctly cast on the snow-covered ground is often the sole cause of its presence being betrayed as it glides silently by the hunter's camp fire in the still hours of a moonlight night. Far oftener, however, it announces itself by the unearthly wierd call notes peculiar to this species, which surpass in startling effect those of all other Owls with which I am familiar. It is necessary to listen to such a vocal concert to fully appreciate its many beauties, (!) as it is impossible to give an accurate description of the sounds produced when a pair or more of these birds try to outdo one another.

Mr. J. W. Preston, of Baxter, Iowa, writes me as follows on this subject: "Their notes are variable and sometimes easily mistaken for those of the the human voice. The base upon which they work is a 'hoo-hoo,' or 'too-too, but these syllables are modified and interchanged at pleasure. Here are a few samples, taken down as uttered by one of these birds while close to it: 'Hoo-hoo, ho-ho-ho-ho-ho, hoo-hoo-too-too, to-to, too-o.' Another call is somewhat like this, 'too-too, to-to, to-to, o-o' and still another, 'haw-haw, hoo-hoo.' Now and then a coarse mocking laugh, very humanlike, may be heard, and again this is changed to a mournful wail. Sometimes their notes are all cut up and uttered in great haste, with seemingly no cause for these violent outbursts. I disturbed a young bird once, causing one of its parents great uneasiness. It is impossible to describe all the notes uttered by it at this time; they were rendered in a subdued muttering and complaining strain, parts of which sounded exactly like 'old-fool, old-fool, don't-do-it, don't-do-it.'"

The Barred Owl is not infrequently heard calling in the daytime, more particularly during cloudy weather before a storm, these sounds emanating usually from the most dismal and gloomy parts of the forest. In the early evening and throughout the night, especially during the mating season, their vocal efforts are so perceptibly increased that two or three individuals may succeed in producing such a startling variety of sounds as to lead to the belief that a small army of these birds had been turned loose in the neighborhood.

While the Barred Owl is not quite as harmless a bird as the three preceding species, when impartially judged it does far more good than harm; and many of the depredations of the Great Horned Owl are erroneously charged to it, owing to the slight difference in size between these birds. From a careful examination of ninety-five stomachs, made under the direction of Dr. C. Hart Merriam, in charge of the Division of Ornithology and Mammalogy, U. S. Department of Agriculture, it appears that only three contained poultry; twelve, other birds, among them two Screech and one Saw-whet Owl, but no game birds; while forty contained mice; fifteen, other mammals, among the latter but very few rabbits and squirrels; four, frogs; nine, crawfish; two, fish, and sixteen were empty. This is certainly not a bad record.

Mr. Frank Bolles, who recently published an exceedingly interesting and instructive article on "Barred Owls in Captivity," drawing attention to peculiarities of these birds, which only a thorough lover of nature would notice, says: "They feed also on snakes, earth worms, grasshoppers, different species of beetles and flies. They not only drank water freely, but took prolonged baths whenever fresh water was given them. Their tank was one foot and a half long, a foot wide, and ten inches deep. Their reflection in this comparatively deep and dark pool greatly amused them for a time. * * * With great interest in the result I placed nine live perch and bream in the Owls' tank one morning when they were about three months old. They had never seen fish before. As the light played upon the red fins and bright scales, the birds' excitement was amusing to see. In a very short time, however, they plunged feet foremost into the water, and with almost unerring aim lanced the victims with their talons and flew out with them. Then the head was crushed at its juncture with the backbone, the spines were bitten into jelly and the fish was swallowed."[1] Every page of this paper is full of interesting information, but want of space forbids additional quotations therefrom.

Except during the mating and breeding season the Barred Owl is an unsocial bird, spending the greater part of the year in solitude and resenting all intrusion of its kind. Throughout the greater part of its range the mating season begins about the first week in February, occasionally a week or two earlier or later, according to latitude and season.

The Barred Owl generally nests in natural cavities of trees, preferably in dense and swampy woods, and such as are in the vicinity of water courses, chestnuts, poplars, oaks, American lindens, elms, walnut, sycamores, maples, willows, birch, and sweet gum trees being most often selected for this pur-

[1] Auk, Vol. VII, April, 1890, pp. 101-114.

pose. A suitable site once taken possession of is tenaciously held from year to year, no matter how persistently the birds may be robbed of their eggs or young. Only when such a site is not to be obtained in the vicinity of their chosen home is an old Hawk's or Crow's nest made use of. If this is not sufficiently bulky it is deepened by throwing out the inner and finer material, as well as the lining, should there be any. They rarely add anything to a nest, unless such feathers as fall from the body of the incubating bird can be called additions. The Barred Owl rarely builds a nest of its own; a cavity being selected the eggs are simply deposited on the rubbish, chips of decayed wood, and dried leaves which may have accumulated therein.

In the more southern parts of its range it usually commences laying in February, while in the Middle States it generally begins from about the second week in March to the first week in April. It is sometimes influenced by the condition of the weather. Even in the southern New England States, near Norwich, Connecticut, Mr. C. L. Rawson has occasionally found them nesting in February, and has taken their eggs lying on a solid cake of ice, both from holes and open nests in trees when the ground was covered with a foot of snow.

But a single brood is raised in a season. Should the first set of eggs be taken, a second, and even a third, is occasionally laid, the last generally smaller in number than the first. The female seems to attend mostly to the duties of incubation, which lasts from three to four weeks. When so engaged she is loth to leave her treasures, and occasionally has to be dislodged by force, uttering her protests either by an angry snapping of her mandibles or by a hissing noise from a limb close by.

The number of eggs laid varies from two to four; sets of three seem to predominate slightly over the smaller number, while four are rare. The U. S. National Museum collection contains a so-called set of five eggs collected in Iowa in April, 1869. In three of these specimens the holes are much larger than in the remaining two, and if really all taken at one time the first eggs were probably laid at a considerable interval from the last. Ordinarily they are deposited every third day. Like all Owls' eggs they are pure white in color, the shell is more or less granulated, slightly rough to the touch, and not very glossy. In shape they are oval or rounded oval.

The average measurement of eighty-two of these eggs is 49.5 by 42.5 millimetres; the largest specimen measures 55.5 by 44, the smallest 41 by 37.5 millimetres. The latter from a set of two (the other being of the normal size, 49 by 43 millimetres), was collected by Mr. C. L. Rawson in New London County, Connecticut, on March 29, 1890, and is really a runt egg, but contained a well formed yolk.

The type specimen, No. 20633 (Bendire collection), from a set of three, was taken by Mr. G. Peck in Black Hawk County, Iowa, March 2, 1878. It is figured on Pl. 12, Fig. 4.

115. Syrium nebulosum alleni Ridgway.

FLORIDA BARRED OWL.

Strix nebulosa alleni Ridgway, Proceedings U. S. National Museum, III, March 27.
1880, 8.
(B —, C —, R 397a, C 477, U 368a.)

GEOGRAPHICAL RANGE: South Atlantic and Gulf coast region of the United States, from eastern South Carolina, Georgia, and all of Florida west to southeastern and central Texas.

The Florida Barred Owl, a slightly darker colored bird than the preceding, was first described from specimens taken in Florida, and is now known to be common in the heavily wooded bottom lands of southeastern and central Texas (Lee County), and it has also been taken at Gainesville, Cooke County, near the northern border of that State. All the Barred Owls found along the intervening Gulf coast, in the southern portions of Alabama, Mississippi, and Louisiana, south of latitude 31°, are doubtless referable to this race. Its known range has recently been likewise extended along the Atlantic coast. Mr. J. E. Benedict shot a specimen, which is now in the U. S. National Museum collection, on June 4, 1891, near Georgetown, South Carolina, in the northeastern part of the State, which is perfectly typical, and he says that this Owl is quite common in that vicinity. It may, therefore, be looked for throughout eastern Georgia and the greater portion of eastern South Carolina as well, and as these birds are constant residents wherever found it is reasonable to presume that they also breed there.

The habits of the Florida race appear to be very similar to those of the common Barred Owl; like the latter, it frequents the densest forests of the bottom lands, nesting, however, almost exclusively in hollow trees and stumps, such as oak, pine, cypress, and gum trees, while in Florida they make use occasionally of hollow cabbage palmettos, the eggs being laid on such rubbish as may be found at the bottom of the cavity or on the bare wood; open nests in trees are only used on very rare occasions. These sites are usually situated at a height of from 15 to 50 feet from the ground and rarely higher.

Dr. William L. Ralph writes me: "The Florida Barred Owl, though much less numerous now than formerly, is still very abundant in the unfrequented districts, and not uncommon even in the more settled portions of St. Johns and Putnam Counties in that State. During the first few years in which I visited these localities they were so abundant that at times—when they were mating, I think —I have heard nearly a hundred calling at once. The call notes of the Florida Barred Owl are about the same as those of the northern bird. These notes consist of three syllables, which sound like 'who,' 'ah,' and 'whack,' but they are often given in such different tones that it appears as if they uttered more. Their usual note is a loud single 'who-ah.' Where several birds are together, sometimes in the midst of almost a perfect silence, one would begin with 'who-

who-who-who, who-who-who-who-who, who-ah.' Then another would answer
in the same note, and perhaps several others in turn. After this note had been
given by several birds in succession, another would utter a call like 'ah-ah-ah-
who-ah,' or perhaps 'who-ah, whack-whack-whack, who-oo-ah,' which would
hardly be begun before others would join in successively, some uttering the
first notes and some the second, until it would seem as if every tree in the
neighborhood held one of these Owls. After a few seconds' continuance at its
greatest height, this racket would gradually die away until there was almost a
perfect silence again, which would last for a few minutes, and then the Owls
would begin to call once more. I have never heard anything that could equal
one of these Owl concerts of former days, and never expect to again. It was
not the degree of noise that made them peculiar, for I have heard as much or
more in the breeding places of water birds, but a wierdness that is indescrib-
able. It was not a common occurrence, however, to hear so many of these
birds calling together, but only at times during the mating season, and in
isolated places. They are particularly active and call more on bright moonlight nights than in dark ones. It also seems as if all the eggs of these birds
whose nests I have found were laid during the moonlight nights of January
and February.

"This Owl is the noisiest of any with which I am familiar, much more so
than its northern relative, and is oftenest heard during the mating season,
rarely while incubating, and not very often during the remainder of the year.
Formerly these birds were very tame, and during my first visits to this locality,
some years ago, I have known them to utter their calls from the roof of the
house in which I lived as unconcernedly as they do now in the most isolated
swamps. They seem to be more sociable than the northern bird, but this may
be only because they are more common. As with most birds of prey, they do
not nest near one another, although they seem not to mind the proximity of
nests occupied by other birds. They are constant residents and become just as
much attached to their nesting places as do the Great Horned Owls. Like these
they are most active by night, but cover a much wider range than the latter,
often leaving the woods and hunting in open fields, and formerly among the
houses of small settlements. From their habit of living and breeding in swamps
they are called by the natives 'Swamp Owls.'

"They nearly always nest in cavities in trunks or large limbs of trees, and
this, with their retiring habits, makes it rather difficult to find their nesting sites
before the eggs are hatched. The cavities they choose for nesting sites are
of all sizes and shapes. I have seen some so large that a person could easily
stand in one of them, others so small that the birds could with difficulty squeeze
through the openings, and again others so shallow that the tail of the sitting
bird could be seen projecting from them. I have never known these birds to
take any material to the cavities in which they nest, there being always more
or less trash in these places, and this, with the feathers from the birds themselves, often makes a soft, thick, and usually a very filthy bed for the eggs to

lie on, especially after such a cavity has been used several years in succession. I have found quite a number of their nests with young in such situations, and only a single instance has come under my observation where a pair of these birds made use of an open nest, presumably one occupied the previous season by a pair of Harlan's Hawks, which were seen about this nest when it was first found on February 3, 1891. This nest was situated in a large pine tree, 62 feet from the ground, at a point where the trunk divided into several large limbs, and it was placed in the forks thereby formed. It was composed of sticks and Spanish moss, and lined with small twigs, Spanish moss, and feathers from the sitting birds. It was found in a wild and desolate spot about 6½ miles south of San Mateo, the tree containing the nest standing on the edge of a small but dense cypress swamp. When examined, on February 16, two young Owls were found, one about a day, the other three days old. Both parents were seen, and the female, which was on the nest, after being driven from it, kept on returning to it.

"Another nest with eggs was found on February 17 in the hollow top of a broken cypress tree standing near the edge of a small swamp, 4½ miles south of San Mateo. The eggs, two in number, were about three-fourths incubated; they measured 53.5 by 46 and 52 by 44 millimetres. The cavity containing the eggs was 56 feet above the ground and so shallow that the tail of the sitting bird could be seen projecting from it. The parent, on being driven from the nest, hooted several times; and it was the first time I ever heard an Owl call when it had eggs. The eggs were lying on pieces of rotten wood and feathers from the birds. After the climber descended from the tree the bird returned to its nest and remained there while our party was in the neigborhood, about fifteen minutes.

"Another nest, first found on February 18, when it contained a single egg, was left until February 23, in hopes that another might be added to it, in which I was disappointed. The nesting site was a hole in the side of a cypress tree about 28 feet from the ground and about 10 inches deep. The cavity was lined with cypress bark, small twigs, rotten wood, and feathers from the sitting birds. One of the birds was driven from the nest both times it was visited, when it would alight on a tree near by, where it stayed for a little while and then disappeared from sight, remaining away during the time our party was in the vicinity. The parent of another set of two eggs, taken on the same date, remained in the hole while the eggs were being taken. These eggs were within a day or two of hatching.

"Another set of two was found February 24; one of these eggs had been incubated about six days, the other was quite fresh. These are small and measure only 47.5 by 40.4 and 47.3 by 40.6 millimetres. The nest was in a hole in the side of a very large pine tree, 21 feet from the ground. The cavity was 18 inches in diameter; the eggs were placed on the side opposite the opening, and, indeed, so were all the other eggs when the nest proper was smaller than the cavity. This nest had the bones of small mammals and fish, together with several pine knots lying around it. The eggs were deposited on some rotten wood and feathers from the sitting birds. I think the bird which laid these eggs is a very old one, and the nest has been occupied for several years. The latest date on which I found eggs of these birds was on March 10, when a set of two fresh eggs were taken, one, of rather large size, measuring 53.3 by 42.2, the other a runt egg, which measures 39.1 by by 34.5 millimetres. I believe this to be a second laying. The runt egg contained a yolk and would probably have hatched. The site was in the hollow of a partly decayed pine stump, about 15 feet from the ground. The cavity was about 1 foot deep and 10 inches in diameter. A number of other nests of this subspecies were examined by me, most of which contained young, and the nesting sites were similar to those already described.

"I found these birds already mated on my arrival at San Mateo, on January 13, 1891, and they were making a great deal of noise every night, which they kept up until the nights began to grow dark during the last of the month, when they gradually ceased, and for about ten days previous to February 14, they were hardly heard. On that date they began calling again, though not so much as at first, and since then they have made a good deal of noise on some nights, while on others they hardly uttered a note.

"The mating season of the Florida Barred Owl begins usually about the 1st of January, perhaps a week or two earlier in some seasons, and their eggs are generally deposited during February. According to my observation they always lay one or two eggs, usually the latter number. Both sexes assist in incubation and in the care of the single brood raised. As a rule they are not very solicitous about their homes, but fully as much, if not more so, at least so far as the young are concerned, as the Great Horned Owls. The young owlets when first hatched are covered with a fluffy white down and grow rapidly. When about a week old the first pinfeathers commence to show on the wings and back, while at the age of a month they are fairly well feathered throughout, the plumage being of a very loose and soft structure; the primaries being about half the normal size, but the tail feathers are correspondingly much shorter and still hidden by the down. I believe they are able to leave the nest when about six weeks old.

"Their food consists of the smaller mammals, such as rabbits, squirrels, gophers, rats and, mice, as well as fish and crawfish, and I believe they sometimes catch snakes and frogs. They rarely capture birds or poultry, and are far more beneficial than otherwise."

In southern Florida nidification occasionally takes place in December, but more frequently in January, Mr. J. F. Menge writing me that he had found two nests with young on February 15, 1890, in the vicinity of Myers, Lee County, Florida. One of these contained a single young bird, the other two, nearly full grown and about ready to leave the nest.

In southern Louisiana Mr. G. E. Beyer found Barred Owls, which are probably referable to this race, nesting early in March, and laying from two to four eggs, while in Lee County, Texas, according to Mr. J. A. Singley, they are the commonest Owls found in that part of the State. Here it nests usually in the latter half of February or the first week in March, and it seems to lay but two eggs, corresponding in this respect to the Florida birds. In seventeen nests found by him, none contained more than that number of eggs or young. The cavities in which these Owls nested were of various depths and dimensions, some of them fully 2 feet deep and 15 inches in diameter, others quite shallow, not over 10 inches deep. The eggs are deposited at intervals of several days, and incubation seems to begin with the first egg laid.

The eggs of the Florida Barred Owl are very similar in color and shape to those of the common Barred Owl; one of the specimens before me is a perfect ovate. The average measurement of eight eggs of this subspecies from Florida, all collected by Dr. Ralph, is 51.1 by 42.7 millimetres, the largest measuring 53.5 by 46, the smallest 47.3 by 40.6, while the average measurement of fifteen eggs of this race from Texas, is 48.5 by 41.5 millimetres, and judging from the small number of Florida specimens before me, these, in this instance at least, average larger than the eggs of the common Barred Owl.

116. Syrnium occidentale XANTUS.

SPOTTED OWL.

Syrnium occidentale XANTUS, Proceedings Academy Natural Sciences, Phila., 1859, 193.
(B —, C 324, R 398, C 478, U 369.)

GEOGRAPHICAL RANGE: Highlands of Mexico; north to southern Colorado, New Mexico, Arizona, California, and Lower California.

The range of *Syrnium occidentale*, the western representative of the Barred Owl, and a somewhat darker colored bird, as far as is known at present, extends through the mountain regions of California, south to Arizona, thence east to New Mexico, and north to southern Colorado. It has also been met with in the mountains of Lower California and extends southward over the higher table-lands of Mexico to Guanajuato, latitude 21°.

Mr. Xantus, one of the pioneer naturalists of the Pacific coast, discovered the Spotted Owl in the vicinity of Fort Tejon, California, in the southern Sierra Nevada Mountains, and described it in the Proceedings of the Academy of Natural Sciences of Philadelphia in 1859. This specimen remained unique until I found the bird again in the spring of 1872, in the vicinity of Whipple's Station, some 10 miles northwest of Tucson, Arizona. Since then it has been met with in other places in the West, but it does not appear to be an abundant species anywhere, except perhaps in Calaveras County, California, where Mr. L. Belding reports it common in summer and perhaps in winter. According to his observations it frequents the densest parts of the fir forests. On June 13, 1882, a male and female were shot together in the early evening.[1] Mr. R. B. Herron also obtained a specimen midway between San Diego and Riverside, in southern California in the fall of 1885.

[1] Belding's Birds of the Pacific District, 1890, p. 49.

The Spotted Owl is mainly nocturnal in its habits, and is rarely seen in the daytime. It may be much more common in suitable localities than is generally supposed, and as it seems to be almost entirely confined to the thinly settled mountain regions of the West, its present rarity is easily accounted for.

Very little is as yet known of its general habits. I accidentally stumbled on this bird and its nest while en route from Picacho Peak to my camp on Rillitto Creek, near Whipple's Station, on April 17, 1872. My attention was first drawn to the nest by one of my men, who noticed a bird sitting on it. Rapping on the trunk of the tree it flew into the branches of another close by, from which I shot it. On picking it up I supposed it to be a common Barred Owl, and only on my return to camp did I realize the prize I had secured; it was too much mutilated, however, to make a good skin. The nest appeared to me to be a new one, built by the birds themselves; it was about 30 feet from the ground and placed in a fork close to the trunk of a large and bushy cottonwood tree standing in the midst of a dense grove of younger trees of the same species. It was composed of sticks, twigs, and the dry inner bark of the cottonwood, lined with some dry grass and a few feathers. The inner cavity was about 2 inches deep, and the nest itself about the size of that of the larger Hawks. It was readily seen from below, but not easily observed a little distance away, the foliage of the tree hiding it pretty effectually. Some few weeks later I saw a pair of these birds in the same vicinity and secured another.

I several times heard the calls of what I supposed to be this species, in March of the following year, among the tall timber near my camp, and I believe the Spotted Owl is not uncommon in that part of Arizona. As near as I remember, their call notes are very similar to those of the Barred Owl, consisting of a series of continuous and far reaching hootings.

I believe it nests in cavities of trees as well as in open nests, and that like the Barred Owl, it lays from two to four eggs to a set. Nidification seems to take place in April, somewhat later than that of the latter species.

Mr. O. C. Poling writes me: "I discovered a nest and four newly hatched young of the Spotted Owl in the foothills among the oaks at the northern end of the Huachuca Mountains in Arizona. This was on May 23, 1890. Both parents were close to the nest and took little notice of me as I approached close to them. The nest was simply a large cavity in an oak about 10 feet from the ground."

The single egg before me, taken from the nest found on April 17, 1872, already mentioned, was fresh; it is oval in shape and pure white in color. The shell is slightly granulated and shows but little gloss. It measures 52 by 45.5 millimetres, and, as it is very similar to the egg of the Barred Owl, is not figured.

117. Scotiaptex cinerea (GMELIN).

GREAT GRAY OWL.

Strix cinerea GMELIN, Systema Naturæ, I, i, 1788, 291.
Scotiaptex cinerea SWAINSON, Classification of Birds, II, 1837, 217.
(B 53, C 322, R 399, C 471, U 370.)

GEOGRAPHICAL RANGE: Northern North America, south in winter to the northern border of the United States.

The breeding range of the Great Gray Owl is confined principally to the more northern regions of the North American continent, from the shores of Hudson Bay north to the limit of timber in about latitude 68°. It is reported common throughout the fur country and the interior of Alaska, and has also been met with by Drs. Cooper and Newberry in different parts of the Pacific coast. The former observed it near the mouth of the Columbia River in June, and obtained a specimen at the time, which would certainly indicate that it might breed in the vicinity, while the latter is said to have observed it as far south as Sacramento Valley in California, as well as in the Cascade Mountains, the Des Chutes Basin, and on the Columbia River, in Oregon. During an extended tour of duty in various parts of Oregon, Washington, and Idaho, I failed to meet with this species at any time of the year. It is more or less migratory during the winter, and much more abundantly met with along our northern border in some seasons than others. The winter of 1889-'90 was a notable instance, and quite a number of these birds were captured or seen in various parts of the New England States. Nothing new respecting the nesting habits of this Owl has been recorded within recent years.

In the "History of North American Birds," 1874 (Vol. III, pp. 32, 33), it is stated: "On the 23d of May, Dr. Richardson discovered a nest of this Owl built on the top of a lofty balsam poplar, composed of sticks with a lining of feathers. It contained three young birds covered with whitish down. * * *

"Mr. Donald Gunn writes that the Cinereous Owl is to be found both in summer and winter throughout all the country commonly known as the Hudson Bay Territory. He states that it hunts by night, preys upon rabbits and mice, and nests in tall poplar trees, usually quite early in the season."

Mr. R. MacFarlane, in his "Notes on the Land and Sea Birds of the Lower Mackenzie River District," says: "I should not say that this Owl was in 'great abundance' in the Anderson River region, as inadvertently stated on page 33, volume III, of the 'History of North American Birds.' We certainly obtained very few specimens, and we found but one nest (which is referred to in the same paragraph), on the 19th of July, 1862, near the Lockhart River on the route to Fort Good Hope; it was built on a spruce pine tree at a height of about 20 feet and was composed of twigs and mosses, thinly lined with feathers and down. It contained two eggs and two young, both of which had lately died. The female left the nest at our approach and flew to another tree at some distance, where she was shot."

From the limited information we possess about the nesting habits of this species it appears that in Alaska these birds nest sometimes as early as April, and in the interior as late as the middle of June. From two to four eggs seem to be laid to a set, and these are small for the size of the bird. The body of the Great Gray Owl is, however, much smaller than that of the Great Horned Owl, in fact but little larger than that of the Barred Owl. The long tail and the loose fluffy plumage of the bird make it look much larger than it really is.

The eggs are dull white in color with but little luster, and the shell is roughly granulated; in shape they are broad elliptical oval, not as rounded as eggs of species belonging to the genera *Syrnium* and *Bubo*.

The average measurement of nine eggs in the U. S. National Museum collection is 55 by 44 millimetres, the largest egg measuring 57 by 44.5 the smallest 53.5 by 42 millimetres.

The type specimen, No. 10277 (Pl. 12, Fig. 5), from a set of two, was collected by Mr. J. Sibbiston, near Fort Yukon, Alaska, in April, 1864.

118. Scotiaptex cinerea lapponica (Retzius).

LAPP OWL.

Strix lapponica Retzius, Fauna Suecica, 1800, 79.
Scotiaptex cinerea lapponicum Ridgway, Manual North American Birds, 1887, p. 260.
(B —, C —, R 399a, C 475, U 370a.)

GEOGRAPHICAL RANGE: Northern portions of Europe and Asia, straggling to western Alaska (shores of Norton Sound).

The Lapp Owl, a lighter colored bird than the Great Gray Owl, claims a place in our avifauna on the strength of a single specimen, an adult female, brought to Mr. L. M. Turner, from the Yukon Delta, April 15, 1876, while he was on duty as United States Signal observer at St. Michael, Alaska. It was said to be quite rare, but he failed to learn anything regarding its habits, and it may, possibly, breed in small numbers in the northern portions of Alaska.

Dresser, in speaking of this species, says: "This, one of the rarest of the Owls inhabiting the Palæarctic region, is almost entirely confined to the more boreal districts, where it is a resident in the upper portions of the forest belt, but rarely straggling down into the northern parts of central Europe. * * *

"The first published notice of the breeding of the present species appears to be that communicated by Mr. C. G. Löwenhjelm, who states (K. Vet. Ak. Handl. 1843, p. 389) that 'the nest was in a dense pine wood on a stump, about 3 ells high, in the top of which a hollow had been formed by the wood having rotted. In the nest was one white egg about the size of that of the Eagle Owl; and at the foot of the stump another egg was on some moss, quite uninjured.'

"Subsequently, Mr. Wolley obtained its eggs in Kemi Lappmark in 1856, and says (Oöth. Wolleyana, p. 173) that 'the nest was on the top of a broken trunk of a Scotch fir, the main part of which hung down; but from the description, Piety thinks there was some old nest there. He does not remember seeing any nest made. It was not high up, some 2 fathoms, perhaps; but those which he has seen before were not more than 1 fathom high. The top of the tree where it was broken off was not level, but it had a great splinter on one side. The birds are very bold at the nest and the cry of the cock attracts people to the nest. The cry is three notes drawn out, the first hardest, the second lighter and short, the third lightest and longest of all, "hu, hu, hu-u-u."' Another nest, taken at Muoniovara on the 5th of April, 1857, was, he adds, 'made of sticks and all kinds of stuff inside, about 3½ fathoms high up in a large Scotch fir, where it is divided into several great forks.' It was not like a new nest, and he describes it 'as about 2 feet in thickness.'

"I possess several eggs of this Owl received from Mr. Wolley's collectors in Lapland, which are pure white and resemble the eggs of the Snowy Owl, but appear a trifle less smooth on the surface of the shell. In size they vary from 2⅒ by 1⅞ to 2⅒ by 1⅔ inches [equal to about 51.5 by 42.7 and 54 by 43.9 millimetres]." [1]

The number of eggs laid at a clutch is not stated, but is probably from two to four.

A set of two eggs, No. 5322, U. S. National Museum collection, taken in 1861, near Tepasto, Finland, were presented to the Museum by the eminent English naturalist, Mr. A. Newton. These specimens are nearly oval in shape, pure white in color, slightly glossy, and the shell is roughly granulated. They measure 51 by 43.5 and 51.5 by 42 millimetres. None of the eggs are figured, as they resemble those of the preceding species very closely, except that they are a little smaller.

119. Nyctala tengmalmi richardsoni (BONAPARTE).

RICHARDSON'S OWL.

Nyctale richardsoni BONAPARTE, Geographical and Comparative List, 1838, 7.
Nyctale tengmalmi var. *richardsoni* RIDGWAY, American Naturalist, VI, 1872, 283.
(B 55, C 327, R 400, C 482, U 371.)

GEOGRAPHICAL RANGE: Northern North America; south, in winter, to northern border of the United States.

The little Richardson's Owl is an inhabitant of the more boreal portions of the North American continent, and, with the exception of Alaska, breeds, as far as known at present, only north of the United States. Even in winter

---

[1] Dresser's History of the Birds of Europe, Vol. V, pp. 282, 283.

it must be considered as rather a rare visitor to our borders.   Mr. Charles B. Cory reports it as breeding on the Magdalene Islands, Gulf of St. Lawrence, and Mr. Thompson gives it as resident in Manitoba.[1]

In eastern North America it is possibly a very rare resident from latitude 46° N., and northward, becoming more abundant as higher latitudes are reached. In the interior it is reported as common on the banks of the Saskatchewan, in latitude 53°, and reaches thence northward at least to Fort Simpson on the Mackenzie River, and its range is doubtless coextensive with the timber belt. It appears to be very common about Great Slave Lake, specimens having been received from all the different Hudson Bay Company posts located on its shores.

Mr. R. MacFarlane states: "This Owl, or a bird closely answering to the description, was repeatedly observed in the country between Fort Good Hope and the Anderson River, Arctic America."

Mr. Dall obtained a female specimen of this Owl at Nulato, Alaska, April 28, where it was not uncommon. It was often heard crying in the evenings almost like a human being, and was quite fearless. It could be readily taken in the hand without its making any attempt to fly away, but it had a habit of biting viciously.   According to the Indians it generally nests in holes in dead trees, and lays six spherical white eggs.[2]

From the foregoing accounts it appears to have pretty much the same habits as its near relative, *Nyctale acadica*. I have picked up one of these birds while perched in a wild rose thicket near Camp Harney, Oregon. It was fat and in a fine condition, and did not appear to see me approaching. Several others were caught by some of my men under similar circumstances, and I can only attribute it to their poor eyesight during the daytime. From its strictly nocturnal habits as well as its small size it is very apt to be overlooked, even in localities where it may be rather common. It is a hardy bird and well adapted to endure excessive cold.

Nelson says of Richardson's Owl: "In one instance, while at the Yukon mouth, I heard them uttering a peculiar grating cry on a cloudy morning in the middle of May. A fur trader from Kotlick brought me a set of four fresh eggs of this bird, taken from a nest in a bush near the Yukon mouth on the 1st of June. These eggs were white and round, as are most Owls' eggs. The man who brought them unfortunately neglected to bring the nest, but he told me that it was rather a small structure of twigs and grass. It was probably a deserted nest of the common Rusty Blackbird or of the Gray-cheeked Thrush, both of which nest commonly in that vicinity. The eggs were found in the midst of a dense thicket.

"Dall and others tell us that this bird generally nests in a hole in a tree, but the lack of trees at the Yukon mouth and the presence of bushy thickets may lead this bird to even build a nest for itself, and the fur trader insisted that the eggs above mentioned were in a nest of the bird's own construction."[3]

[1] Chamberlain's Canadian Birds, 1887, p. 61.
[2] History of North American Birds, 1874, Vol. III, p. 42.
[3] Report of Natural History Collections made in Alaska, 1887, No. 3, p. 151.

Three of the eggs referred to by Mr. Nelson are now in the U. S. National Museum collection, and measure, respectively, 39 by 32.5, 39 by 31.5, and 38.5 by 32.5 millimetres.

The eggs of the Old World form, *Nyctale tengmalmi*, a species closely related to Richardson's Owl, but a trifle lighter colored, and which, like the Lapp and Hawk Owls, is also likely to occur in Alaska, measure, according to Seebohm, from 31.7 to 33 millimetres in length and from 26.6 to 27.9 millimetres in breadth. There should be little or no difference in the relative size of the eggs of these birds. Like ours, Tengmalm's Owl is principally confined to timbered regions, and is said to nest in natural hollows of trees or the excavations of the Black Woodpecker.

The information obtained by Dall from Indians, whose testimony is not usually very reliable, is in this instance probably entirely correct, and I think if, as Mr. B. R. Ross states, this Owl nests in trees "that is, in an open nest," it will be found the exception and not the rule.

The only genuine eggs of Richardson's Owl in the U. S. National Museum collection are three collected by Mr. Ross, at Fort Simpson on the Mackenzie River, in latitude 62° N., on May 4, 1861. As there appear to be no memoranda showing the exact manner in which these eggs were obtained by him, it is likely that they were either taken from a cavity in, or from an open nest on, a tree.

The eggs described by Mr. Nelson, and previously referred to, are unquestionably those of the American Hawk Owl; they are absolutely indistinguishable and altogether too large for the species under consideration. How close they come to the eggs of the former will be seen by a comparison of the average measurement of thirty-two specimens now in the U. S. National Museum collection, which is 39.5 by 31.5 millimetres.

Seebohm, in speaking of Tengmalm's Owl, says: "This bird is a very early breeder, and even in latitude 67° N., Wheelwright's eggs were all taken between the 2d and 13th of May; whilst at Muoniovara, a degree still farther to the north, Wolley obtained eggs between the 18th of May and 2d of June, and received them from a little farther north between the 1st and 27th of June. Wheelwright describes its call note as a very musical soft whistle, never heard except in the evening and at night. Its food consists of mice, beetles, and small birds."[1]

Like the preceding, Richardson's Owl is an early breeder and nests probably at about the same time in corresponding latitudes. The number of eggs laid is likely the same, from three to seven. These are pure white in color, oval in shape; the shell is smooth, close grained, and shows but little luster.

The three specimens in the U. S. National Museum collection measure, respectively, 33 by 27.5, 32 by 27.5, and 31 by 27 millimetres.

The type, No. 5098 (Pl. 12, Fig. 6), is one of the above mentioned specimens.

---

[1] History of British Birds, Seebohm, 1883, Vol. I, pp. 164, 165.

120. Nyctala acadica (Gmelin).

SAW-WHET OWL.

Striz acadica Gmelin, Systema Naturæ, I, 1788, 296.
Nyctale acadica Bonaparte, Geographical and Comparative List, 1838, 7.
          (B 56, 57, C 328, R 401, C 483, U 372.)

Geographical range: Northern United States and British provinces, rarely south of 40° in eastern portions, but in mountainous western districts south to southern Mexico.

The breeding range of the Saw-whet or Acadian Owl, so far as known, extends east of the Mississippi Valley through the northern portions of the United States from about latitude 40° (Carroll County, Indiana), through the southern parts of the Dominion of Canada to latitude 51° 30' (Moose Factory), near the southern shores of James Bay. In the central Rocky Mountain region it ranges south to at least latitude 35° in northern New Mexico and Arizona, and on the Pacific coast north through the mountains of California, Nevada, Idaho, Oregon, and Washington into British Columbia; it probably breeds throughout all the higher mountain ranges of western North America.

It is more than probable that it will yet be found breeding throughout the more mountainous regions of eastern North America south of latitude 40°, but on account of its small size as well as its nocturnal habits it is seldom noticed, and while it may be rather irregularly distributed and even entirely absent in certain localities, it is nevertheless a far more common species than is generally supposed. It is a constant resident throughout the greater portion of its range within the United States, only migrating from its more northern breeding grounds, and passing the winter season mainly in the Middle States, where it is met with at times in considerable numbers. Mr. W. E. D. Scott took not less than twenty-one specimens during December, 1878, in a cedar grove on a side hill with a southerly exposure, near Princeton, New Jersey. He found some of them very tame and unsuspicious, allowing themselves to be taken by hand; I have also found them equally stupid in the vicinity of Camp Harney, Oregon. Each winter one or more specimens were brought to me alive by some of my men, who found them sitting in the shrubbery bordering a little creek directly in rear of their quarters, where they usually allowed themselves to be taken without making any effort to escape. I thought at first that they were possibly starved, and on that account too weak to fly, but on examination found them mostly in good condition and fairly fat. They seem to be especially fond of dense evergreen thickets in swampy places or near water courses.

Although the little Saw-whet Owl seems to be exceedingly gentle in captivity and of an amiable disposition, it does not appear to lack strength or courage, and is able to capture and kill rodents considerably larger than itself. Mr. George G. Cantwell informs me that he found a nest of this species in a small hollow white oak tree, the cavity, which had evidently been an old squirrel's nest, being filled with leaves and other trash to within a few inches of the entrance, which was about 15 feet up. The tree stood in the midst of a dense piece of timber bordering a small lake near Minneapolis, Minnesota. This nest contained on June 1, 1885, four young owlets of different ages and an added egg, as well as portions of two rats. Mice and other small rodents appear to furnish the principal part of its food, and occasionally a bird is captured. No doubt frogs and insects of various kinds are fed on as well. This bird must certainly be counted among the beneficial species and deserves protection. It seeks its humble prey mostly by night, while the days are passed in hollows in trees or in the gloomiest and darkest portions of the forests.

It is only within the last ten years that anything reliable has been learned about the nesting habits of this species, or fully identified eggs been taken, and perhaps none are better qualified to speak of the nesting habits of these quaint and interesting little Owls than Dr. William L. Ralph and Mr. Egbert Bagg of Utica, New York, who have devoted considerable time to the study of these birds and have taken five sets of their eggs. The former writes me as follows: "We found these birds quite common in Oneida County, New York, especially in the northern and eastern parts. Their nests are not very hard to find, and it seems strange that so few have been taken. Those found by Mr. Bagg and myself were all in the deserted holes of Woodpeckers and the eggs were laid on the fine chips found in such burrows without much of an attempt at making a nest. They were all in woods, wholly or in part swampy, such situations being particularly congenial to these birds, who usually frequent them throughout the year.

"Just before and during the mating season these little Owls are quite lively; their peculiar whistle can be heard in almost any suitable wood, and one may by imitating it often decoy them within reach of the hand. Upon one occasion, when my assistant was imitating one, it alighted on the fur cap of a friend that stood near him. They may often be seen at this season during the day sitting in trees or bushes and sometimes even hunting; but after their eggs are laid they are usually if not always silent, and are then very hard to find, as they keep in holes most of the time, except during the night. They are not at all suspicious and I have more than once stroked one with my hand as it was roosting sleepily in some bush or tree.

"Their call is a frequently repeated whistle, sometimes uttered in a high and again in a low key, and given in either a slow or rapid cadence. Generally it is commenced slowly and gradually becomes faster and faster till it ends quite rapidly. This call, which is the only one I have ever heard them give, sounds not unlike the noise made during the operation of filing a saw, and it is easily imitated.

"Both sexes, I think, assist in incubation, and the only apparent concern they show when their eggs or young are disturbed is the persistency with which they cling to their nests. They are constant residents of this county and seem not to decrease any in numbers. I have often found mice of different species and sometimes moles in and about their nests, but never any other kind of food.

"The first nest was taken near Holland Patent, New York, on April 7, 1886. It was situated 22 feet above the ground in a dead maple stump, and contained seven eggs ranging from fresh to slightly incubated. The second was found near the same place on April 21, 1886, also in a dead stump 40 feet above ground. It contained five young birds and an egg on the point of hatching. The third was found on the same day near Trenton Falls, New York, likewise in a dead stump 20 feet above the ground. It contained seven eggs which were heavily incubated. The fourth was found at Gang Mills, Herkimer County, New York, April 30, 1886, in a dead stump 50 feet above ground, and likewise contained seven eggs on the point of hatching. The fifth and last was taken near Holland Patent, New York, April 30, 1889, and was situated in the dead top of a maple tree 63 feet above the ground, and contained four eggs ranging from fresh to slightly incubated. I believe they lay their eggs at intervals of about two days."

Mr. William Brewster mentions the acquisition of a fully identified set of these eggs, taken by Mr. W. Perham, at Tyngsboro, Massachusetts, April 5, 1881, who, he says, takes many eggs of the Mottled Owl by hanging up artificial nests in suitable places in the woods. These nests are made from sections of hollow trunks boarded up at the open ends, with entrance holes cut in the sides, and the Owls apparently find them quite to their taste, for they freely appropriate them, both as roosting and nesting places. Sometime late in March of the above mentioned year a pair of Saw-whets took possession of one which was nailed against the trunk of an oak in an extensive piece of woodland. No nest was made, the eggs being simply laid on a few leaves which squirrels had taken in during the winter. There were four eggs on April 4, and, as the number was not increased, the following day Mr. Perham decided that the set was complete and accordingly took the parent birds with their clutch. He writes me that he made many unsuccessful attempts to catch the female on her eggs. She invariably flew out when he began to climb the tree, and he was at length obliged to shoot her. This behavior is strikingly different from that of the Mottled Owl under similar circumstances, for the sitting female of the latter species can always be taken off her nest by hand,

and even when pulled out of the hole rarely makes any attempt to escape. The male Saw-whet was shot while sitting on a branch near the nesting hole.[1]

In the same number of the Nuttall Bulletin, page 185, Mr. N. A. Francis, of Brookline, Massachusetts, makes the following statement: "On June 4, 1880, I found a nest of the Acadian Owl (*Nyctale acadica*), containing five nearly fledged young, in a cedar tree, in the midst of a dense swamp in Braintree. The nest was an old one of a Night Heron, repaired with a few leaves and feathers. Close to this nest of the Acadian Owl was found one belonging to a Long-eared Owl."

The little Saw-whet is not uncommon in the southern Rocky Mountains in Colorado, where Mr. Denis Gale found it nesting near Gold Hill, and Mr. William G. Smith in Estes Park. It does not appear to breed as early there as in the East, and the usual number of eggs laid by it is also less. Their nests were all found in abandoned Woodpecker excavations in old cottonwood trees, among the dense thickets along the water courses, and situated from 10 to 30 feet from the ground. The number of eggs or young averaged from three to five, never more.

Mr. Gale says that when disturbed in its nest it utters occasionally a peculiar cry, very similar to that of a startled Robin. Its food seemed to consist of mice and other small rodents, and no feathers were found in any of the cavities examined by him. He thinks that incubation lasts about three weeks, and that the eggs are covered continuously from the time the first one is laid, and are deposited at intervals of two or three days. In a set of three eggs taken June 3, one was fresh, the other two containing embryos in different stages of development. A second nest, found early in July, contained three young of different ages and two sterile eggs. The eldest young, which was more than half feathered, had a uniform chestnut colored breast. Mr. Gale raised these birds, and after they had attained their growth kindly presented them to me. They are deposited in the zoölogical collection now forming in Washington, District of Columbia, and attract considerable attention. They are cleanly, amiable creatures, and make exceedingly interesting pets, performing all sorts of curious antics in the early evening and throughout the night, while during the day they are rather sleepy and quiet, and sit close together in their cage.

Mr. W. G. Smith found a nest of this species in a gloomy ravine near water at an altitude of about 7,000 feet, on June 6, 1890. The eggs, four in number, were partly incubated and placed in a hole in an old cottonwood stump about 10 feet high and 8 inches from the aperture. The nesting material was scant and consisted of a few domestic chicken feathers. The female flew out when disturbed and was secured; the male was not seen. On June 8 he found a second nest containing three young birds already well feathered and within a mile of the former site and in a similar location; it appeared to be an old squirrel's nest.

---

[1] Bulletin Nuttall Ornithological Club, Vol. vi, July, 1881, pp. 143, 144.

One of my men also found a nest in the Blue Mountains, near Camp Harney, Oregon, May 2, 1881. The nest was in a deserted Woodpecker's hole in a pine stump standing in a dense thicket of young pines on the side of a cañon, and contained five nearly fresh eggs. The hole was 8 feet from the ground and about 10 inches deep, the eggs lying on the rubbish in the bottom of the cavity; the female was at home and was secured.

From the foregoing it appears that the Saw-whet Owl usually nests in hollow trees, in old squirrels' nests, or in the abandoned excavations of Woodpeckers, occasionally in artificial nests, as well as in open ones. Such instances as the latter are probably of rare occurrence, however.

The number of eggs laid to a set varies from three to seven, the latter number appearing to prevail in northern New York, where nidification usually commences early in April, while in the West, where the sets of eggs range only from three to five, the nesting season seems to commence either late in May or the beginning of June.

The eggs are pure white in color, with little or no gloss, usually oval in shape, and occasionally a set is found that is slightly ovate.

The average measurement of thirty specimens in the U. S. National Museum collection is 30 by 25 millimetres, the largest egg measuring 31 by 25, the smallest 29 by 25 millimetres.

The egg described in the "History of North American Birds," (Vol. III, p. 47), No. 14538, U. S. National Museum collection, and said to have been taken by R. Christ, Nazareth, Pennsylvania, April 25, 1867, is certainly not an egg of the Saw-whet Owl, being much smaller, very glossy, and measuring only 24 by 21.5 millimetres. From the texture of the shell I should call it a Woodpecker's egg.

The type specimen, No. 23889 (Pl. 12, Fig. 7), from a set of three eggs, was taken June 5, 1889, by Mr. Denis Gale near Gold Hill, Boulder County, Colorado.

121. Megascops asio (LINNÆUS).

SCREECH OWL.

Strix asio LINNÆUS, Systema Naturæ, ed. 10, I, 1758, 92.
Megascops asio STEJNEGER, Auk, II, April, 1885, 184.
(B 49, part. C 318, R 402, C 465, U 373.

GEOGRAPHICAL RANGE: Eastern United States and the southern border of the British provinces, except lower portions of the South Atlantic and Gulf States; west to the Great Plains.

The breeding range of the common Screech Owl, also known as the Mottled and sometimes, according to its plumage, as the Red or Gray Owl, extends through the eastern United States, west to middle Louisiana, excepting southern South Carolina, southern Georgia, all of Florida and the Gulf coast, where it is replaced by the slightly smaller Florida Screech Owl. North and West it ranges well into the Great Plains, having been taken in central Texas at Long Point, and it is reported as abundant in Kansas, not uncommon in Nebraska, and has been met with near Vermillion, Clay County, South Dakota, where it is said to be rare. In northern Minnesota and northern Maine it is rather rare; it is also found along the southern border of the Dominion of Canada. Mr. T. McIlwraith, in his "Birds of Ontario," says: "It is the most abundant of the Owls in this part of the country, yet, like the others, it is of very irregular occurrence. I have met with it once or twice in the woods in summer, but it is most frequently seen in winter when the ground is covered with snow."

Mr. M. Chamberlain, in his "Catalogue of Canadian Birds," says: "It occurs from Lake Huron to the Atlantic, though rare in the maritime provinces." Professor Macoun reports having taken one example, at Birtle, in northwestern Manitoba, in about latitude 50° 30'; but I believe this specimen is more likely to be *Megascops asio maxwelliæ*.

It is well known that the Screech Owl is subject to two distinct variations of plumage, the rufous and the gray, and young birds in both phases are often found in the same nest, even when the parents are of the same color. It is not as yet well understood to what this difference in coloration is due; certain climatological conditions may, however, be a considerable factor. While in some sections the two phases are nearly equally common, in others one or the other predominates. From what I have been able to ascertain it appears that the rufous phase is most frequently found in the Mississippi Valley, while the reverse seems to be the case near the Atlantic seacoast.

The Screech Owl is strictly nocturnal in its habits and is rarely seen in the daytime, which it passes in a hollow tree or in some dense thicket which the sunlight rarely penetrates. From such retreats it is often flushed by the attacks of the Blue Jay, which is the bane of its existence, and by other birds as well, which annoy it persistently wherever met with and generally cause it to seek safety in flight. The attention of the collector is often drawn to it by the noisy scoldings heard at such times.

It is a constant resident wherever found. Although often living in the immediate vicinity of farmhouses (old orchards being especially favored by them), yet on account of its retiring and unobtrusive habits it is seldom seen during the greater part of the year, though often enough heard in the early evenings throughout the mating season, when its doleful call notes are sure to be heard. In the winter when snow covers the ground, the trees denuded of their foliage, and food is scarce in the forests, many of these Owls that spent the summer in such localities now seek some dark and secluded nook in the outbuildings, corncribs, and haylofts about farmhouses in the vicinity, and are then more readily observed.

The Screech Owl, although considered by not a few superstitious and illiterate persons as a bird of ill omen, portending bad luck to the people whose home it frequents, is one of the most profitable and useful birds a farmer can have about his place, as it lives almost exclusively on mice and other rodents, of which it destroys large numbers yearly, and noxious insects as well, seldom molesting a bird of any kind, and occasionally fish, crawfish, and frogs are also eaten. One of these birds kept in confinement by Mr. O. B. Zimmermann seemed to be fond of caterpillars. The indigestible portions of the food used, such as bones and the fur of the rodents killed, are thrown up in the shape of pellets about eight hours after having been eaten.

I could not, if I wished, describe their call notes. It is beyond my ability to do justice to them on paper; the words "lugubrious" and "uncanny" express their sound as well as any, and it is mainly these doleful moans which prejudice many otherwise intelligent persons against these useful birds. Sometimes, however, their notes are rather clear and soft and not unpleasant.

I believe the Screech Owl remains paired through life. The mating period begins usually about the 1st of March, occasionally a little earlier or later, according to the season.

Mr. Lynds Jones, of Grinnell, Iowa, writes me: "I saw this species mating once. The female was perched in a dark leafy tree apparently oblivious of the presence of her mate, who made frantic efforts through a series of bowings, wing-raisings, and snappings to attract her attention. These antics were continued for some time, varied by hops from branch to branch near her, accompanied by that forlorn, almost despairing wink peculiar to this bird. Once or twice I thought I detected sounds of inward groanings, as he, beside himself with his unsuccessful approaches, sat in utter dejection. At last his mistress lowered her haughty head, looked at and approached him. I did not stay to see the sequel."

In the southern and central parts of its range nidification begins about the latter half of March or the first part of April; in the New England and Northern States, usually somewhat later, between the middle of April and the beginning of May.

The nesting site chosen is nearly always a hollow in a tree, either an abandoned Woodpecker's hole, or oftener a natural cavity, varying in height from 3 to 40 feet from the ground, usually at distances from 10 to 20 feet up. Occasionally it nests in boxes nailed up in trees or in some dark corner under the eaves of a barn or outbuilding, in dovecots, etc. Hollow apple trees and oaks seem to be favorite sites with them, and in Kansas and Nebraska, cottonwoods. Mr. Oliver Davie mentions his having found several nests between the broken siding of ice houses along streams. Mr. C. S. Brimley found a set of three eggs of this species placed in a cavity of a stump, the bottom of which was below the level of the ground outside; they were taken on April 27, 1885, and were much incubated when found.

Strictly speaking, the Screech Owl makes no nest, the eggs being laid in the bottom of the cavity on such rubbish as naturally accumulates therein, such as bits of rotten wood, a few dry leaves, and the feathers dropped from the birds during incubation. Four or five eggs are usually laid to a set, and occasionally six, and even seven are sometimes found, but the latter number is rare. Mr. Oliver Davie, in his "Nests and Eggs of North American Birds, 1889," page 196, states that they frequently lay eight, rarely nine, and says that in April, 1885, a farmer brought him nine young, with the parent birds, which he had taken from a hollow tree. The largest set I have any knowledge of, or have been able to fully verify, is one of seven. These eggs were slightly incubated when found and were taken by Mr. W. E. D. Scott, near New Brunswick, New Jersey, April 11, 1879.

Mr. Lynds Jones writes me: "Both parents are generally found near the nest, and not infrequently sitting on the eggs at the same time. In a number of instances I have taken the two from well incubated eggs, but have never flushed both from a fresh set. Between the interval when the first egg is laid and the set is completed, the male may be found in a hollow tree near by and cannot be flushed, while the female watches the nest and flushes easily When incubation begins the male will flush readily for a time, the female, however, remaining. Later, both birds must be dislodged by force. If the cavity is large enough to admit of it, both birds will lie over the eggs; if, however, it be small, the female covers the eggs and the male either wedges himself down by her side or lies on top of her, and sometimes finds a lodgment somewhere higher up in the hole, which, however, is rarely the case."

When suddenly disturbed in their hole they frequently utter a hissing noise and snap their mandibles together, producing a kind of rattling sound. In disposition these birds are rather unsociable, seldom more than a pair being seen together at any time, except when still caring for their young in the summer and early fall.

Incubation lasts about three weeks, or perhaps a few days longer. The young, according to Mr. Lynds Jones, are blind when first hatched, and soon develop an astonishing appetite, keeping both parents busy to provide the needed food for the rapidly growing family. The foraging is done almost entirely after sundown and throughout the night, and a supply of food to last during the day is usually stored away in their nesting site. Their flight is noiseless and easy, enabling them to drop down silently and quietly on their unsuspecting victims.

Though small, the Screech Owl does not lack courage, and will frequently attack animals much larger and heavier than itself, and has been known to kill large rats.

Mr. J. L. Davidson publishes the following remarkable incident in Forest and Stream of March 19, 1885:

"LOCKPORT, NEW YORK, February 26.

"On Saturday last I received a box containing a live Screech Owl (Scops asio), from a young farmer friend, and on Tuesday received the following letter from him:

'Friday morning I found a large Plymouth Rock rooster with his head and neck badly torn and covered with blood, and, after some search, I saw a small Owl up in the barn. I caught it and found fresh blood on the feathers around its beak. To be sure it was the Owl that did the mischief, I put both the fowl and the bird in a darkened place and was at once treated to a surprise, for the Owl flew at the cock and lit on his neck, and began to pick at his head in a very furious manner. Being sure that I had the culprit, I boxed it up and sent it to you. The rooster weighed 9 pounds alive, rather large prey for such a small bird.'"

They are easily tamed and make rather amusing pets, are cleanly in their habits, and very fond of bathing.

The eggs of the Screech Owl are pure white in color, usually oval or nearly globular in shape, and moderately glossy. In the majority of specimens the shell is smooth and finely granulated, while in a few it is rough to the touch. They are deposited at intervals of a day or two.

The average measurement of fifty-six specimens in the U. S. National Museum collection is 35.5 by 30 millimetres, the largest egg of the series measuring 38 by 31, the smallest 32 by 29.5 millimetres.

The type specimen, No. 20571 (Pl. 12, Fig. 8), selected from a set of four, Bendire collection, was taken near Orange, New Jersey, April 20, 1875.

122. Megascops asio floridanus (RIDGWAY).

FLORIDA SCREECH OWL.

Scops asio var. *floridanus* RIDGWAY, Bulletin Essex Institute, December, 1873, 200.
Megascops asio floridanus STEJNEGER, Auk, II, April, 1885, 184.
(B —, C 318c, R 402a, C 469, U 373a.)

GEOGRAPHICAL RANGE: Southern South Carolina and Georgia, Florida and the Gulf coast to southern Louisiana.

The breeding range of the Florida Screech Owl, a much darker and smaller bird than the preceding, and like it, found in both the red and gray phases of plumage, extends from southern South Carolina, southern Georgia and Florida, westward along the Gulf coast to southern Louisiana, where, according to Dr. A. K. Fisher, it is not uncommon in the vicinity of New Orleans. A Screech Owl has also been reported as breeding at Houma, Terre Bonne County, Louisiana, which is unquestionably referable to this race, and it probably reaches west to southern Texas. It is a constant resident wherever found.

The general habits of the Florida Screech Owl are similar to those of the preceding subspecies. Like it, it never constructs a nest of its own, but lays its eggs in the abandoned excavations of the larger Woodpeckers or in natural hollows in trees, old stumps, cabbage palmettos, and occasionally in a dovecot, and in the latter case usually dispossessing the rightful owners.

Dr William L. Ralph says: "They are not at all particular as to the height at which they nest. I have found them occupying holes anywhere from 8 to 80 feet from the ground. They nest frequently in rotten stumps at such heights as to make it dangerous, if not impossible, to reach them. I remember one pair that nested near the house where I boarded, in a hole at least 80 feet above the ground, near the top of a very large rotten stump which towered above the tops of a clump of trees among which it was standing. Every time during the breeding seasons of two years that I would go near this stump one of the pair, whichever might be sitting, would look out of the hole in a most provoking manner, for I wanted a set of eggs of this subspecies very much at that time, but the stump was not climable. Usually it is a hard matter to make these birds show themselves; this pair, however, seemed to know that they were perfectly safe, and never hesitated to make their appearance.

"Like the northern Screech Owls, they utter their call notes with a peculiar tremulous sound, which is hard to describe, and from which they get their local name of 'Shivering Owls.' They feed on mice, moles, insects, and other small members of the animal kingdom.

"The mating season of the Florida Screech Owl begins sometime in March, usually in the first half of the month, and their eggs are deposited in April, generally between the 10th and 20th. I found my first nest April 29, 1883, near San Mateo, Florida, in a rotten stump, in a hole 8 feet above the ground. The hole was 8 inches deep, and the two about half incubated eggs which it contained were deposited on a few rotten chips, leaves, and feathers that had collected in it. The bird on the nest was taken from it by hand, in fact I could not get it to leave in any other way. It was a male in the red plumage, and its mate was not seen. A second nest, found on May 6, 1884, in the same vicinity, was in a pine stump, and the cavity was about 12 inches deep and 8 feet above the ground. It contained three eggs well advanced in incubation. The parent had to be taken out by force; it was in the gray plumage, sex unknown, and its mate not seen."

In southern Florida (near Myers, Lee County), Mr. J. F. Menge found the Florida Screech Owl nesting sometimes as early as March 1. He says: "They feed their young to a great extent on lizards and grasshoppers, like the Florida Burrowing Owl."

From two to four eggs are laid to a set, three being most often found, and as far as known but a single brood is raised in a season. Both sexes assist in incubation. The eggs are similar in color and shape to those of the common Screech Owl, but average somewhat smaller. Occasionally an egg is found which is almost spherical.

The average measurement of ten eggs in the U. S. National Museum collection is 32.5 by 28.3 millimetres. The largest of these eggs measures 34 by 29.5, the smallest 32 by 27.5 millimetres. One nearly round measures 31 by 29 millimetres. None of these eggs are figured.

123. Megascops asio mccallii (Cassin).

TEXAN SCREECH OWL.

Scops mccallii CASSIN, Illustrated Birds of California, Texas, etc., July, 1854, 184.
Megascops asio maccalli STEJNEGER, Auk, ii, April, 1885, 184.
(B 50, C 318b, R 402b, C 468, U 373b.)

GEOGRAPHICAL RANGE: Southern Texas and eastern Mexico; south to Guatemala.

The Texan Screech Owl, a slightly smaller race than *Megascops asio*, and, like this, found in both the red and gray phases, is, according to Mr. R. Ridgway, distinguishable from the latter in having the light mottlings on upper parts much coarser and more conspicuous. It is a resident of the semi-tropical portions of southern Texas, especially of the valley of the Lower Rio Grande, and eastern Mexico, and breeds wherever found.

But little is as yet known about the general habits and food of this form, but they are doubtless very similar to those of the rest of this genus. Like them it breeds in cavities in trees and lays from two to five eggs. Mr. George B. Sennett found a nest of this race on April 23, 1877, situated in a dead stub, about 9 inches in diameter, and so weak and rotten that it could easily have been pushed over; it contained three fresh eggs; location about 4 miles from Hidalgo, up the river and within about one-fourth of a mile of its banks.[1] Another set of five eggs found about April 1, 1878, were sent to him by Mr. Bourbois, who took them from a nest on his ranch at Lomita.

Dr. James C. Merrill, assistant surgeon U. S. Army, says with reference to this Owl: "Common resident. Near Hidalgo on May 6, I captured a female of this race on her nest in an old hollow stump about 5 feet from the ground. There were two eggs nearly hatched, placed on a few chips at the bottom of the hole."[2]

There seems to be no especial difference in the time of nidification, and the eggs of the Texan Screech Owl are similar in every respect to those of the common form, excepting that they are a trifle smaller.

The average measurement of twelve specimens of this race in the U. S. National Museum collection is 33.5 by 29 millimetres, the largest egg measuring 35.5 by 30 the smallest 32.5 by 28 millimetres. No specimen has been figured.

[1] Bulletin U. S. Geological Survey, 1878, p. 40.
[2] Proceedings of the U. S. National Museum, Vol. i, 1878, p. 151.

124. Megascops asio bendirei (Brewster).

CALIFORNIA SCREECH OWL.

Scops asio bendirei Brewster, Bulletin Nuttall Ornithological Club, vii, January, 1882, 31.
Megascops asio bendirei Stejneger, Auk, ii, April, 1885, 184.
(B —, C —, R —, C —, U 373c.)

Geographical range: California.

The range of the California Screech Owl, a brownish gray and somewhat larger bird than any of the preceding races of *Megascops*, and only found in this phase, is confined, so far as known, to the State of California, and principally if not wholly to that portion lying west of the Sierra Nevada Mountains. It has also been reported from southwestern Oregon, but I believe that the birds found there, and even those of the coast regions of northern California, in Del Norte and Humboldt Counties, are not typical, but intermediate between this race and *Megascops asio saturatus*, a new subspecies recently described by Mr. William Brewster in The Auk (Vol. viii, April, 1891, p. 141). Our present knowledge regarding these different subspecies is still too limited to define their ranges with any degree of accuracy. It is well known that a Screech Owl is also found on the peninsula of Lower California that has not as yet been fully identified, though it may prove to be this form, but is much more likely to be *Megascops asio trichopsis*, or a new race. They are constant residents and breed wherever found.

Their general habits are essentially like those of the eastern Screech Owls, except that, according to Mr. Charles A. Allen, of Nicasio, Marin County, California, their call notes are slightly different and more prolonged. He also informs me that they are very fond of a cold bath in the early mornings, and literally soak themselves, and that he has shot them in the act. He also says they are the commonest Owl in that vicinity.

Mr. W. Otto Emerson, of Haywards, California, nailed some boxes up in a small grove of Australian gum trees on March 14, 1885. Examining these on April 10, he found one occupied by a pair of these Owls which contained two eggs, the parent flushing and flying into a tree close by. Two days later another egg had been laid, but no more were deposited after that date. Incubation, he says, did not commence until the 17th. The male was only seen once during incubation, when he appeared instantly at the call of the female, which had also left the box. As he climbed up to examine it the male snapped his mandibles at him from a limb overhead. On May 10 he found three white downy owlets in the nest, one not yet quite dry. He says: "The mother seemed to think I was getting too free with her household, for when I took one out she flew about, calling the male. They both came near, snapping their bills all the while. The young could not hold up their heads, their eyes were round and full, but not open. On taking them out they gave a slight peep, like a young chicken just hatched. They grew very fast, and in eight days commenced to show signs of pinfeathers and had strength enough to snap their bills. About this time, May 18, I found one of the young with its head missing, which I could not account for unless food got short, and rather than see them all starve the birds fed one to the others. Next day the body had disappeared. I put all the remains of birds skinned on the top of the box; they made way with these and the young grew very fast. By the 25th, fifteen days after hatching, they had their eyes open. They would now back up into a corner, snapping their bills on my trying to get them out. * * * They were still in the box on June 14, when I went to the Farallon Islands, so do not know how they got on after that. * * *

"I found one the past spring that had taken up quarters in an old wood rat's nest placed 'on' a limb of a bay tree some 30 feet from the ground. A large mass of dead leaves from the tree had been put together and a hollow formed in the center, lined with feathers of fowls and birds."[1]

Mr. B. T. Gault writes me as follows: "On May 12, 1883, while riding through the timber along the Santa Anna River, near Riverside, California, a nest of this subspecies was unexpectedly discovered in an old abandoned Woodpecker's hole, presumably that of *Colaptes cafer*, in a cottonwood tree. It seemed remarkable that such a small hole would so easily accommodate a bird of this size, as we had to enlarge it to enable me to insert my hand in order to examine its contents, and I was not a little surprised in doing so to find it occupied by a family of these Owls, consisting of one of the parent birds and four young, perhaps ten days old. The mother bird appeared dazed when brought to the light, and singularly enough in taking her from the nest the entire brood was also removed at the same time, she having instinctively grasped one of the young, that one another, and so on until they all became attached, and they certainly presented a ludicrous sight as they came dangling out of the hole, each retaining a firm hold of the other, but the young finally dropped off and tumbled to the ground."

Their favorite nesting sites are old Woodpeckers' holes or natural cavities in oak and cottonwood trees, generally not over 15 feet from the ground. According to Mr. Emerson's observations, incubation lasts about twenty-three days, and I believe he is the first ornithologist who published the fact that these Owls were hatched blind. Mr. Lynds Jones confirms this fact, and has observed it in the young of *Megascops asio*.

In California nidification usually commences about the middle of April, occasionally in the latter part of March, and again, in late and cold seasons, not until the beginning of May. The eggs, numbering from three to six, generally four, and occasionally five, are deposited at intervals of one or two days. Incubation ordinarily does not begin until the set is complete.

---

[1] Ornithologist and Oölogist, Vol. X, 1885, No. 2, pp. 173, 174.

The eggs of the California Screech Owl are similar to those of the common eastern bird, but, notwithstanding that it is somewhat larger than the latter, their eggs average a trifle smaller.

The average measurements of twenty-six specimens in the U. S. National Museum collection is 35 by 30 millimetres, the largest egg measuring 36 by 32, the smallest 32 by 28 millimetres. As there is so little difference in the eggs of the various races of this genus, I have not figured one in every case.

125. Megascops asio kennicottii (ELLIOT).

KENNICOTT'S SCREECH OWL.

Scops kennicottii ELLIOT, Proceedings Academy Natural Sciences Phila., 1867, 69.
Megascops asio kennicottii STEJNEGER, Auk, II, April, 1885, 184.
(B —, C 318a, R 402d, C 466, U 373d.)

GEOGRAPHICAL RANGE: Northwest coast region, southern Alaska, from Sitka south probably to northern British Columbia.

Mr. William Brewster has recently rearranged the genus *Megascops*, and described three new subspecies in the Auk (Vol. VIII, No. 2, April, 1891, pp. 139–144), which I include. According to this new arrangement Kennicott's Screech Owl, an extremely dark tawny colored race, is confined to the southern coast regions of Alaska from the vicinity of Sitka, south probably to northern British Columbia.

The type, No. 59847, U. S. National Museum collection, a male, was taken by Mr. Ferdinand Bishop, at Sitka, March, 1866, and remains, so far as I am aware, unique in collections.

It is likely to be a constant resident wherever found, and there is probably no difference in its nesting habits and eggs from those of other members of this genus. It is one of the largest of the several races of Screech Owls.

126. Megascops asio maxwelliæ (RIDGWAY).

ROCKY MOUNTAIN SCREECH OWL.

Scops asio var. *maxwelliæ* RIDGWAY, Field and Forest, June, 1877, 210, 213.
Megascops asio maxwelliæ STEJNEGER, Auk, II, April, 1885, 184.
(B—, C —, R 402c, C 467, U 373e.)

GEOGRAPHICAL RANGE: Foothills and adjacent plains of the eastern Rocky Mountains from Colorado north to Montana.

The Rocky Mountain Screech Owl is the lightest colored and, in my opinion, the handsomest of the geographical races of Screech Owls found in the United States. Comparatively little is as yet known regarding the extent of its breeding range. As far as I am aware it has only been found along the foothills among the cottonwood timber of the creek bottoms on the eastern slopes of the Rocky Mountains, from central Colorado, in the vicinity of Colorado Springs, where it is believed to be a winter visitor only, to the northern parts of the State, in Boulder and Larimer Counties, where it is known to breed, rarely, however, reaching a greater altitude than 6,000 feet; and also near Fort Custer, in southeastern Montana, where I first met with it in the winter of 1884–'85, and which point, I believe, still marks the northern boundary of its known range.

As Screech Owls are usually residents wherever found, rarely straggling to any great distance from their summer ranges, it is reasonable to presume that this subspecies breeds more or less commonly along the lower foothills and plains of the eastern slopes of the Rocky Mountains, as well as along the outlying spurs of the same, ranging from the eastern portions of northern Colorado through similar localities in Wyoming, north to southeastern Montana (Fort Custer), and possibly still farther in this direction. It is also likely to be found in the extreme northwestern parts of Nebraska, but it is doubtful whether it occurs on the western slopes of the Rocky Mountains. As yet it has only been found breeding in northern Colorado, where it seems to be tolerably common in suitable localities.

The following notes on the Rocky Mountain Screech Owl are taken from an article recently published by me:

"The credit of the discovery of the nest and egg of this race, the handsomest of the genus *Megascops*, belongs, I believe, to Mr. A. W. Anthony, one of our younger and most energetic naturalists, who has done excellent work in this line, as well as in other branches of natural history, in various portions of the West, and has generously donated through the writer a number of his rarest and most interesting specimens to the U. S. National Museum collection at Washington. He writes me as follows regarding this species: 'On May 4, 1883, while collecting on the Platte River, about 6 miles from Denver, Colorado, my attention was attracted by the hammering of a Red-shafted Flicker, and pushing my way through a very thick growth of willows and small cottonwoods, I found the bird at work on one of the latter, where he was excavating a nesting site. The tree was a very large one; its top had been partly broken off about 12 feet up, blown over, and some of its limbs rested on the ground. As I climbed up via the leaning top to the *Colaptes* burrow, which was located in that part of the trunk of the tree still standing upright, a Rocky Mountain Screech Owl flew out from a knot hole not before noticed and dashed almost in my face, lit on a tree within 6 feet of me, and, after staring at me in amazement for a few minutes, dropped down and out of sight in the dense undergrowth in the neighborhood. The two burrows were about 4 feet apart, nearly on a level with each other, but on opposite sides of the tree.'

"The Owl's nest was in an old knot hole about 15 inches in length, and judging from a rough sketch sent me by Mr. Anthony at the time, the base of the nest was almost on a level with the entrance. It contained three young about a week old and an addled egg. This egg was not found until his return to the nest a second time a few hours afterward, when one of the parents was caught and a careful examination of the nest made. This, if it can be called a nest, was composed of bits of rotten wood, a few feathers of small birds, and a good many fish scales. The tree was standing within 100 yards of the river. Fish of various species seem to form no inconsiderable portion of the diet of other small Owls as well, as I have found on more than one occasion good sized brook trout (*Salmo purpuratus*) in the burrows used by Kennicott's Screech Owl (*Megascops asio kennicottii*) in Washington Territory.[1] Just how they manage to catch an active fish like a brook trout, if they take them alive, which, I must confess, is very questionable, would be interesting to know.[2]

"Mr. Anthony thinks that the Rocky Mountain Screech Owl breeds also in old abandoned nests of the Black-billed Magpie (*Pica pica hudsonica*), and he writes me that he has often found them roosting in them both in winter and spring, and has found the American Long-eared Owl (*Asio wilsonianus*) breeding in such nests.

"Mr. Denis Gale, of Gold Hill, Colorado, has taken several nests of this bird during the last three years, and finds them not at all uncommon in that vicinity. A set of four eggs now before me was found by him on April 20, 1886, on Boulder Creek, near Boulder City, Colorado. He writes me regarding these eggs as follows: 'Judging from the different stages of advancement in the embryos, I am inclined to think they were laid at intervals of from forty-eight to seventy-two hours, and that the eggs were covered continuously from the time the first one was laid. The burrow used for a nesting site by this pair of birds was an old Flicker's hole, in a cottonwood tree, about 20 feet from the ground. There was nothing between the eggs and the bare wood bottom on which they lay that bore the semblance of a nest, excepting a little wood dust and a few wing and tail feathers of the Arctic Bluebird and several species of Sparrows. These feathers were without doubt the remnants of birds fed to the sitting female by her mate, the soiled and stained eggs showing plainly their coming in contact with the mangled food devoured over them.' The female covered her eggs with great persistency and was only removed off them by force, snapping her bill and using her sharp claws with great energy when handled. Mr. Gale tells me that besides small birds, several species of the smaller rodents, frogs, and crawfish also form part of their bill of fare.

"Mr. Gale writes further as follows: 'Rarely does this species follow the creeks far into the foothills; I have not observed them at 6,000 feet altitude. Like others of their genus they seem to delight in a sheltered, shady location, close to a pond or creek where they select a domicile, either in a

---

[1] Recently described as *Megascops asio macfarlanei*, BREWSTER.
[2] Since this article has been written, I notice that this fishing propensity is not confined to the two races mentioned therein, but is common to the eastern form as well. (See M. A. Frazar, in Bull. Nutt. Ornith. Club, Vol. II, No. 3, July, 1877, p. 80, and Willard E. Treat, in Auk, Vol. VI, April, 1889, p. 189.)

natural tree hole or in a Flicker's old nest site. If for any reason the Flicker wishes to retain his previous year's nest site, and Scops is in possession, strife is carried on between them with great vigor, ending as often in favor of one as the other, judging from the broken eggs upon the ground ejected by the victor. The Flicker dares not enter to turn Scops out, but if the premises are vacated for ever so short a time, he enters and holds them against all comers. His formidable bill pointing out at the door is sufficient apology for leaving him in quiet possession.'

"About the middle of April is the usual period for the eggs, which are from three to five in number. The nest is usually a sparse gathering of wing and tail feathers of small birds; in some instances no litter of any kind is present. As a rule, the first two or three eggs are laid on consecutive days, with intervals of two, three, or more days between the third and last one or two, as the case may be. The female is always in charge, and at no time leaves the nest while sitting or while her brood is very young. She is waited upon and fed by the male, who, being a skillful hunter, provides liberally for her wants. Searching for nests I have sometimes discovered the male hidden in a tolerably well stocked larder, in close proximity to the nest site. In one cache were portions of a Bluebird, a mouse, and a frog; in another a Junco, a Tree Sparrow, and a minnow 3½ inches long; claws and legs of crawfish were also present. In a few cases I have discovered the male sitting upon a bough close to the stem of a cottonwood tree, perfectly motionless, with eyes almost closed as if asleep, the pupil of the eye closed to the merest slit, but with ears erect, and all alive to the danger threatening his sitting mate close by; in this well selected position, his colors and markings so nearly resembling the rough corrugated bark of the tree, he seemed to have the fullest assurance of security against observation.

"The female is a close sitter. To induce her to leave her nest is a difficult matter unless she has been frequently disturbed and understands what is meant when she hears the tree grappled in climbing it. She will then fly out. Otherwise you will have to take her off her eggs. In some instances she will feign dead and lie on her back in your open palm with her eyes shut. Immediately you throw her off, however, she will right herself on wing, and gaining a bough on a neighboring tree will crouch forward, bending her eartufts back and look very spiteful and wicked. At other times when removed from her eggs she will snap her bill, moan slightly, and show fight. Both male and female indulge in the screech, which differs but little from that of their eastern cousins. Its sharp distressing notes can be heard of a still night a mile distant.

"The lately hatched young are clothed in beautiful white down. In the latter part of June, before they are well able to fly, they may be seen sitting side by side, perfectly motionless, upon a limb close by the nest site. The young and their parents seem to desert their holes and live among the **trees for the** balance of the summer; but when the cold winds strip the **leaves from the** trees in the fall suitable tree holes are selected for their **winter** quarters.

"While stationed at Fort Custer, Montana, during the winter of 1884–'85, I took five of these birds, but was unable to find their nests. I discovered their presence there quite accidentally. On December 1, 1884, while out hunting Sharp-tailed Grouse in a bend of the Big Horn River, a few miles south of the post, as I was walking by a thick clump of willows I indistinctly noticed a whitish looking object dropping on the ground, apparently out of the densest portion of the thicket and on the opposite side from where I was standing at the time, and simultaneously heard several plaintive squeaks from that direction. Carefully skirting around the thicket, which was some 20 yards long and perhaps 5 yards wide, I saw the object of my search savagely engaged in killing a meadow mouse which it had just captured. I promptly shot it. It proved to be a female and excessively fat; in fact all the specimens I secured subsequently showed conclusively that they managed to secure an abundance of food in that Arctic winter climate, and that a portion of this at least seems to be obtained in the daytime. The four other specimens collected by me were all obtained in similar locations. I have no doubt that it breeds in the vicinity of Fort Custer, but I lost trace of these birds in the spring months and failed **to** bear their love notes at that time. It is possible that they retire a little **nearer** to the mountains to breed. This is, up to date, the most northerly local- **ity** recorded at which the Rocky Mountain Screech Owl has been obtained."[1]

In Colorado full sets of eggs are sometimes found by April 1, and again **as late as May** 25, the latter probably a second **laying,** the first having been destroyed. **But a** single brood is raised **in a season. Hollow** cottonwood trees furnish **their** favorite nesting sites, **at distances from** 4 to 25 **feet** from the ground, but **according** to Mr. Gale they **also** breed at times in natural cavities in box elders **and** black willows, and occasionally a pair will make use of the abandoned nest of the Black-billed Magpie, as Mr. Anthony sur- mised. Mr. William G. Smith, of Loveland, Colorado, has since informed me that he found a set of eggs in such a situation in the spring of 1890. He also states that remains of young cotton-tail rabbits are conspicuous near their nesting sites while rearing their broods. He has found them nesting in the timber along the creek bottoms, fully 20 miles out on the plains, away from the foothills, and they probably reach points still farther east of the mountains.

The eggs of the Rocky Mountain Screech Owl are pure white in color and moderately glossy; the shell is smooth and finely granulated. In shape they vary from oval to a broad elliptical oval, some being decidedly more elongated **than any** other eggs of the genus *Megascops* I have seen. The usual number **of** eggs **to a set** is four. Of ten sets taken by Mr. Gale, three con- tained three eggs **each,** six contained four, and one set five eggs. They are usually deposited **at** intervals of a couple of days, and incubation sometimes begins with the first egg laid.

[1] Auk, Vol. vi, 1889, No. 4, pp. 298–302.

The average measurement of nine of those eggs in the U. S. National Museum collection, all but one presented by Mr. Gale, is 36 by 29.5 millimetres, the largest egg measuring 40 by 29.5, the smallest 34 by 29.5 millimetres. Thirty-eight eggs taken by Mr. Gale average 36.5 by 30 millimetres.

The type specimen, a single egg, taken on the Platte River near Denver, Colorado, on May 4, 1884, by Mr. A. W. Anthony, No. 22450, is figured on Pl. 12, Fig. 10.

127. Megascops asio trichopsis (WAGLER).

MEXICAN SCREECH OWL.

Scops trichopsis WAGLER, Isis, 1832, 276.
Megascops asio trichopsis RIDGWAY, Proceedings U. S. National Museum, VIII, 1885, 355.

(B —, C —, R 403, C 470, U 373*f*.)

GEOGRAPHICAL RANGE: Northwestern Mexico and the contiguous border of the United States, in Arizona and New Mexico (Lower California?).

The Mexican Screech Owl is a slightly smaller race of the common form, and distinguished from it by its pure ashy gray plumage, with the middle streaks black and in strong contrast; it is only found in this phase. Its range, as far as known, extends from central Arizona and New Mexico south to northwestern Mexico, and probably Lower California, where a Screech Owl is known to occur, but of which no adult birds have as yet been taken. A pair of Screech Owls obtained by Mr. Vernon Bailey near Oracle, Pinal County, Arizona, on June 11, 1889, are, according to Mr. William Brewster, intermediate between this race and the newly described *Megascops asio aikeni*, from southern Colorado.

Its habits are similar to the other members of this extensive genus, and, like them, it nests in hollow trees or sahuaras (the giant cactus) in abandoned excavations of the larger Woodpeckers, or in natural cavities, generally at no great distance from the ground.

I first met with this little Screech Owl among the shrubbery in the Rillitto Creek bottom, near Tucson, Arizona, in 1872; it appeared to be rather rare here and I saw but few. During February and March their weird, mournful notes were frequently heard shortly after sundown. On March 10, I finally obtained one, which was sitting in a bushy willow overrun with wild grape vines. It was perched on a small limb close to the main trunk and did not appear to have seen me before I noticed it sitting in a drowsy attitude, and as soon as it saw me it straightened up, sat very erect, with all its feathers pressed close to the body, making it appear nearly as long again, and it might then have been easily mistaken for a slender stump of the limb on which it was resting. I have no doubt but that it was done for the purpose of mimicry and in which it is frequently successful.

On March 26, 1872, I found one of their nests in an old Woodpecker's hole in a willow stump not more than 7 inches in diameter and about 6 feet from the ground. The cavity was slightly over 2 feet deep, and the four eggs it contained, which had been incubated for a few days, were lying on bits of rotten wood and a few dead leaves, not sufficient to call a nest. The female was at home and had to be taken out forcibly, protesting and uttering a hissing sound, and, after being turned loose, snapping her mandibles rapidly together from her perch on a small walnut tree, into which she had flown. I was in hopes she might continue to use the same site again, but was disappointed in this.

Mr. Herbert Brown, of Tucson, finds this race nesting frequently, in that vicinity at least, in the sahuara cactus (*Cereus giganteus*), which are often used as nesting sites by both the Gila Woodpecker (*Melanerpes uropygialis*), and especially by the Gilded Flicker (*Colaptes chrysoides*), being easily excavated and affording both roomy and secure homes in which to rear their families. I believe such cavities are selected in preference to any others by the Mexican Screech Owl, as well as the more numerous Elf Owls. In a recent letter received from him, he writes me: "I have found the Mexican Screech Owl nesting in holes of sahuaras within 4 feet from the ground, and from that distance up to almost the extreme top of the plant. The sahuaras along the river bottoms, and on the mesas bordering them, are their favorite nesting grounds. At the head of the pass leading through the Tucson Mountains, about 2½ miles from the Santa Cruz River, there is a large forest of immense sahuaras, some of them 60 feet high and both body and limbs bored full of Woodpeckers' holes. As nearly all were too large to climb, I cut several down, but failed to obtain a single Owl of either this subspecies or the more common Elf Owl, and at the Quijota, 90 miles southwest of Tucson, I also cut down a number of these in a similar location some distance from water, with the same result, and I have come to the conclusion that these Owls only occupy the sahuaras growing in the lowlands and not those in the higher hills or out in the deserts.

"They are pugnacious and wicked little fellows, who will use their claws, and bite as well, on the least provocation. Small birds, kangaroo rats, gophers, different species of mice, lizards, scorpions, grasshoppers, and beetles are their staple articles of diet. From the lateness of the season in which I have occasionally found them nesting, I believe that two broods are sometimes raised in a year; still I have no actual proof of this."

In the oak regions of southern Arizona they nest in the natural cavities of these trees, most of which are hollow. Mr. O. C. Poling found two of their nests in such trees, one on May 1, the other on May 5, 1890.

Nidification begins occasionally in the last half of March, continuing through April and the first week in May, according to altitude; birds breeding in the hot valleys nesting fully a month earlier than those living in the mountain regions.

26957—Bull. 1——24

The number of eggs laid is usually three or four, rarely five. They are similar in shape and color to those of the rest of this genus. Now and then a set is found which is so badly stained by the excrement of fleas inhabiting their burrows in large numbers that the eggs, judging by their color, might be taken for those of the Sparrow Hawk.

The average measurement of twenty specimens is 34 by 29.5 millimetres. The largest egg of this series measures 36 by 31, the smallest 32.5 by 27.5 millimetres.

The type specimen, No. 20569 (Pl. 12, Fig. 11), U. S. National Museum collection, selected from a set of four from the Bendire collection, was taken by the writer near Tucson, Arizona, on March 26, 1872.

128. Megascops asio aikeni Brewster.

AIKEN'S SCREECH OWL.

Megascops asio aikeni Brewster, Auk, VIII, April, 1891, 139.
(B —, C —, R —, C —, U 373*g*.)

GEOGRAPHICAL RANGE: Southern Colorado (south probably to central New Mexico and northeastern Arizona).

According to Mr. William Brewster, who recently described this new race, Aiken's Screech Owl is of about the same size as the California Screech Owl, and resembles *Megascops asio trichopsis* somewhat in coloration, but with the mesial streaks and stripes of dull black coarser and more conspicuous, thus giving it a darker appearance.

The type was obtained by Mr. Charles E. Aiken in El Paso County, Colorado, on May 29, 1872, and according to his observations it is not uncommon among the cottonwood timber along the creeks in El Paso and Fremont Counties, in central Colorado. It inhabits similar localities as the Rocky Mountain Screech Owl, and the latter is frequently found, in the winter at least, in the same groves with this subspecies, whose range it overlaps here.

It is questionable if Aiken's Screech Owl occurs at any points north of Douglas County, Colorado, but it probably inhabits all the more open country along the foothills of the southern Rocky Mountains south at least to central New Mexico and northeastern Arizona.

Dr. R. W. Shufeldt, U. S. Army, found a Screech Owl breeding near Fort Wingate, New Mexico, which I think is referable to this subspecies. He took three well incubated eggs on April 18, 1887, from a cavity in an oak tree 10 feet from the ground, capturing alive both parents at the same time. I have seen photographs of these specimens, and they show every indication that the originals belonged to this race. The eggs are now in the U. S. National Museum collection, and resemble those of the other members of this genus in shape and color, and average slightly larger than the eggs of the Mexican Screech Owl, measuring 36.5 by 31, 36 by 30.5, and 34.5 by 31 millimetres. A pair of birds from Oracle, Pinal County, Arizona, seem to be intermediate between this race and the slightly smaller and lighter colored Mexican Screech Owl. The habits of Aiken's Screech Owl are similar to those of the other geographical races and the number of eggs laid to a set varies probably from three to five, and are deposited during the months of April and May, according to latitude. It is probably a constant resident wherever found.

129. Megascops asio macfarlanei BREWSTER.

MACFARLANE'S SCREECH OWL.

Megascops asio macfarlanei BREWSTER, Auk, VIII, April, 1891, 140.
(B —, C —, R —, C —, U 373h.)

GEOGRAPHICAL RANGE: Southeastern Washington to western Montana (and probably through the entire intervening region from the eastern slope of the Cascade Mountains in Washington, south to central Oregon, east through northern and central Idaho to the eastern foothills of the Bitter Root Mountains in western Montana, north into southeastern British Columbia).

According to Mr. William Brewster, this recently described race (named in honor of Mr. R. MacFarlane, of the Hudson Bay Company, to whom we are indebted for a great deal of valuable information about the nesting habits of so many of our birds), is similar to the California Screech Owl in coloration, but much larger, resembling in this respect the type of Kennicott's Screech Owl, which appears to be equally large.

MacFarlane's Screech Owl is a constant resident wherever found and its habitat as far as known seems to be restricted to the timbered bottom lands of the lower sagebrush and bunch grass covered valleys and plains of the dry interior portions of the States above mentioned. It seems to avoid the mountains, and I do not believe that it is found at much greater altitudes than 4,000 feet. Its general habits are in no way different from those of the other members of the genus *Megascops*, excepting that on account of its larger size it is compelled to nest entirely in natural cavities of trees, the excavations made by the larger Woodpeckers breeding in the same localities, like *Melanerpes torquatus* and *Colaptes cafer*, being too small to accommodate them.

I found my first nest of MacFarlane's Screech Owl in southeastern Oregon on April 16, 1877, but referred it at the time to the California race, as the female caught in the cavity corresponded with the latter in plumage, but it unquestionably belonged to the race now under consideration. This nest was found in a hollow willow stump, in a small grove of these and cottonwood trees among which I camped while on a hunt after waterfowl on Lower Silvies River, near Malheur Lake, 20 miles southwest of Camp Harney, Oregon. The hole was about 5 feet from the ground, 18 inches deep, and contained six partly incubated eggs. There was no nest, the eggs lying on some rubbish which had accumulated in the hole; the female was caught on the nest, and beyond snapping her mandibles made no resistance; the male was not seen.

I did not hear the tremulous notes of this bird about Camp Harney nor at Fort Klamath, Oregon, and do not believe it is found in close proximity to the mountains. At Fort Walla Walla, Washington, it is not uncommon, and I there secured a number of specimens. I also found two of its nests in the year 1881, and three more in 1882. Here the mating season began early in March, and their doleful notes could be heard every evening, shortly after sundown, throughout the month. After incubation commenced the birds were silent. In 1881 a pair of MacFarlane's Screech Owls nested in a natural cavity of a good sized cottonwood tree, about 25 feet from the ground, and within 100 yards of my quarters, giving me ample opportunity to watch them. Whenever I rapped on the tree the occupant would stick its head out and look about, but did not fly away. When first examined on March 29, I found a single egg and a dead mouse in the hole. On April 7 the set was apparently complete, consisting of four eggs. These were laid on small pieces of decayed wood and a few dead leaves. The parent remained in the hole while the eggs were being removed. I was in hopes of securing another set from this pair later in the season, but they abandoned the locality. A pair of Sparrow Hawks (*Falco sparverius*), had taken possession of an old Woodpecker's hole only about 2 feet above the one occupied by the Owls, and seemed to live in harmony with them.

The earliest date on which I took a full set of their eggs was March 26, 1882. This nest was also in a hollow cottonwood, the cavity being about 2 feet in depth and 15 feet from the ground; it contained five fresh eggs and two black spotted brook trout (*Salmo purpuratus*), as well as the parent. The latest date on which I took eggs was April 30, 1882. This nest was found on April 24, when it contained three eggs, which were left, and on reëxamination a week later, the nest was abandoned and no other eggs had been added.

All the nests found near Walla Walla, Washington, were placed in natural cavities in cottonwood trees, from 15 to 30 feet from the ground, and invariably near water. In two of the holes occupied by them I found trout from 6 to 8 inches long and a small whitefish (*Coregonus williamsonii*) about 10 inches long. It still puzzles me to know just how they manage to catch such active fish, but believe that, where obtainable, these as well as frogs form no inconsiderable portion of their daily fare, while the smaller rodents and grasshoppers supply the remainder. I do not believe it catches birds to any extent, and must be considered an eminently useful species.

But a single brood seems to be raised in a season. The eggs, which are deposited in the latter part of March or in the first two weeks in April, vary from three to six in number, usually four or five, and are deposited at intervals of one or two days. Incubation lasts from three to four weeks, both sexes assisting in this duty. The young when first hatched are covered with fine white down, and if they are born blind I failed to notice the fact.

The eggs like those of all Owls are pure white in color, rather glossy, and mostly oval in shape; some are nearly spherical; the shell is smooth and closely granulated. While not quite as large as the average egg of the American Long-eared Owl, their capacity, the shape being a more perfect oval as a rule, is very nearly the same, and the contrast in size between them and the eggs of all the other members of this family found with us (with the exception of Kennicott's Screech Owl, perhaps) is especially noticeable when placed side by side.

The average measurement of twenty-seven speimens in the U. S. National Museum collection is 37.5 by 32 millimetres, the largest egg measuring 39 by 33.5, the smallest 35 by 31.5 millimetres.

The type specimen, No. 20563, selected from a set of five eggs (Bendire collection), was taken by the writer on April 12, 1881, near Fort Walla Walla, Washington. It is figured on Pl. 12, Fig. 9.

130. Megascops asio saturatus, Brewster.

PUGET SOUND SCREECH OWL.

Megascops asio saturatus Brewster, Auk, VIII, April, 1891, 141.
(B —, C —, R —, C —, U 373i.)

GEOGRAPHICAL RANGE: Shores and islands of Puget Sound, Washington, and southward along or near the coast to central Oregon, west of the Cascade Mountains; north to Vancouver Island and southern British Columbia.

According to Mr. William Brewster, this newly described race, which he names the "Puget Sound Screech Owl," is somewhat smaller than Kennicott's Owl, with the general coloring darker and less tawny, and the face and under parts much more white. It is found in two phases of plumage, gray and ferruginous.

The habits of the Puget Sound Screech Owl appear to be very much like those of the California and MacFarlane's Screech Owls. Like these it inhabits the timbered bottom lands in close proximity to water courses, nesting in natural cavities in trees, generally in oaks and cottonwoods, and raising usually a single brood in a season. In Marion County, northwestern Oregon, nidification commences usually in the latter part of April and is frequently protracted into May.

Mr. C. L. Keller informs me that the number of eggs laid by this race is usually four, occasionally five, and should the first set be taken a second and smaller one is now and then laid again in the same nest.

The sites found occupied by these Owls varied from 20 to 30 feet in height, and the eggs were laid on very little nesting material, simply a few leaves and feathers. Occasionally he found some of these birds in the fir timber on the outskirts of the forests, but more often among the cottonwoods and oaks close to water. They are constant residents in the Willamette Valley, Oregon. Its food is similar to that of the other members of this genus, and, like them, seems to be fond of fish.

Mr. A. W. Anthony writes me as follows, and his remarks evidently apply to this race: "While living at Beaverton, Washington County, Oregon, I caught a Screech Owl in a steel trap set for beaver in a ditch about 2 feet wide and 6 feet deep. The trap was not baited but placed fully 4 inches under water, and the Owl managed to get both feet well into the jaws of the trap. The bird was evidently after fish, as I have frequently seen trout in this ditch."

The eggs are similar in shape to those of the other members of this genus. A set of three eggs of this subspecies, taken by Mr. C. L. Keller, in Marion County, Oregon, on May 8, 1891, from a cavity in an old cottonwood tree, 31 feet from the ground, contained good sized embryos when found. Another, of two eggs, taken in the same locality on July 13, 1883, from a cavity 20 feet from the ground, in an oak tree, were probably a second clutch. The average measurement of these specimens, which are in the U. S. National Museum collection, is 36.4 by 30.9 millimetres.

131. Megascops flammeolus (KAUP).

FLAMMULATED SCREECH OWL.

Scops flammeola KAUP, Transactions Zoölogical Society London, IV, 1862, 226.
Megascops flammeolus STEJNEGER, Auk, II, April, 1885, 184.
(B —, C 319, R 404, C 471, U 374.)

GEOGRAPHICAL RANGE: Highlands of Guatemala and Mexico, and northward to Colorado and northern California.

The little Flammulated Owl, one of the smallest of our Screech Owls, is a resident of the elevated plateaus of Guatemala and central Mexico, the southern Rocky Mountains in southwestern Colorado, the mountain regions of Arizona and New Mexico to northern California, where the late Capt. John Feilner, U. S. Army, an enthusiastic naturalist, obtained a single specimen, a young bird of the year, near Fort Crook, August 23, 1860, which was evidently raised there.

This little Owl is still one of the rarest birds in North American ornithological collections, and up to the year 1890 very little was known about its breeding habits. The single egg obtained with the female parent from a Woodpecker's excavation in an old pine tree, by Mr. Charles A. Aiken, of Colorado Springs, in Wet Mountain Valley, Colorado, June 15, 1875, and which is now in the U. S. National Museum collection, remained unique, as far as known to me, until the year 1890, when several of their nests and eggs were taken. Mr. William G. Smith, of Loveland, Colorado, well known as a good ornithologist and reliable collector, found three of their nests during that season, and has given me the following description of them: "The first nest was taken on June 2, 1890, in Estes Park, Colorado, at an altitude of probably 10,000 feet. The site, a Woodpecker's hole in a dead aspen, was about 10 feet from the ground and the burrow about 10 inches deep. It contained three fresh eggs. The female, which was in the hole, had to be removed by force, and in doing so one of the eggs was broken; they were lying on a few chips and feathers from the bird.

"On June 4, I found a second nest about a mile from the former site and in a similar situation, a ravine near water. This contained two fresh eggs and an egg of a Flicker (*Colaptes cafer*). They were placed in a Woodpecker's hole in a large aspen, about 8 feet from the ground and 10 inches below the aperture, while about 6 feet above this was a nest of young Flickers. The cavity appeared to have been formerly used by a squirrel and the eggs were deposited on the old nesting material. It also contained a few Flicker's feathers. The female clung tenaciously to her eggs.

"On June 20, I found the third nest, but this time at a considerably less altitude, probably at about 8,000 feet. It was in a pine tree in a Woodpecker's hole about 14 feet from the ground, and contained four partly incubated eggs. On rapping on the tree the old bird flew out and perched on a limb close by while I investigated the nest. This consisted of a few feathers in the bottom of the burrow, which was about 10 inches deep. The bird's stomach contained the remnants of some small mammal. In none of these cases did I see the males, although my son and I searched around the vicinity of each nest thoroughly. I believe these birds are strictly nocturnal, and consequently rarely seen."

This set, with the female parent shot at the same time and purchased, is now in the U. S. National Museum collection. The four eggs measure, respectively, 29.5 by 25, 29 by 25.5, 28 by 25.5, and 28 by 25 millimetres.

Mr. Evan Lewis found a set of three eggs of this species on June 7, 1890, near Idaho Springs, Clear Creek County, Colorado. The nesting site was a Woodpecker's hole in a dead spruce tree about 15 feet from the ground, the eggs lying on a few feathers. Elevation about 8,700 feet. These eggs are now in the collection of Mr. Thomas H. Jackson, West Chester, Pennsylvania.

From what we know of the habits of the Flammulated Owl they seem to vary but little from the other races of the Screech Owl family. They are apparently strictly nocturnal, and their food consists of the smaller mammals, as well as beetles and other insects. The stomach of a specimen killed by Dr. C. Hart Merriam in the Grand Cañon of the Colorado, September 13, 1890, contained a scorpion, some beetles, and other insects.

Three or four eggs are laid to a set, and in the southern Rocky Mountain region in Colorado, the only locality where it has as yet been found nesting, nidification begins either late in May or the first week in June, but probably considerably earlier at less altitudes elsewhere in its range.

The eggs of the Flammulated Screech Owl are white, with a faint creamy tint, and oval in shape. The shell is strong, finely granulated, and slightly glossy. The average measurement of the five eggs in the U. S. National Museum collection is 28.6 by 25 millimetres.

The type specimen, No. 17199, figured on Pl. 12, Fig. 15, was taken by Mr. Charles A. Aiken, in Wet Mountain Valley, Fremont County, Colorado, June 15, 1875.

132. Megascops flammeolus idahoensis, MERRIAM.

DWARF SCREECH OWL.

Megascops flammeolus idahoensis MERRIAM, in North American Fauna, No. 5, 1891, p. 96.

(B —, C —, R —, C —, U 374a.)

GEOGRAPHICAL RANGE: Mountains of central Idaho.

The type of this new subspecies, an adult male, and as far as known the smallest of our Screech Owls, was obtained in the Big Wood River Mountains, near Ketchum, Alturas County, Idaho, on September 22, 1890, during a biological survey of that portion of the State, made under the direction of Dr. C. Hart Merriam, in charge of the Division of Ornithology and Mammalogy, U. S. Department of Agriculture. It was found among the straggling pine timber on the hills bordering Big Wood River.

It is a slightly smaller bird than the Flammulated Screech Owl and much lighter colored. The dark markings are much finer and the ashy gray tints very pronounced. It probably inhabits the mountain regions of the interior of northwestern North America and seems to attain a higher northern range than the Flammulated Screech Owl.

Nothing is as yet known about its breeding habits and eggs, but these are undoubtedly similar to those of the preceding subspecies.

133. Bubo virginianus (GMELIN).

GREAT HORNED OWL.

Strix virginiana GMELIN, Systema Naturæ, i, i, 1788, 287.
Bubo virginianus BONAPARTE, Geographical and Comparative List, 1838, 6.
(B 48, C, 317, R 405, C 462, U 375.)

GEOGRAPHICAL RANGE: Eastern North America; south through eastern Mexico to Costa Rica.

The breeding range of the Great Horned Owl may be defined as follows: It extends over eastern North America from Florida and the Gulf coast, north into the southern portions of the Dominion of Canada to southern Labrador, and thence westward, principally south of latitude 50° to eastern Manitoba. In the United States the Great Plains form its western limit, including eastern

North and South Dakota, eastern Nebraska, eastern and central Kansas, the same parts of the Indian Territory, and the more heavily wooded districts of eastern and central Texas.

The Great Horned Owl, also known as the Cat and Hoot Owl, is the most powerful and destructive bird of this family found within the United States. Although apparently smaller than the Great Gray and Snowy Owls, it is really considerably larger in body than either and correspondingly stronger.

Excepting possibly in the extreme northern portions of its range, it is a constant resident wherever found, and, though mostly nocturnal in its habits, it sees well enough in the daytime and hunts its prey occasionally on cloudy days, especially when it has young to provide for.

Except during the mating and breeding season, it is an unsociable and solitary bird, rarely allowing another of the same species in the vicinity of its range, which is usually some heavily wooded tract near water. As is the case with most Raptores the female is considerably larger than the male, the latter being but a poor match for his spouse at any time, and I have little doubt that he occasionally falls a victim to the churlish and cannibalistic propensities of his stronger mate, which sometimes happens when pairs of these birds are kept in captivity.

It is generally conceded that the Great Horned Owl is by far the most destructive of all our Raptores, and, on the whole, commits more damage than all the other species together. In this instance, at least, actual facts fully bear out this universal supposition.

Dr. C. Hart Merriam states that one of these Owls has been known to decapitate three Turkeys and several Chickens in a single night, leaving their bodies uninjured and fit for the table.[1]

Aside from its frequent depredations in the poultry yard, where it helps itself to anything within reach, and often kills many more victims than it actually requires, such as Turkeys, Geese, Guinea Fowl, Ducks, Chickens, and Pigeons, and even entering coops after them, it is the worst and most relentless enemy our game birds, such as the Wild Turkey, the Ruffed and Pinnated Grouse, the Bob White, and Woodcock, have to contend against, and wherever these valuable birds are still fairly common they furnish a considerable portion of their daily food. Among the mammals, hares, rabbits, squirrels, skunks, opossums, muskrats, and the smaller rodents help to fill out their bill of fare, and if fish are procurable they show an equal fondness for this sort of food.

Whenever provender is plenty they often content themselves with simply eating the heads of their victims, rejecting the remainder, and thus wipe out whole families of birds in a single night; their sight is so keen that few manage to escape. They are generally able to procure an abundance of food even in the coldest weather, and it is the exception and not the rule to find one of these Owls in poor condition at any time of the year.

---

[1] Birds of Connecticut, 1877, p. 67.

The mating season of the Great Horned Owl begins in midwinter when the greater portion of the range it inhabits is still covered with snow and ice. Mr. Lynds Jones, of Grinnell, Iowa, informs me: "I once had the good fortune to steal unnoticed upon a pair of these birds in their love making. The ceremony had evidently been in progress sometime. When discovered the male was carefully approaching the female, which stood on a branch, and she half turned away like a timid girl. He then fondly stroked his mate with his bill, bowed solemnly, touched or rubbed her bill with his, bowed again, sidled into a new position from time to time, and continued his caresses. All these attentions were apparently bashfully received by the female. Soon thereafter the pair flew slowly away side by side. It is at this time that their hootings are frequently heard. The common call which is most often uttered, and I believe that of the male, is a far reaching "to-hoot-to-hoot-to-hooh," while the answering one of the female is shorter, and usually consists simply of an "õõ," or "to-õõ." Aside from these, they have several others, one a cat-like squeal or cry like "waah-hu," and again a series of yelps, similar to the barking of a dog.

None of these calls can be said to be pleasing to the ear at any time, and when suddenly awakened by them from a restful slumber, perhaps while camped in the silent and snow-covered woods on some hunting expedition, they sound uncanny enough to startle even an old woodsman on first hearing them. These Owls are often attracted by the camp-fires of hunting parties, and their flight is so easy and silent that the first notice one has of their presence is their pertinent querry, "who-who-cooks-for-you" from a tree top perhaps directly overhead, undoubtedly uttered as a protest against the invasion of their own favorite hunting grounds.

In the eastern parts of its range, where the ax of the lumberman has nearly succeeded in destroying all the primeval forests, and large hollow trees are now comparatively scarce, the Great Horned Owl breeds at present mostly in open nests, generally those of the larger Hawks, and occasionally that of the common Crow, while in the heavily timbered bottom lands of the Ohio, Missouri, and Mississippi Rivers, the majority of these birds still nest in natural cavities in trees.

Capt. B. F. Goss, of Pewaukee, Wisconsin, who is well qualified to speak on this subject, writes me as follows: "I think the natural breeding place of the Great Horned Owl is in hollow trees, and where suitable cavities can be found they are always selected. In the early settlement of this part of Wisconsin such breeding places were abundant, and I do not remember finding a single pair of these birds nesting in any other location, but with the rapid improvement of the country the large trees were mostly cut down, until now hardly one remains, and these birds are now compelled to resort to other places, and we find them making use of old Hawks' nests."

Prof. D. E. Lantz, of Manhattan, Kansas, informs me that of twelve nests found by him in that vicinity, three were in old Red-tailed Hawks' nests, one in an old Crow's nest, and eight in hollow trees.

Dr. William L. Ralph has kindly furnished me with the following information on this species: "In the Indian River region of Florida, the Great Horned Owl usually lives in the pine wood districts, breeding altogether in these localities, and I have never known it to nest in other situations in any part of this State that I am familiar with. At and in the vicinity of Merritt's Island, where I visited for several winters, these birds were so common that eight of their nests were found in one season while looking for those of the Bald Eagle, but, like most Florida birds, they are gradually decreasing.

"In this region these Owls always deposit their eggs in the nests of the Bald Eagle, and while I think that these are usually, if not always first deserted by the original owners, the natives say that the Owls drive the Eagles from and appropriate them for their own use. One of the reasons why I think the nests taken by the Owls are deserted ones is because nearly all those found occupied by these birds were situated rather near the ground, and these are the ones the Eagles generally abandon first. These nests are originally constructed of large sticks and limbs, lined with dead grasses, palmetto leaves, flags, and weeds—usually with swamp grasses alone —and after being taken by the Owls are always further thickly lined with scales of pine bark, a material I have never found in any quantity in the nests occupied by the Eagles. The amount of this bark in each nest seems to be about the same, which would not likely be the case had it fallen into the nests by chance, which may occasionally happen to a limited extent. In addition to this bark there are always more or less feathers from the birds in this second lining. Many birds of prey line their nests with leaves or bark from resinous trees and they do this as a preventive remedy for parasites, with which they are always more or less troubled. I have never heard or seen this bird in the vicinity of San Mateo, Florida, and while they are common in some parts of the State, they are entirely absent in other sections, although apparently equally suitable.

"These birds become very much attached to certain localities and seldom wander far from them, even in cases of extreme persecution. As a usual thing they will, should their nest be disturbed, take another in the immediate vicinity, and after a season or two return again to the first one; but in this locality I have known one of these Owls to lay a third set of eggs in the same nest from which the first two had been successively taken. In Florida this species usually commences breeding in December. I have taken eggs about one-third incubated December 17, and found nearly fresh ones January 5. These are the earliest and the latest dates of which I have any personal records, and have never found more than two eggs in a nest, and about 60 per cent. of the sets consisted of a single egg.

"The average measurement of a number of specimens taken by me in Florida is 56.4 by 47.7 millimetres. One egg measured only 50.8 by 42.9 millimetres, and I am sure that it was from a first laying, as it was one of the earliest taken. This egg was the only one in the nest, and partly incubated when found. Seven weeks afterward I took another from the same nest, which was still smaller, but so nearly hatched that I did not try to preserve it. The Owl to which these eggs belonged was one of the largest I have ever seen, and believe that their small size was due to the very old age of the bird. It is almost certain that the Great Horned Owl raises but one brood in a season in Florida, where they feed almost entirely on waterfowl and the smaller mammals, such as rabbits, squirrels, gophers, mice, etc. I have never heard of their catching poultry in this region, and believe that they do not."

The Great Horned Owls are early breeders, laying their eggs throughout the greater part of their range in the beginning of February and occasionally even in the latter part of January. There seems to be but little difference in the time of oviposition between some of the more southern localities, Florida excepted, and those considerably farther north, and it also appears that climate has little influence in the matter. In some of the Western States, like Illinois, Iowa, and Missouri, full sets of their eggs are not infrequently found by February 1, while in the southern New England States it is not unusual to find them in the second and third week of that month, mostly however, about the beginning of March, and in New Brunswick, Nova Scotia, and Newfoundland they nest about the latter part of March or the beginning of April. The country is usually still covered with snow and ice when nidification begins, and their eggs are not infrequently frozen by the intense cold prevailing at the time.

In Newfoundland, as well as occasionally in other places, the Great Horned Owl, according to Mr. Henry Reeks, nests sometimes on the ground. In his notes on the "Zoölogy of Newfoundland," in speaking of this species, he says: "The only nest that came under my observation was built on the ground on a tussock of grass in the center of a pond. The same nest had been previously occupied for several years by a pair of Geese (*Bernicla canadensis*)."[1]

Mr. George E. Beyer, of New Orleans, Louisiana, also found a nest of this species, containing three young, in a hollow pine log on the ground. It is a well known fact that the Western Horned Owl resorts to somewhat similar locations to nest, in regions where suitable trees are wanting. Mr. Audubon also says that he has twice found the eggs of the Great Horned Owl in fissures of rocks, and while such nesting sites are perhaps rare with the eastern bird they are by no means uncommon with the Western Horned Owl. Col. N. S. Goss, in his "Birds of Kansas," states that on the plains or treeless portions of the State it likewise nests in fissures of rocks. These birds are poor nest builders, and if they do construct one of their own, it is through necessity and not from choice. In the Eastern States the majority use open nests, generally those of the Red-tailed and Red-shouldered Hawks, the Crows, and sometimes those of the larger Herons, while farther west hollow trees, when procurable, are still, to a considerable extent, resorted to. The trees most frequented by them for

[1] Zoölogist, 2d series, IV, 1869, p. 1614.

purposes of nidification are elms, oaks, chestnuts, ash, maples, pines, spruces, and cedars, and, in the more western parts of their range, sycamores and cottonwood trees.

The height from the ground varies considerably, some being placed not over 10, others fully 90 feet up, generally averaging from 25 to 40 feet. Among peculiar nesting sites the following deserve mention:

Judge John N. Clark, of Saybrook, Connecticut, writes me that he found a pair of these birds nesting in a quadruple fork of a large chestnut tree some 25 feet from the ground, the eggs lying on the bare wood, without any loose material around them whatever, not even a single leaf. Mr. P. W. Smith, jr., found another pair occupying an old soap box which had originally been put up for squirrels in a grove not over 100 yards from a house. The top of the box had blown off and it was nearly filled with dry leaves. In this condition the Owls had taken possession, and had evidently nested in it several years before discovered by him.

Such scanty repairs as may be needed are made to the nest sometime before nidification commences, and perhaps a little lining, consisting of strips of bark and dry grasses, and as incubation advances many of the feathers of the birds are added in the open nests, while if a hollow tree is used, nothing whatever is done, the eggs being deposited on the rubbish, which may have accumulated therein, such as bits of rotten wood, old leaves, and the feathers dropped from the incubating bird.

An unusual cold and wet spell may freeze or spoil the first eggs laid, and a second set is subsequently added, the former, in such case, are often pushed down among the loose rubbish in the nest. This accounts for some of the extra large sets that are sometimes found, which in reality are two sets, laid at different times, one addled the other fertile.

From one to five eggs have been found to a set, but as a rule two or three are all that are laid, the smaller number more frequently. In some sections, however, sets of four eggs are not unusual. Mr. J. W. Preston, of Baxter, Iowa, writes me that this number is found by him about once in three sets, and that in the early part of March, 1875, he found a set of five eggs too far advanced in incubation to disturb them, and which were all hatched later. This unusually large set was found in an open nest in the top of a medium sized black oak in heavy woods.

Capt. B. F. Goss writes me that he never found more than three eggs in a set, and that two are far more common. He says: "I found two nests with four young in each of them, both in hollow trees. In one the tree had been bent over and broken off, leaving a horizontal hole in the end. Two of the young were more than half grown and partly feathered, the remaining two very small and still in the down. There seemed to be a month's difference in their ages, but it occurred to me as being possible that the two large ones got in front and took most of the food, and the other two were dwarfed by starvation. In the second nest the young were of different sizes, the largest nearly full feathered, the smallest still in the down. This regular difference can scarcely be accounted for by inequality of food, but seems to indicate that the eggs were laid at intervals of about two weeks."

I believe that where the Great Horned Owl nests in hollow trees the number of eggs laid by them is usually apt to be larger than where an open nest is used. The young are more secure in such a location and not so likely to fall or be crowded out.

According to the observations of several careful collectors, incubation is said to last only three weeks, but I believe that twenty-eight days comes nearer to the actual time required. Positive assertions in such matters cannot well be made, especially as it appears that the eggs are, sometimes at least, laid at considerable intervals, and in such cases incubation begins with the first one laid. Where sets do not exceed the usual number, two, incubation probably does not begin until the set is completed, and it is not likely that ordinarily a longer interval than three days occurs between the laying of the two eggs.

The Great Horned Owl will sometimes breed in confinement. Professor Lantz, of Manhattan, Kansas, writes me: "A pair kept in a large roomy cage, where they were seen and teased by many people, became very combative. In 1885 the female laid eggs as follows: One on January 14; this was frozen because she would not sit on it. January 29 the nest contained two more eggs, which were taken, and on February 25 two others. No more were laid."

They are not the kind of birds to make pets of. As a rule they are ill tempered, no matter how well treated, and will attack their keeper without any provocation, inflicting severe and sometimes dangerous wounds. One of my correspondents, who raised one of these Owls from the nest and kept it for three years, called it a "veritable feathered tiger," but they do not all deserve quite so bad a name.

I believe the female attends to the duties of incubation almost exclusively, the male providing her with food.

The Great Horned Owl is certainly a diligent, as well as a successful hunter, and an abundance of food is generally found in a nest with the young. Captain Goss found in one nest several partly devoured rabbits and more than a dozen rats, all without their heads, but otherwise untouched. A correspondent of Forest and Stream, in the number of May 4, 1882, writing from Saratoga Springs, New York, under the *nom de plume* of "Hawkeye," states that in a nest he examined, containing two young Owls, he found the following animals: "A mouse, a young muskrat, two eels, four bullheads, a Woodcock, four Ruffed Grouse, one rabbit, and eleven rats. The food taken out of the nest weighed almost 18 pounds. A curious fact connected with these captives was that the heads were eaten off, the bodies being untouched."

Where open nests are resorted to, these are not unfrequently used by two different species in the same year, the Great Horned Owls being the first tenants, and as soon as their young have left it is taken possession of by one of the larger Hawks for the same purpose. Although at all other times unsocial, during the season of reproduction the Great Horned Owls are generally devoted and courageous in the defense of their young, caring for them long after leaving the nest. Collectors have been known to be vigorously attacked and even beaten off by them, and were quite willing to make a hasty retreat in order to keep out of reach of their sharp and powerful talons.

The eggs of the Great Horned Owl, usually two or three in number, are white in color, and show little or no gloss, though there are occasional exceptions; they are rounded oval in shape; the shell is thick and rather coarsely granulated, feeling rough to the touch.

The average measurement of twenty-five specimens is 56 by 46.5 millimetres. The largest of these eggs measures 58.5 by 48.5, the smallest 51 by 44.5 millimetres.

The type specimen, No. 20629 (Pl. 12, Fig. 12), from a set of two, was taken by Capt. B. F. Goss, near Pewaukee, Wisconsin, March 13, 1883.

134. Bubo virginianus subarcticus (Hoy).

WESTERN HORNED OWL.

Bubo subarcticus HOY, Proceedings Academy Natural Sciences Phila., VI, 1852, 211.
Bubo virginianus β subarcticus RIDGWAY, Ornithology of the 40th Par., 1877, 572.
(B 48, part; C 317a, part; R 405a, part; C 463, part; U 375a.)

GEOGRAPHICAL RANGE: Western United States (except northwest coast); eastward across the Great Plains, straggling to northern Illinois, Wisconsin, and western Canada; north to Manitoba; south over the table-lands of Mexico (Lower California ?).

The breeding range of the Western Horned Owl, a lighter gray and buff colored bird than the preceding subspecies, extends from the Mexican table-lands, north through southwestern Texas, New Mexico, Arizona, Colorado, western Kansas, western Nebraska, Utah, Wyoming, Montana, and western South and North Dakota, as well as beyond our border into western Manitoba, Assiniboia, and southern Alberta. On the Pacific coast it is found from (Lower? and) southern California northward through all the intervening States, on both sides of the Sierra Nevada, passing through British Columbia to Alaska, along the Lower Yukon River and shores of Bering Sea, to about latitude 65° N.

According to Mr. William Brewster, the Horned Owls found in Lower California are much smaller in size and darker colored than the Western Horned Owl, resembling the Dusky Horned Owl somewhat in coloration, and they will have to be separated as a new geographical race.

The Western Horned Owl is only found in the lower foothills and more open country throughout the range indicated, while in the higher mountain regions, it is replaced by the Dusky Horned Owl (*Bubo virginianus saturatus*).

There is no perceptible difference in the general habits of the Western Horned Owl from those of its eastern relatives. Their call notes are also similar; and, like it, it is the most destructive and insatiable of all the Raptores found in its range, feeding to a great extent on valuable game birds, especially the Columbian and Prairie Sharp-tailed Grouse, wherever these are abundant, as well as on Ducks, other waterfowl, and the smaller land birds. Among mammals, hares, prairie dogs, polecats, marmots, the different species of tree and ground squirrels, wood rats, and other rodents, contribute to its fare. In the more settled regions poultry yards also suffer, as these Owls rapidly develop a strong taste for such food.

In the choice of nesting sites the Western Horned Owl shows a wider range, however, than the preceding. While perhaps the majority of these birds resort likewise to hollow trees or old nests of the larger Hawks and of the common Crow, quite a number nest in the wind-worn holes in sandstone and other cliffs, small caves in clay and chalk bluffs, in some localities on the ground, and, I believe, even occasionally in badger holes under ground. On the grassy plains in the vicinity of the Umatilla Indian Reservation, in northeastern Oregon, I have several times seen Owls of this race sitting on the little mounds in front of badger or coyote burrows, near the mouths of which small bones and pellets of fur, were scattered about. While unable to assert positively that they do actually breed occasionally in such holes, the indications point that way, and this would not seem to be due to the absence of suitable timber, as an abundance of large trees grow along the banks of the Umatilla River not more than a mile away. When nesting in trees, large cottonwoods, sycamores, willows, pecans, pines, oaks, and firs, are generally preferred. In regions, however, where heavy timber is scarce they content themselves with nests in small mesquite and hackberry trees, frequently placed not over 10 feet from the ground. In Lower California (?) and southern Arizona they also nest occasionally in the sahuara, the giant cactus, so common in those regions. In Colorado, Wyoming, and Idaho, they are known to make use of old Black-billed Magpies' nests, laying their eggs occasionally inside, but more often on the broken down roof of these bulky structures. In the neighborhood of Nueces Bay, in southwestern Texas, they nest in holes in high banks, and in portions of California similar situations are occupied.

Mr. Charles A. Allen, of Nicasio, California, writes me as follows: "On the seacoast near Point Reyes I have found their nests on the ground. All along the coast the water rushing down from the hills during the rainy season has worn and cut channels out of the soft and friable soil to the depth of 100 or 200 feet as it approaches the shore, the sides of these gulches being frequently nearly perpendicular. The slopes of these cuts are in many parts covered with a growth of coarse grass, bullrushes, and tall ferns, and a place among these is usually selected by them for a nesting site. A shallow hole is scratched out next to the bank, and although you may be able to look down into the nests, they are frequently inaccessible. I have, while hunting sea birds, often started Owls from off their nests in such places."

Mr. W. Otto Emerson, of Haywards, California, found a nest of this Owl, containing three young birds, on a sleeper under a railroad bridge, and Lieut. Robert C. Van Vliet, U. S. Army, tells me that he frequently saw these birds flying about within the town limits of Santa Fé, New Mexico, a pair occupying the tower of the cathedral, and he thinks they nested there.

Mr. Denis Gale says: "Each pair of these birds have their particular range, and no amount of harassing or robbing them of their eggs two or three times a year, will induce them to leave a locality once chosen. The food supply, of course, is the chief consideration influencing their choice. In some cases half a mile of creek bottom defines the limit of their preserve or hunting ground, and occasionally it is larger, every square foot of which, in time, becomes familiar by careful watching night and day. No doubt every burrow and hiding place, from that of a mouse to a jack-rabbit, is known to them. * * * A choice of location once made is never abandoned, unless civilization blots out the cover or kills the birds."

The Western Horned Owl is extremely abundant in favorable localities. At Fort Custer, Montana, situated in the angle formed by the confluence of the Big and Little Horn Rivers, I obtained not less than twenty-eight of these Owls in the winter of 1884–'85, and at least a dozen others were killed which I did not receive. All were shot within a radius of 6 miles of the post, among the cottonwood timber on these streams. Every specimen, old or young, was excessively fat, showing that notwithstanding their numbers, they all readily procured an abundance of food even in the severest winter weather. The Columbian Sharp-tailed Grouse appeared to have suffered greatly from their depredations, as fully one-half of the birds secured contained remains of these in their crops. A few of the specimens obtained here were intermediate in plumage between this race and the Arctic Horned Owl, probably migrants from the north.

In the southern parts of their range nidification begins occasionally in the first part of January. Capt. B. F. Goss found a set of their eggs on the 8th of that month. Usually it does not begin much before February 15, and lasts until the middle of March. Climate seems to have little to do with the time of nesting with these birds, as they nest sometimes fully as late in the semi-tropical regions as they do much farther north.

The Western Horned Owl is a persistent layer. Mr. Gale writes me that he has taken three sets of eggs from the same pair of birds in the season of 1889 at intervals of about four weeks. The first set contained four eggs, the second three, and the last two each, and the nesting site was changed each time. Where they use open nests the site is likely to be changed each season, but when a hollow tree or a hole in a cliff is chosen they usually occupy the same from year to year, unless too often disturbed. The old birds can generally be found in the vicinity of their breeding ground throughout the year. Mr. Gale believes that these Owls do not breed until the second year, and as a rule only a single brood is raised; but the fact that young birds not yet able to fly are occasionally found so late in the season, and sometimes in localities where they certainly had not been disturbed previously, it would appear as if a second might now and then be reared. A few days after my arrival at Fort Klamath, Oregon (June 18, 1882), one of my men brought me a young Owl of this subspecies which he had caught alive in the pine forest south of the post. It could barely fly at the time, and if not from a second brood the eggs must have been laid several weeks later than usual.

The eggs number two or three to a set, occasionally four, and sets of three are about as often found as the smaller number, while those of four are not especially rare. Mr. Charles F. Morrison reports taking one of six in Wyoming, an extremely large set, and Mr. Charles C. Neale writes me that he took a set of five eggs from a nest in an oak tree in the mountains in Plumas County, California.

They are deposited generally at intervals of two or three days, the female attending to the duty of incubation exclusively, I believe, and which lasts about four weeks. The male supplies his mate with the necessary food while she is so engaged, and when not hunting is usually found in close proximity to the nest. The eggs are similar to those of the Great Horned Owl.

The average measurement of fifteen specimens in the U. S. National Museum collection is 55.5 by 47 millimetres. The largest egg of this series measures 58.5 by 48.5, the smallest 53.5 by 45 millimetres. None are figured.

135. Bubo virginianus arcticus (Swainson).

ARCTIC HORNED OWL.

Strix (Bubo) arctica Swainson, Fauna Boreali Americana, II, 1831, 86, Pl. 30.
Bubo virginianus var. *arcticus* Cassin, Illustrated Birds of California, etc., 1854, 178.
(B 48, part; C 317a, part; R 405b, C 463, part; U 375b.)

GEOGRAPHICAL RANGE:- Arctic America, chiefly the interior; south in winter to the Great Plains (the two Dakotas, Montana, etc.).

The breeding range of the Arctic Horned Owl, a much lighter colored race than the two preceding forms, is confined, as far as known at present, to those parts of the interior of British North America situated between James Bay (Moose Factory), the west shores of Hudson Bay, and the eastern slopes of the Rocky Mountains, north of latitude 51°, and extending in a northwesterly direction to northern Alaska, where a specimen was obtained by Mr. C. L. Mackay on the Attoknagik River, August 24, 1881. Like the Western Horned Owl it inhabits the more open country throughout its range, more especially along the shores of the numerous lakes and streams found in those inhospitable regions. In winter it migrates southward, though rarely entering our borders. As yet I have not seen a specimen of this race obtained within the limits of the United States that could be called typical. While stationed at

Fort Custer, Montana, I received several Owls which approached this form, being intermediate between it and the Western Horned Owl, but none were perfect types of either.

Mr. R. MacFarlane met with the Arctic Horned Owl in the country between Fort Good Hope on the Lower Mackenzie and the Anderson River region, within the Arctic circle, and in a collection of birds and eggs recently received from him is a very light colored female of this race, a perfectly typical specimen from Moose Lake, eastern Saskatchewan, shot in May, 1890, which probably marks nearly the southern limit of its breeding range. These birds feed on the numerous waterfowl, Ptarmigan, and the Arctic hares inhabiting these regions, and are probably common enough in suitable localities, where an abundance of food is easily obtainable.

Nothing is as yet known about their mode of nesting or their eggs, which are not likely to differ from those of the preceding races.

136. Bubo virginianus saturatus RIDGWAY.

DUSKY HORNED OWL.

Bubo virginianus saturatus RIDGWAY, Ornithology of the 40th Par., 1877, 572, footnote.

(B 48, part; C 317b, R 405c, C 464, U 375c.)

GEOGRAPHICAL RANGE: From Labrador and Hudson Bay; west through the interior to Alaska, and south probably through all the higher regions of the Rocky and Sierra Nevada Mountains; south to Arizona (San Francisco Mountain).

The range of the Dusky Horned Owl, the darkest colored of the different races of the genus *Bubo*, has until recently been supposed to be confined to the coast regions of Oregon, Washington, British Columbia, and Alaska Territory. As it is well known to occur also in Labrador—and, furthermore, to breed there, showing that it is not an accidental straggler—it probably also inhabits the wooded regions of the interior, covered with hardly spruce and pine forests, which connect these widely separated points, and reach from the North Atlantic Ocean nearly to Bering and the Arctic Sea. As yet, however, no specimens of this race have been obtained from the interior of British North America. This is not surprising when we consider the fact that this large bird has until very recently been overlooked in regions far more accessible than the so-called "fur countries."

During a biological survey, conducted under the direction of Dr. C. Hart Merriam, Chief of the Division of Ornithology and Mammalogy, U. S. Department of Agriculture, made in August and September, 1889, and which resulted in some extremely interesting discoveries, a specimen of this dark colored race was shot on September 14, in the pine belt on San Francisco Mountain, central Arizona. Another was seen at the same time, and they are reported as tolerably common in that vicinity. This extends the range of this race south to latitude 35° N.

Since it occurs in Arizona it will doubtless be found to inhabit all the higher timbered ranges and spurs of both the Rocky and Sierra Nevada Mountains, within the United States, and possibly the Sierra Madre of northern Mexico as well, and breeding perhaps entirely in the fir and spruce belt above an altitude of 8,000 feet, where they find a summer climate similar to that of the higher latitudes they inhabit in the Arctic regions. On the approach of winter they probably all leave the mountain summits and descend to the foot-hills.

I am well aware of the fact that the Dusky Horned Owl is a migrant, at least in some parts of its range, having shot quite a number of these birds in different parts of Oregon and Washington during the winter months. At Fort Walla Walla, Washington, situated in the fertile valley bearing the same name, near the northern slopes of the Blue Mountains, I found this race especially common at that season. Of the eighteen birds obtained there, twelve were referable to this form; three were intermediate, and three were typical *Bubo virginianus subarcticus*, the latter breeding there regularly, while the Dusky Horned Owls seem to retire to the higher timbered mountains at the approach of spring; at any rate, none of the dark birds were seen during the nesting season.

The late Mr. Robert Kennicott found a nest and two eggs of the Dusky Horned Owl near Fort Yukon, Alaska, April 16, 1862. The female parent (No. 27075, U. S. National Museum collection), procured at the same time, is one of the darkest colored specimens of the entire series of skins of this race in the collection. The nest is described as a large structure, made of dry branches, and placed in the top of a spruce standing in a dense grove of trees of the same species; the inner cavity was shallow and simply lined with a few feathers. It contained two fresh eggs, and another, fully formed but broken, was found in the oviduct of the female, also a smaller ovum about the size of a musket ball.

Mr. H. Connelly found the Dusky Horned Owl breeding at Fort Nis-copec, Labrador, in the spring of 1863. A single egg and the female parent (No. 34958) are now in the National Museum collection. This specimen is the darkest colored bird of the entire series.

Mr. L. M. Turner, of the U. S. Signal Service, procured several specimens of this Owl while stationed at Fort Chimo, Ungava. In his manuscript on the "Birds of Labrador and Ungava," he says: "From intelligent and trustworthy sources I have learned that this species of Owl is quite common near the head of Hamilton Inlet and the southern portion of Labrador. The char-acter of that region would indicate a greater abundance of birds of prey than in the sparsely wooded district of Ungava. That the bird is resident and breeds in Ungava is attested by specimens of both young and old."

The Dusky Horned Owl is evidently only a resident of the heavier timbered portions of British North America, the pine and spruce forests of these regions, and the higher mountain ranges of the more temperate zone.

Its general habits and food are similar to those of the preceding races. In the northern parts of its range nidification, as far as known, begins about the middle of April, and from two to four eggs are probably laid, and these are indistinguishable from those of the other Horned Owls.

Of the three specimens previously mentioned, the two taken by Mr. R. Kennicott, near Fort Yukon, Alaska, April 16, 1862, measure 55 by 47.5 and 52.5 by 48 millimetres, and the other, from Fort Niscopee, Labrador, taken by Mr. H. Connelly, measures 57.5 by 48.5 millimetres. None are figured.

137. Nyctea nyctea (LINNÆUS.)

SNOWY OWL.

Strix nyctea LINNÆUS, Systema Naturæ, ed. 10, I, 1758, 93.
Nyctea nyctea LICHTENSTEIN, Nomenclator Musco Berolinensis, 1854, 7.

(B 61, C 325, R 406, C 479, U 376.)

GEOGRAPHICAL RANGE: Extreme northern portions of northern hemisphere in summer; migrating southward in winter in North America, almost across the United States, and even reaching, accidentally, the Bermudas.

The breeding range of the Snowy Owl in North America extends from about latitude 53° in Labrador north to the Arctic Sea, and it has been observed at the highest latitudes our Arctic explorers have as yet been able to reach. It is likewise common in Greenland during the breeding season, but much more so in the northern than the southern portions. Both Downes and Reeks report it abundant in Newfoundland during the greater part of the year, but I cannot find any positive records that it has actually been found breeding on this island, though it may do so in limited numbers in the less often visited parts of the interior. It is very doubtful if it nests at any time south of latitude 53° N., although it has been reported as nesting occasionally in Nova Scotia, New Brunswick, and northern Maine. Mr. Le Grand T. Meyer records it as breeding in Manitoba, the exact locality not stated, where he says he found a nest containing six more or less incubated eggs, February 26, 1879. He states: "I learned that a pair had used the same nest for two years. * * * The nest, aggregated by the several additions, was about 18 inches above the level of the prairie, composed of hay, grass, and sticks, warmly lined with feathers from their breasts."[1]

Gen. A. W. Greely, Chief Signal Officer, U. S. Army, mentions a nest of this species taken near Fort Conger, Grinnell Land, May 25, 1882, and young birds on July 8. He says: "The Snowy Owls bred abundantly in the vicinity of Fort Conger, and as many as fifteen or twenty-five young birds were raised in 1882, and kept by us until approaching winter compelled us to release them. A nest near Fort Conger resembled that described by Major Feilden, which was a mere hollow scooped out of the earth and

[1] Oölogists Exchange, Vol. I, No. 4. I give this record, but have not been able to verify it.—C. B.

situated on the summit of an eminence which rose from the center of the valley." In this case a few feathers and a little grass were present. The nest found by Major Feilden in Grinnell Land was obtained June 20, 1876, in latitude 82° 40′, probably the most northerly point this species has been found nesting."[1]

On the Pacific coast it is a resident, and breeds throughout northern Alaska, both in the interior and near the seacoast, and has there been met with by Dall, Turner, Nelson, Murdoch, and others.

Mr. R. MacFarlane states that the Snowy Owl is not numerous in the Anderson River country, and though every effort was made to secure its eggs and nest, the search was unsuccessful. He says: "On one occasion we noticed a White Owl hunting marmots (*Arctomys empetra*), in the barren grounds; and there can be no doubt that this and other Owls sometimes rob Ptarmigan and Ducks of their eggs."

While a few Snowy Owls remain throughout the year, braving the severe Arctic winters, even in high latitudes, by far the greater number of these birds perform annually extended migrations southward. Some winters they appear in considerable numbers in the northern parts of the United States, and a few even straggle into the Southern States. On the Pacific coast they do not come nearly as far south, and are rarely seen in the more open portions of Washington and Oregon, and I believe my records are the first from those regions. I observed it on three occasions in the vicinity of Camp Harney, Oregon, in January, 1875, and December, 1876, and at Fort Walla Walla, Washington, one each on December 1, 1880, and November 10, 1881. I did not succeed in getting any of these birds at Harney, as they were excessively shy and wide awake, but the two last mentioned were secured. One of these is now in the collection of Mr. William Brewster, Cambridge, Massachusetts.

The home of the Snowy Owl is on the immense moss and lichen covered tundras of the boreal regions, where it leads an easy existence and finds an abundant supply of food during the short Arctic summers. It hunts its prey at all hours and subsists principally on the lemming, and is said to be always abundant wherever these mammals are found in any numbers. Small rodents are also caught, as well as Ptarmigan, Ducks, and other waterfowl, and even the Arctic hare, an animal fully as heavy again as these Owls, is said to be successfully attacked and killed by them.

Mr. L. M. Turner, in his "Notes on the Birds of Labrador and Ungava," says: "This bird never seizes its prey except while the latter is in motion, except in the case probably of fish, for it is said to be an expert fisher, seeking the places overgrown with seaweed to seize any sculpin (*Cottus*) that may be lurking among the crevices of the rocks. The hares are chased and seized near the lumbar region and held by the bird, which spreads its wings and partly lifts the animal from the ground, thus depriving it of the power to use its strong hinder parts. * * * The natives assert that when a Ptarmigan is sighted, the Owl endeavors to start the bird into a run and is then seized, though it adopts different tactics in this case; the prey is crushed to the ground, the outspread wings of the captor preventing those of its prey from allowing it to rise."

The nests of the Snowy Owl are ordinarily placed on the ground, usually on the highest and driest point in the surrounding tundra. Occasionally a nesting site on a rocky ledge or a cliff is chosen. In either case the nest is but a flimsy affair at best, consisting, if on the ground, of a slight hollow scratched out by the birds, and this is usually lined with a little moss and a few feathers; if on top of a ledge or a cliff, the eggs frequently lie on the bare rock, with just enough material around them to keep them in place and prevent them from rolling about. Incubation begins with the first egg laid, which are deposited at irregular intervals. Young birds are often found in the same nest with nearly fresh eggs. Mr. Turner says that the old birds, especially the female, are very fierce in the defense of their young. They fly close to the head of the intruder with their talons fully spread and snapping their mandibles. The natives in such instances seize one of the young and make it struggle, which causes a charge of the parent on the intruder, who holds up the stock of his gun and lets the bird dash against it, which usually stops further action on their part.

From three to ten eggs are laid by this species, usually from five to seven. Mr. Collett states that in 1871 he found as many as ten in several instances in northern Norway. They are deposited about the latter part of May, and in its extreme northern range not until June. All of the eggs of this species in the U. S. National Museum collection appear to have been taken in June. Some of these come from Labrador, others from Fort Churchill, Hudson Bay, latitude 59° N., a set of six from Repulse Bay, Melville Sound, Arctic America, latitude 66° N., and a set of three from the Yukon Delta, Alaska. But one brood is raised in a season.

The eggs of the Snowy Owl are white in color, with a faintly perceptible creamy tint in some instances, and oblong oval in shape. None of these eggs are as round as those of the genus *Bubo*, and their shell is roughly granulated and without luster. A few corrugated lines starting a trifle beyond the center of the egg and running to the longer axis are noticeable in the majority of specimens examined by me.

The average measurement of fifteen specimens in the U. S. National Museum collection is 57 by 45 millimetres; the largest egg of the series measuring 60 by 47, the smallest 55 by 44 millimetres.

The type specimen, No. 13041 (Pl. 12, Fig. 19), selected from a set of six eggs, was obtained by Capt. C. F. Hall, during his Arctic expedition in the *Polaris*, in June, 1867, at Repulse Bay, Melville Sound, latitude 66° N., while in search of the remains of Sir John Franklin's party.

138. Surnia ulula (Linnæus).

HAWK OWL.

Strix ulula Linnæus, Systema Naturæ, ed. 10, 1, 1758, 93.
Surnia ulula Bonaparte, Catalogo Metodico degli Uccelli Europei, 1842, 22.

(B —, C —, R 407a, C 481, U 377.)

Geographical range: Northern portions of eastern hemisphere, from Norway to Kamchatka, and more northern Asiatic shores of Bering Sea (Plover Bay), accidental in western Alaska (St. Michael).

The Hawk Owl, another Old World species, likewise claims a place in our avifauna on the strength of several specimens obtained in the vicinity of St. Michael, Alaska, by the indefatigable Mr. L. M. Turner, while on duty there in connection with the U. S. Signal Service during the years 1874 to 1881.

It is slightly larger than the American Hawk Owl and readily recognized by its much lighter coloration. It seems to be a stupid bird, at times at least, as Mr. Turner caught one alive in his hands while it was sitting in a clump of rank grass. He says: "The natives assert that it is a resident, and breeds in the vicinity of St. Michael; also that it is a coast bird, *i. e.*, not going far into the interior, and that it can live a long time in winter without food, as it remains for days in the protection of the holes about the tangled roots of the willow and alder patches."[1]

Mr. Henry Seebohm, in speaking of this species, says: "It breeds throughout the pine forests of Scandinavia and North Russia, occasionally reaching as high as the birch region on the confines of the tundra. * * * The principal food of the Hawk Owl is mice and lemmings; and the bird follows the migratory parties of the last named little mammal to prey upon them. From its indomitable spirit, however, few birds of the forest are safe from its attacks. * * *

"The breeding season of the Hawk Owl apparently commences in the middle of April and lasts to the end of June. As this bird possesses the habit, in common with many of its congeners, of laying their eggs at intervals and sitting on them as soon as laid, they may be found as late as the third week in June. It makes no nest, and the eggs are usually laid in the hole of a decayed pine tree and rest on the powdered wood alone, as is the case with those of the Woodpeckers. Collett mentions a nest of this Owl in Norway, on the top of a broken pine trunk, some 6 feet below which was a Golden-eye Duck sitting on her nest. Wolley mentions a similar instance in Lapland. This Owl will also frequently take possession of the nest boxes placed by the peasants for Ducks and rear its young in them. The eggs of the Hawk Owl are from five to eight in number, white in color, smooth, and possess considerable gloss. They measure from 1.65 to 1.55 inches in length, and from 1.25 to 1.17 inches in breadth [equal to about 41.9 to 39.4 in length and 31.7 to 29.7 millimetres in breadth].

---

[1] Contributions to Natural History of Alaska, Turner, 1886, p. 164.

"Both birds sit upon the eggs and are sometimes found on them in company. While the female is upon her charge the male bird will perch close at hand, ready to do battle with any intruder, not even excepting man himself. Numerous instances are recorded of this bird's dauntless courage when its nest is assailed. It strikes at the intruder again and again, seeming not to care for its own safety, and but too often pays the price of its temerity with its life."[1]

Ten eggs of this species in the U. S. National Museum collection from Lapland and Finland give an average measurement of 39.5 by 31.5 millimetres. No specimen is figured, as the eggs are indistinguishable from those of the American Hawk Owl.

139. Surnia ulula caparoch (Müller).

AMERICAN HAWK OWL.

Strix caparoch MÜLLER, Systema Naturæ Supplement, 1776, 69.
Surnia ulula caparoch STEJNEGER, Auk, 1, October, 1884, 363.
(B 62, C 326, R 407, C 480, U 377a.)

GEOGRAPHICAL RANGE: Northern North America; south, in winter to northern border of the United States; British Islands.?

The breeding range of the American Hawk Owl is principally confined to the "fur country," the Hudson Bay territory, and the timbered districts of Alaska. It is said to breed in the interior of Newfoundland, and thence north and westward from latitude 48° N., through Labrador, the northeast and northwest territories of British North America to the end of the timber zone in about latitude 68° N.

According to Mr. G. A. Boardman, the American Hawk Owl is a rare resident in the vicinity of Calais, Maine. The U. S. National Museum collection contains a couple of eggs of this subspecies, obtained through Mr. Boardman, which are said to have been collected on the shores of the Gulf of St. Lawrence in the spring of 1861. This, if correct, is the most southern breeding record of this species known to me. The nest from which these eggs were taken was placed in the top of a thick fir tree. Dr. C. Hart Merriam also states:[2] "The Hawk Owl unquestionably breeds in northern Idaho. August 11, 1872, I shot one on Madison River, Montana, only a few miles from the Idaho boundary."

The American Hawk Owl is diurnal in its habits, hunting its prey to a great extent by daylight, generally early in the morning or in the evening, being often seen at such times, and on that account considered more common than other Owls inhabiting like regions, and at the same time easily obtained; its habits are fairly well known, and is sure to be noticed wherever it occurs. Its food is said to consist principally of small rodents, insects, and an occasional bird.

[1] History of British Birds, Seebohm, 1883, Vol. 1, pp. 184, 185.
[2] North American Fauna, No. 5, 1891, p. 96.

According to Mr. Turner, it is a very common resident in the Yukon district, and also quite abundant near the coast. He says: "They usually seclude themselves in the willow or alder patches, and are frequently startled from some grass-covered bank of a lake; they fly equally well by night or by day. I once observed a bird of this species sitting during a bright day on a post, and approached to within a few feet of the bird; it squatted, then stood up, and seemed ready to fly at any moment; I went within 6 feet of it, and it then settled down as if to take a nap; I retired and threw a stick at it to make it fly; I shouted and made other noises, and only after several attempts to dislodge it did it fly. When taking flight from an elevated position they invariably drop to within a few feet of the earth and sail away rapidly."[1]

Mr. W. H. Dall, of the U. S. Coast Survey, found a nest of this species, containing six eggs, on the top of an old birch stub about 15 feet from the ground, near Nulato, Alaska, May 5, 1868. The eggs were lying directly on the rotten wood, and the male was sitting on them. Climbing to the nest, the bird dashed at him and knocked off his cap. These eggs are now in the U. S. National Museum collection.

Mr. R. MacFarlane says: "The Hawk Owl is not uncommon in the region of Anderson River. Four nests of this species were discovered and the eggs taken therefrom. All of these were built in pine trees at a considerable height from the ground. One was actually placed on the topmost boughs, and, like the others, constructed of small twigs and sticks, and lined with hay and moss. This nest contained two young birds, one apparently ten days and the other three weeks old, together with an addled egg. Two of the other nests contained six eggs and one seven. The parents always disapproved of our proceedings."[2]

One of these nests was found on April 28, another on May 2, and the one containing young on June 20, 1863. A single egg, taken by Mr. MacFarlane, near Fort Providence, Great Slave Lake, on April 14, 1885, shows that the American Hawk Owl breeds very early, even in high latitudes, and that some winter there also.

According to Mr. B. R. Ross, it nests occasionally in cliffs, but its usual nesting sites are probably natural cavities in trees, where they are obtainable, but when such are wanting open nests placed on the decayed tops of stumps or among the limbs of thick and bushy conifers are used.

Mr. Turner found the American Hawk Owl to be rare in southern Labrador. In his "Notes on the Birds of Labrador and Ungava," he says: "In the latter part of June, 1884, an Indian brought me two young of this species just emerging from the downy stage and not yet able to fly. As I desired to make a study of these young birds I kept them for several days and fed them on the carcasses of birds, mice, and young Ptarmigan. The quantity of food which these small Owls could dispose of was astonishing. They

[1] Contributions to the Natural History of Alaska, 1886, Vol. II, p. 16.
[2] From R. MacFarlane's Manuscript Notes on Land and Water Birds Nesting in British America.

were very nice pets, and in the course of a few days came to recognize me whenever I came to them, and always greeted me with a whistling note of plaintive tone."

The eggs of the American Hawk Owl are from three to seven in number, and nidification commences frequently long before the disappearance of the ice and snow. Like the Hawk Owl of the Old World, it lays at irregular intervals and commences to incubate as soon as the first egg is deposited, both sexes taking part in these duties. Eggs may be looked for from the latter part of April through the month of May; these vary from oval to oblong oval in shape, are pure white in color, and somewhat glossy; the shell is smooth and fine grained. They resemble the eggs of the Short-eared Owl very closely and are scarcely distinguishable from them.

The average measurement of thirty-eight specimens in the U. S. National Museum collection is 39.5 by 31.5 millimetres, the largest egg measuring 43 by 32, the smallest 36.5 by 30 millimetres.

The type specimen, No. 14564 (Pl. 12, Fig. 18), selected from a set of six eggs, was taken by Mr. William H. Dall, U. S. Coast Survey, near Nulato, Alaska, May 5, 1868.

140. Speotyto cunicularia hypogæa (BONAPARTE).

BURROWING OWL.

Strix hypogæa BONAPARTE, American Ornithology, I. 1825, 72.
Spheotyto cunicularia var. *hypogæa* RIDGWAY, in COUES'S Key to North American Birds, 1872, 208.

(B 58, 59, C 332, R 408, C 487, U 378.)

GEOGRAPHICAL RANGE: Western North America; north to and beyond the northern boundary of the United States; east throughout the Great Plains; south to Guatemala; accidental in New York (city) and Massachusetts.

The breeding range of the little Burrowing Owl includes the prairie regions west of the Mississippi and Missouri Rivers, the Great Plains of the United States from northern Texas north, through the Indian Territory, Kansas, Nebraska, North and South Dakota, to about latitude 48° N. It also occurs in small numbers in western Minnesota (Swift County). As far as I am aware, it has not as yet been noticed in the southern parts of the provinces of Manitoba and Assiniboia, in the Dominion of Canada, but probably occurs there also in favorable localities. Thence it is found westward throughout the intervening States and Territories, the timbered and mountainous regions excepted, to the Pacific coast, where it reaches the southern border of British Columbia in about latitude 50°. Prof. J. Macoun found it very abundant at Kamloops, British Columbia, in 1889, and took a specimen at Revels Lake in 1890.

It is common in various localities in Idaho, Washington, and Oregon, east of the Cascade range of mountains, and it is well known to occur in considerable numbers in California, especially the southern portion of the State, while in Arizona and Lower California it is somewhat rarer. It passes south, thence into Mexico and Central America.

A good deal of nonsense has found its way into print about the life history of this Owl, and the sentimental story of its living in perfect harmony with prairie dogs and rattlesnakes, both of which inhabit a considerable portion of the range occupied by these Owls, was for years accepted as quite true and furnished the ground work for many an interesting tale. Dr. Elliott Coues was perhaps the first naturalist who showed the fallacy of this generally accepted fact in an interesting article on this species in his "Birds of the Northwest." From an extended acquaintance with the habits of the Burrowing Owl, lasting through a number of years' service in the West, I can most positively assert, from personal experience and investigation, that there is no foundation based on actual facts for these stories, and that no such happy families exist in reality. I am fully convinced that the Burrowing Owl, small as it is, is more than a match for the average prairie dog, and the rattlesnake as well; it is by no means the peaceful and spiritless bird that it is generally believed to be, and it subsists to some extent at least on the young dogs, if not also on the old ones.

In California, Oregon, Washington, and Idaho, where I believe the true prairie dog is not found, they occupy the burrows of the numerous spermophiles (such as Douglas's, Townsend's, and Beechey's) infesting these States, and which are the greatest pests the farmers have to contend against in these regions. The birds enlarge these burrows to suit their needs and live on both the old and young squirrels. Wherever they abound a colony of these Owls is sure to be found also, but I doubt if they ever dig burrows themselves. At Fort Custer, in southeastern Montana, where prairie dogs are common, the young dogs furnish a considerable portion of their fare, as it was found from examination of the stomachs of two specimens that both had been feeding on such; and, judging from the ease with which they dispatch an adult ground squirrel, I have no doubt that a prairie dog would stand but little chance to hold its own against one of these Owls. Their food is quite varied, and consists principally of rodents, such as young prairie dogs, the different species of ground squirrels already mentioned, chipmunks, pocket gophers, mice, as well as shrews, small hares (cotton tails), frogs, fish, lizards, snakes, and insects of different kinds, such as grasshoppers, the large and exceedingly destructive black crickets (*Anabus simplex*) of the Great Basin, beetles, and scorpions. Birds are also said to be caught by them, and such may sometimes be the case, although I have never found any of their remains in their burrows, and I have examined quite a number.

East of the Rocky Mountains this little Owl is said to be a constant resident, even in some of the more northern portions of its range, in South Dakota, for instance,[1] and while this may possibly be true in these regions, I do not

[1] Report on Bird Migration, Cooke, Bulletin No. 2, U. S. Department of Agriculture, 1888, p. 124.

believe the same conditions hold good west of these mountains, where the winter climate is much milder in corresponding latitudes. In Washington, Idaho, and Oregon, they appear to migrate about the beginning of November and sometimes earlier, returning to their summer homes in the early part of March. At any rate, without actually examining any of their burrows during the winter months, to ascertain their presence, I never saw one of these birds, as far as I can remember, sitting in front of these at such times, and I have lived where they were very common and would certainly have noticed one occasionally if actually about. The belief that these Owls migrate regularly seems to be confirmed by the observations made by Dr. James C. Merrill, U. S. Army, in the vicinity of Brownsville, Texas, who, in his "Notes on the Ornithology of Southern Texas," makes the following statement:[1] "The Burrowing Owl is rather abundant during the winter months, but I do not think that any remain to breed." The fact that Mr. George B. Sennett, who collected in the same region for several seasons, failed to detect this Owl during the breeding season tends still more to confirm that they are only winter visitors, and consequently migrants from the North. Dr. Merrill tells me that even as far south as Fort Reno, Indian Territory, he failed to notice any of these birds during the winter, although common enough at all other seasons. That they hybernate, as some observers suppose, I do not believe for an instant.

In the vicinity of Fort Walla Walla, Washington, and Fort Lapwai, Idaho, they usually made their appearance in the first week of March, about the same time the earliest migrants arrived, and by the middle of the month they were abundant.

These birds are diurnal in their habits, and may be seen sitting in front of their burrows at any hour of the day. Where not unduly molested they are not at all shy, and usually allow one to approach them near enough to note their curious antics. Their long slender legs give them rather a comical look, a sort of top-heavy appearance, and they are proverbially polite, being sure to bow to you as you pass by. Should you circle around them they will keep you constantly in view, and if this is kept up it sometimes seems as if they were in danger of twisting their heads off in attempting to keep you in sight. If you venture too close they will rise and fly a short distance and generally settle down near the mouth of another burrow close by, uttering at the same time a chattering sort of note, and repeat the bowing performance. Occasionally, when disturbed, they alight on a small sage bush, probably to get a better view of the surroundings.

They hunt their prey mostly in the early evening and throughout the night, more rarely during the daytime. As soon as the sun goes down they become exceedingly active, and especially so during the breeding season. At such times they are always busy hunting food, and go and come constantly, and they may often be seen hovering suspended in the air like the Sparrow Hawk,

---

[1] Proceedings U. S. National Museum, Vol. 1, 1878, p. 151.

locating their prey or darting down noiselessly and swiftly, and grasping it with their talons without arresting their flight an instant. The actual amount of food a pair of these birds require to bring up their numerous family, generally averaging eight or nine, is something enormous. Each Owl will eat fully its own weight in twenty-four hours, if it can get it.

I have, at different times, kept some of these birds in confinement for a week or more and fed them on Townsend's ground squirrels, an animal weighing more than this Owl. These were caught alive, and absolutely uninjured, in wire traps baited with carrots, and turned loose in the room where the Owls were kept; first, to see if they could actually kill rodents of this size, and, second, to find out how much they would eat in a day. To test both, I fed a pair of these Owls four live full-grown Townsend's ground squirrels in one day, besides the carcasses of five small birds which had been skinned, and was astonished at the ease and celerity with which these rodents were killed and the small amount of resistance they made. I watched the proceedings through a small hole in the door. As soon as a squirrel was turned loose in the room with the Owls, one of them would pounce on it, and, fastening its sharp talons firmly in the back of the squirrel, spread its wings somewhat, and with a few vigorous and well-directed blows of its beak break the vertebræ of the neck, and before it was fairly dead it commenced eating the head. This was always eaten first and is the favorite part. Next morning there was but little left of squirrels or birds, and the two Owls had certainly eaten considerably more than their own weight in the twenty-four hours. It actually kept one busy to supply them with the necessary food they would consume, which gives a fair idea of how much a family of half-grown young must require. As nearly all the food used by them consists of noxious vermin, it readily appears what an immensely beneficial bird the Burrowing Owl is, considered from an economic point of view, and deserving of the fullest protection.

They appear to be mated when they make their first appearance in the early spring, and I believe remain paired through life. At this season, where they are abundant, and they are generally found in little colonies of several pairs at least, their peculiar love note can be heard on all sides about sundown; it reminds me more of the call of the European Cuckoo (*Cuculus canorus*) than anything else, a mellow sonorous and far-reaching "coo-c-o-o," the last syllables somewhat drawn out, and this concert is kept up for an hour or more. These notes are only uttered when the bird is at rest, sitting on the little hillock surrounding its burrow. While flying about, a chattering sort of note is used, and when alarmed a short shrill "tzip-tzip." When wounded and enraged it utters a shrill scream and snaps its mandibles rapidly together, making a sort of rattling noise, throws itself on its back, ruffles its feathers, and strikes out vigorously with its talons, and with which it can inflict quite a severe wound.

Preparations for nidification begin in the latter part of March and continue well into April. When not disturbed, the same burrow is used from year to year; in such a case it is cleaned out and repaired, if necessary. In different localities their choice in the selection of nesting sites varies somewhat. At Fort Lapwai, Idaho, they generally selected a burrow on a hillside with a southerly exposure, while at Walla Walla their nests were always found in burrows on level ground. At Camp Harney, Oregon, where the Burrowing Owls were not very common, one under a large basaltic bowlder seemed to be a favorite site with them, and here they encroached upon the timber in the foothills of the Blue Mountains. At Fort Custer, Montana, I found them mostly on level ground, generally bottom lands, and always at the outskirts of a prairie dog village. On the Pacific coast the burrows of the ground squirrels are more often used for nesting sites, and occasionally those of badgers, which are quite common in some sections. If one of the former is selected, it has first to be considerably enlarged, and which requires a good deal of patient labor on the part of the Owls to accomplish. While stationed at Fort Lapwai I had an opportunity to see an Owl at work enlarging and cleaning out a burrow. The loosened dirt was thrown out backward with vigorous kicks of the feet, the bird backing gradually toward the entrance and moving the dirt outward in this manner as it advanced. These burrows vary greatly in length and depth, and are rarely less than 5 feet in length and frequently 10 feet and over. If on level ground they usually enter diagonally downward for 2 or 3 feet, sometimes nearly perpendicularly for that distance, when the burrow turns abruptly, the nesting chamber being always placed above the lowest part of the burrow. If in a hillside it will frequently run straight in for a few feet, and then make a sharp turn direct to the nesting chamber. At other times the burrow follows the curves of a horseshoe, and I have more than once found the eggs in such a burrow lying within 2 feet of the entrance and close to the surface of the hill on a trifle higher level; where, had it been known they could have been reached with little trouble. These burrows are generally about 5 inches in diameter, and the nesting chamber is usually from 1 foot to 18 inches wide. After the burrow is suitably enlarged, especially at the end, dry horse and cow dung is brought to the entrance of it, where it is broken up in small pieces, carried in and spread out in the nesting chamber which is usually lined with this material to a thickness of 1 or 2 inches, and I have never found any other material in the nest. In California, however, they are said to line them occasionally with dry grasses, weed stalks, feathers, and similar materials. On one thing most observers agree, namely, that their burrows invariably swarm with fleas.

In southern California the Burrowing Owl commences laying about the beginning of April; in Oregon, Washington, and Idaho, rarely before the 15th of the month, and usually about the latter part of it; in Kansas and northern Texas it begins about the same time; in Utah, fresh eggs have been found as late as June 15, and at Fort Collins, Colorado, on July 1.

Although incubation does not appear to begin until the clutch is nearly completed, I always found one of the parents at home, even if there was but a single egg in the nest. The old bird is courageous in the defense of its

domicile, and as a rule will not leave it, although the way may be left clear
for it to do so. Backing up to the extreme end of its burrow, it will strike
with beak and claws in defense of its nest. Frequently when within a foot
or two of the nest proper, and before it was yet visible the occupant made
a rattling noise, produced by the rapid movement of its mandibles, which
sounded very much like the warning of the rattlesnake when disturbed; this
would easily impose on the average investigator, and proceeding out of the
burrow, somewhat muffled and subdued, is very similar indeed to the rattle
of the latter.

The number of eggs laid by the Burrowing Owl varies from six to
eleven. From seven to nine are more often found, while sets of ten and
eleven are not especially rare, and Mr. Walter E. Bryant, of Oakland, California, found one of twelve near Carson, Nevada. The eggs are usually
found in a single layer and disposed in the form of a horseshoe. On two
occasions in extra large sets, I found them placed on top of each other. It
is astonishing how they manage to cover them all, but they do, and it is
rare to find an addled egg. Both parents assist in incubation, which lasts
about three weeks, and but a single brood is raised in a season. A second,
and somewhat smaller set is frequently laid in the same burrow or in another
close by, if the first eggs are taken.

The eggs of the Burrowing Owl, after they are washed, are pure white in
color, but as taken from the burrow they are usually much soiled by the
excrement of the numerous fleas inhabiting these domiciles, and bear then
no resemblance to white. They are much more glossy than most Owls' eggs
and are usually rounded ovate in shape. The shell is close grained and
rather smooth, but in some sets it is strongly granulated.

The average size of a fine series of these eggs is 31 by 25.5 millimetres,
the largest egg of the series measuring 34 by 27 millimetres, the smallest 28 by
25 millimetres.

The type specimen, No. 20578 (Pl. 12, Fig. 14), from the Bendire collection, selected from a set of nine eggs, was taken near Fort Walla Walla,
Washington, April 21, 1881.

141. Speotyto cunicularia floridana RIDGWAY.

FLORIDA BURROWING OWL.

Speotyto cunicularia var. floridana RIDGWAY, American Sportsman, v, July 4, 1874,
216.

(B —, C —, R 408a, C 488, U 378a.)

GEOGRAPHICAL RANGE: Florida and the adjacent Bahama Islands (New Providence).

The breeding range of the Florida Burrowing Owl, a variety somewhat
lighter colored than the preceding, seems to be confined to the State of
Florida and to some of the Bahama Islands. It is an inhabitant of the vast
prairie found north of Lake Okeechobee and similar localities in other parts
of the State.

Mr. Walter Hoxie, in an interesting article on the breeding habits of this species, says: "Although in the west the Burrowing Owl usually inhabits the deserted domicile of some animal, this does not seem to be the case with the Florida Burrowing Owl (*Speotyto cunicularia floridana*). There are no animals in the country the latter bird inhabits that make such holes as they require, and I am assured that every hole is occupied by a pair of Owls in the spring, and when one is caved in by the feet of cattle or horses, its occupants at once proceed to excavate a new one. The Indians say that they use their feet for this purpose, and dig pretty fast, too. Their imitation of the cry of the bird was very much like the notes of the Cuckoo, and not at all Owl like in its tone. * * *

"The burrows are found either in the very highest parts of the prairie or in the thickest vegetation, and occupy a peculiar sort of sandy flat ground, which, however, is covered with a good tough turf. They are about 5 inches wide and 3½ high, and extend under ground, on the average, less than 6 feet. A few found were 8 feet and over and only one less than 4 feet. The superincumbent soil is from 8 inches to 1 foot thick, and the chamber at the extremity in which the nest is placed is quite circular, and not less than 1 foot in diameter. It is higher than the passageway leading to it, and being likewise slightly domed, brings the top quite near to the surface of the ground. It is this part of the habitation which is most often caved in by the feet of passing cattle. The sand that is thrown out at the mouth of the burrow makes quite a conspicuous mound in the open prairie, but in the 'roughs,' or those places that have not been burned over for some years, the weeds and grass are rich and rank about it, doubtless fertilized by the droppings and castings of the inhabitants. This hides the burrow pretty effectually from the casual observer, but after a little experience these circular patches of richer vegetation were quite valuable as guides in my search. I found no very large towns, the usual number of burrows being five or six. The largest number found together was eleven, and the smallest three. The holes open to all points of the compass, although one of the oldest settlers in the region assured me that they always extended south under ground. They seldom make much of a turn. When one hole was found I always looked for others within at least a rod, and occasionally they were not more than a yard apart."[1]

Mr. J. F. Menge writes me from Myers, Florida, as follows: "The Florida Burrowing Owl nests on the high open prairies in this vicinity, beginning to lay about March 15 and up to the middle of May. The eggs range from four to eight to a set, usually six. The burrows vary greatly both in length and depth. I have found nests all the way from 6 to 16 feet from the entrance, and only 5 inches to 3 feet under ground. The nest is always placed from 2 to 4 inches above the level of the passageway in a circular chamber near the end of the burrow. The eggs are laid on a

[1] Ornithologist and Oölogist, Vol. XIV, 1889, No. 3, pp. 33, 34.

bed of dry cow dung and grass roots. The food of these Owls consists mostly of beetles, grasshoppers, small snakes, and frogs. Many of their nests are yearly destroyed by skunks and opossums, who seem to be very fond of the eggs. During the spring of 1890 I secured several full sets in the following manner. By not disturbing the entrance, simply inserting a pliable stick as far as it would go in the hole, withdrawing it and measuring, then digging a small hole down to the passageway and using the stick again as before until I came to the nest, I could examine it, and, in case the set of eggs was not complete, cut a turf to fit each hole made, taking care to shut out all light, and thus found that the birds would continue laying as if nothing had disturbed them. A few days later I went back to the nests so treated, and lifting the turf carefully, invariably found the bird at home. She always made a rattling sound like that of the rattlesnake, so much so, indeed, that the party who was with me jumped back and said, 'Lookout, there is a rattlesnake!' These birds are persistent layers. I robbed a pair three times, first of eight eggs, then of five, and the last time of four. These were all laid in the same burrow, but each set was a little deeper in. One of their burrows found by me described almost a complete circle, the nesting chamber being situated within 2 feet of the entrance hole."

After the breeding season is over, the Florida Burrowing Owl is said to disappear for a time from its usual haunts, but where it goes is not positively known. On the whole, its habits are very similar to those of the common Burrowing Owl.

Mr. Menge, from whom I obtained a set of six eggs, sent the nesting material on which they were found. In this instance it consisted entirely of the burnt ends of grass stalks and the charred roots, evidently pulled up by the birds, and mixed with this material were some few breast feathers of the Ground Dove (*Columbigallina passerina*).

The eggs are indistinguishable from those of the preceding subspecies, and their shells are equally smooth and glossy; their shape is also similar.

The average measurement of twenty-three specimens is 31.5 by 26.5 millimetres, the largest egg of the series measuring 33 by 29, the smallest 30 by 25.5 millimetres.

The type specimen, No. 18192 (Pl. 12, Fig. 16), was obtained from Dr. J. W. Velie, and taken in southern Florida in the spring of 1880.

142. Glaucidium gnoma WAGLER.

PYGMY OWL.

Glaucidium gnoma WAGLER, Isis, 1832, 275.
(B 60, C 329, R 409, C 484, U 379.)

GEOGRAPHICAL RANGE: The interior and mountainous parts of western North America, from British Columbia; south through the United States to the table-lands of Mexico, excepting the Pacific coast region; east to the eastern slopes of the Rocky Mountains.

The breeding range of the Pygmy Owl, as far as known, extends through the timbered regions of western North America, from the southern Rocky Mountains in Colorado, New Mexico, and Arizona, westward to eastern California, eastern Oregon, and eastern Washington, north into eastern British Columbia, and south into Mexico.

The following is quoted from an article written by me and published in The Auk, October, 1888 (Vol. v, pp. 366–372), with a few verbal changes:

"The general habits of the Pygmy Owl are by this time pretty well known, and there remains little for me to add to its life history that is really new. It is a well established fact that it is quite diurnal and hunts its prey, to a great extent at least, during the daytime; its food consists not alone of grasshoppers and other insects, as some of the earlier naturalists surmised, but also of birds and the smaller rodents, some of the latter considerably heavier than itself.

"I presume that it is not at all uncommon throughout the entire mountainous and timbered portions of the West; but from its small size and retiring habits, generally being hidden in dense evergreen trees, it is not often noticed by the naturalist, and usually only by accident. I have taken it personally in the Blue Mountains in Washington and in several places in Oregon, but have never met with more than one at a time. My specimens were, with but a single exception, found in or near pine timber. While hunting Sage Fowl on the morning of February 5, 1875, in the vicinity of Camp Harney, Oregon, I shot a female Pygmy Owl at least 5 miles from the nearest timber. It was perched on a large bowlder lying at the foot of a basaltic cliff, and allowed me to approach quite closely. It had just about finished breakfasting upon a Western Tree Sparrow, as indicated by the feathers scattered about and on the rocks; it was in prime condition and exceedingly fat.

"The first of these little Owls coming under my observation was shot by Sergeant Smith, who used frequently to hunt with me. On the morning of December 14, 1874, we were out hunting Sooty Grouse along the southern slopes and among the foothills of the Blue Mountains, a few miles north of Camp Harney, and had been quite successful. The sergeant was walking along the edge of a mesa, while I was about 100 yards below him hunting among some service berry bushes growing about half way up the slope of the hill, and in which Grouse were usually found feeding at that time of the year. Hearing the sergeant fire (he could not be seen from where I was standing), I called to him and asked what he had shot. His reply seemed at the instant rather strange to me. It was, 'Captain, I shot a baby Owl riding on a rat; I have got them.' Had I not known the sergeant to be a strictly sober man, not at all addicted to drinking, I should have readily agreed with him 'that he had them,' and laid it to overindulgence in something stronger than water on that particular morning; but when I climbed up to where he was standing the matter was fully explained.

"It appears that a tall old pine tree had been uprooted years ago by some of the heavy wind storms that occasionally sweep over that region, and the roots of it were lying partly under a younger and bushy tree of the same species that was taking the place of the older one in the course of nature. The massive trunk of the old tree was free from limbs for about 40 feet, and was slowly but surely decaying. A large sized gopher, which perhaps, found a congenial home amidst the roots of the old tree, on hearing the noise the sergeant made in his approach, had climbed up on to the trunk of the tree, possibly to get a good view of the intruder and to warn the balance of his family, when, quick as a flash, a little Pygmy Owl that had been securely hidden among the branches of the growing pine, dropped down with unerring aim on its victim and fastened its sharp little talons securely into the astonished gopher's back. Sergeant Smith's attention was drawn to the performance by a squeak from the gopher, which, in trying to escape, ran along on top of the fallen pine almost its entire length, making rather slow progress, however, hampered as it was by the Owl, when the sergeant fired, killing both. During this time, nearly a couple of minutes, the Owl sat upright on the gopher's back, never letting go its hold an instant, twisting its head nearly off the body in trying to keep an eye on the sergeant, who was rapidly approaching, but apparently showing no uneasiness whatever. He told me that the whole thing was done in such a business-like manner that it was evidently not the first ride of the kind this little Owl had so taken. It held on to its prey even in death. (I published a short account of this occurrence at the time in the Proceedings of the Boston Society of Natural History (Vol. xviii, October 6, 1875). Both specimens are in the U. S. National Museum collection.)

"I also met with the Pygmy Owl on several occasions at Fort Klamath, Oregon, and remember quite distinctly seeing one (presumably the same individual) several times at various hours of the day, sitting patiently, but wide awake, on a single long and slender willow branch overhanging Fort Creek, but a little distance from the post. I refrained from shooting it as I suspected it nested in the vicinity, and it would also have been rather difficult to secure. I cannot say positively, but think that it used that particular perch for no other purpose than to catch frogs. The willow overhung a marshy, reed-covered spot, where the water was rather shallow, and which seemed to be a favorite resort for numbers of these batrachians. Small birds, of which there were numbers about in the vicinity in the willow thickets bordering the stream, did not seem to resent the presence of the little Owl, and paid no attention whatever to it.

"Its call notes may often be heard during the early spring months while mating, and usually shortly after sundown. Its love notes are by no means unmusical. They somewhat resemble the cooing of the Mourning Dove (*Zenaidura macroura*), like 'coohuh, coohuh,' softly uttered, and a number of times repeated. Although I have not positively seen this bird while in the act of calling its mate, am quite certain that the notes emanated from this little Owl and no other. I am familiar with the notes of the Acadian and MacFarlane's Owls (*Nyctala acadica* and *Megascops asio macfarlanei*), the only other of the small Owls at all likely to be found there, but their notes are different, and they were not heard by me while stationed at Fort Klamath, Oregon.

"Mr. Henshaw found the Pygmy Owls quite numerous in the southern Rocky Mountains, and states that they are rather sociable in disposition, especially during the fall months. He says he has imitated their call and readily lured them up close enough to be seen.[1] I am inclined to think that they are much more common there than farther north. * * *

"During an absence once from Fort Klamath on official matters, one of my men found on June 10, 1883, a burrow occupied as a nest by the true *Glaucidium gnoma*, which at the time it was first discovered must have contained eggs. The nest was not disturbed till the day after my return to the post, June 25, when he showed it to me. The nesting site used was a deserted Woodpecker's excavation, in a badly decayed but still living aspen tree and was about 20 feet from the ground; the cavity was about 8 inches deep and 3½ wide at the bottom. This tree, with two others of about the same size, stood right behind, and but a few feet from a target butt on the rifle range, which had been in daily use since May 1, target firing going on three or four hours daily. All this shooting did not seem to disturb these birds, for the first egg must have been deposited some two or three weeks after the target practice season began, but the strangest thing is that the Owls were not discovered long before, as two men employed as markers were constantly behind the butt in question during the firing and directly facing the entrance hole of the burrow. When the nest was shown me I had it examined, and, much to my disgust, found it to contain, instead of the much coveted eggs, four young birds about a week or ten days old. I took these; two of them are now in the U. S. National Museum, the remaining two in Mr. William Brewster's collection at Cambridge, Massachusetts. The cavity was well filled with feathers of various kinds, and contained besides the young, the female parent and a full grown Say's chipmunk (*Tamias lateralis*), that evidently had

[1] Auk, Vol. III, January, 1886, p. 79.

just been carried in, as it had not been touched. The cavity was almost entirely filled up by the contents mentioned."[1]

Mr. William G. Smith, of Loveland, Colorado, well known as a reliable naturalist and collector, writes me that he found a nest of this species, on May 31, 1890, containing three young birds apparently about two days old and a single egg which was on the point of hatching, in a ravine near Estes Park at an altitude of about 10,000 feet. The site was in an old Woodpecker's hole in a dead aspen about 14 feet from the ground. The nest, if it can be called such, was composed of a few feathers and rubbish. The female was in the hole, and in trying to take her out the egg was broken; the male was perched on a tree close by and was likewise secured. Mr. Smith also states in one of his letters to me that one of these little Owls made itself quite familiar in the town of Loveland, Colorado, during the winter (1890), flying about the houses during the daytime for several days, until shot. He regards them as rare in that vicinity.

Mr. Charles F. Morrison reports having taken four sets of their eggs in La Plata County, Colorado, during 1886 and 1887, but fails to describe or to give measurements of these eggs, and I have been unable to borrow any of these specimens for examination. He states that they nested in deserted Woodpeckers' holes and hollow stubs on the sides of gulches grown up with pine timber. He gives the earliest date of nesting as June 1, and the latest June 22. The nesting sites were from 8 to 20 feet up. The only note he heard them utter was a faint squeak.

The food of the Pygmy Owl consists principally of the smaller rodents and birds, some considerably larger than itself. It is a decidedly savage little fellow and a courageous one as well. In speaking of their food, my friend Dr James C. Merrill, U. S. Army, makes the following statement: "One captured February 21, had just struck at a Robin, and was struggling with it on the ground. It is said to be especially abundant in summer at Modoc Point, Klamath Lake, Oregon, and to feed upon a lizard that is common there. I have also found fragments of field mice in their stomachs. Insects, however, and especially grasshoppers, constitute the greater part of their food, when they can be obtained. When the Owl is searching for these, the smaller birds pay but little attention to it, even if it happens to alight near them."[2]

Judging from the date of nesting but a single brood is raised in a season, and incubation appears to begin with the first egg laid. They are constant residents wherever found.

I regret that I am unable to give a detailed description and measurement of their eggs, which appear to be usually four in number, but they will unquestionably prove indistinguishable from those of its near relative, the California Pygmy Owl, which are known.

[1] Auk, Vol. v, No. 4, October, 1888, pp. 366–371.
[2] Auk, Vol. v, No. 2, April, 1888, p. 146.

143. Glaucidium gnoma californicum (Sclater).

CALIFORNIA PYGMY OWL.

Glaucidium californicum Sclater, Proceedings Zoölogical Society London, 1857, p. 4.
Glaucidium gnoma californicum Bendire, Auk, Vol. v, October, 1888, p. 366.
(B 69, part; C 329, part; R 409, part; C 484, part; U 379a.)

Geographical range: Coast region of California, Oregon, Washington, and southern British Columbia.

The California Pygmy Owl, a darker colored race than the preceding, is a resident of and breeds in the timbered regions adjacent to the Pacific coast from about latitude 37° N. in middle California, through western Oregon, Washington, and southern British Columbia, where Mr. Clark P. Streator reports it as common in the vicinity of Mount Lehman, in latitude 49° N., and breeding. In the drier climate in the interior of these States it is replaced by the true *Glaucidium gnoma*, and its habits generally are very similar to that subspecies. I believe the credit of discovery of the nest and eggs of this Owl belongs to Mr. George H. Ready, of Santa Cruz, California, who found a nest on June 8, 1868, containing three eggs, in a deserted Woodpecker's excavation in an old and isolated poplar tree growing on the banks of the San Lorenzo River, near the above mentioned locality. The cavity was 75 feet from the ground. A short account of this find was published by me in the Proceedings of the Boston Society of Natural History (Vol. xix, March 21, 1879, p. 132), and a somewhat fuller description, by W. C. Cooper, can be found in the Bulletin of the Nuttall Ornithological Club (Vol. iv, April, 1879, p. 86).

The two eggs which Mr. Cooper has made drawings of, and which I have before me, are nearly ovate in shape, and said to be dull white in color, with a scarcely perceptible yellowish tinge. Their surface is described as quite smooth, and to have the appearance of having been partly punctured with a fine needle point over the whole egg. Judging from the drawings they are decidedly pointed for Owls' eggs, and perhaps somewhat abnormal in this respect, resembling the eggs of the Burrowing Owl in shape. Their size is given as 1.18 by 0.90 and 1.17 by 0.87 inches (about 30 by 22.9 and 29.7 by 22.1 millimetres).

I recently had an opportunity to examine two sets of eggs, now in the oölogical collections of Messrs. Samuel B. Ladd and Joseph Hoopes, of West Chester, Pennsylvania, which are said to belong to this subspecies. Both were collected in Benton County, Oregon, and found in Woodpeckers' excavations in dead trees in swampy woods. One of these sets, taken May 20, 1889, containing four eggs, and now in Mr. Ladd's collection, measures, respectively, 27 by 23, 26.5 by 23, 26 by 23, and 26 by 22.5 millimetres. The shells of these eggs are smooth, close grained, and very thin, almost semitranslucent. In color they are dull milky white, with a very faint creamy tint, and show no luster, but have the peculiar pittings or punctures already referred to, and which seem to be characteristic of the eggs of this Owl. They are rounded oval in shape, about the size of the well known egg of the little Whitney's or Elf Owl, but can be readily distinguished from these by the different texture of the shells.

Mr. Charles A. Allen, writing me about the California Pygmy Owl from Nicasio, California, says: "In this section it inhabits the heavy coniferous forests, and is rather difficult to obtain at any time. Its love note is a soft low musical 'toot-toot,' repeated at intervals of a few seconds between each call. During the mating season, which commences here about the last of February, these birds can be heard any still morning up to 8 or 9 o'clock, and if it be dull and cloudy to nearly 11 o'clock; after that hour they remain silent until sundown. I consider them good weather prognosticators, as I hear them invariably just before a storm. After the breeding season is over and during the summer months they remain silent. The male when calling usually selects one of the tallest trees and perches near the topmost branches. Their food consists of small birds and mammals, and I have found them feeding on *Pipilo oregonus* and *Habia melanocephala*, and also saw one pounce down on a *Tamias townsendi* sitting on a log, seize it and fly up into a large red-wood tree, where I found it feeding three young which were sitting on a limb close beside it, and crowding one another to obtain the food as fast as the parent could tear it up into suitable pieces for them."

Prof. O. B. Johnson, in his "List of the Birds of the Willamette Valley, Oregon," referring, I think, to this bird, says: "Quite common; I have not seen the nest. They are savage little fellows and will attack cage birds in daylight, and I know of two that suffered death thereby."

As with the preceding subspecies, nidification commences in May, and occasionally not before June. The number of eggs to a set seems to be three or four, and probably but a single brood is raised in a season. As the egg of the California Pygmy Owl is indistinguishable from that of the Ferruginous Pygmy Owl, as far as size and shape are concerned, and of which a specimen is figured, I have not illustrated it.

144. Glaucidium gnoma hoskinsii BREWSTER.

HOSKIN'S PYGMY OWL.

Glaucidium gnoma hoskinsii BREWSTER, Auk, v, April, 1888, p. 136.
(B —, C —, R —, C —, U 379b.)

GEOGRAPHICAL RANGE: Lower California.

Mr. William Brewster, who first described this new subspecies, speaks of it as being "similar to *Glaucidium gnoma californicum*, but smaller and grayer, the forehead and fascial disc with more white and the upper parts less distinctly spotted."

The type specimen, an adult male, was taken in the Sierra de la Laguna, Lower California, May 10, 1887, by M. Abbott Frazar; and two others were obtained in the same locality June 2 and 4, respectively.   Mr. Walter E. Bryant took another male near Comondu, Lower California, March 22, 1889.

Nothing is as yet known about the breeding habits and eggs of this new subspecies, but there is no reason to believe that they differ in any respect from those of the two preceding geographical races.

145. Glaucidium phalænoides (DAUDAIN).

FERRUGINOUS PYGMY OWL.

Strix phalænoides DAUDAIN, Traité d'Ornithologie, II, 1800, 206.
Glaucidium phalænoides CABANIS, Journal für Ornithologie, 1869, 208.
(B —, C 330, R 410, C 485, U 380.)

GEOGRAPHICAL RANGE: Whole of tropical America (except West Indes); north to southwestern border of the United States (southern Texas to Arizona).

The Ferruginous Pygmy Owl is a resident of the southern border of the United States, breeding in the valley of the Lower Rio Grande in Texas, and in southern Arizona.   It doubtless occurs in the intervening regions as well.

Mr. George B. Sennett found it not uncommon at Lomita, Texas, and says: "From its small size it is not readily seen in heavy timber.   Its note, a clear whistle, quite difficult to follow, was often heard during April and May.   A female captured April 9 contained eggs nearly ready to be laid."

I am able to add but little more relating to the life history of this Owl than what I have already published in the Auk (Vol. v, No. 4, October, 1888, pp. 371, 372), which is here appended, as follows: "This widely distributed species was first described by Prince Max z. Wied in 1820.   It inhabits the whole of tropical America (the West Indies excepted), and is found to the northward along the southwestern border of the United States, occurring in southern Texas and Arizona.   It was first added to our fauna by the writer, who took several in 1872 in the heavy mesquite thickets bordering Rillitto Creek, near the present site of Camp Lowell, in the vicinity of Tucson, Arizona. The first specimen was taken January 24, 1872, showing that it is a resident throughout the year; other specimens were obtained during the following spring and summer.   Unfortunately, I was not then an adept in taxidermy; the skins made by me in those days looked as if they had passed through the jaws of a hungry coyote, and they were only useful in determining species.   Like *Glaucidium gnoma*, this little Owl is quite diurnal in its habits.   Its call, according to my own notes, is 'chu, chu, chu,' a number of times repeated, and most frequently heard in the evening.   According to Mr. F. Stephens, its note is a loud 'cuck,' repeated several times as rapidly as twice each second.   He further states that at each utterance the bird jerked its tail and threw back its head.   Occasionally a low 'chuck,' audible for only a short distance, replaced the usual call.   Mr. Stephens's notes come perhaps nearer the mark than my own; I know him to be an exceedingly careful, conscientious, and reliable observer.   According to

Prince Max z. Wied, in Burmeister's 'Thiere Brasilien's' (Vol. II, 1856, p. 142), its call is said to be 'keck, keck, keck.' The best account of the life history of this little Owl is found in the Journal für Ornithologie (Vol. XVII, 1869, pp. 244, 245), under 'Notes on the Natural History of the Birds of Brazil,' by Carl Euler.

"According to this authority, small as the Ferruginous Pygmy Owl is, it has been known to carry off young chickens, and he was informed by the natives that it even attacked Jacú hens (*Penelope*), a bird of greater size than domestic fowls. It was stated to him that the little Owl fastened itself under the wings of the latter, gradually tearing it to pieces, and wearing it out and eventually killing it. I am aware, from personal observations, that some of our small Owls are the peer, as far as courage is concerned, of the noblest Falcon ever hatched, but I should not quite care to father that story. Carl Euler says, further, that in captivity, when fed on birds, it always carefully removed all the larger feathers from the carcass before beginning its meal. Also that it was not at all afraid of light, and that he met with it several times during bright sunshiny days, sitting on perfectly bare and leafless trees. He gives its call note as 'khiu, khiu.' Apparently none of us here mentioned agree on the call note of this Owl, and I leave it to the reader to take his choice.

"Euler surmises that it rears two broods a season, one in October the other in December. He once met in March a family of four, two adults and two young, sitting close together on a limb of a tree, waiting, as he says, for twilight. The nest is said to be made in hollow trees, but no mention is made of the eggs having been found, however, and I cannot find any description of them in any of the works accessible to me.

"A nest containing two fully fledged young found by me in a hole in an old mesquite tree in the spring of 1872, in a chaparral thicket near Camp Lowell, and referred to at the time as being that of *Micropallas whitneyi*, may possibly, and probably, have been one of this little Owl, as the Elf Owl seems to confine itself in its nesting sites mainly to excavations in giant cactus (*Cereus giganteus*), so far as known."

Since this account was published, Mr. George B. Sennett has secured the eggs of this species. He says: "On May 2, 1888, my collector took an adult female and an egg of this Owl at Cañon del Caballeros, near Victoria, Tamaulipas, Mexico. The locality is high and at the base of the more precipitous mountains. The nest was in a hollow tree and contained but a single fresh egg, which is white, shaped like that of a *Megascops*, measures 1.05 by 0.90 inches [about 26.7 by 22.9 millimetres], and is now in my collection with the parent bird. It will be observed that in size it is very close to the egg of *Micropallas whitneyi*."[1]

Since Mr. Sennett published the above he has received a set of four eggs of this species taken within our southern border, which, as well as the first mentioned egg, he has kindly allowed me to examine. The latter were

found on May 3, 1890, and were apparently fresh. The nesting site was a
Woodpecker's hole in a mesquite tree, about 10 feet from the ground, in thick
woods near Brownsville, Texas. The eggs are oval in shape and measure,
respectively, 28 by 23.5, 28 by 23.5, 27.5 by 24, and 27 by 23 millimetres.

Compared with the eggs of *Glaucidium gnoma californicum*, the shells are
apparently much thicker, and are rather coarsely granulated, considering their
small size, considerably more so than the egg of *Micropallas whitneyi*, and
they are not as glossy as the latter. The texture of the shells is decidedly
different from that of the eggs said to be those of the California Pygmy
Owl. In none of the specimens before me are the peculiar punctures or pittings noticeable and purporting to be characteristic of the eggs of the preceding species. In fact, the reverse is rather the case, most of the specimens
showing a few slight protuberances on their surface.

The type specimen figured on Pl. 12, Fig. 17, is the one taken near
Victoria, Mexico, and now in Mr. G. B. Sennett's collection.

146. Micropallas whitneyi (COOPER).

ELF OWL.

Athene whitneyi COOPER, Proceedings of California Academy Sciences. 1861, 118.
Micropallas whitneyi SENNETT, Auk, Vol. VI, 1889, 276.
(B —, C 331, R 411, C 486, U 381.)

GEOGRAPHICAL RANGE: Lower California and southwestern United States, from
southeastern California and southern Arizona to southern Texas; south, to southern
Mexico (Pueblo and Guanajuato).

The range of the Elf Owl, the smallest of our Owls found within the
United States, extends from southeastern California (Mojave) southeastward
through southern Arizona, where it is the commonest Owl in that Territory.
Mr. George B. Sennett obtained a single specimen from Hidalgo County,
Texas, taken on April 5, 1889, extending its range considerably in this direc-
tion. Mr. L. Belding reports it as not uncommon in the vicinity of Miraflores,
Lower California, where he took several specimens in April, 1882, and it is
likewise an inhabitant of the greater portion of Mexico.

Although probably a constant resident, and breeding wherever found, I
believe its eggs have as yet only been taken in the vicinity of Tucson, southern Arizona.

Since Dr. J. G. Cooper obtained the type specimen near Mojave, California, April 26, 1861, no others have been taken in this State, and this
point probably marks the extreme western and northern limit of its range.
The type remained unique for eleven years, until I met with this little Owl
again in April, 1872, in the vicinity of Tucson, Arizona.

The Elf Owls are nocturnal in their habits and are seldom seen moving
about in the daytime, which they pass either in abandoned excavations of

Woodpeckers in trees and giant cactus, or in the densest thickets in the creek bottoms. I first met with the Elf Owl in April, 1872, my attention being drawn to it by its peculiar call notes, resembling the syllables "cha-cha, cha-cha," frequently repeated in different keys, sometimes quite distinct and again so low that they could not be heard more than 20 yards off. Although uttered in a rather tremulous tone there is nothing unpleasant in the sound, in fact it is rather the reverse. They become active shortly after sundown and were probably attracted to the vicinity of my camp by the guard fire, which was usually kept up all night. This also attracted numerous insects, on which these little fellows feed to a great extent.

On April 20, 1872, while pushing my way through a dense mass of willows in the Rillitto Creek bottom, I saw one of them perched in the thicket and shot it. Although I had made considerable noise, it allowed me to approach quite close and did not seem to be disturbed by my intrusion into its retreat. I took several others subsequently, most of them shortly after sundown, by carefully watching for the point from which they uttered their call notes. When they find themselves observed they sit quite erect and perfectly motionless, and may in such a position be easily mistaken for a part of the limb on which they are perched.

Mr. William Brewster published the following notes made by Mr. F Stephens, while collecting for him in 1882. Referring to this species, he says: "I was walking past an elder bush in a thicket when a small bird started out. Thinking it had flown from its nest, I stopped and began examining the bush, when I discovered a Whitney's Owl sitting on a branch with its side toward me and one wing held up, shield fashion, before its face. I could just see its eyes over the wing, and had it kept them shut I might have overlooked it, as they first attracted my attention. It had drawn itself into the smallest possible compass so that its head formed the widest part of its outline. I moved around a little to get a better chance to shoot, as the bush was very thick, but whichever way I went the wing was always interposed, and when I retreated far enough for a fair shot, I could not tell the bird from the surrounding bunches of leaves. At length, losing patience, I fired at random and it fell. Upon going to pick it up I was surprised to find another which I had not seen before, and which must have been struck by a stray shot. Rather curiously both these specimens proved to be adult males. It is by no means positive, however, that the males are not, to a certain extent, gregarious during the breeding season, for on another occasion two more were killed from a flock of five which were sitting together in a thick bush. They had several different call notes, one of which sounded like the syllable 'churp,' while another was a low 'tro-jur-rrr.' These cries were heard at all times of the night, but more often in the early evening and again at daybreak."[1]

Mr. Herbert Brown writes me from Tucson, Arizona, as follows: "Their food consists largely of ants and beetles. I have examined more than a

---

[1] Bulletin Nuttall Ornithological Club, Vol. VIII, 1883, p. 28.

dozen of their stomachs, always with the same result, and am of the opinion that they do not prey upon either birds or mammals, however small. In dozens of excavations occupied by them I have failed to find a vestige of fur or feathers. In the spring of 1885 I raised five of their young. When I first took them they were little downy cottony things, blind, and not larger than the end of an ordinary sized man's thumb. I kept them until full grown and then sent three to the Zoölogical Garden in Philadelphia and two to the one in Cincinnati. They were fed with raw meat and did well. They are beautiful little creatures and perfectly harmless. I have taken dozens out of their holes, and, so far as I can remember, but one snapped its beak at me. When taken out they offer no resistance and make not the least attempt to get away, but will lie in the closed hand apparently dead. Release the grasp, however, but for one moment and they are gone. The first hostile demonstration I saw one make was a few days ago. I had taken two, a male and female, and placed them in a box, but shortly after, on adding another, the male previously captured raised his feathers on end and looking as wicked as it was possible for the little fellow to do, he began in true Owl fashion to sway his body from side to side and kept it up as long as I watched him. He so frightened the newcomer that it tried to escape, but failing in this, huddled down in the opposite corner of the box in apparent terror. This convinced me that notwithstanding their wonderful meekness the little fellows fight among themselves. During the rainy season they leave the sahuaras and take to the bush, at least none are found occupying such burrows at that time of the year, as I have repeatedly looked for them, but without success."

Two stomachs of the Elf Owl, kindly sent by Mr. Brown and turned over by me to Dr. C. Hart Merriam for examination, contained the following remains of food: One, taken on March 30, 1890, the remains of twenty beetles (*Trimytis*) and fragments of two ribs of a small mammal. Another, taken April 20, 1890, the remains of nineteen beetles and one grasshopper.

In southern Arizona nidification rarely begins before May 10, and lasts through this month and the first week in June. The favorite nesting sites of the Elf Owls are old Woodpeckers' holes in giant cactus, the sahuara of the natives. These grow frequently to a height of 30 and 40 feet, and the main trunk as well as the candelabra-like arms or branches are sometimes fairly riddled with these holes, many of which are undoubtedly bored simply for amusement, as they are easily excavated, and also furnish safe and cosy homes, a fact which these little Owls seem to have found out. Some of these holes are occasionally found as low as 4 feet, and again near the very tops of these curious plants, but more often from 10 to 20 feet from the ground. In very rare instances they nest in holes of trees, such as the mesquite and cottonwood.

As a rule only such cactus as grow along the lowlands of the river bottoms and the table-lands bordering them are selected to breed in. Mr.

Brown has repeatedly examined sahuaras out in the deserts without finding a single Owl in any of them, although they were full of suitable excavations. In a single instance he found one of these birds some distance away from the lowlands, near the summit of the Quijotoa Range, 90 miles southwest of Tucson, Arizona.

From two to five eggs are laid to a set, but the most common number found is three. Of thirty-eight sets taken by Mr. F. Stephens, who found the first eggs of this species, twenty-four sets contained three each, twelve sets contained four, and two sets five eggs. The cavities in which they are deposited are usually about 10 inches deep. There is no nest, the eggs simply lying on the dry chips in the bottom of the hole. One of the parents is always at home after the first egg is laid, and frequently both. The male assists in incubation, which lasts about two weeks. The eggs of the Elf Owl are pure white in color and oval in shape, the shell is finely granulated, and while some specimens are rather glossy, the majority are only moderately so.

The average measurement of twenty-three specimens in the U. S. National Museum collection, all taken by Mr. E. W. Nelson, is 27 by 23 millimetres. The largest of these eggs measures 28.5 by 24, the smallest 25 by 22.5 millimetres.

The type specimen, No. 22134, was taken by the above mentioned gentleman near Fort Lowell, Arizona, May 25, 1884. The set contained three eggs. The specimen is figured on Pl. 12, Fig. 13.

Fig.   1. Colinus virginianus, *Linnæus*. Bob White.
Figs.   2, 3. Oreortyx pictus plumiferus, Gould. Plumed Partridge.
Figs.   4, 5. Callipepla squamata, Vigors. Scaled Partridge.
Figs.   6, 7. Callipepla squamata castanogastris, Brewster. Chestnut-bellied Scaled Partridge.
Figs. 8, 9, 10. Callipepla californica, Shaw. California Partridge.
Figs. 11, 12, 13, 14. Callipepla gambeli, Nuttall. Gambel's Partridge.
Fig.   15. Cyrtonyx montezumæ, Vigors. Massena Partridge.
Figs. 16, 17, 18, 19. Dendragapus obscurus fuliginosus, Ridgway. Sooty Grouse.
Figs. 20, 21, 22, 23. Dendragapus canadensis, Linnæus. Canada Grouse.

416

PLATE I

Fig.   1. Bonasa umbellus, Linnæus.   Ruffed Grouse.
Fig.   2. Bonasa umbellus togata, Linnæus.   Canadian Ruffed Grouse.
Fig.   3. Bonasa umbellus umbelloides, Douglas.   Gray Ruffed Grouse.
Fig.   4. Bonasa umbellus sabini, Douglas.   Oregon Ruffed Grouse.
Figs.   5, 6, 7, 8, 9, 10. Lagopus lagopus, Linnæus.   Willow Ptarmigan.
Figs. 11, 12, 13, 14, 15. Lagopus rupestris, Gmelin.   Rock Ptarmigan.
Figs.   16, 17. Lagopus leucurus, Swainson.   White-tailed Ptarmigan.
Figs.   18, 19, 20. Tympanuchus americanus, Reichenbach.   Prairie Hen.
Fig.   21. Zenaida zenaida, Bonaparte.   Zenaida Dove.
Fig.   22. Engyptila albifrons, Bonaparte.   White-fronted Dove.
Fig.   23. Melopelia leucoptera, Linnæus.   White-winged Dove.
Fig.   24. Columbigallina passerina pallescens, Baird.   Mexican Ground Dove.
Fig.   25. Scardafella inca, Lesson.   Inca Dove.
Fig.   26. Geotrygon montana, Linnæus.   Ruddy Quail Dove.

PLATE II.

EXPLANATION TO PLATE III.

Fig.   1. Tympanuchus pallidicinctus, Ridgway.   Lesser Prairie Hen.
Fig.   2. Tympanuchus cupido, Linnæus.   Heath Hen.
Figs.   3, 4, 5. Pediocætes phasianellus, Linnæus.   Sharp-tailed Grouse.
Figs.   6, 7, 8. Pediocætes phasianellus columbianus, Ord.   Columbian Sharp-tailed Grouse.
Figs.   9, 10. Pediocætes phasianellus campestris, Ridgway.   Prairie Sharp-tailed Grouse.
Figs. 11, 12, 13. Centrocercus urophasianus, Bonaparte.   Sage Grouse.
Fig.   14. Meleagris gallopavo, Linnæus.   Wild Turkey.
Fig.   15. Meleagris gallopavo mexicana, Gould.   Mexican Turkey.
Fig.   16. Ortalis vetula maccalli, Baird.   Chachalaca.
Fig.   17. Columba fasciata, Say.   Band-tailed Pigeon.
Fig.   18. Columba fasciata vioscæ, Brewster.   Viosca's Pigeon.

420

EXPLANATION TO PLATE IV.

Figs. 1, 3. Carthartes aura, Linnæus. Turkey Vulture.
Fig. 2. Columba flavirostris, Wagler. Red-billed Pigeon.
Fig. 4. Columba leucocephala, Linnæus. White-crowned Pigeon.
Fig. 5. Pseudogryphus californianus, Shaw. California Vulture.
Fig. 6. Ectopistes migratorius, Linnæus. Passenger Pigeon.
Figs. 7, 10. Catharista atrata, Bartram. Black Vulture.
Figs. 8, 9. Zenaidura macroura, Linnæus. Mourning Dove.

422

PLATE IV

EXPLANATION TO PLATE V.

Figs.   1, 2. Elanoides forficatus, Linnæus.   Swallow-tailed Kite.
Figs.   3, 4. Elanus leucurus, Vieillot.   White-tailed Kite.
Fig.   5. Ictinia mississippiensis, Wilson.   Mississippi Kite.
Figs.   6, 7. Rostrhamus sociabilis, Vieillot.   Everglade Kite.
Figs.   8, 9, 10. Circus hudsonius, Linnæus.   Marsh Hawk.
Figs. 11, 12, 13.
        14, 15, 16.
            17. Accipiter velox, Wilson.   Sharp-shinned Hawk.
Figs. 18, 19, 20. Accipiter cooperi, Bonaparte.   Cooper's Hawk.

424

PLATE V

EXPLANATION TO PLATE IX.

Figs. 1, 2, 4. Archibuteo ferrugineus, Lichtenstein. Ferruginous Rough-legged Hawk.
Figs. 3, 5. Aquila chrysaëtos, Linnæus. Golden Eagle.
Figs. 6, 8, 9. Falco rusticolus gyrfalco, Linnæus. Gyrfalcon.
Fig. 7. Haliæetus leucocephalus, Linnæus. Bald Eagle.

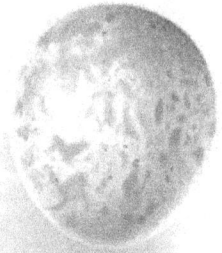

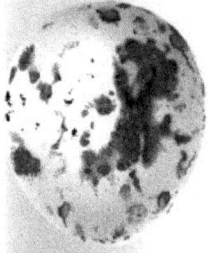

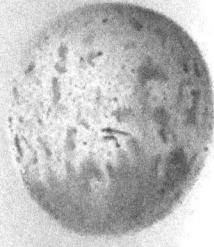

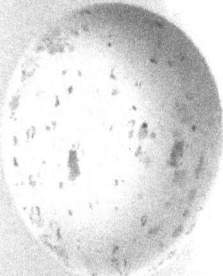

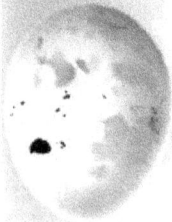

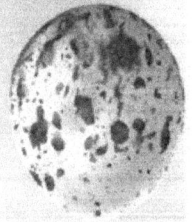

Fig. 1. Accipiter atricapillus, Wilson. American Goshawk.
Fig. 2. Accipiter atricapillus striatulus, Ridgway. Western Goshawk.
Figs. 3, 4. Parabuteo uuicinctus harrisi, Audubon. Harris's Hawk.
Figs. 5, 6. Buteo borealis, Gmelin. Red-tailed Hawk.
Figs. 7, 8. Buteo borealis calurus, Cassin. Western red-tailed Hawk.
Fig. 9. Buteo lineatus elegans, Cassin. Red-bellied Hawk.

PLATE VIII.

Figs. 1, 2, 3, 4, 5. Buteo lineatus, Gmelin. Red-shouldered Hawk.
Fig.   6. Buteo abbreviatus, Cabanis. Zone-tailed Hawk.
Fig.   7. Asturnia plagiata, Schlegel. Mexican Goshawk.
Figs.   8, 9. Buteo albicaudatus, Vieillot. White-tailed Hawk.
Figs. 10, 11, 12. 13. Buteo latissimus, Wilson. Broad-winged Hawk.

PLATE IX.

Lith. by Ketterlinus. Phila.

434

PLATE X

EXPLANATION TO PLATE XI.

Figs. 1, 2, 3, 4. Polyborus cheriway, Jacquin. Audubon's Caracara.
Figs. 5, 6, 7, 8, 9. Pandion haliaëtus carolinensis, Gmelin. American Osprey.

PLATE XI.

Fig. 1. Strix pratincola, Bonaparte. American Barn Owl.
Fig. 2. Asio wilsonianus, Lesson. American Long-eared Owl.
Fig. 3. Asio accipitrinus, Pallas. Short-eared Owl.
Fig. 4. Syrnium nebulosum, Forster. Barred Owl.
Fig. 5. Scotiaptex cinerea, Gmelin. Great Gray Owl.
Fig. 6. Nyctala tengmalmi richardsoni, Bonaparte. Richardson's Owl.
Fig. 7. Nyctala acadica, Gmelin. Saw-whet Owl.
Fig. 8. Megascops asio, Linnæus. Screech Owl.
Fig. 9. Megascops asio macfarlanei, Brewster. MacFarlane's Screech Owl.
Fig. 10. Megascops asio maxwelliæ, Ridgway. Rocky Mountain Screech Owl.
Fig. 11. Megascops asio trichopsis, Wagler. Mexican Screech Owl.
Fig. 12. Bubo virginianus, Gmelin. Great Horned Owl.
Fig. 13. Micropallas whitneyi, Cooper. Elf Owl.
Fig. 14. Speotyto cunicularia hypogæa, Bonaparte. Burrowing Owl.
Fig. 15. Megascops flammeolus, Kaup. Flammulated Screech Owl.
Fig. 16. Speotyto cunicularia floridana, Ridgway. Florida Burrowing Owl.
Fig. 17. Glaucidium phalænoides, Daudain. Ferruginous Pigmy Owl.
Fig. 18. Surnia ulula caparoch, Müller. American Hawk Owl.
Fig. 19. Nyctea nyctea, Linnæus. Snowy Owl.

438

PLATE XII.

ALPHABETICAL INDEX.